Springer Proceedings in Advanced Robotics 4

More information about this series at http://www.springer.com/series/15556

Jadran Lenarčič · Jean-Pierre Merlet
Editors

Advances in Robot Kinematics 2016

Springer

Editors
Jadran Lenarčič
Jožef Stefan Institute
Ljubljana
Slovenia

Jean-Pierre Merlet
INRIA Sophia Antipolis
Sophia Antipolis Cedex
France

ISSN 2511-1256 ISSN 2511-1264 (electronic)
Springer Proceedings in Advanced Robotics
ISBN 978-3-319-56801-0 ISBN 978-3-319-56802-7 (eBook)
DOI 10.1007/978-3-319-56802-7

Library of Congress Control Number: 2017940802

Printed on acid-free paper

This Springer imprint is published by Springer Nature
The registered company is Springer International Publishing AG
The registered company address is: Gewerbestrasse 11, 6330 Cham, Switzerland

Foreword

Robots! Robots on Mars and in oceans, in hospitals and homes, in factories and schools; robots fighting fires, making goods and products, saving time and lives. Robots today are making a considerable impact from industrial manufacturing to health care, transportation, and exploration of the deep space and sea. Tomorrow, robots will become pervasive and touch upon many aspects of modern life.

The *Springer Tracts in Advanced Robotics (STAR)* was launched in 2002 with the goal of bringing to the research community the latest advances in the robotics field based on their significance and quality. During the latest fifteen years, the STAR series has featured publication of both monographs and edited collections. Among the latter, the proceedings of thematic symposia devoted to excellence in robotics research, such as ISRR, ISER, FSR, and WAFR, have been regularly included in STAR.

The expansion of our field as well as the emergence of new research areas has motivated us to enlarge the pool of proceedings in the STAR series in the past few years. This has ultimately led to launching a sister series in parallel with STAR. The *Springer Proceedings in Advanced Robotics (SPAR)* is dedicated to the timely dissemination of the latest research results presented in selected symposia and workshops.

This volume of the SPAR series brings the proceedings of the fifteenth edition of ARK on Advances in Robot Kinematics, whose proceedings have been previously published by Kluwer and Springer since 1991. This edition took place in Grasse, France, from June 27 to June 30, 2016. The volume edited by Jadran Lenarčič and Jean-Pierre Merlet contains 46 scientific contributions, revised and extended after the meeting. This collection focuses on mechanism and kinematics with special emphasis on parallel robots, control, and singularities.

From its excellent technical program to its warm social interaction, ARK culminates with this unique reference on the current developments and new advances in robot kinematics—a genuine tribute to its contributors and organizers!

Naples, Italy
Stanford, CA, USA
March 2017

Bruno Siciliano
Oussama Khatib
SPAR Editor

Preface

Kinematics, the motion of mechanisms, is one of the most fundamental aspects of robot design, analysis, and control, but it is also relevant to other scientific domains, such as biomechanics, molecular biology, and others. This series of books on Advances in Robot Kinematics, which reports the latest achievements in the field, has a long history, as the first symposium was organized in 1988, and the first book was published by Springer in 1991. Since then, a new issue has been published every two years. Each book is linked to a single-track symposium in which the participants exchange their results and opinions in a meeting that brings together the world's best researchers and scientists as well as young students. Since 1992, these symposia have come under the patronage of the International Federation for the Promotion of Machine Science (IFToMM).

In 2016, the symposium related to this book was organized by the French National Research Institute in Computer Science and Control Theory (INRIA) in Grasse, France. We are grateful to the authors for their contributions and to the large team of reviewers for their critical and insightful recommendations. The papers in this book show that robot kinematics is an exciting domain with an enormous number of research challenges that go well beyond the field of robotics. We are also indebted to the members of the HEP-HAISTOS team of INRIA for their help in organizing the symposium. The articles from this symposium were first published in a green open-access archive to favor the free dissemination of the results. We are grateful to Y. Papegay for putting the edition together.

The current book is the 13th in the series of Springer (and Kluwer) and is the result of a peer-review process intended to select the newest and most original achievements in this field. The book was published after the conference. This was unusual for the series from the symposia Advances in Robot Kinematics, because the books are typically released before the conference. However, this circumstance allowed the authors to have their manuscripts further improved and to take into account the opinions and constructive criticisms of the conference participants. Some authors even made re-calculations and produced new and more valuable results.

First and foremost, we are grateful to the authors who participated in this project with all their enthusiasm and commitment. We are grateful to Springer, to the whole team, but especially to Nathalie Jacobs and Cynthia Feenestra, who have made this publication possible. Above all, we are grateful to our younger colleague Tadej Petrič, Ph.D., whose assistance was crucial in the technical production of the book. Without him, things would not have taken place as efficiently and rapidly.

We hope that our book will again reach the shelves of scholars, researchers, and students around the world who are attracted to the unique field of robot kinematics.

Ljubljana, Slovenia
September 2016

Jadran Lenarčič
Jean-Pierre Merlet

Contents

Mass Equivalent Pantographs for Synthesis of Balanced Focal Mechanisms

Volkert van der Wijk

Abstract Force balance is an important property in the design of high-speed high precision machinery to reduce base vibrations and also for the design of inherently safe large movable structures. This paper presents the synthesis of inherently balanced overconstrained focal mechanisms with mass equivalent pantographs. It is shown how pantograph linkages can be combined into an overconstrained but movable linkage by connecting them in their similarity points. With mass equivalent modeling the force balance conditions are derived for which the common center of mass is in the focal point for any motion. As examples, Burmester's focal mechanism is investigated for balance and a new balanced focal mechanism of three mass equivalent pantographs is presented.

1 Introduction

In robotics, dynamic (shaking) force balance is an important property for high-speed motion with minimal base vibrations [4]. Since force balanced mechanisms are statically balanced too, it is also a useful property for large moving structures for save motion with minimal effort.

A problem of common approaches to balance pre-existing mechanisms is that generally a multitude of counter-masses is required [1, 9], leading to unpractical designs with a significant increase of mass and inertia [6]. Instead, a reversed approach was presented where balanced mechanisms are synthesized from inherently balanced linkage architectures [4]. These linkage architectures consist solely of the essential

V. van der Wijk (✉)
Mechatronic System Design, Department of Precision and Microsystems Engineering, Faculty of Mechanical, Maritime, and Materials Engineering, Delft University of Technology, Mekelweg 2, Delft, The Netherlands
e-mail: V.vanderWijk@tudelft.nl; Volkert.vanderWijk@kcl.ac.uk

V. van der Wijk
Centre for Robotics Research, Department of Informatics, Faculty of Natural and Mathematical Sciences, King's College London, Strand, London, UK

J. Lenarčič and J.-P. Merlet (eds.), *Advances in Robot Kinematics 2016*, Springer Proceedings in Advanced Robotics 4, DOI 10.1007/978-3-319-56802-7_1

kinematic relations for balance. With this method a variety of new advantageous inherently balanced mechanism solutions were found among which the first high-speed dynamically balanced parallel manipulator that was successfully built and tested [8].

With inherent balancing it is also possible to synthesize balanced mechanism solutions from *overconstrained* inherently balanced linkage architectures [7]. These architectures have more links than kinematically needed. This gives the designer the freedom to select links to keep or eliminate to obtain a normally constrained balanced mechanism solution. Also more solutions can potentially be found.

The goal of this paper is to investigate focal mechanisms, which are overconstrained and movable, for inherent balance. The focal mechanism of Burmester [2] - the cognate of Kempe's focal mechanism - can be regarded a combination of two pantographs [3]. It is shown how these two pantograph linkages can be combined by connecting them in their similarity points. For force balance the two pantographs need to be mass equivalent with a model of which the common center of mass (CoM) is in the focal point. The conditions for this are derived. In addition also a new inherently balanced focal mechanism of three combined pantographs is presented at the end.

2 CoM in Focal Point of Burmester's Focal Mechanism

Figure 1 shows Burmester's focal mechanism which consists of the two pantograph linkages $P_1A_1P_2S$ - with link lengths l_1, l_2, a_1, and a_2 - and $P_3A_3P_4S$ - with link lengths l_3, l_4, a_3, and a_4 - that are connected with revolute pairs in the similarity points A_0, A_2, and S. This linkage is two times overconstrained yet movable since both pantographs are similar, i.e. elements $A_0A_1P_1 \sim A_1A_2P_2 \sim A_0A_3P_3 \sim A_3A_2P_4$ with angles β_1 and β_2. These four triangular elements are also similar to triangle A_0A_2S for

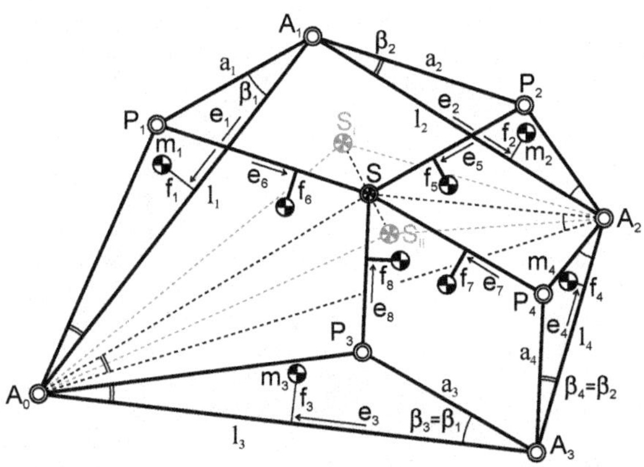

Fig. 1 Burmester's focal mechanism of two pantograph linkages connected in their similarity points A_0, A_2, and S. S is the focal point and is the common CoM of all elements for force balance

any motion of the mechanism. Both pairs of opposite internal four-bars are reflected similar to one another, with one pair being parallelograms.

When, for example, for the upper pantograph a_1, a_2, l_1, and β_1 are given, l_2 and β_2 can be calculated as

$$\lambda_1^S = 1 - \frac{a_1}{l_1}\cos\beta_1, \quad \lambda_2^S = \frac{a_1}{l_1}\sin\beta_1$$
$$\beta_2 = \tan^{-1}\frac{\lambda_2^S}{\lambda_1^S}, \quad l_2 = \frac{a_2}{\lambda_1^S}\cos\beta_2 = \frac{a_2}{\lambda_2^S}\sin\beta_2 \tag{1}$$

with λ_1^S and λ_2^S the constant similarity parameters of the four triangular elements and triangle A_0A_2S. When, subsequently, for the lower pantograph l_3 and l_4 are given, a_3 and a_4 can be calculated as

$$a_3 = (1-\lambda_1^S)\frac{l_3}{\cos\beta_1} = \lambda_2^S\frac{l_3}{\sin\beta_1}, \quad a_4 = l_4\frac{\lambda_1^S}{\cos\beta_2} = l_4\frac{\lambda_2^S}{\sin\beta_2} \tag{2}$$

These parameters can also be obtained from the similarity conditions of the four triangular elements which write

$$\frac{a_1}{l_1} = \frac{a_3}{l_3}, \quad \frac{a_2}{l_2} = \frac{a_4}{l_4} \tag{3}$$

In Fig. 1 each of the eight links i has a mass m_i of which the CoM is defined with parameters e_i and f_i as illustrated. The aim is to design the mechanism such that the common CoM of all elements is in focal point S for any motion. Then the mechanism is inherently force balanced with respect to the focal point.

The force balance conditions describe how the CoM parameters of each element are related for balance. These conditions can be found by mass equivalent modeling with real and virtual equivalent masses [4, 5]. With mass $m_I = m_1 + m_2 + m_5 + m_6$ of upper pantograph $P_1A_1P_2S$ and mass $m_{II} = m_3 + m_4 + m_7 + m_8$ of lower pantograph $P_3A_3P_4S$ the total mass of the focal mechanism can be written as $m_{tot} = m_I + m_{II}$. The common CoM of the upper pantograph is denoted S_I and the common CoM of the lower pantograph is denoted S_{II}. With similarity points A_0 and A_2 these two points form two triangles as well which also have to remain similar for any motion. For force balance then each pantograph is mass equivalent to a 2-DoF mass equivalent model with the conditions [5]

$$m_I^a = m_I(1-\lambda_1^I), \quad m_I^b = m_I\lambda_1^I, \quad m_I^c = m_I\lambda_2^I$$
$$m_{II}^a = m_{II}(1-\lambda_1^{II}), \quad m_{II}^b = m_{II}\lambda_1^{II}, \quad m_{II}^c = m_{II}\lambda_2^{II} \tag{4}$$

with λ_1^I and λ_2^I the similarity parameters of triangle $A_0A_2S_I$, λ_1^{II} and λ_2^{II} the similarity parameters of triangle $A_0A_2S_{II}$, and real equivalent masses m_j^a and m_j^b and virtual equivalent mass m_j^c of each pantograph j. For the upper pantograph in Fig. 2a, b shows the 2-DoF mass equivalent model adapted from [5]. Essentially the virtual

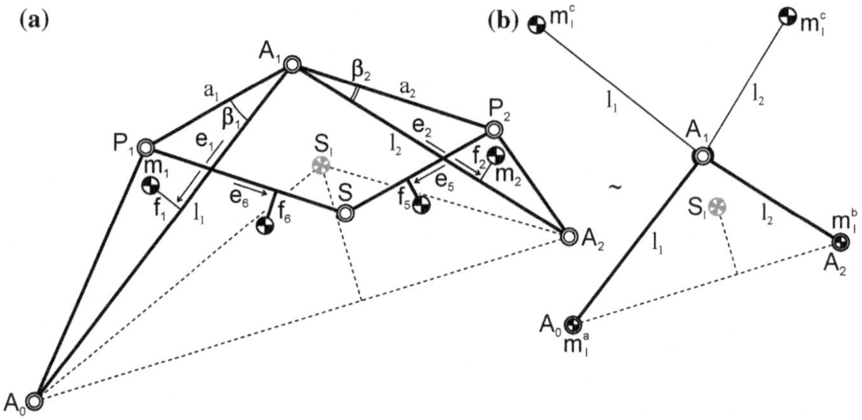

Fig. 2 For force balance (**a**) each pantograph must be mass equivalent to the (**b**) 2-DoF mass equivalent model, here shown for the upper pantograph with CoM in S_I

equivalent mass determines the link CoMs relative to the lines connecting the joints, i.e. the values of parameters f_i, whereas the real equivalent masses determine the link CoMs along the lines connecting the joints, i.e. the values of parameters e_i.

To have the common CoM in the focal point, the sum of the mass equivalent models of the two pantographs should equal the mass equivalent model of the complete mechanism. This can be written as $m_I\lambda_1^I + m_{II}\lambda_1^{II} = m_{tot}\lambda_1^S$ and $m_I\lambda_2^I + m_{II}\lambda_2^{II} = m_{tot}\lambda_2^S$. The resulting model is similar to Fig. 2b but with each equivalent mass replaced with the sum of the equivalent masses of the two pantograph models as $m^a = m_I^a + m_{II}^a, m^b = m_I^b + m_{II}^b$, and $m^c = m_I^c + m_{II}^c$. The conditions for the mass equivalent model of the complete mechanism then are written as

$$m^a = m_{tot}(1 - \lambda_1^S), \quad m^b = m_{tot}\lambda_1^S, \quad m^c = m_{tot}\lambda_2^S \tag{5}$$

The force balance conditions for each pantograph can be derived from the linear momentum equations of each DoF individually where the linear momentum of the mass equivalent model must equal the linear momentum of the real pantograph, similar as for the dyads in [5]. Figure 3a shows the mass motions of DoF 1 of the upper pantograph where link A_1A_2 is fixed and link A_0A_1 rotates about A_1 with angle θ_{I1}. The mass motion of the pantograph for this DoF is shown on the right with a compact Equivalent Linear Momentum System (ELMS) where all masses are projected on element A_0A_1. Figure 3b shows the mass motions of DoF 2 where link A_1A_0 is fixed and link A_1A_2 rotates about A_1 with angle θ_{I2}. Also here the mass motion of the pantograph for this DoF is shown on the right with a compact ELMS where all masses are projected on element A_1A_2. The linear momentum L_1 and L_2 of these individual motions can be written with respect to their relative reference frames $x_{I1}y_{I1}$ and $x_{I2}y_{I2}$, which are aligned with lines A_0A_1 and A_2A_1, respectively, as

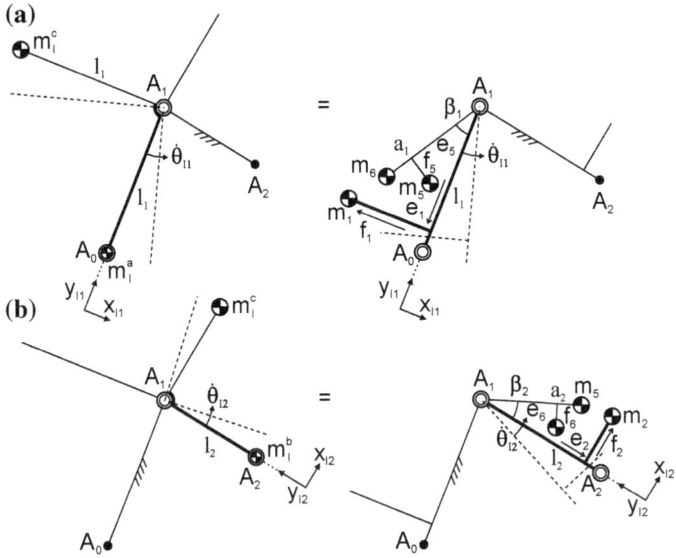

Fig. 3 The force balance conditions are derived from the linear momentum equations of each DoF individually which are equal for the mass equivalent model (*left*) and the real pantograph (*right*, here shown as compact Equivalent Linear Momentum Systems)

$$\frac{\overline{L}_1}{\dot{\theta}_{I1}} = \begin{bmatrix} m_I^a l_1 \\ -m_I^c l_1 \end{bmatrix} = \begin{bmatrix} m_1 e_1 + m_5(e_5 \cos\beta_1 + f_5 \sin\beta_1) + m_6 a_1 \cos\beta_1 \\ -m_1 f_1 - m_5(e_5 \sin\beta_1 - f_5 \cos\beta_1) - m_6 a_1 \sin\beta_1 \end{bmatrix} \quad (6)$$
$$\frac{\overline{L}_2}{\dot{\theta}_{I2}} = \begin{bmatrix} m_I^b l_2 \\ m_I^c l_2 \end{bmatrix} = \begin{bmatrix} m_2 e_2 + m_5 a_2 \cos\beta_2 + m_6(e_6 \cos\beta_2 + f_6 \sin\beta_2) \\ m_2 f_2 + m_5 a_2 \sin\beta_2 + m_6(e_6 \sin\beta_2 - f_6 \cos\beta_2) \end{bmatrix}$$

These equations result in the four force balance conditions

$$m_I^a l_1 = m_1 e_1 + m_5(e_5 \cos\beta_1 + f_5 \sin\beta_1) + m_6 a_1 \cos\beta_1 \quad (7)$$
$$m_I^c l_1 = m_1 f_1 + m_5(e_5 \sin\beta_1 - f_5 \cos\beta_1) + m_6 a_1 \sin\beta_1 \quad (8)$$
$$m_I^b l_2 = m_2 e_2 + m_5 a_2 \cos\beta_2 + m_6(e_6 \cos\beta_2 + f_6 \sin\beta_2) \quad (9)$$
$$m_I^c l_2 = m_2 f_2 + m_5 a_2 \sin\beta_2 + m_6(e_6 \sin\beta_2 - f_6 \cos\beta_2) \quad (10)$$

For the other pantograph the force balance conditions can be derived similarly as

$$m_{II}^a l_3 = m_3 e_3 + m_7(e_7 \cos\beta_1 + f_7 \sin\beta_1) + m_8 a_3 \cos\beta_1 \quad (11)$$
$$m_{II}^c l_3 = m_3 f_3 + m_7(e_7 \sin\beta_1 - f_7 \cos\beta_1) + m_8 a_3 \sin\beta_1 \quad (12)$$
$$m_{II}^b l_4 = m_4 e_4 + m_7 a_4 \cos\beta_2 + m_8(e_8 \cos\beta_2 + f_8 \sin\beta_2) \quad (13)$$
$$m_{II}^c l_4 = m_4 f_4 + m_7 a_4 \sin\beta_2 + m_8(e_8 \sin\beta_2 - f_8 \cos\beta_2) \quad (14)$$

These are the 8 general force balance conditions of the focal mechanism in Fig. 1 for which the common CoM is in the focal point S. For example, from the first four equations the equivalent masses m_I^a, m_I^b, and m_I^c may be found to subsequently calculate with Eq. (5) the equivalent masses m_{II}^a m_{II}^b, and m_{II}^c to be used in the latter four balance conditions. It is also possible to initially choose values for m_I^a m_I^b, and m_I^c. Then for instance from the first four equations e_5, f_5, e_6, and f_6 can be derived as

$$e_5 = \frac{\sin\beta_1(m_I^c l_1 - m_1 f_1 - m_6 a_1 \sin\beta_1) + \cos\beta_1(m_I^a l_1 - m_1 e_1 - m_6 a_1 \cos\beta_1)}{m_5}$$

$$f_5 = \frac{\sin\beta_1(m_I^a l_1 - m_1 e_1 - m_6 a_1 \cos\beta_1) - \cos\beta_1(m_I^c l_1 - m_1 f_1 - m_6 a_1 \sin\beta_1)}{m_5}$$

$$e_6 = \frac{\sin\beta_2(m_I^c l_2 - m_2 f_2 - m_5 a_2 \sin\beta_2) + \cos\beta_2(m_I^b l_2 - m_2 e_2 - m_5 a_2 \cos\beta_2)}{m_6}$$

$$f_6 = \frac{\sin\beta_2(m_I^b l_2 - m_2 e_2 - m_5 a_2 \cos\beta_2) - \cos\beta_2(m_I^c l_2 - m_2 f_2 - m_5 a_2 \sin\beta_2)}{m_6}$$

3 Focal Mechanism of Three Pantographs

In general it is possible to synthesize a variety of inherently force balanced focal linkages by combining multiple mass equivalent pantographs in the same way as in the previous section. Figure 4 shows a new focal mechanism that is composed of the three pantographs $P_1A_1P_2S$, $P_3A_3P_4S$, and $P_5A_5P_6S$ which are connected in similarity points A_0, A_2, A_4, and S where S is the focal point. The resulting linkage is four times overconstrained yet movable. Also here each pantograph has similar triangular elements and a similar triangle of the similarity points. However in this case

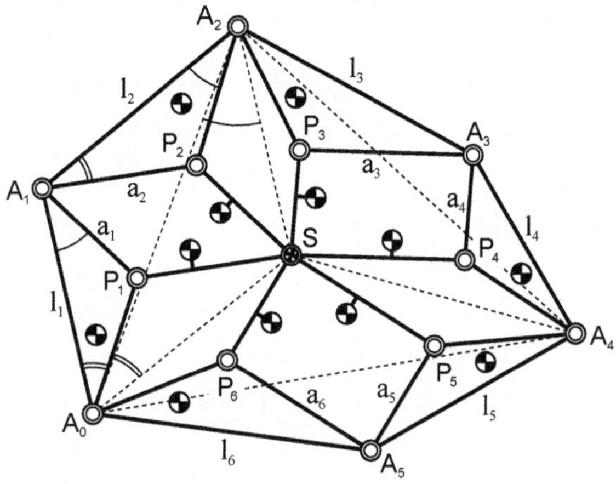

Fig. 4 Focal mechanism of three pantograph linkages connected in their similarity points A_0, A_2, A_4, and S. S is the focal point and is the common CoM of all elements for force balance

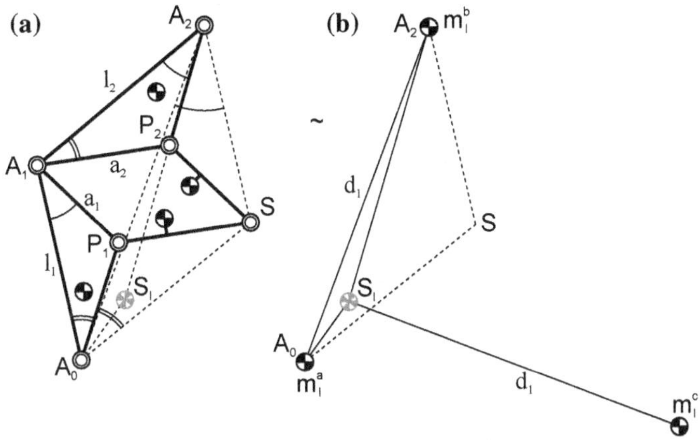

Fig. 5 (**a**) A pantograph with CoM in S_I can be modeled as a (**b**) mass equivalent single element with equivalent masses m_I^a, m_I^b, and m_I^c

the pantographs differ from one another, e.g. the triangular elements of pantograph $P_1A_1P_2S$ are not similar to the triangular elements of the other pantographs. In fact the focal mechanism is a combination of the three different triangles A_0A_2S, A_2A_4S, and A_0A_4S that together form the triangle $A_0A_2A_4$. For each pantograph the dimensions of the elements can be calculated with Eq. (1) with for each pantograph different λ^S parameters. For two pantographs the λ^S parameters can be chosen independently such that with the triangle $A_0A_2A_4$ the third is determined.

The approach to derive the force balance conditions for which the common CoM is in focal point S is similar to Burmester's focal mechanism. Here the mechanism can be considered a combination of three mass equivalent models with each a mass m_I, m_{II}, and m_{III} with CoMs in S_I, S_{II}, and S_{III}, respectively. For each pantograph the force balance conditions can be found with Eq. (6). The equivalent masses m_j^a, m_j^b, and m_j^c of each mass equivalent model are defined according to Eq. (4).

To find the mass equivalent model of the complete focal mechanism it is possible to model each pantograph as a mass equivalent single element as shown for pantograph $P_1A_1P_2S$ in Fig. 5. In Fig. 5b triangle $A_0A_2S_I$ can be regarded a rigid element with a real equivalent mass m_I^a in joint A_0, a real equivalent mass m_I^b in joint A_2, and a virtual equivalent mass m_I^c located at a distance d_1 from S_1, the CoM of the model, as illustrated. d_1 is equal to the distance between A_0 and A_2.

Combining the mass equivalent single elements of the three pantographs then results in the model in Fig. 6. This mass equivalent model of the complete focal mechanism has real equivalent masses $m_I^a + m_{III}^b$ in A_0, $m_I^b + m_{II}^a$ in A_2, and $m_{II}^b + m_{III}^a$ in A_4 and it has virtual equivalent masses m_I^c about S_I, m_{II}^c about S_{II}, and m_{III}^c about S_{III} as illustrated. Figure 7 shows the unified mass equivalent model of the complete focal mechanism. The difference with Fig. 6 is that here all the three virtual equivalent masses are located about S. From this model the conditions for which S is the CoM can be derived as

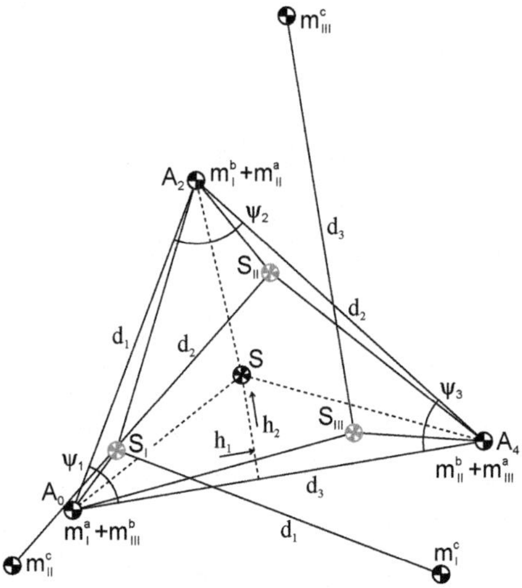

Fig. 6 Combination of the three mass equivalent single models with their equivalent masses. The common CoM of the focal mechanism is the CoM of this combined mass equivalent model

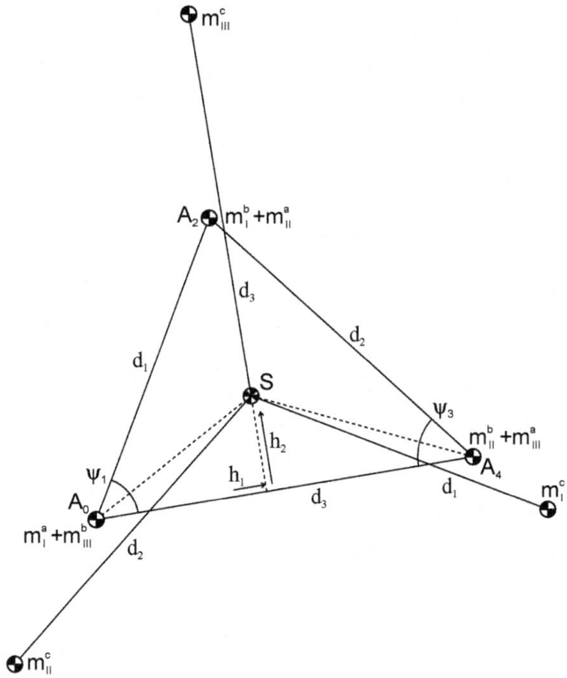

Fig. 7 The unified mass equivalent model of the focal mechanism in Fig. 4 with the CoM in *S* about which all three virtual equivalent masses are located

$$
\begin{aligned}
&(m_{II}^b + m_{III}^a)d_3 + (m_I^b + m_{II}^a)d_1 \cos\psi_1 = m_{tot}h_1 \\
&(m_I^b + m_{II}^a)d_1 \sin\psi_1 = m_{tot}h_2 \\
&m_{III}^c d_3 - m_{II}^c d_2 \cos\psi_3 - m_I^c d_1 \cos\psi_1 = 0 \\
&m_I^c d_1 \sin\psi_1 - m_{II}^c d_2 \sin\psi_3 = 0
\end{aligned}
\tag{15}
$$

with total mass $m_{tot} = m_I + m_{II} + m_{III}$ and with the CoM in S defined with respect to A_0A_4 by h_1 and h_2. The meaning of the first two conditions is that S is the CoM of the real equivalent masses, while the meaning of the last two conditions is that S is the CoM of the virtual equivalent masses.

4 Discussion and Conclusion

The inherent force balance of Burmester's focal mechanism was investigated and the force balance conditions were derived. It was shown that for balancing the focal mechanism can be considered composed of two mass equivalent pantographs. Combination of the mass equivalent models of the pantographs then results in one mass equivalent model of which the center of mass is in the focal point.

It was also shown how with three mass equivalent pantographs a new focal mechanism could be designed. In general, by combining multiple mass equivalent pantographs a variety of inherently balanced focal mechanisms can be synthesized. Unifying the mass equivalent models of all pantographs then results in a single mass equivalent model of which the center of mass is in the focal point.

Parameters a_i are the principal dimensions of the focal mechanism when its common center of mass is in the focal point. When the center of mass of an individual pantograph is in the focal point, then a_i are also the principal dimensions of this individual pantograph.

Although in Burmester's focal mechanism the two pantographs are in opposite branch, this is not required from the force balance conditions. This means that for force balance one of the pantographs or both of them may also be in the other branch, which means that they could also appear as being on top of one another.

Acknowledgements This publication was financially supported by the Niels Stensen Fellowship.

References

1. Briot, S., Bonev, I.A., Gosselin, C.M., Arakelian, V.: Complete shaking force and shaking moment balancing of planar parallel manipulators with prismatic pairs. Multi-Body Dyn. **223**(K), 43–52 (2009)
2. Burmester, L.: Die brennpunktmechanismen. Zeitschrift für Mathematik und Physik **38**, 193–223 and 3 appendices (1893)
3. Dijksman, E.A.: Motion Geometry of Mechanisms. Cambridge University Press (1979)

4. Van der Wijk, V.: Methodology for analysis and synthesis of inherently force and moment-balanced mechanisms - theory and applications (dissertation). University of Twente (free download: http://dx.doi.org/10.3990/1.9789036536301) (2014)
5. Van der Wijk, V.: Mass equivalent dyads. In: Bai, S., Ceccarrelli, M. (eds.) Recent Advances in Mechanism Design for Robotics. MMS 33, pp. 35–45. Springer, Berlin (2015)
6. Van der Wijk, V., Demeulenaere, B., Gosselin, C., Herder, J.L.: Comparative analysis for low-mass and low-inertia dynamic balancing of mechanisms. Mech. Robot. **4**(3), 031008 (2012)
7. Van der Wijk, V., Herder, J.L.: Inherently balanced 4R four-bar based linkages. In: Lenarčič, J., Husty, M. (eds.) Latest Advances in Robot Kinematics, Proceedings of the IFToMM 13th International Symposium on Advances in Robot Kinematics, pp. 309–316. Springer, Berlin (2012). ISBN 978-94-007-4619-0
8. Van der Wijk, V., Krut, S., Pierrot, F., Herder, J.L.: Design and experimental evaluation of a dynamically balanced redundant planar 4-RRR parallel manipulator. Int. J. Robot. Res. **32**(6), 744–759 (2013)
9. Wu, Y., Gosselin, C.M.: Design of reactionless 3-DOF and 6-DOF parallel manipulators using parallelepiped mechanisms. IEEE Trans. Robot. **21**(5), 821–833 (2005)

Compliant Serial 3R Chain with Spherical Flexures

Farid Parvari Rad, Rocco Vertechy, Giovanni Berselli and Vincenzo Parenti-Castelli

Abstract A spherical flexure is a special kind of compliant hinge specifically conceived for spherical motion. It features an arc of a circle as centroidal axis and an annulus sector as cross-section, circle and annulus having a common center coinciding to that of the desired spherical motion. This paper investigates a compliant spherical 3R open chain that is obtained by the in-series connection of three identical spherical flexures having coincident centers and mutually orthogonal axes of maximum rotational compliance. The considered spherical chain is intended to be used as a complex flexure for the development of spatial parallel manipulators. The compliance matrix of the proposed chain is first determined via an analytical procedure. Then, the obtained equations are used in a parametric study to assess the influence of spherical flexure geometry on the overall stiffness performances of the considered 3R open chain.

1 Introduction

Compliant mechanisms (CMs) are a special kind of articulated systems in which motion, force or energy are transferred or transformed through the deflection of flexible members (hereafter briefly referred to as "flexures" or "flexural hinges") [10]. Thanks to the absence (or reduced use) of traditional kinematic pairs, which are

F. Parvari Rad (✉) · R. Vertechy · V. Parenti-Castelli
Department of Industrial Engineering, University of Bologna, Bologna, Italy
e-mail: farid.parvarirad2@unibo.it

R. Vertechy
e-mail: rocco.vertechy@unibo.it

V. Parenti-Castelli
e-mail: vincenzo.parenti@unibo.it

G. Berselli
Department of Mechanical, Energy, Management and Transportation Engineering,
University of Genova, Genova, Italy
e-mail: giovanni.berselli@unige.it

J. Lenarčič and J.-P. Merlet (eds.), *Advances in Robot Kinematics 2016*,
Springer Proceedings in Advanced Robotics 4, DOI 10.1007/978-3-319-56802-7_2

instead based on mating surfaces, CMs are almost not affected by wear, friction and backlash, and only require minimal maintenance with no need of lubrication. Due to their hinge-less nature, CMs can be manufactured in a single piece (for instance via laser or water jet cutting, electrical discharge machining or additive manufacturing), thereby reducing number of parts, assembly needs and, thus, manufacturing costs. With the above-mentioned features, CMs are ideal to work in vacuum, contamination-free, wet or dirty environments and in devices requiring resistance to shocks and silent operation. Common applications of CMs span high-precision manufacturing [27, 36], minimally invasive surgery [9, 18] and micro-electromechanical systems (MEMS) [1, 30].

As regards the existing literature, several studies have been devoted to the design, the characterization and the comparative evaluation of different flexure geometries and CMs formed therewith (see e.g. [14, 21, 31, 35]). In particular, most of these devices have been specifically conceived for the generation of planar motions only, out-of-plane displacements being regarded as *parasitic effects* to be minimized when possible [13]. On the other hand, despite the huge potentialities, exploitation and study of CMs specifically conceived for spatial motions have been much more rare (see e.g. [3–5, 20, 26, 28, 32, 34, 39]). Within this scenario, the development of Spherical CMs (SCMs) has recently attracted the attention of several researchers. SCMs are an important class of flexure-based spatial CMs in which all points of the end-link are ideally constrained to move on concentric spherical surfaces that are fixed with respect to the grounded link. In particular, the in-series ensemble of two or three compliant revolute (R) joints (of either planar notch, planar leaf spring or straight torsion beam type) with orthogonal and intersecting axes has been proposed in [8, 17, 19, 37] to conceive compliant spherical 2R or 3R serial chains to be used as compliant universal or spherical joints for the development of Cardan's [33] and Double-Hooke's couplings [17] and of spatial parallel manipulators [4, 5, 20, 26, 28, 32, 34, 39]. In these applications, the use of compliant spherical 2R or 3R serial chains in place of the axial-symmetric notch primitive flexure is usually preferred owing to the more limited ranges of motions and larger stress concentrations of this latter. The connection of four, five, six or eight bars with an equal number of compliant revolute joints (of either straight crease or lamina emergent torsional type) with intersecting axes has been considered in [6, 7, 38] for the development of 4R, 5R, 6R or 8R closed single-loop lamina-emergent SCMs, as well as arrays thereof (including the six bar Watt's and Stephenson's linkages), to be used in origami-inspired foldable systems such as pop-up books, industrial packaging and deployable devices. Planar notch and straight torsion beam flexures have been used in [19] to develop an actuated miniature 3-CRU (C and U denoting cylindrical and universal joints respectively) spherical parallel CM for the orientation of parts and tools in space. The in-parallel connection of three symmetrically placed spherical 3R serial chains employing either lamina emergent straight torsion beam or notch flexures has been proposed in [11, 29] for the development of 3-(3R) spherical parallel CMs with flat initial state to be used in compact pointing devices such as in MEMS beam-steering mirrors or medical instruments.

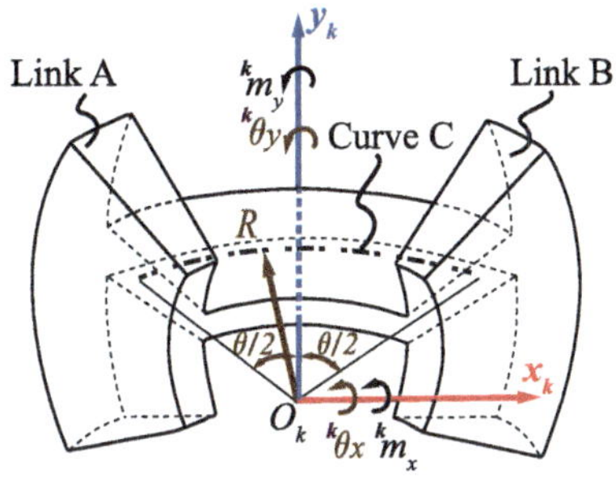

Fig. 1 Spherical flexures

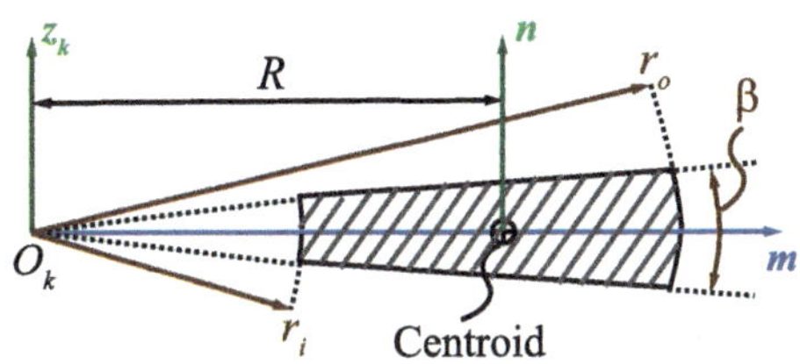

Fig. 2 Cross section properties of SFs

In all the above-mentioned studies, the considered SCMs have been obtained by employing compliant revolute flexures specifically conceived for planar motion applications. In contrast to this, Circularly-Curved Beam Flexures (CCBFs) with constant cross-section and featuring lower rotational rigidity along the radial direction have been proposed in [12, 22] for the development of SCMs with improved spherical motion capabilities. Among these CCBFs, those with annulus sector cross-section as depicted in Figs. 1 and 2, hereafter referred to as Spherical Flexures (SFs), have recently been demonstrated among the most effective ones in reducing the drift of the desired center of spherical motion under the combined action of torques and forces [25].

In this context, this paper investigates the use of SFs for the development of compliant spherical 3R serial chains to be used as SCMs or as spherical complex flexure components for spatial CMs with either serial or parallel architecture. As depicted in Fig. 3, the considered spherical chains are obtained by the in-series connection of three identical SFs that are arranged in space so as to share the same center of curvature and have mutually orthogonal axes of maximum rotational compliance. In particular, analytical results are provided to characterize the compliance behavior of the considered chain in 3D space as a function of flexure geometric parameters.

2 Formulation

A spherical flexure connecting the rigid links A and B is depicted in Fig. 1. It is a solid of revolution characterized by an annulus sector cross-section with inner and outer radii, r_i and r_o, and subtended angle β (see Fig. 2), an axis of revolution z_k

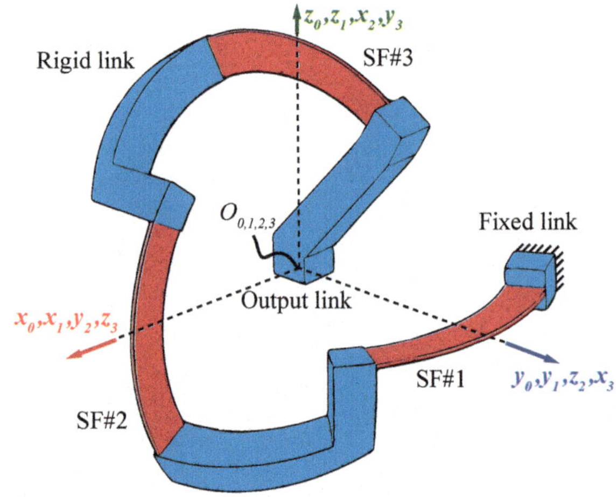

Fig. 3 SF-based compliant spherical 3R open chain

passing through the center O_k of the annulus and orthogonal to the cross-section axis of symmetry, **m**, (see Fig. 1), and revolution angle θ (which describes the flexure length). Cross-section dimensionless parameters, β and w^* ($w^* = \frac{r_o - r_i}{r_o}$), are such that its smaller area moment of inertia is in the direction of the **m** axis. Assuming link A being clamped and B free and loaded, the small deflection behavior of the flexure about its unloaded configuration can be described by the following relation [23]:

$$ {}^k\mathbf{s} = \begin{bmatrix} {}^k\mathbf{u} \\ {}^k\boldsymbol{\theta} \end{bmatrix} = \begin{bmatrix} {}^k\mathbf{C}_{uf} & {}^k\mathbf{C}_{um} \\ {}^k\mathbf{C}_{\theta f} & {}^k\mathbf{C}_{\theta m} \end{bmatrix} \cdot \begin{bmatrix} {}^k\mathbf{f} \\ {}^k\mathbf{m} \end{bmatrix} = {}^k\mathbf{C} \cdot {}^k\mathbf{w} \tag{1} $$

where ${}^k\mathbf{s}$ is composed of an incremental translation ${}^k\mathbf{u} = [{}^k u_x \;\; {}^k u_y \;\; {}^k u_z]^T$ and an incremental rotation ${}^k\theta = [{}^k\theta_x \;\; {}^k\theta_y \;\; {}^k\theta_z]^T$, ${}^k\mathbf{w}$ is composed of an incremental force ${}^k\mathbf{f} = [{}^k f_x \;\; {}^k f_y \;\; {}^k f_z]^T$ and an incremental torque ${}^k\mathbf{m} = [{}^k m_x \;\; {}^k m_y \;\; {}^k m_z]^T$, whereas ${}^k\mathbf{C}_{uf}$, ${}^k\mathbf{C}_{um}$, ${}^k\mathbf{C}_{\theta f}$, ${}^k\mathbf{C}_{\theta m}$ are three-dimensional matrices composed of entries with dimensions [m/N], [1/N], [rad/N], and [rad/Nm] respectively.

As a consequence, ${}^k\mathbf{C} \equiv {}^k C_{ij}$ is a 6×6 matrix with entries of non uniform physical dimensions, the submatrices ${}^k\mathbf{C}_T = [{}^k\mathbf{C}_{uf} \;\; {}^k\mathbf{C}_{um}]$ and ${}^k\mathbf{C}_R = [{}^k\mathbf{C}_{\theta f} \;\; {}^k\mathbf{C}_{\theta m}]$ relating the external wrench to the resulting translations and rotations respectively.

The expression of Eq. 1 is frame dependent. For any SF intended for spherical motion about the center of its centroidal axis circle, a suitable frame is S_k that features center at O_k and orthogonal axes x_k, y_k and z_k respectively lying on centroidal axis plane, on beam symmetry plane and along the intersection of these two planes (see Fig. 1). In this frame, indeed, sub-matrices ${}^k\mathbf{C}_{uf}$ and ${}^k\mathbf{C}_{\theta m}$ are diagonal (meaning that x_k, y_k and z_k are along the principal directions of rotational and translational compliance of the flexure), and the components of ${}^k\mathbf{C}_{uf}$ and ${}^k\mathbf{C}_{um}$ (or ${}^k\mathbf{C}_{\theta f}$) indicate how the desired center of spherical motion drifts as a consequence of applied external forces and torques.

Knowing matrix ${}^{k}\mathbf{C}$ for a single spherical flexure, the compliance matrix of the in-series ensemble of any number n of identical flexures can be obtained with the following formula [2]:

$$ {}^{0}\mathbf{C} = \sum_{k=1}^{n} {}^{0}\mathbf{T}_{k}^{-T} \cdot {}^{k}\mathbf{C} \cdot {}^{0}\mathbf{T}_{k}^{-1} = \sum_{k=1}^{n} {}^{k}\mathbf{T}_{0}^{T} \cdot {}^{k}\mathbf{C} \cdot {}^{k}\mathbf{T}_{0} \quad (2) $$

where ${}^{k}\mathbf{T}_{0}$ is a 6×6 matrix to transform the components of the stiffness matrix ${}^{k}\mathbf{C}$ of the $k{-}th$ flexure from the local frame S_k to a ground frame S_0. In particular, the expression of ${}^{k}\mathbf{T}_{0}$ is:

$$ {}^{k}\mathbf{T}_{0} = \left[\begin{array}{c:c} {}^{k}\mathbf{R}_{0} & \mathbf{0} \\ \hdashline {}^{k}\tilde{\mathbf{r}}_{0} \cdot {}^{k}\mathbf{R}_{0} & {}^{k}\mathbf{R}_{0} \end{array} \right] = \left[\begin{array}{c:c} {}^{0}\mathbf{R}_{k}^{T} & \mathbf{0} \\ \hdashline \left({}^{0}\tilde{\mathbf{r}}_{k} \cdot {}^{0}\mathbf{R}_{k}\right)^{T} & {}^{0}\mathbf{R}_{k}^{T} \end{array} \right] \quad (3) $$

where ${}^{k}\mathbf{R}_{0}$ denotes the rotation matrix of frame S_0 with respect to frame S_k and ${}^{0}\tilde{\mathbf{r}}_{k}$ indicating the skew symmetric matrix of the position vector ${}^{k}\mathbf{r}_{0}$, which locates the origin of frame S_0 with respect to frame S_k.

For the compliant spherical 3R chain shown in Fig. 3, made by three identical spherical flexures with coincident centers O_k and mutually orthogonal axes, the overall compliance matrix expressed with respect to the reference frame of the first spherical flexure (namely, $S_0 \equiv S_1$) results as [24]:

$$ {}^{0}\mathbf{C}^{3\mathrm{R}} = \begin{bmatrix} C_{x,f_x} & 0 & 0 & 0 & C_{x,m_y} & C_{x,m_z} \\ 0 & C_{y,f_y} & 0 & C_{y,m_x} & 0 & C_{y,m_z} \\ 0 & 0 & C_{z,f_z} & C_{z,m_x} & C_{z,m_y} & 0 \\ 0 & C_{\theta_x,f_y} & C_{\theta_x,f_z} & C_{\theta_x,m_x} & 0 & 0 \\ C_{\theta_y,f_x} & 0 & C_{\theta_y,f_z} & 0 & C_{\theta_y,m_y} & 0 \\ C_{\theta_z,f_x} & C_{\theta_z,f_y} & 0 & 0 & 0 & C_{\theta_z,m_z} \end{bmatrix}_{3\mathrm{R}} \quad (4) $$

where:

$$ \begin{aligned} C_{x,f_x} &= C_{y,f_y} = C_{z,f_z} = \frac{R\theta\left(I_n GJ + GJR^2A + R^2EAI_n\right)}{EAI_n GJ} = C_t \\ C_{\theta_x,m_x} &= C_{\theta_y,m_y} = C_{\theta_z,m_z} = \frac{R\theta\left(I_n EI_m + I_n GJ + GJI_m\right)}{GJEI_m I_n} = C_r \\ C_{x,m_y} &= C_{y,m_z} = C_{z,m_x} = C_{\theta_x,f_z} = C_{\theta_y,f_x} = C_{\theta_z,f_y} = \frac{-2R^2\sin(\theta/2)}{GJ} = C_{tr_1} \\ C_{x,m_z} &= C_{y,m_x} = C_{z,m_y} = C_{\theta_x,f_y} = C_{\theta_y,f_z} = C_{\theta_z,f_x} = \frac{2R^2\sin(\theta/2)}{EI_n} = C_{tr_2} \end{aligned} \quad (5) $$

In Eq. 5, E and G are the Young's and shear moduli of the employed material. A, R, I_m, I_n and J are, respectively, cross section area, centroidal axis radius, area moments of inertia and torsional constant of the flexure cross section (refer to Fig. 2) that read as follows [25]:

$$A = \frac{r_o^2\beta}{2} - \frac{r_i^2\beta}{2} = \frac{(r_o^2 - r_i^2)\beta}{2} \quad (6)$$

$$R = \frac{4}{3}\frac{(r_o^3 - r_i^3)\sin\beta/2}{(r_o^2 - r_i^2)\beta} \quad (7)$$

$$I_m = \frac{1}{8}(r_o^4 - r_i^4)(\beta - \sin\beta) \quad (8)$$

$$I_n = \frac{1}{8}(r_o^4 - r_i^4)(\beta + \sin\beta) - \frac{8}{9}\frac{(r_o^3 - r_i^3)^2\sin^2(\beta/2)}{(r_o^2 - r_i^2)\beta} \quad (9)$$

$$J = \frac{2}{3}\sin^3(\beta/2)(r_o^4 - r_i^4) - 16\sin^4(\beta/2)(V_L r_o^4 + V_S r_i^4) \quad (10)$$

where:

$V_L = 0.10504 - 0.2\sin(\beta/2) + 0.3392\sin^2(\beta/2) - 0.53968\sin^3(\beta/2) + 0.82448\sin^4(\beta/2)$
$V_S = 0.10504 + 0.2\sin(\beta/2) + 0.3392\sin^2(\beta/2) + 0.53968\sin^3(\beta/2) + 0.82448\sin^4(\beta/2)$

Equation 10 is the formula for the torsional constant firstly proposed by J.B. Reynolds to account for the warping of annulus sector cross-sections [15, 16].

As one can notice from Eqs. 4 and 5, the compliance matrix of the compliant spherical 3R chain with respect to frame S_0 still retains diagonal translational and rotational sub-matrices (${}^0\mathbf{C}_{uf}$ and ${}^0\mathbf{C}_{\theta m}$), and is only a function of four independent factors: C_t, C_r, C_{tr_1} and C_{tr_2}. C_r is the primary rotational compliance of the 3R chain, which should be as high as possible to minimize resistance to desired spherical motions. C_t is a secondary translational compliance, which should be as close as possible to zero to minimize drift of the desired center of spherical motion (O_0) under the action of the force vector ${}^0\mathbf{f}$ applied on the end-link. C_{tr_1} and C_{tr_2} are secondary coupled rotational-translational compliances, which should be as close as possible to zero to minimize spherical motion center drift under the action of the torque vector ${}^0\mathbf{m}$ applied on the end-link.

3 Parametric Evaluation of the Compliant Spherical 3R Chain

This section investigates the influence of flexure geometry on the ability of the considered 3R chain in the generation of spherical motions. The study is performed by evaluating the following three indices:

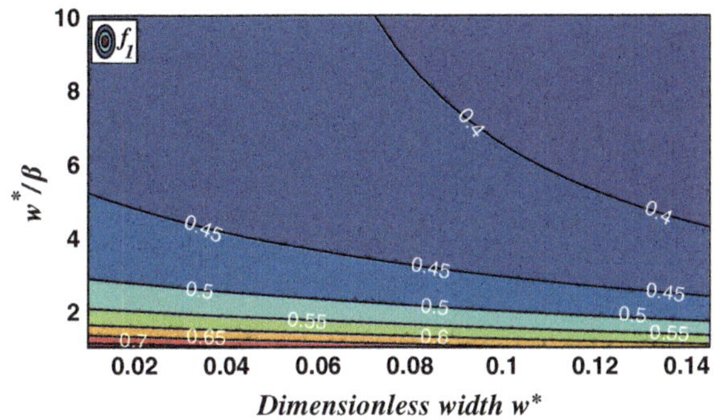

Fig. 4 The influence of varying w^* and w^*/β on f_1

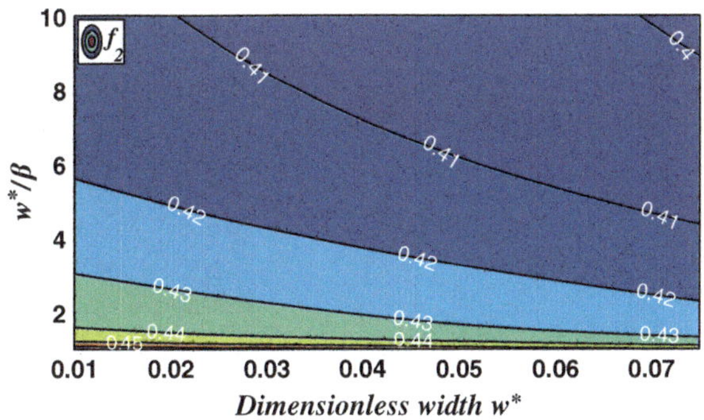

Fig. 5 The influence of varying w^* and w^*/β on f_2 for $\theta = 45°$

$$f_1 = \left| \frac{1}{r_o^2} \frac{C_t}{C_r} \right| \qquad f_2 = \left| \frac{1}{r_o} \frac{C_{tr_1}}{C_r} \right| \qquad f_3 = \left| \frac{1}{r_o} \frac{C_{tr_2}}{C_r} \right| \tag{11}$$

that represent the dimensionless ratios of the translational and coupled translational-rotational compliances of a generic compliant spherical 3R chain to the rotational counterpart. In the definition of these indices, the curvature radius r_o of the SF is used as characteristic size to obtain scale-independent expressions that only depend on the flexure shape dimensionless parameters w^*, β and θ. In particular, f_1 is only a function of w^* and w^*/β, whereas f_2 and f_3 also depend on θ. Among the possible choices, r_o has been chosen as characteristic length since it describes the overall encumbrance of the 3R chain, which is often the most important application constraint in the design optimization process. Plots of Eq. 11 are reported in Figs. 4, 5, 6, 7 and 8 as a function of the SF aspect ratios w^* and w^*/β. The dependency of f_2 and f_3 on θ is shown by comparing Figs. 5 and 7 (for $\theta = 45°$) to Figs. 6 and 8 (for $\theta = 90°$). In addition, the contour plot of the size independent factor $C_r^* = C_r * r_o^3/\theta$ (which is constant irrespective of the value of θ and only dependent on the cross section aspect ratios w^* and w^*/β) is reported in Fig. 9. As figures show, maximization of

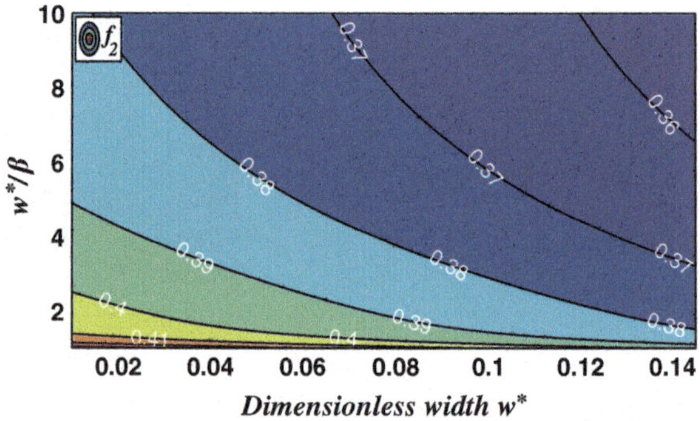

Fig. 6 The influence of varying w^* and w^*/β on f_2 for $\theta = 90°$

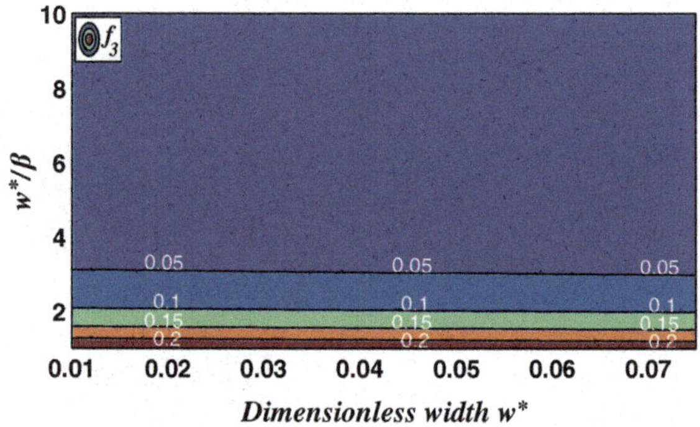

Fig. 7 The influence of varying w^* and w^*/β on f_3 for $\theta = 45°$

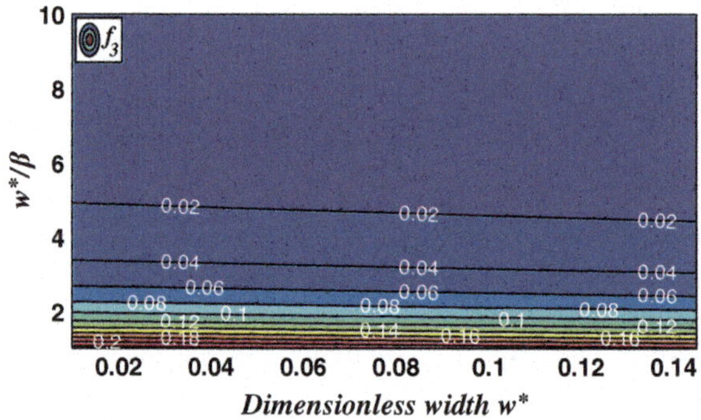

Fig. 8 The influence of varying w^* and w^*/β on f_3 for $\theta = 90°$

the spherical motion generation capabilities of the considered compliant 3R chain (that is, minimization of secondary to primary compliance ratios) can be obtained by adopting the largest possible values for θ and w^*/β (within the limits of physical realizability) as well as for w^* (within the limit of validity of the slender beam approximation; namely $w^* < 0.1\theta$).

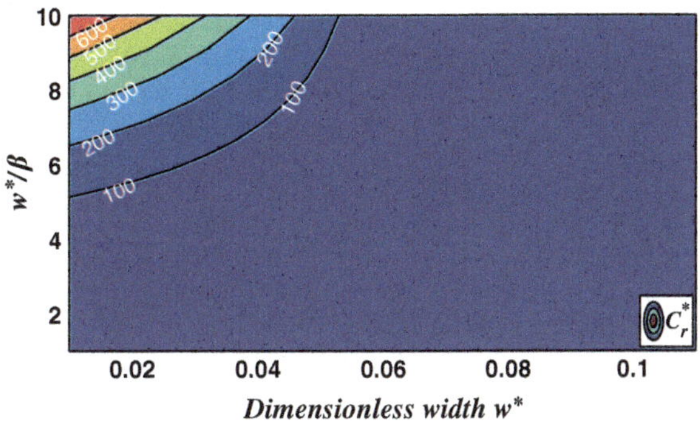

Fig. 9 The influence of varying w^* and w^*/β on C_r^*

4 Conclusions

A compliant open chain featuring three in-series connected identical primitive spherical flexures with coincident centers of curvature and mutually orthogonal axes of principal compliance, is introduced and analyzed for application in spherical compliant mechanisms. First, the closed form compliance equations of the proposed spherical chain are presented as a function of flexure dimensions and employed material. The obtained equations are then used to study the influence of flexure dimensions on spherical chain parasitic motions. The study is performed by evaluating three dimensionless ratios of the translational and coupled translational-rotational compliances of a generic compliant spherical 3R chain to the rotational counterpart. The results show that maximization of the spherical motion generation capabilities of the considered compliant 3R chain (that is, minimization of secondary to primary compliance ratios) can be obtained by adopting the largest feasible values for θ and w^*/β (within the limits of physical realizability) as well as for w^* (within the limit of validity of the slender beam approximation; namely $w^* < 0.1\theta$). Future activities will be devoted to the study of the compliant spherical chains in the case of stockier flexures as well as in the large deformation range.

References

1. Belfiore, N.P., Balucani, M., Crescenzi, R., Verotti, M.: Performance analysis of compliant MEMS parallel robots through pseudo-rigid-body model synthesis. In: ASME ESDA 11th Biennial Conference on Engineering Systems Design and Analysis, pp. 329–334 (2012)
2. Carter Hale, L.: Principles and techniques for designing precision machines. Ph.D. thesis, Department of Mechanical Engineering, MIT, Cambridge, MA (1999)
3. Chen, S., Culpepper, M.L.: Design of a six-axis micro-scale nanopositioner μhexflex. Precis. Eng. **30**(3), 314–324 (2006)
4. Dong, W., Sun, L., Du, Z.: Stiffness research on a high-precision, large-workspace parallel mechanism with compliant joints. Precis. Eng. **32**(3), 222–231 (2008)

5. Dunning, A., Tolou, N., Herder, J.: A compact low-stiffness six degrees of freedom compliant precision stage. Precis. Eng. **37**(2), 380–388 (2013)
6. Greenberg, H., Gong, M., Magleby, S., Howell, L.: Identifying links between origami and compliant mechanisms. Mech. Sci **2**(2), 217–225 (2011)
7. Hanna, B.H., Lund, J.M., Lang, R.J., Magleby, S.P.: Waterbomb base: a symmetric single-vertex bistable origami mechanism. Smart Mater. Struct. **23**(9), 094009 (2014)
8. Hesselbach, J., Wrege, J., Raatz, A., Becker, O.: Aspects on design of high precision parallel robots. Assem. Autom. **24**(1), 49–57 (2004)
9. Hong, M.B., Jo, Y.H.: Design and evaluation of 2-DOF compliant forceps with force-sensing capability for minimally invasive robot surgery. IEEE Trans. Robot. **28**(4), 932–941 (2012)
10. Howell, L.L.: Compliant Mechanisms. Wiley, New York (2001)
11. Jacobsen, J.O., Chen, G., Howell, L.L., Magleby, S.P.: Lamina emergent torsional (LET) joint. Mech. Mach. Theory **44**(11), 2098–2109 (2009)
12. Li, G., Chen, G.: Achieving compliant spherical linkage designs from compliant planar linkages based on prbm: a spherical young mechanism case study. In: 2012 IEEE International Conference on Robotics and Biomimetics (ROBIO), pp. 193–197, IEEE (2012)
13. Lobontiu, N.: Compliant Mechanisms: Design of Flexure Hinges. CRC Press (2002)
14. Lobontiu, N., Paine, J., Garcia, E., Goldfarb, M.: Corner-filleted flexure hinges. J. Mech. Des. **123**(3), 346–352 (2001)
15. Love, A.: A Treatise on the Mathematical Theory. Dover Public (1944)
16. Lyse, I., Johnston, B.: Structural beams in torsion, 1934. Fritz Laboratory Reports (1934)
17. Machekposhti, D.F., Tolou, N., Herder, J.: A review on compliant joints and rigid-body constant velocity universal joints toward the design of compliant homokinetic couplings. J. Mech. Des. **137**(3), 032301 (2015)
18. Moon, Y., Choi, J.: A compliant parallel mechanism for needle intervention. In: 2013 35th Annual International Conference of the IEEE Engineering in Medicine and Biology Society (EMBC), pp. 4875–4878 (2013)
19. Palmieri, G., Palpacelli, M.C., Callegari, M.: Study of a fully compliant u-joint designed for minirobotics applications. ASME J. Mech. Des. **134**(11), 111003(9) (2012)
20. Parlaktaş, V., Tanık, E.: Single piece compliant spatial slider-crank mechanism. Mech. Mach. Theory **81**, 1–10 (2014)
21. Paros, J.: How to design flexure hinges. Mach. Des. **37**, 151–156 (1965)
22. Parvari Rad, F., Berselli, G., Vertechy, R., Parenti-Castelli, V.: Evaluating the spatial compliance of circularly curved-beam flexures. In: Computational Kinematics, pp. 329–336. Springer (2013)
23. Parvari Rad, F., Berselli, G., Vertechy, R., Parenti-Castelli, V.: Stiffness analysis of a fully compliant spherical chain with two degrees of freedom. In: Advances in Robot Kinematics, pp. 273–284. Springer (2014)
24. Parvari Rad, F., Berselli, G., Vertechy, R., Parenti-Castelli, V.: Design and stiffness analysis of a compliant spherical chain with three degrees of freedom. Precis. Eng. **47**, 1–9 (2017)
25. Parvari Rad, F., Vertechy, R., Berselli, G., Parenti-Castelli, V.: Analytical compliance analysis and finite element verification of spherical flexure hinges for spatial compliant mechanisms. Mech. Mach. Theory **101**, 168–180 (2016)
26. Pham, H.H., Chen, I.M.: Stiffness modeling of flexure parallel mechanism. Precis. Eng. **29**(4), 467–478 (2005)
27. Polit, S., Dong, J.: Development of a high-bandwidth XY nanopositioning stage for high-rate micro-/nanomanufacturing. IEEE/ASME Trans. Mechatron. **16**(4), 724–733 (2011)
28. Ratchev, S.: Precision assembly technologies for mini and micro products. In: Proceedings of the IFIP TC5 WG5, 5 Third International Precision Assembly Seminar (IPAS'2006), 19–21 February 2006, Bad Hofgastein, Austria, vol. 198. Springer Science & Business Media (2006)
29. Rubbert, L., Renaud, P., Gangloff, J.: Design and optimization for a cardiac active stabilizer based on planar parallel compliant mechanisms. In: ASME 2012 11th Biennial Conference on Engineering Systems Design and Analysis, pp. 235–244. American Society of Mechanical Engineers (2012)

30. Sauceda-Carvajal, A., Kennedy-Cabrera, H.D., Hernndez-Torres, J., Herrera-May, A.L., Mireles, J.: Compliant MEMS mechanism to extend resolution in fourier transform spectroscopy. In: Proceedings of SPIE, Micromachining and Microfabrication Process Technology XIX **8973**, 89,730S–89,730S–9 (2014)
31. Schotborgh, W., Kokkeler, F., Tragter, H., van Houten, F.: Dimensionless design graphs for flexure elements and a comparison between three flexure elements. Precis. Eng. **29**(1), 41–47 (2005)
32. Su, H.J.: Mobility analysis of flexure mechanisms via screw algebra. J. Mech. Robot. **3**(4), 041010 (2011)
33. Tanık, Ç.M., Parlaktaş, V., Tanık, E., Kadıoğlu, S.: Steel compliant cardan universal joint. Mech. Mach. Theory **92**, 171–183 (2015)
34. Teo, T.J., Chen, I.M., Yang, G.: A large deflection and high payload flexure-based parallel manipulator for uv nanoimprint lithography: Part ii. Stiffness modeling and performance evaluation. Precis. Eng. **38**(4), 872–884 (2014)
35. Tian, Y., Shirinzadeh, B., Zhang, D., Zhong, Y.: Three flexure hinges for compliant mechanism designs based on dimensionless graph analysis. Precis. Eng. **34**(1), 92–100 (2010)
36. Tian, Y., Zhang, D., Shirinzadeh, B.: Dynamic modelling of a flexure-based mechanism for ultra-precision grinding operation. Precis. Eng. **35**(4), 554–565 (2011)
37. Trease, B., Moon, Y., Kota, S.: Design of large-displacement compliant joints. J. Mech. Des. **127**(4), 788–798 (2005)
38. Wilding, S.E., Howell, L.L., Magleby, S.P.: Spherical lamina emergent mechanisms. Mech. Mach. Theory **49**, 187–197 (2012)
39. Wu, T.L., Chen, J.H., Chang, S.H.: A six-dof prismatic-spherical-spherical parallel compliant nanopositioner. IEEE Trans. Ultrason. Ferroelectr. Freq. Control **55**(12), 2544–2551 (2008)

Combining Tube Design and Simple Kinematic Strategy for Follow-the-Leader Deployment of Concentric Tube Robots

Cédric Girerd, Kanty Rabenorosoa and Pierre Renaud

Abstract Concentric tube robots show promising performances for many medical applications. A particularly useful but challenging deployment of these robots, called "follow-the-leader" deployment, consists in the robot following the path traced out by its tip. In this paper, we propose to combine a simple and analytical kinematic approach combined with now possible tube design to offer efficient follow-the-leader behavior. The approach is presented and then assessed with promising performances using a realistic scenario in the context of human nose exploration.

1 Introduction

Concentric tube robots (CTR) constitute a class of continuum robots that is of particular interest in the medical context [3]. The displacements of the robot end-effector are then obtained by relative translations and rotations of precurved elastic tubes with diameters that can be typically below 3 mm [5, 10]. Complex shapes of robots can be generated using remote actuation, that is particularly relevant for navigation in constrained anatomical areas. CTR kinematics are however complex because of the mechanical interactions between the tubes. Several models have been derived that now include in particular the impact of tube torsion which occurs in the general case [2, 6]. Complementary work is now focused on the design of CTR tubes, with local modifications of their shape and structure to obtain anisotropic behavior and hence to modify the relative importance of bending and torsion in the tubes. Until now, it is however still difficult to build a CTR and its control to satisfy a so-called follow-the-leader approach, where the CTR body only occupies the volume swept by its tip during the deployment. Even though this approach can be mandatory from an application point-of-view, limited solutions have indeed been reported to choose

C. Girerd (✉) · P. Renaud
ICube, UDS-CNRS-INSA, 300 bd Sébastien Brant, Illkirch, France
e-mail: cedric.girerd@icube.unistra.fr

K. Rabenorosoa
FEMTO-ST, AS2M, Univ. Bourgogne Franche-Comté, UFC/ENSMM/CNRS,
Besançon, France

J. Lenarčič and J.-P. Merlet (eds.), *Advances in Robot Kinematics 2016*,
Springer Proceedings in Advanced Robotics 4, DOI 10.1007/978-3-319-56802-7_3

accordingly a robot design, i.e. the number and geometry of precurved tubes, and the deployment sequence [6].

In this paper we propose a simple-to-use and yet efficient approach to CTR design, with the corresponding deployment strategy, that allows an approximate follow-the-leader behavior. In a realistic case study, i.e. the context of human nose exploration, it is shown to be of adequate performance. The design and deployment strategy combines the now existing possibility of tube design with a kinematic approach. For planar trajectories, the goal of the paper, analytical formulation of robot geometry, defined by number of tubes and their curvatures, and path following sequence are introduced. Required modifications of tubes are then determined from the task simulation. The method we propose consists first in path generation and robot design. Their descriptions are introduced in Sects. 2 and 3 respectively with illustration of the considered application. In Sect. 4, the determination of tube modifications is performed and the method assessed through simulation of torsion impact as well as control errors during the deployment. Conclusions and perspectives are finally given in Sect. 5.

2 Trajectory Generation

In [6], the conditions for exact follow-the-leader deployment have been investigated. To be admissible, a trajectory must not induce tube torsion. From a design point-of-view, tubes with precurved helical shapes are identified as candidates for such deployments, but their manufacturing remains delicate. The other option, for planar paths, uses planar precurved tubes with constant curvatures in the same or opposite configurations. As a first step of our method, the CTR path is hence determined by including this latter constraint. It can be easily demonstrated that a set of tubes with planar configurations and constant curvatures form a CTR with 2D shape and constant curvature by sections. The first step is therefore to identify a trajectory that is admissible from an application perspective and constituted by portions of constant curvatures.

In the following of the paper, an application is used for evaluation. We consider the deployment of a CTR to reach the olfactory cleft, located in the upper part of the nose. Figure 1a represents the 3D shape of the area reconstructed from CT images. The goal is to reach the olfactory cleft starting from the nostril. A planar path with constant curvature by sections can be identified to join the two regions, as illustrated in Fig. 1a, b. A point-cloud extraction and circle-fitting algorithm based on least-squares optimization, not detailed for the sake of compactness, is used to identify the path parameters. Three sections are determined, with length and curvature for each one being equal to: 17.32, 14.27, 14.20 mm and 0.0353, 0.0646 and 0.0592 mm^{-1} respectively. The resulting trajectory is a key element that will be used as a reference to quantify deployment errors during the robot insertion.

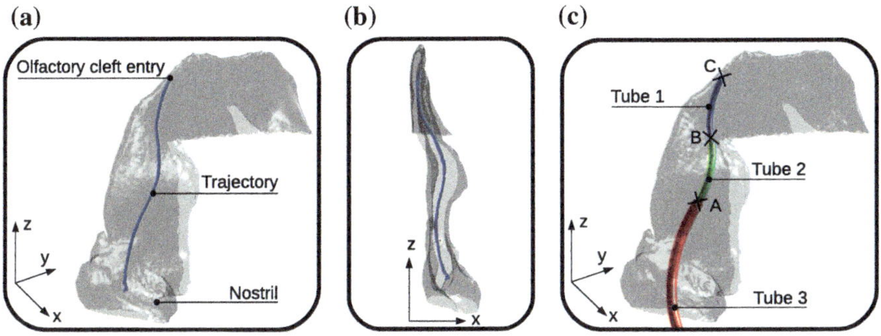

Fig. 1 Front view and perspective view of the nasal cavity with suitable trajectory (**a**, **b**) and CTR with 3 tubes deployed (**c**)

3 Robot Design and Deployment

The CTR is considered to be composed of planar precurved tubes of constant curvatures, located in the same plane, with aligned or opposite curvatures. In this situation indeed, no torsion occurs in the tubes. If the tubes have aligned curvatures, one can easily imagine that the resulting robot will be in a stable configuration. On the contrary, depending on tube mechanical properties, opposite curvatures can make the configuration unstable. This can be of course a main issue for the deployment. To handle this situation, modification of the tube structure as described in [1, 7] is considered. This does not affect the robot synthesis, described below, with determination of the number of tubes and their overall shapes, lengths and curvatures. Robot design is thus described below, tube design being introduced in Sect. 4.

Let n be the number of tubes of the CTR. Each tube numbered i is described by its precurvature $\boldsymbol{u}_i^{*F_i(s)}(s) = [u_{ix}^*(s)\ u_{iy}^*(s)\ u_{iz}^*(s)]^T$ expressed in its cross section material coordinate frame $F_i(s)$. The stiffness properties of the tube are expressed by the frame-invariant stiffness tensor $K_i = \text{diag}(E_i I_i, E_i I_i, G_i J_i)$, with E_i, G_i respectively its Young and Shear modulus, and I_i, J_i respectively its cross section area and polar moment of inertia. The curvature resulting from the combination of the n tubes is then given by [5]:

$$\boldsymbol{u}^{F_0(s)}(s) = \left(\sum_{i=1}^{n} K_i\right)^{-1} \sum_{i=1}^{n} K_i \boldsymbol{u}_i^{*F_0(s)}(s) \tag{1}$$

with $F_0(s)$ the cross-section reference frame which experiences no twist when translated along the robot's centerline.

In order to determine the robot geometry, we propose to invert Eq. (1) in order to solve for the tube curvatures. As the tubes are considered to have a constant curvature along their length, without any presence of torsion, their curvatures can then be written as $\boldsymbol{u}_i^* = [\kappa_i\ 0\ 0]^T$ or $\boldsymbol{u}_i^* = [0\ \kappa_i\ 0]^T$. The path sections are characterized by their lengths $(s_1, \ldots, s_n)$ and curvatures $(1/r_1, \ldots, 1/r_n)$, starting from the path end.

The number n of tubes is immediately determined to correspond to the number of constant curvature sections on the path generated in the previous section. The tube lengths $(l_1, \ldots, l_n)$ and curvatures $(\kappa_1, \ldots, \kappa_n)$ can then be determined using Eq. (2), which has to be solved first for tube 1, the inner tube, up to tube n, the outer tube, in increasing index order, and using positive or negative path and robot curvatures depending on the curvature direction:

$$\begin{aligned} l_j &= \sum_{i=j}^{n} s_i \\ \kappa_1 &= \frac{1}{r_1} \quad \text{and} \quad \kappa_{j,j>1} = \frac{1}{E_j I_j}\left(\sum_{i=1}^{j} \frac{E_i I_i}{r_j} - \sum_{i=1}^{j-1} E_i I_i \kappa_i\right) \end{aligned} \tag{2}$$

In (2), the values of Young's modulus and cross section area moments of inertia I_j are chosen from characteristics of circular Nitinol tubes, the standard choice for CTR because of the material superelasticity. Using diameters of commercially-available circular tubes, the lengths and curvatures are obtained from Eq. (2) and indicated in Table 1.

By extending the strategy described in [6], the computed CTR geometry can be deployed with a follow-the-leader approach. As a first step, the n tubes are inserted altogether by pure translation until the tube n reaches the end of its stroke (Fig. 2).

Table 1 Robot parameters after synthesis with the proposed method for the considered application

Tube index	Young's modulus (GPa)	Shear modulus (GPa)	Inner diameter (mm)	Outer diameter (mm)	Length (mm)	Curvature (mm^{-1})
1	80	30	0.880	1.200	45.8	0.059
2	80	30	1.296	1.524	31.6	0.136
3	80	30	1.760	2.184	17.3	0.066

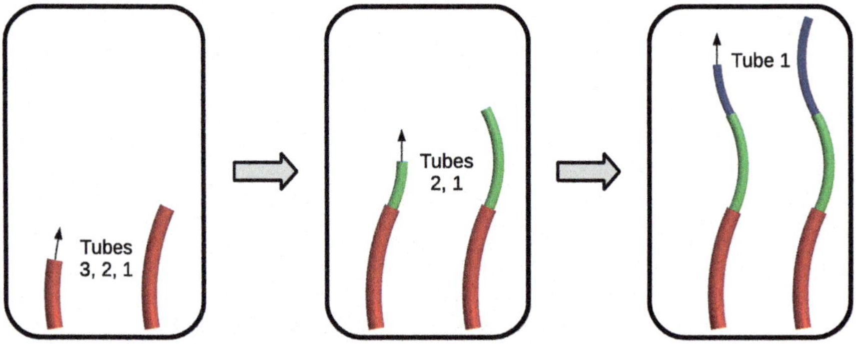

Fig. 2 Deployment sequence of CTR with three tubes ($n = 3$)

Then, the $n-1$ tubes are inserted together, with the tube n remaining fixed, until tube $n-1$ reaches the end of its stroke. The procedure is repeated until the robot is fully deployed.

4 Tube Design and Deployment Assessment

We have introduced in Sects. 2 and 3 how it is possible to generate a CTR geometry and its deployment strategy for planar paths composed of constant curvature sections. It was previously outlined that the use of concentric tubes with opposite curvatures can lead to unstable configurations. To make use of a CTR for medical purposes, it is obvious that only stable equilibrium positions should be used. In this section, we propose to design tubes with anisotropic properties by local structure modification to handle this particular issue and make the deployment strategy safe and accurate.

4.1 Robot Kinematic Model

The evaluation of the influence of tube properties is conducted using the application data. The presented application context will also be used to assess the deployment accuracy. Therefore, a model is needed to analyze the impact of torsion on the interactions between the tubes, and possible deviations between stable and unstable configurations. We here use the torsional model developed in [4, 9], based on an energy method. For a set of n tubes experiencing bending and torsion, the energy stored in the tubes is given by Eq. (3) if the tubes overlap continuously for $s \in [\sigma_1, \sigma_2]$.

$$E = \frac{1}{2}\int_{\sigma 1}^{\sigma 2} (\boldsymbol{u}(s) - \boldsymbol{\alpha}(s))^T K(\boldsymbol{u}(s) - \boldsymbol{\alpha}(s)) + C ds, \tag{3}$$

with

$$\begin{aligned} \boldsymbol{\alpha}(s) &= K^{-1}\sum_{i=1}^{n} K_i \overline{\boldsymbol{u}}_i^*(s), \quad K = \sum_{i=1}^{n} K_i, \quad \theta_i(s) = \psi_i(s) - \psi_1(s), \\ \overline{\boldsymbol{u}}_i^*(s) &= R_{\theta_i}\boldsymbol{u}_i^* - \dot{\theta}_i(s)\boldsymbol{e}_3, \quad C(s) = \sum_{i=1}^{n} \overline{\boldsymbol{u}}_i^{*T}(s) K_i \overline{\boldsymbol{u}}_i^*(s) - \boldsymbol{\alpha}^T(s) K \boldsymbol{\alpha}(s), \end{aligned} \tag{4}$$

In these equations, $\boldsymbol{u}(s)$ represents the equilibrium curvature vector for the robot, and R_{θ_i} the rotation matrix of angle θ_i about $\boldsymbol{e}_3 = [0\ 0\ 1]^T$. The angles $(\psi_1, \ldots, \psi_n)$ designate the absolute angular variables of each tube. As neither $C(s)$ nor $\boldsymbol{\alpha}(s)$ depends on $\boldsymbol{u}(s)$, the minimal energy is obtained for $\boldsymbol{u}(s) = \boldsymbol{\alpha}(s)$. In order to compute $\boldsymbol{\alpha}(s)$, the variables $\psi_i(s)$ that describe the evolution of torsion along the tubes have to be estimated. For tubes with constant curvature of the form $\boldsymbol{u}_i^* = [\kappa_i\ 0\ 0]^T$

or $\boldsymbol{u}_i^* = [0\ \kappa_i\ 0]^T$, as previously considered, functions $\psi_i(s)$ can be expressed by solving the system (5) of two first order differential equations [8]:

$$\begin{aligned}\dot{\psi}_i(s) &= u_{iz}(s)\\ \dot{u}_{iz}(s) &= \frac{E_i I_i}{EIG_i J_i}\sum_{j=1}^{n} E_j I_j \kappa_i \kappa_j \sin(\psi_i(s) - \psi_j(s))\end{aligned} \tag{5}$$

where $EI = \sum_{j=1}^{n} E_j I_j$. The system (5) can be easily used for a CTR composed of several sections by writing it for each set of overlapped tubes and adding continuity constraints to the solution.

The set of equations representing torsion along the entire length of the robot is a boundary value problem that is solved numerically using a finite difference code as implemented in the bvp5c function in Matlab (The MathWorks Inc., Natick, USA). Boundary conditions are $\psi_i(0)$, the known angles at the tubes insertion points, and $\dot{\psi}_i(l_i) = 0$ as tubes can not apply axial moments at their distal ends. Finally, the shape of the robot can be determined using Eq. (6), with superscript symbol "hat" denoting the conversion of an element of $\mathbb{R}^3$ to an element of $\mathfrak{so}(3)$, the Lie algebra of Lie group SO(3).

$$\begin{aligned}\dot{\boldsymbol{p}} &= \boldsymbol{R}\boldsymbol{e}_3\\ \dot{\boldsymbol{R}} &= \boldsymbol{R}\hat{\boldsymbol{u}}\end{aligned} \tag{6}$$

To solve Eq. (6), $\boldsymbol{u}(s)$ can be approximated by constant values over a given step size, and then a Runge-Kutta method can be used for the resolution.

4.2 *Tube Design*

At this point, all the tube characteristics are known except the polar moments of inertia J_j, $j \in [1, n]$. To select their adequate values, that can be adjusted by local tube modification as described in [7], the ratio between bending and torsion stiffnesses is being varied in simulation. The ratio λ is equal to $E_j I_j / G_j J_j$ [7] and is considered identical for all the tubes. As expected, Fig. 3 shows that lowest values of λ minimize the torsional effects and finally the position errors in stable configurations (Table 2). However, it is challenging to obtain experimentally very low values for λ. In [7], the authors have successfully reached a value of 0.344. For our application, a ratio $\lambda = 0.348$ results in a tip error of 2.51 mm, which is acceptable for our application. This value is therefore selected, which ends the robot design with the proposed approach.

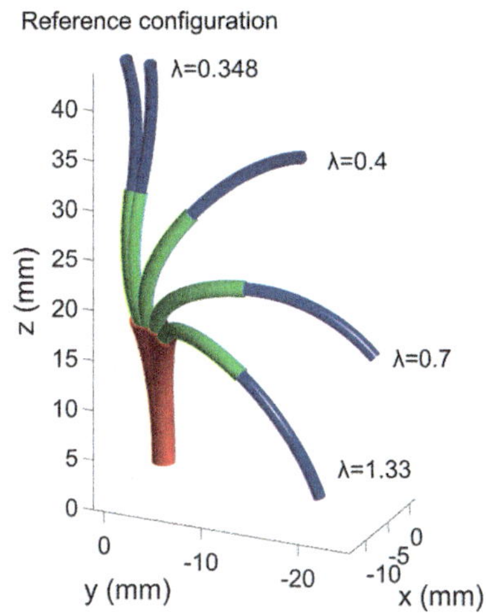

Fig. 3 Robot positions for different λ values

Table 2 Tip position errors as a function of λ. λ = 1.33 corresponds to standard non-modified circular tubes

λ	Tip position error (mm)
0.348	2.51
0.4	19.24
0.7	36.49
1.33	44.96

4.3 Follow-the-Leader Behavior Assessment

We now investigate precisely the follow-the-leader deployment errors by measuring tip position error and root mean square error along the robot's centerline for different deployment stages. The tubes are considered held at their insertion point. Measurements are reported in Table 3 with two indicators: the maximum error (Max), which is equal to the tip error, and the RMS value of deviation along the deployed robot. Three configurations during the deployment are considered: after deployment of the 3 tubes altogether (Point A, Fig. 1c), of tubes 1 and 2 altogether (Point B, Fig. 1c), and finally of tube 1 (Point C, Fig. 1c). At point A, errors are equal to zero. Further analysis shows that anisotropy leads to the existence of only one configuration, superimposed with the desired deployed geometry, a situation previously observed

Table 3 Position errors during the follow-the-leader deployment. Maximum errors are in plain letters, RMS values are in bold

Situation	A	B	C
Without error (mm)	0 (**0**)	0.79 (**0.27**)	2.51 (**0.95**)
With error (mm)	0 (**0**)	1.05 (**0.36**)	3.31 (**1.26**)

for two tubes in [9]. Results obtained are suitable for our application in terms of tip deviation.

In order to go further and explore the robustness of the follow-the-leader behavior, we introduce angular errors in the control of the tubes at their base. A value of 0.005 degree, achievable with standard encoders and transmissions, is chosen. Results in Table 3 indicate the maximum errors in presence of tube angular errors. Those results still remain acceptable for our application, which is encouraging and show the interest of the proposed robot design and deployment method.

5 Conclusion

In this paper, we have proposed a method for CTR design and deployment. It combines a simple kinematic approach and tube design to achieve approximate follow-the-leader deployment. We have shown that interesting accuracy can be obtained for a medical application. The effect of the actuation control errors on the robot deployment have also been studied, and we have demonstrated that such errors remain acceptable for our application. The interest of tube modification is outlined, with high sensitivity of the robot behavior to these modifications. Further work will now be focused on selecting the best tube patterning techniques to obtain anisotropic properties. Evaluation of sensitivity to design parameters and extension to other 3D deployment situations will also be considered.

Acknowledgements This work was supported by the French National Agency for Research within the Biomedical Innovation program (NEMRO ANR-14-CE17-0013), and the Investissements d'Avenir (Robotex ANR-10-EQPX-44, Labex CAMI ANR-11-LABX-0004 and Labex ACTION ANR-11-LABX-0001-01)

References

1. Azimian, H., Francis, P., Looi, T., Drake, J.: Structurally-redesigned concentric-tube manipulators with improved stability. In: IEEE/RSJ International Conference on Intelligent Robots and Systems, pp. 2030–2035 (2014)
2. Bergeles, C., Gosline, A., Vasilyev, N., Codd, P., del Nido, P., Dupont, P.: Concentric tube robot design and optimization based on task and anatomical constraints. IEEE Trans. Robot. **31**(1), 67–84 (2015)
3. Burgner-Kahrs, J., Rucker, D.C., Choset, H.: Continuum robots for medical applications: a survey. IEEE Trans. Robot. **31**(6), 1261–1280 (2015)
4. Dupont, P., Lock, J., Butler, E.: Torsional kinematic model for concentric tube robots. In: IEEE International Conference on Robotics and Automation, pp. 3851–3858 (2009)
5. Dupont, P., Lock, J., Itkowitz, B., Butler, E.: Design and control of concentric-tube robots. IEEE Trans. Robot. **26**(2), 209–225 (2010)
6. Gilbert, H., Neimat, J., Webster, R.: Concentric tube robots as steerable needles: achieving follow-the-leader deployment. IEEE Trans. Robot. **31**(2), 246–258 (2015)

7. Lee, D., Kim, J., Kim, J., Baek, C., Noh, G., Kim, D., Kim, K., Kang, S., Cho, K.: Anisotropic patterning to reduce instability of concentric-tube robots. IEEE Trans. Robot. **31**(6), 1311–1323 (2015)
8. Rucker, D.C.: The mechanics of continuum robots: model–based sensing and control. Ph.D. thesis, Vanderbilt University (2011)
9. Rucker, D.C., Webster I, R.J., Chirikjian, G.S., Cowan, N.J.: Equilibrium conformations of concentric-tube continuum robots. Int. J. Robot. Res. **29**, 1263–1280 (2010)
10. Webster I, R.J., Romano, J.M., Cowan, N.J.: Mechanics of precurved-tube continuum robots. IEEE Trans. Robot. **25**, 67–78 (2009)

A Screw-Based Dynamic Balancing Approach, Applied to a 5-Bar Mechanism

Jan de Jong, Johannes van Dijk and Just Herder

Abstract Dynamic balancing aims to reduce or eliminate the shaking base reaction forces and moments of mechanisms, in order to minimize vibration and wear. The derivation of the dynamic balance conditions requires significant algebraic effort, even for simple mechanisms. In this study, a screw-based balancing methodology is proposed and applied to a 5-bar mechanism. The method relies on four steps: (1) representation of the links' inertias into point masses, (2) finding the conditions for these point masses which result in dynamic balance in one given pose (instantaneous balance), (3) extending these conditions over the workspace to achieve global balance, (4) converting the point mass representation back to feasible inertias. These four steps are applied to a 5-bar mechanism in order to obtain the conditions which ensure complete force balance and additional moment balance over multiple trajectories. Using this methodology, six out of the eight balancing conditions are found directly from the momentum equations.

1 Introduction

The ever increasing demands on the throughput of robots requires reduction of their cycle times without compromising the accuracy and the lifetime. Higher velocities induce stronger base reaction forces and moments which in turn cause frame vibration and wear of the manipulator [8]. Dynamic balancing aims to design the kinematics and the mass distribution of the manipulator such that both the changing base reaction forces and moments are eliminated [10]. With force balancing, only the changing reaction forces are considered [1].

J. de Jong (✉) · J. van Dijk · J. Herder
University of Twente, Enschede, Netherlands
e-mail: j.j.dejong@utwente.nl

J. van Dijk
e-mail: j.vandijk@utwente.nl

J. Herder
e-mail: j.h.herder@utwente.nl

J. Lenarčič and J.-P. Merlet (eds.), *Advances in Robot Kinematics 2016*,
Springer Proceedings in Advanced Robotics 4, DOI 10.1007/978-3-319-56802-7_4

Dynamic balancing often involves the addition of linkages and masses - such as counter masses and/or counter rotations - which in general leads to higher complexity and higher motor torques [8]. For parallel mechanisms, the closure equations supports finding dynamic balance without additional linkages or counter-rotations [5]. However, for mechanisms with more DOFs, the dynamic balance conditions become increasingly difficult to find as the number of bodies increase and the kinematic closure equations become more complicated. To overcome this, several synthesis methods are presented; such as stacking of dynamically balanced 4-bar linkages [10], and synthesis based on principal vector linkages [7]. Nevertheless, these synthesis methodologies do not cover all the possible solutions and require considerable effort to find the balancing conditions.

In this paper, a screw theory-based, four step methodology is presented to simplify the process of finding the dynamic balance conditions for planar mechanisms, with a potential extension to spatial mechanisms. The methodology relies on two insights. Firstly, the geometric screw theory gives the conditions for the direct calculation of a subset of the balancing conditions without differentiation or solving the kinematic closure equations. Secondly, the dynamics equations are simplified using an inertia decomposition method derived from Foucault and Gosselin [2]. This approach is illustrated by applying it to a 5-bar mechanism to obtain complete force balance (similar to [4]) with additional moment balance over multiple trajectories (similar to the Dual V [9]). First the kinematic model of a 5-bar mechanism (Sect. 2.1), and the screw dynamics (Sect. 2.2) are described, based on which the four steps are illustrated (Sects. 2.3–2.6).

2 Method

2.1 Kinematic Model of a 5-Bar Mechanism

The 5-bar mechanism under investigation consists of two RR linkages connected by a revolute joint at $\boldsymbol{x}$ (see Fig. 1). To each body a reference frame (ψ_i) is associated in the joint as seen in the figure. The base reference frame is placed arbitrarily. The frame in which a point is represented is denoted with a superscript (e.g. $\boldsymbol{a}^i$).

The velocity of a body in space is described by a 6D twist vector ($\boldsymbol{t}_k^j$), which is the general global velocity of frame ψ_k expressed in ψ_j. The angular velocity is denoted by $\boldsymbol{\omega}$ and the linear velocity by $\boldsymbol{v}$. The coordinate transformation matrix ($\boldsymbol{X}_i^j$) changes the expression of a twist from frame ψ_i to ψ_j. The matrix consists of a rotation matrix ($\boldsymbol{R}_i^j$) and a translation vector ($\boldsymbol{o}_i^j$)[1]:

[1]The $[\boldsymbol{o}_i^j \times]$ is a skew symmetric form of vector $\boldsymbol{o}_i^j$.

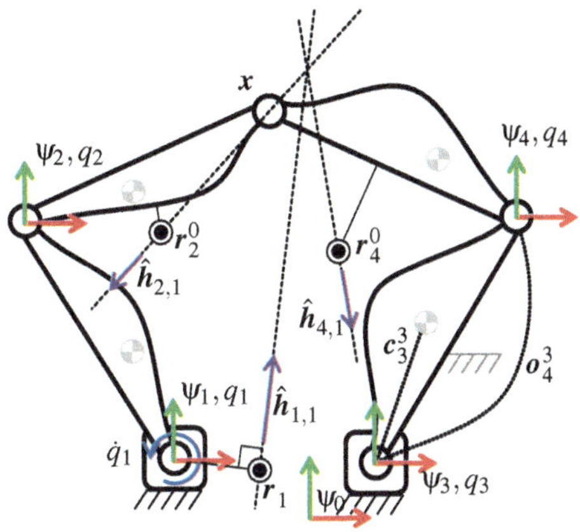

Fig. 1 Kinematic model and instantaneous balance of a 5-bar mechanism. The momentum wrenches sum to zero for a pure motion of joint 1

$$t_k^j = \begin{bmatrix} \omega_k^j \\ v_k^j \end{bmatrix} = X_i^j t_k^i \qquad X_i^j = \begin{bmatrix} R_i^j & 0 \\ [o_i^j \times] R_i^j & R_i^j \end{bmatrix} \tag{1}$$

The body Jacobian (J_i) relates the joint velocities to the twist (t_i^0) of each body. As input joint velocities ($\dot{q}$) we choose the base joints. This makes bodies 1 and 3 the active, and 2 and 4 the passive (non-actuated) bodies. The Jacobian of the mechanism can be found using methods such as presented by Zoppi et al. [12]. The body Jacobians are concatenated such that the total mechanism Jacobian (J) becomes:

$$t_i^0 = J_i \dot{q}_f \qquad J = \begin{bmatrix} J_1 \\ \vdots \\ J_4 \end{bmatrix} = \begin{bmatrix} X_1^0 \hat{t} & 0 \\ X_1^0 \hat{t} + d_1 X_2^0 \hat{t} & d_2 X_2^0 \hat{t} \\ 0 & X_3^0 \hat{t} \\ d_3 X_4^0 \hat{t} & X_3^0 \hat{t} + d_4 X_4^0 \hat{t} \end{bmatrix} \tag{2}$$

in which $\hat{t} = \left[(n_z)^T \; 0^T\right]^T$ is the local twist axis, in which n_z is the unit vector in z direction. The Jacobian coefficients are:

$$d_1 = -\frac{x^4 \cdot R_1^4 [n_z \times] o_2^1}{x^4 \cdot R_1^4 [n_z \times] R_2^1 x^2} - 1 \qquad d_2 = \frac{x^4 \cdot R_3^4 [n_z \times] o_4^3}{x^4 \cdot R_3^4 [n_z \times] R_2^3 x^2} \tag{3}$$

$$d_3 = \frac{o_2^1 \cdot R_2^1 [n_z \times] x^2}{x^4 \cdot R_1^4 [n_z \times] R_2^1 x^2} \qquad d_4 = -\frac{o_4^3 \cdot R_2^3 [n_z \times] x^2}{x^4 \cdot R_3^4 [n_z \times] R_2^3 x^2} - 1 \tag{4}$$

Note that the denominators of all the coefficients are equal, yet, we write them differently in terms of the rotation matrix in the nominator for later use (Eq. 11).

2.2 Dynamics

The conditions for dynamic balance are usually derived from the momentum equations. Shaking forces and moments are the derivate of momentum. When assuming zero initial velocity, the shaking forces and moments are zero when the momentum is zero for all motions.

In screw theory, this momentum is seen as wrench [6]; the momentum wrench ($\boldsymbol{h}$), a concatenation of the angular momentum ($\boldsymbol{\xi}$) and the linear momentum ($\boldsymbol{p}$). The momentum wrench can be expressed in another frame using a second coordinate transformation matrix. The momentum generated by a body, is calculated from the twist or the Jacobain of that body and the inertia matrix $\boldsymbol{M}_i$, which is given later (Eq. 7).

$$\boldsymbol{h}^j = \begin{bmatrix} \boldsymbol{\xi}^j \\ \boldsymbol{p}^j \end{bmatrix} = (\boldsymbol{X}^i_j)^T \boldsymbol{h}^i \qquad \boldsymbol{h}^0_i = \boldsymbol{M}_i \boldsymbol{t}^0_i = \boldsymbol{M}_i \boldsymbol{J}_i \dot{\boldsymbol{q}} \tag{5}$$

The admissible momentum wrench of a mechanism is defined by its momentum span. As the 5-bar mechanism is a 2 DOF mechanism, the dimension of the span is maximally two. In the current study we choose the bases of this momentum span (indicated with a hat) as the momenta generated by unit velocity of the two base joints.

$$\begin{bmatrix} \hat{\boldsymbol{h}}^0_1 \; \hat{\boldsymbol{h}}^0_3 \end{bmatrix} = \sum_i^4 \boldsymbol{M}_i \boldsymbol{J}_i = 0 \tag{6}$$

When the momentum span is only zero for a certain pose we have obtained a local momentum equilibrium or instantaneous balance, this is a necessarily but not sufficient condition for dynamic balance. For global dynamic balance, these instantaneous conditions have to be extended over the complete workspace.

2.3 Step 1. Inertia Decomposition

Wu and Gosselin [11] used the property that the inertia of a body can be represented as a collection of point masses to study the dynamic equivalence of robotic platforms. Continuing on that, we recognize that the inertia of a planar body can be sufficiently represented by two point masses. For a given center of mass (COM), inertia and mass, four equations have to be satisfied [2]. As two point masses give six variables, the location of one point mass can be chosen freely, fixing the location of the other mass and the mass distribution over the two points.

When the free point mass is placed on the revolute joint, this joint has no influence on the motion of the free point. Therefore this point can also be regarded to be fixed to the connecting body. This leaves the initial body with one point mass representation $(\boldsymbol{r}_i, m_i)$. This inertia decomposition can be applied throughout the whole mechanism, such that the inertia properties of each body are characterized by a single point mass. This reduces the number of dynamic parameters from 4 to 3 per body. This finally gives the inertia matrix for a planar body:

$$\boldsymbol{M}_i = m_i \begin{bmatrix} -[\boldsymbol{r}_i^0 \times]^2 & [\boldsymbol{r}_i^0 \times] \\ -[\boldsymbol{r}_i^0 \times] & \boldsymbol{I}_3 \end{bmatrix} \tag{7}$$

2.4 Step 2. Instantaneous Balance

Dynamic balancing occurs when the location and mass of these points are such that the momentum span reduces to zero. For a 5-bar, the mechanism's momentum basis is defined by the motion of one joint while the other joint (and body) is fixed. This means that only three bodies contribute to each mechanism's momentum basis.

$$-\hat{\boldsymbol{h}}_{1,1}^0 = \hat{\boldsymbol{h}}_{2,1}^0 + \hat{\boldsymbol{h}}_{4,1}^0 \qquad -\hat{\boldsymbol{h}}_{3,3}^0 = \hat{\boldsymbol{h}}_{2,3}^0 + \hat{\boldsymbol{h}}_{4,3}^0 \tag{8}$$

The three body momentum bases are represented as wrenches (see to Fig. 1). For force balance, the vector sum of the linear momenta has be to zero. For additional moment balance, the three wrenches have to intersect at one point. A momentum wrench generated by rotation of a point mass around an axis passes trough the point mass in a direction perpendicular to the point and the axis location. Therefore it follows that the point mass of the base body has to be on the intersection point of a line perpendicular to the wrench line $(\hat{\boldsymbol{h}}_{1,1})$ and the axis of rotation, as indicated in Fig. 1. The mass to be located at this point is given by ratio of linear and angular momentum.

$$\boldsymbol{r}_1^1 = \frac{1}{m_1}[\boldsymbol{n}_z \times]\left(\hat{\boldsymbol{p}}_{2,1}^1 + \hat{\boldsymbol{p}}_{4,1}^1\right) \qquad m_1 = -\frac{\|\hat{\boldsymbol{p}}_{2,1}^1 + \hat{\boldsymbol{p}}_{4,1}^1\|^2}{\boldsymbol{n}_z \cdot \left(\hat{\boldsymbol{\xi}}_{2,1}^1 + \hat{\boldsymbol{\xi}}_{4,1}^1\right)} \tag{9}$$

If similar conditions are imposed on the second base link ($\boldsymbol{r}_3^3$, and m_3), we have obtained six instantaneous balance conditions.

2.5 Step 3. Global Force Balance

Global force balance is obtained when the sum of the linear momentum span of the passive bodies (2 and 4) - expressed in the base bodies reference frames - is constant over the workspace. This is required to enforce a pose independent solution for Eq. 9. Therefore, the global force balance conditions only depends on the dynamic properties of the passive bodies. After coordinate transformation of the linear part of Eq. 8, the following constraint equation is obtained:

$$-\hat{\boldsymbol{p}}^1_{1,1} = m_2[\boldsymbol{n}_z\times]\boldsymbol{o}^1_2 + m_2(1+d_1)[\boldsymbol{n}_z\times]\boldsymbol{R}^1_2\boldsymbol{r}^2_2 + m_4 d_3[\boldsymbol{n}_z\times]\boldsymbol{R}^1_4\boldsymbol{r}^4_4 = const. \quad (10)$$

Inspection shows that Eq. 10 and the terms d_1 and d_3 of Eqs. 3 and 4 are only written in terms of the variables $\boldsymbol{R}^1_2(q_{2,1})$ and $\boldsymbol{R}^1_4(q_{4,1})$. For global force balance, the derivative of Eq. 10 with respect to these two angles should remain zero:

$$\frac{\delta}{\delta q_{1,4}}\left(\hat{\boldsymbol{p}}^1_{1,1}\right) = \frac{d_3}{\boldsymbol{x}^4\cdot\boldsymbol{R}^4_1[\boldsymbol{n}_z\times]\boldsymbol{R}^1_2\boldsymbol{x}^2}\left(m_2\|\boldsymbol{x}^4\|^2[\boldsymbol{n}_z\times]\boldsymbol{R}^1_2\boldsymbol{r}^2_2\right.$$
$$\left. + m_4\left(\left(\boldsymbol{x}^4\cdot\boldsymbol{x}^2\right) + \left(\boldsymbol{x}^4\cdot[\boldsymbol{n}_z\times]\boldsymbol{x}^2\right)\boldsymbol{I}_3\right)\boldsymbol{R}^1_2\boldsymbol{r}^4_4\right) = 0 \quad (11)$$

From this derivative, the following global force balance conditions is obtained.

$$\boldsymbol{r}^2_2 = \frac{m_4}{m_2}\frac{1}{\|\boldsymbol{x}^4\|^2}\left(\left(\boldsymbol{x}^4\cdot[\boldsymbol{n}_z\times]\boldsymbol{x}^2\right)[\boldsymbol{n}_z\times] - \left(\boldsymbol{x}^4\cdot\boldsymbol{x}^2\right)\boldsymbol{I}_3\right)\boldsymbol{r}^4_4 \quad (12)$$

This constraint equation can also be obtained when differentiating Eq. 10 to the other angles ($q_{1,2}$), and from the derivatives of ($\hat{\boldsymbol{p}}^3_{3,3}$). The implications of global force balance and additionally instantaneous dynamic balance on reactionless trajectories are discussed in Sect. 2.7. This global solution step still requires considerable algebraic effort.

2.6 Step 4. Inertia Recomposition

The resulting balancing conditions are written in terms of the point masses, describing a range of inertias. To select a proper inertias, we recognize that at the joints ($\boldsymbol{u}_{i,j}$) - connecting body i with j - a point mass ($a_{i,j}$) can be exchanged between the bodies (see to Fig. 2). The mass which is added to one link has to be subtracted from the connecting link ($a_{i,j} = -a_{j,i}$). In such a way the inertia (g_i) and COM ($\boldsymbol{c}_i$) of mass ($m_{t,i}$) can be selected which satisfy the balance conditions.

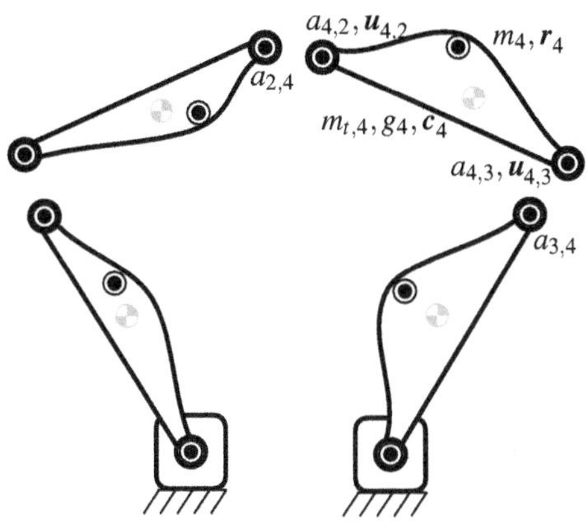

Fig. 2 Inertia decomposition of the 5 bar mechanism. The COM ($\boldsymbol{c}_4$), inertia (g_4), and mass (m_4) of body 4 is given by sum of three point masses ($\boldsymbol{u}_{4,2}$, $\boldsymbol{u}_{4,3}$, and $\boldsymbol{r}_4$)

$$m_{t,i} = m_i + \sum_{j=1}^{n} a_{ij} \quad m_{t,i}\boldsymbol{c}_i = m_i\boldsymbol{r}_i + \sum_{j=1}^{n} a_{ij}\boldsymbol{u}_{ij}$$

$$g_i + m_{t,i}\|\boldsymbol{c}_i\|^2 = m_i\|\boldsymbol{r}_i\|^2 + \sum_{j=1}^{n} a_{ij}\|\boldsymbol{u}_{ij}\|^2 \tag{13}$$

The mechanism can be built as long as the inertia and mass are positive. This precludes a range selectable inertia distributions.

2.7 *Reactionless Trajectories*

With global force balance the dimension of momentum for planar mechanism is reduced to one. Since a 5-bar mechanism has 2 DOF, there exist a velocity vector in each pose for which the momentum is zero. This null space motion of Eq. 5 is numerically integrated to form a reactionless trajectory. In the instantaneous balance poses, the momentum for both directions is always zero. This implies that in these poses, two reactionless trajectories meet.

3 Results

To evaluate the presented method, a geometry is selected, as depicted in Fig. 3a and Table 1. The dynamic balance of the mechanism is evaluated using multibody software package Spacar [3]. The mechanism moves over two reactionless trajectories (red and blue) and one arbitrary unbalanced trajectory (yellow).

The maximal shaking forces of all the trajectories are in the order of computation accuracy (max: 1.07^{-09} N), confirming that the mechanism is force balanced. Also the shaking moments are approximately zero (max: 3.01^{-04} N m) for the two balanced trajectories. For the unbalanced trajectory a maximal shaking moment of 6.69 N m is found (Fig. 3b).

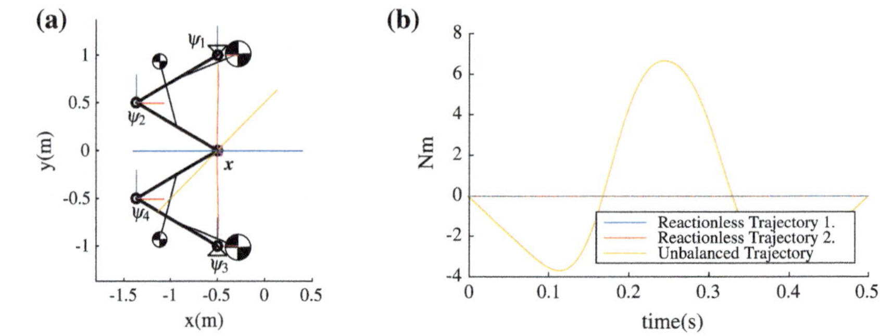

Fig. 3 **a** Geometry and trajectories. **b** Shaking moments

Table 1 Geometrical and dynamic parameters

Joint position [m]		COM [m]		Mass [kg]		Inertia [kgm^2]	
$\boldsymbol{o}_1^0$	[0, 1]	$\boldsymbol{c}_1^1$	[0.220, 0.013]	$m_{t,1}$	1.0155	g_1	0.0240
$\boldsymbol{o}_2^1$	[-0.866, 0.5]	$\boldsymbol{c}_2^2$	[0.250, 0.433]	$m_{t,2}$	0.2000	g_2	0.0500
$\boldsymbol{o}_3^0$	[0, -1]	$\boldsymbol{c}_3^3$	[0.220, -0.013]	$m_{t,3}$	1.0155	g_3	0.0240
$\boldsymbol{o}_4^3$	[-0.866, -0.5]	$\boldsymbol{c}_4^4$	[0.250, -0.433]	$m_{t,4}$	0.2000	g_4	0.0500

4 Discussion and Conclusion

Using the presented method a simplification of the balancing process is obtained, such that six (Eq. 9 for $\boldsymbol{r}_1^1, \boldsymbol{r}_3^3, m_1$, and m_3) out of eight conditions for dynamic balance can be calculated directly from the momentum equations without manipulation of the kineto-dynamic relationships. Furthermore, the two remaining conditions for global balance (Eq. 12 for $\boldsymbol{r}_2^2$) are found to be only dependent on the dynamic properties of the passive bodies. However, these last two conditions require effort in taking derivatives and manipulation of the momentum equations. The applicability of this balance methodology to more complex planar and ultimately spatial mechanisms is under investigation.

In this paper, a screw-based balancing method is presented and applied to a 5-bar mechanism. The balancing conditions are found for force balance over the complete workspace and additional moment balance over multiple trajectories, as shown by simulation results. To arrive at these dynamic balance conditions, a screw-based approach was presented. It consists of four steps. In the first step it was recognized that the dynamic properties of a planar mechanism with revolute joints can be simplified to one point mass per body. In the second step, instantaneous balance was found by placing the point masses of the base links orthogonal to the momentum wrench line generated by the rest of the mechanism. In the third step, the place of the remaining point masses was calculated such that the force balance extends over the workspace. In the last step, the resulting point masses where converted into actual inertias such that the mechanism can be built.

References

1. Berkof, R., Lowen, G.: A new method for completely force balancing simple linkages. J. Eng. Ind. **91**(1) (1969)
2. Foucault, S., Gosselin, C.: On the development of a planar 3-DOF reactionless parallel mechanism. In: ASME 2002 IDETC, pp. 1–9 (2002)
3. Jonker, J.B., Meijaard, J.P.: SPACAR — computer program for dynamic analysis of flexible spatial mechanisms and manipulators. Multibody Systems Handbook, pp. 123–143. Springer, Berlin (1990). http://www.utwente.nl/ctw/wa/software/spacar/
4. Ouyang, P., Li, Q., Zhang, W.: Integrated design of robotic mechanisms for force balancing and trajectory tracking. Mechatronics **13**(8–9), 887–905 (2003)
5. Ricard, R., Gosselin, C.M.: On the development of reactionless parallel manipulators. In: ASME 2000 DECTC, Baltimore, Maryland, vol. 1, pp. 1–10 (2000)
6. Stramigioli, S., Bruyninckx, H.: Geometry of dynamic and higher-order kinematic screws. In: Proceedings 2001 IEEE ICRA, pp. 3344–3349 (2001)
7. Van der Wijk, V.: Shaking-moment balancing of mechanisms with principal vectors and momentum. Front. Mech. Eng. **8**(1), 10–16 (2015)
8. Van der Wijk, V., Herder, J.: Guidelines for low mass and low inertia dynamic balancing of mechanisms and robotics. Advances in Robotics Research, pp. 21–30. Springer, Berlin (2009)
9. Van der Wijk, V., Krut, S., Pierrot, F., Herder, J.L.: Design and experimental evaluation of a dynamically balanced redundant planar 4-RRR parallel manipulator. Int. J. Robot. Res. **32**(6), 744–759 (2013)
10. Wu, Y., Gosselin, C.M.: Synthesis of reactionless spatial 3-DoF and 6-DoF mechanisms without separate counter-rotations. Int. J. Robot. Res. **23**(6), 625–642 (2004)
11. Wu, Y., Gosselin, C.M.: On the dynamic balancing of multi-DOF parallel mechanisms with multiple legs. J. Mech. Design **129**(2), 234 (2007)
12. Zoppi, M., Zlatanov, D., Molfino, R.: On the velocity analysis of interconnected chains mechanisms. Mech. Mach. Theory **41**(11), 1346–1358 (2006)

A Novel S-C-U Dual Four-Bar Linkage

Pierre Larochelle and Sida Du

Abstract This paper presents the kinematic architecture, analysis, and simulation of the S-C-U Dual Four-Bar Linkage (Larochelle and Du, Dual four-bar linkage mechanism, 2015, [4]). The S-C-U Dual Four-Bar Linkage, or SCUD Linkage, is a biologically inspired design for articulating the leg in a mechanical walking machine. The SCUD Linkage consists of dual planar four-bar mechanisms used to support and generate the desired motion of a rod in three-dimensional space. The rod is supported by two joints; each of these binary joints connecting the rod to a coupler point on a planar four-bar mechanism. At one end of the rod a universal joint connects it to a coupler point. At the rod's midpoint, a combination of a cylindrical joint and a spherical joint are used to connect it to the other supporting coupler point. The result is a two degree of freedom closed kinematic chain with two parallel sub chains. The end-effector or workpiece of the SCUD Linkage is affixed to the free end of the rod. Here, the concept of the SCUD Linkage as well as its kinematic analysis and simulation are presented. The analysis of an example SCUD Linkage is included.

1 Introduction

The motivation for the creation of the SCUD Linkage came from the same source, independently, to each of the authors before they had met. For both authors the source of the inspiration was identical; it was the motion of the front legs of a tortoise. In 1999 the first author acquired as a family pet a *geochelone sulcata* or African Spurred tortoise named Chomper, see Fig. 1. The complex motion of the front legs of the tortoise and their ability to support the relatively large mass of the tortoise inspired the first author in 2006 to pursue the design of kinematic closed chains that could replicate the motion of the front foot. Similarly, the second author, while pursuing his bachelors degree in 2009, was inspired by the front leg motion of tortoises while

P. Larochelle (✉) · S. Du
Florida Institute of Technology, Melbourne, FL, USA
e-mail: pierrel@fit.edu

S. Du
e-mail: sdu2013@my.fit.edu

J. Lenarčič and J.-P. Merlet (eds.), *Advances in Robot Kinematics 2016*,
Springer Proceedings in Advanced Robotics 4, DOI 10.1007/978-3-319-56802-7_5

Fig. 1 A Geochelone sulcata or African spurred tortoise

working on the design of a mobile robot for the Biomimetic Robotics Competition organized that year by the Harbin Institute of Technology. Independently, each author explored the use of open kinematic chains to generate the desired motion however they found that four or more degrees of freedom were required for each foot. Simpler, lower degree of freedom, solutions were sought. Once the authors met in Fall 2013 they began collaborating to identify kinematic architectures that would effectively and efficiently replicate the front leg motion of the geochelone sulcata. The SCUD Linkage presented here is one of the promising architectures that the authors have designed.

Human beings have long been curious about the behavior of the world's wonderful creatures and have tried to understand and imitate them. The earliest walking machines were mechanical toys. Their legs were driven by cranks or cams from a source of rotary power, usually clockwork, and executed a fixed cycle [8]. The first documented walking mechanism appeared in about 1870 and was based on a four-bar mechanism invented by the Russian mathematician P.L. Chebyshev as an attempt to imitate natural walking (Artobolevsky, 1964) [3]. In 1893 the first patents for legged systems were registered with the US Patent Office [3].

Based on the number of legs the robot has, there are bipeds e.g. humans or birds, quadrupeds e.g. mammals and reptiles, hexapods e.g. insects, and octopods e.g. spiders [3]. A hexapod robot is a mechanical device that walks on six legs. One example is RHex, a biologically inspired hexapod with compliant legs [1, 7]. The SphereWalker is another biologically inspired hexapod device [5]. The SphereWalker was developed by the first author and colleagues and was also inspired by the front legs of a tortoise. The SphereWalker is composed of three spherical four-bar linkages each connected to base plates that are serially connected to each other using two universal joints. Each linkage in SphereWalker is identical and each has two feet attached to its extended coupler link.

2 The SCUD Linkage

To fulfill the desired characteristics and performance attributes discussed above a novel linkage was designed. The S-C-U Dual Four-Bar Linkage, or SCUD Linkage, consists of dual planar four-bar mechanisms used to support and generate the desired motion of a rod in three-dimensional space. The rod is supported by two joints; each of these binary joints connecting the rod to a coupler point on a planar four-bar mechanism. These coupler points support the rod at its midpoint and at one of its ends. At one end of the rod a universal joint connects it to a coupler point. Along the rod's span a combination of a cylindrical joint and a spherical joint are used to connect the rod to the other supporting coupler point. The result is a two degree of freedom spatial closed kinematic chain with two parallel sub chains. The SCUD Linkage may also be classified as a parallel robot [6]. The end-effector or workpiece of the SCUD Linkage is affixed to the free end of the rod. A kinematic diagram of a SCUD Linkage is shown in Fig. 2. The two planar closed chain four-bar linkages generate motion in either parallel or intersecting planes. These mechanisms transmit motion and torque from their driving motors thru their couplers to the rod. Each of the planar four-bar mechanisms drives their coupler point along coplanar coupler curves and these curves determine the spatial motion of the end-effector rod. One end of the rod is connected to one of the couplers with a universal joint. This end of the rod is named the head of the rod and this four-bar mechanism is referred to as the U-linkage. Its plane of motion is the U-plane and the curve generated by the coupler point of the U-linkage is the U-curve. Along the rod is a second joint that

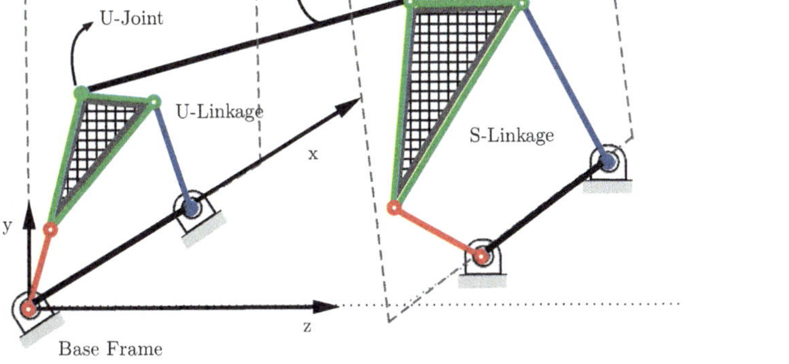

Fig. 2 An example SCUD linkage

connects it to the other four-bar mechanism. This joint consists of a combination of a spherical joint and a cylindrical joint. This linkage is named the S-linkage, due to the spherical joint, its plane is the S-plane, and its coupler curve is the S-curve. The two intersection points between the rod and the U and S planes move along the U-curve and S-curve and determine the spatial motion of the end-effector rod.

3 Kinematic Analysis

Here we present the kinematic analysis of the SCUD Linkage. First, Cartesian coordinate frames are affixed to the links. Then, the transformation operators, velocity vectors and acceleration vectors of the coupler points of the planar linkages and that of the end-effector are derived to describe their motion with respect to the inputs. Next, the parameters of the SCUD Linkage are used to build models both in Matlab and in Creo [2] to generate and analyze the workspace of this SCUD Linkage. These models are utilized to visualize the kinematic capabilities and limitations of the SCUD Linkage.

Consider the SCUD Linkage whose U-linkage and S-linkage are each crank-rocker four-bar mechanisms as shown in Fig. 3. The kinematic analysis is performed in 3 steps:

1. The derivation of the planar four-bar motions. In this first step, a four-bar position analysis is performed to determine the relative angle (ϕ) of the coupler with respect to the crank as a function of the input angle (θ). Next the two coupler point frames relative to the base frame are determined; i.e. the frames $\{U\}$ and $\{S\}$.
2. The derivation of the rod's spatial motion operator. In this second step, the angles λ_x and λ_y are solved from a set of non-linear equations that describe the geometric relationship between the $\{U\}$ and $\{S\}$ frames. Next, the frame $\{U^{'}\}$ is derived by rotating the frame $\{U\}$ by λ_x and λ_y and then the end-effector frame $\{E\}$ is obtained by translating the frame $\{U^{'}\}$ along the operator rod.
3. The derivation of the kinematic derivatives. The velocity vectors and the acceleration vectors of the coupler points and the end-effector are obtained by computing the derivatives of the transformation operators obtained in steps 1 and 2.

Step 1 can be performed using the known position analysis of the planar four-bar crank-rocker mechanism and applying coordinate transformations to yield the frames $\{U\}$ and $\{S\}$. Step 2 requires the angles λ_x and λ_y that represent the motion at the U joint. These angles are determined by performing an analysis of the orientation of frame $\{U\}$ with respect to the $\{S\}$ frame. In this problem, the unknown frame $\{U^{'}\}$ has the same position as the known frame $\{U\}$ and is rotated with respect to $\{U\}$ such that its z axis is directed towards the origin of $\{S\}$. Utilizing this geometric relationship, the angles λ_x and λ_y are determined. The known frames $\{U\}$ and $\{S\}$ and the angles λ_x and λ_y are then used to obtain the frames $\{U^{'}\}$, $\{S^{'}\}$, and $\{E\}$. See Fig. 3. The details follow.

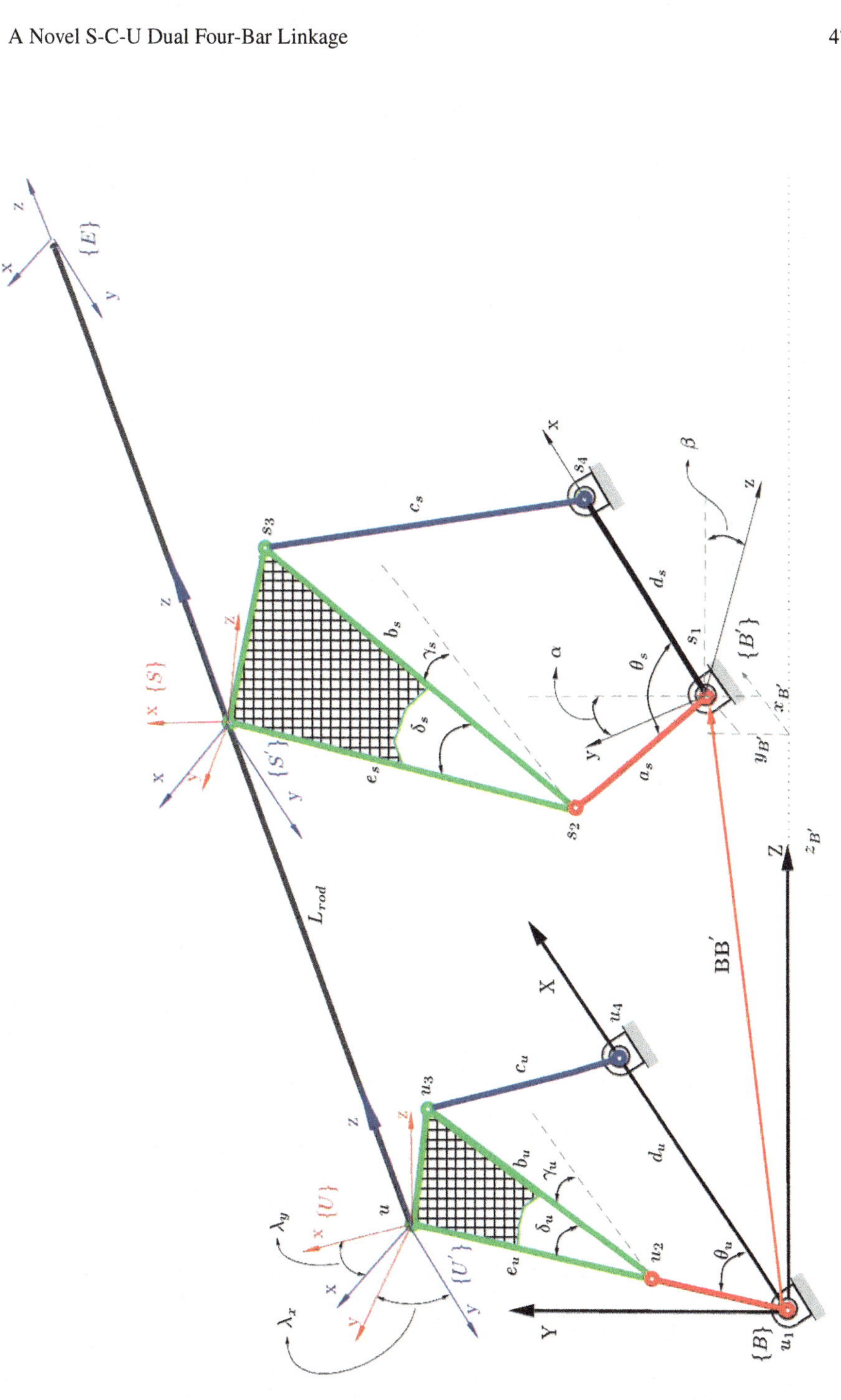

Fig. 3 The example SCUD linkage with parameters and frames defined

Frame $\{U^{'}\}$ is defined as having origin coincident with $\{U\}$ and z-axis directed along the rod toward $\{S\}$. Equating the direction of the z-axis of the $\{U^{'}\}$ and a vector along the rod from $\{U\}$ toward $\{S\}$ yields,

$$\begin{aligned} p_x &= \lambda(\cos \Gamma_u \sin \lambda_y + \sin \Gamma_u \sin \lambda_x \cos \lambda_y) \\ p_y &= \lambda(\sin \Gamma_u \sin \lambda_y - \cos \Gamma_u \sin \lambda_x \cos \lambda_y) \\ p_z &= \lambda(\cos \lambda_x \cos \lambda_y). \end{aligned} \tag{1}$$

where $\mathbf{p}$ is the known vector from the origin of the $\{U\}$ frame to the origin of the $\{S\}$ frame, $\Gamma_u = \gamma_u + \delta_u$, and λ is an arbitrary scale factor. The maximum possible range of motion of the universal joint is $[-\frac{\pi}{2}, \frac{\pi}{2}]$. These bounds are in the range of the arctan function therefore the angles λ_x and λ_y are also in this range and can be computed from Eq. 1 by eliminating λ and using arctan,

$$\lambda_x = \arctan\left(\frac{-(p_y \cos \Gamma_u - p_x \sin \Gamma_u)}{p_z}\right) \tag{2}$$

$$\lambda_y = \arctan\left(\frac{\cos \lambda_x p_x - p_z \sin \Gamma_u \sin \lambda_x}{p_z \cos \Gamma_u}\right). \tag{3}$$

Having found the λ_x and λ_y angles a straight-forward kinematic analysis yields $\{U^{'}\}$ and $\{S^{'}\}$ as well as the end-effector frame $\{E\}$. Step 3 requires differentiating the transformation operators. Though tedious, the process is straight-forward.

3.1 Workspace

The workspace of a SCUD Linkage is defined as the set of all reachable points of the origin of the end-effector frame $\{E\}$. The SCUD Linkage possesses 2 degrees of freedom therefore its workspace is a 2 dimensional spatial surface. A visualization of the workspace of a SCUD Linkage is generated in 3 steps. Step 1: discretize the working range of θ_u. Step 2: for each value of θ_u vary the input angle θ_s throughout its range of motion. Determine the origin of $\{E\}$ for each θ_s. This results in a curve that is on the surface of a sphere whose center is at the U joint and whose radius is the length of the rod L_{rod}. Step 3: repeat Step 2 for each value of θ_u. The resulting curves represent the discretized 2 dimensional workspace. The discrete points that represent the workspace of the SCUD Linkage may be visualized as a surface by utilizing MATLAB's surface rendering function *surf*, see Fig. 4. Moreover, the workspace of the SCUD Linkage may be visualized as a set of curves by utilizing the *tracecurve* function in Creo, see Fig. 5.

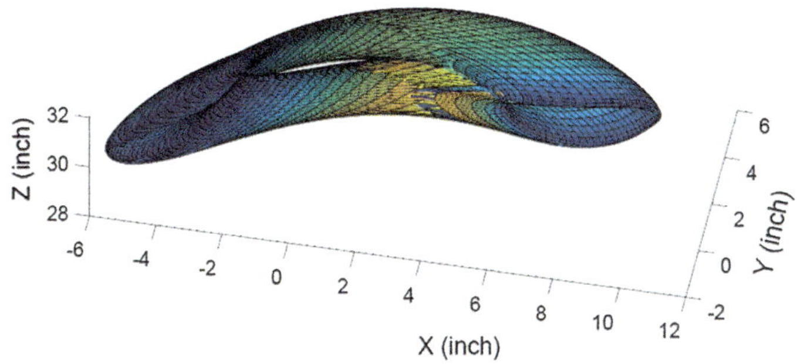

Fig. 4 An example SCUD linkage workspace rendered in MATLAB

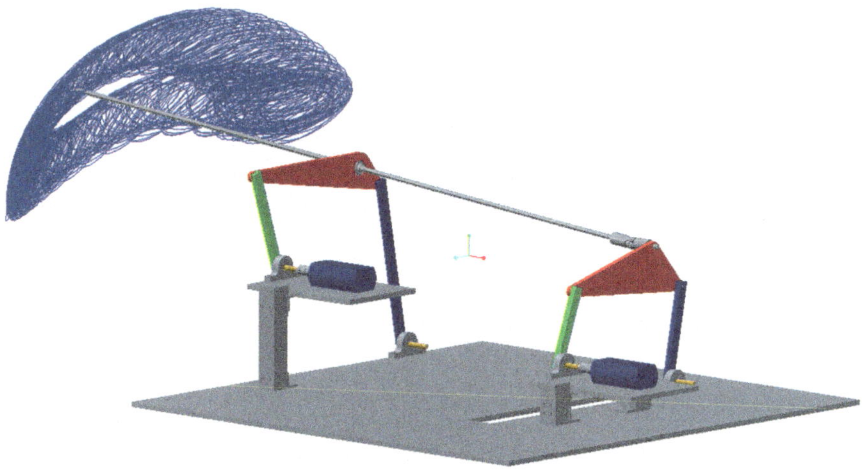

Fig. 5 An example SCUD linkage and workspace rendered in Creo

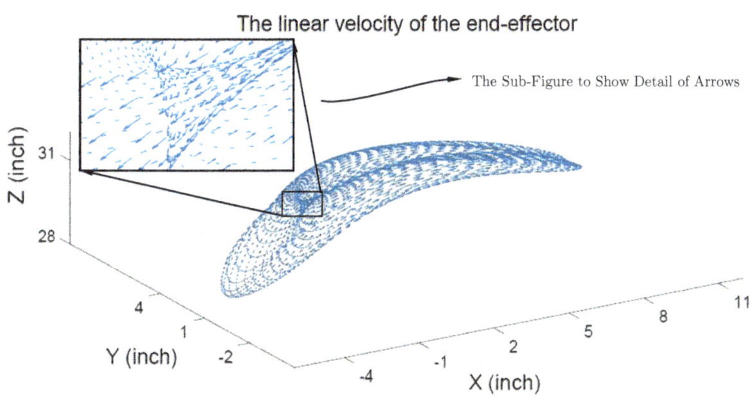

Fig. 6 An example SCUD linkage linear velocity field: $\dot{\theta}_u = 0$

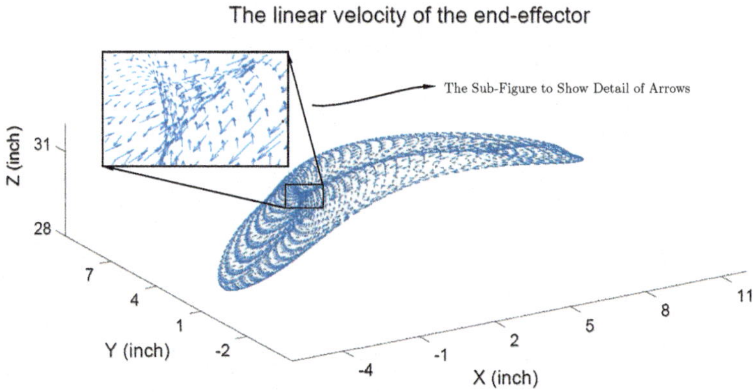

Fig. 7 An example SCUD linkage linear velocity field: $\dot{\theta}_s = 0$

3.2 Velocity

Results of a linear velocity analysis of the example SCUD Linkage are shown in Figs. 6 and 7.

References

1. Altendorfer, R., Moore, N., Komsuoglu, H., Buehler, M., Brown, H.B., McMordie, D., Saranli, U., Full, R., Koditschek, D.: Rhex: a biologically inspired hexapod runner. Auton. Robot. **11**, 207–213 (2001)
2. Chang, K.H.: Mechanism Design and Analysis Using PTC Creo Mechanism 3.0. SDC Publications, Kansas (2015)
3. Gonzalez de Santos, P., Garcia, E., Estremera, J.: Quadrupedal Locomotion: An Introduction to the Control of Four-legged Robots. Springer, London (2006)
4. Larochelle, P., Du, S.: Dual four-bar linkage mechanism. US Provisional Patent App. No. 62196504, 2015
5. Larochelle, P., Zimmerman, O., Mori, J., Scully, C., Lucas, C., Sleight, J., Lee, G.: Robotic walking apparatus. U.S. Patent No. 9162721, 2015
6. Merlet, J.P.: Parallel Robots, 2nd edn. Springer, Berlin (2010)
7. Saranli, U., Buehler, M., Koditschek, D.: Rhex: a simple and highly mobile hexapod robot. Int. J. Robot. Res. **20**(7), 616–631 (2001)
8. Todd, D.: Walking Machines: An Introduction to Legged Robots. Anchor Press Ltd., New York, Kogan Page Publishing, London (1985)

Inverse Kinematics Analysis of a P2CuP2Cu Concentric Tube Robot with Embedded Micro-actuation for 3T-1R Contactless Tasks

Mohamed Taha Chikhaoui, Kanty Rabenorosoa and Nicolas Andreff

Abstract This paper introduces a novel kinematic structure based on the concentric tube robot (CTR) paradigm, augmented with embedded soft micro-actuation. The latter allows to replace troublesome R-joints in CTR with 3 tubes by active tube curvatures (Cu-joints). First, the forward kinematic model is derived. Furthermore, the inverse kinematic problem is partially solved by restricting it to 3-translations/1-rotation movements. Finally, the inverse model is used to perform path planning schemes in medical scenarios.

1 Introduction

Flexible and miniaturized instruments are widely used for minimally invasive interventions. In this scope, continuum robots provide doctors with a controllable small device offering high precision navigation inside the human body for both therapeutic and diagnosis purposes [4]. Particularly, concentric tube robots (CTR) have proven their efficiency for a wide range of medical applications [2, 10, 11, 13, 15]. CTR are formed by several pre-shaped tubes nested in each other that can translate and rotate relatively in a telescopic way. The improved efficiency is due, among other reasons, to the use of tubes of small diameters ranging from 3 mm down to 0.8 mm [16] with a free lumen able to embed different medical tools. Despite all these advantages, CTR suffer from mechanical limits such as snapping explained in [16] and caused by the high torsional energy involved when rotating the tubes relatively. In fact, the rotation input at the tube base is all the more different from its output angle at the tip as its length and/or curvature are higher. Another issue with CTR is that once the tubes are assembled, the curvature of each component is a fixed parameter and

M.T. Chikhaoui (✉)
Laboratory for Continuum Robotics, Leibniz Universitaet Hannover, Appelstrasse 11, 30167 Hannover, Germany
e-mail: chikhaoui@lkr.uni-hannover.de

K. Rabenorosoa · N. Andreff
AS2M Department, FEMTO-ST Institute, UBFC/UFC/CNRS/ENSMM/UTBM, 24 Rue A. Savary, Besançon, France

J. Lenarčič and J.-P. Merlet (eds.), *Advances in Robot Kinematics 2016*,
Springer Proceedings in Advanced Robotics 4, DOI 10.1007/978-3-319-56802-7_6

cannot be modified afterward, which narrows the possible workspace to cover as demonstrated in [6–8]. Furthermore, in order to perform the relative movements, each tube is attached to the rotation stage, which can be connected to the translation stage [3, 10, 18]. For n concentric tubes, one needs $2n$ actuators that should overcome the frictional and torsional efforts of the tubes and thus are of a consequent size compared to the effector size. With a 3-tubes CTR, 6 actuators are used implying the possibility to control the 6 degrees of freedom (DoF) of the robot end-effector.

The aim of this work is to propose an equivalent structure of a standard 3-tubes CTR but augmented with embedded soft micro-actuation based on 2 tubes only and considered as a P2CuP2Cu. P denotes the prismatic joints, equivalent to the translational movements; and the 2 active tube curvatures in 2 orthogonal directions for each tube are denoted by Cu for each bending. This structure is intended to keep the same performances with less mechanical constraints and an embedded soft actuation scheme able to be easily integrated into an actual operating room. If the forward kinematic model is rather easy to derive, inversion of the robot model in a geometrical way is quite challenging but useful for stable path planning including obstacle avoidance, tissue and organ examination with imaging systems, and full robot shape monitoring in constrained environments. From a control point of view, this solution is proposed here in contrast with (i) Jacobian-based inverse-kinematics that require a full knowledge of the kinematics, an important computational time and that are tributary to the non-singular configurations [10], and with (ii) the kinematic-equivalent model based inverse kinematics [14] that monitor exclusively the position of the robot tip (added to the overall shape) by solution exploration without controlling its orientation.

For the sake of the completeness of this introduction, a few words are needed on technology. The comparison of the available micro-actuators such as shape memory alloys (SMA) and piezoelectric materials (PEM) presents the electro-active polymers (EAP) as the best candidates for this application. In fact, EAP-based actuators do not produce additional heating in contrast with SMA, require very low voltages (<2 volts for some ionic EAP) in contrast with PEM, and do not alter the structure mechanics because of their lightweight [6]. Biocompatibility with a medical device is also confirmed and a relatively high strain (more than 20%) should be noted. Section 2 presents the combination of these promising technologies and the forward modeling as the first work led to the best of our knowledge. Furthermore, the inversion of these models is developed in a geometrical way in Sect. 3 for the P2CuP2Cu, it stems for the computation of two tangentially connected circular arcs of controllable curvature, orientation (hence, arc planes), and length. Simulations of path planning results are also presented based on the developed models.

2 Concentric Tube Robot with Embedded Soft Micro-actuation

2.1 Embedded Soft Micro-actuation in Concentric Tube Robots

Adding micro-actuation to a concentric tube robot provides several improvements. First, the intrinsic curvature of each tube is accessible when activating the EAP-based soft micro-actuators deposed as 4 electrodes onto each tube as described in Fig. 1a. In fact, each pair of electrodes generates antagonistic efforts due to an electro-chemo-mechanical conversion. When one positively activated electrode expands in volume, its diametrically opposite one (negatively activated) shrinks which leads to the bending of the substrate tube along the plane containing these electrode central lines. This emerging technology is under continuous improvement and has shown promising results that required expertise in chemistry, micro-fabrication and clean room developments [1, 9]. Moreover, with this 4-electrodes configuration, the rotation motors are not mandatory which reduces substantially the actuation unit size. Furthermore, biocompatible flexible tubes are used and thus small efforts are required to deploy them telescopically. Smaller translation stages are adequate and replace for the high-torque-requiring motors used in standard CTR actuation. The thickness of such actuators varies between 10 and 30 microns and thus does not alter the concentric tube approach and saves the free-lumen configuration, in contrast with cable-driven continuum robots [5]. Finally, in order to preserve the accessible 6 DoF of the standard CTR, we introduce the P2CuP2Cu: a configuration with 2 concentric tubes augmented with 4 EAP electrodes each in order to control their bending in 2 orthogonal directions added to their telescopic deployment.

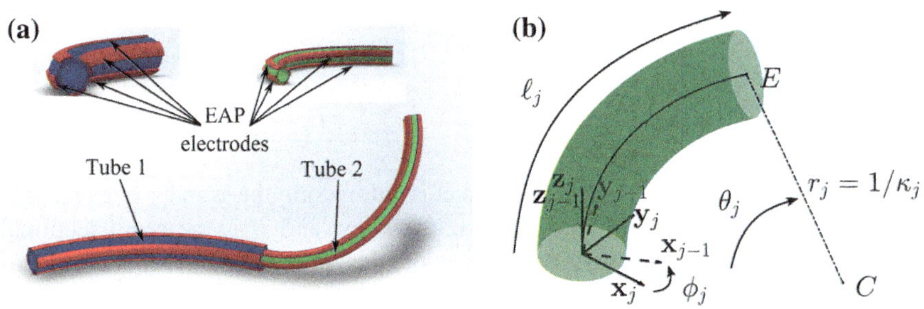

Fig. 1 **a** CAD design of the proposed P2CuP2Cu robot based on embedded soft micro-actuators at an arbitrary configuration, **b** Schematic description of an arc of a circle in 3D

2.2 Forward Kinematic Model

The forward kinematic model first describes a single section of a CTR considering its actuators. The modeling is based on the piece-wise constant curvature assumption [12, 17]. Indeed, an arc of a circle in 3D is defined, following Fig. 1b, by its curvature κ_j (inverse of the radius of curvature r_j), its length ℓ_j, and the angle of the plane in which it is located ϕ_j. Thus, the transformation matrix from the arc origin to its tip is:

$$^{j-1}\mathbf{T}_j = {}^{j-1}\mathbf{T}_j(\phi_j)^{j-1}\mathbf{T}_j(\kappa_j, \ell_j) = \begin{bmatrix} \mathbf{R}_z(\phi_j) & 0 \\ 0 & 1 \end{bmatrix} \begin{bmatrix} \mathbf{R}_y(\theta_j) & \mathbf{p}_j \\ 0 & 1 \end{bmatrix} \tag{1}$$

where $\theta_j = \kappa_j \ell_j$ is the bending angle and $\mathbf{p}_j = [r_j(1 - \cos\theta_j), 0, r_j \sin\theta_j]^T$. The arc variables are directly linked to the robot actuators, depending on the overlapping of the n tubes constituting m sections. For the considered P2CuP2Cu with 2 tubes, the actuator space is constituted of $\mathbf{q} = [v_{1x}\ v_{1y}\ v_{2x}\ v_{2y}\ \rho_1\ \rho_2]^T$ where $v_{ix,y}$ is the supplied voltage to the ith tube according to its x and y axes respectively, and ρ_i is its translation for $i = \{1, 2\}$. The intrinsic curvatures of each tubes are denoted $\kappa_{ix,y_{in}} = C_{EAP_i} v_{ix,y}$ noting that C_{EAP_i} is the EAP electro-chemical constant of the ith tube electrodes. The intrinsic arc variables for the 2 tubes ($i = \{1, 2\}$) are given by:

$$\begin{cases} \kappa_{i_{in}} = \sqrt{\kappa_{ix_{in}}^2 + \kappa_{iy_{in}}^2} \\ \phi_{i_{in}} = atan2(\kappa_{iy_{in}}, \kappa_{ix_{in}}) \end{cases} \tag{2}$$

The second section (containing only tube 2) variables are directly identified as $\kappa_2 = \kappa_{2_{in}}$ and $\phi_2 = \phi_{2_{in}}$. However, for the first section subject to the mechanical interaction of 2 tubes, one must compute the first section variables by:

$$\begin{cases} \kappa_1 = \sqrt{\kappa_{1x}^2 + \kappa_{1y}^2} \\ \phi_1 = atan2(\kappa_{1y}, \kappa_{1x}) \end{cases} \tag{3}$$

where $\kappa_{1x} = \dfrac{\sum_{i=1}^{2} E_i I_i \kappa_{i_{in}} \cos\phi_{i_{in}}}{\sum_{i=1}^{2} E_i I_i}$, and $\kappa_{1y} = \dfrac{\sum_{i=1}^{2} E_i I_i \kappa_{i_{in}} \sin\phi_{i_{in}}}{\sum_{i=1}^{2} E_i I_i}$. Note that κ_{ix} and κ_{iy} are the decomposition of the main curvature along the x and y axes respectively for the jth section, E_i is the elastic modulus, and I_i is the cross sectional moment of inertia of the ith tube. We should note that $\kappa_{i_{in}}$ is constant for the standard CTR when they are directly accessible in the proposed P2CuP2Cu with the pair of electrodes. Finally, considering the initial pose of the robot where all the tubes are withdrawn, the link lengths are such that $\ell_j = 0, \forall j = \{1, 2\}$. Whenever the tubes are deployed, the link lengths are computed as $\ell_1 = \rho_1$ and $\ell_2 = \rho_2 - \rho_1$.

3 A Geometrical Approach to Kinematic Model Inversion for 3T-1R Contactless Planning

3.1 *Closed-Form Planar Solution to the Translation Part*

The model is inverted geometrically in order to find a closed-form planar solution to properly control the robot end-effector position as a first step. For standard CTR with 3 tubes, straightforward computation of the exact inverse kinematic model is very arduous due to the non-linear equations involved and require challenging inverse kinematics or heuristic methods which are often limited by singularity issues, demand significant computational resources and a consequent execution time. In summary, the proposed algorithm computes the inverse kinematic model of the P2CuP2Cu in 3D for 4 DoF including the three position components and one in-plane orientation as described below.

3.1.1 Plane Definition

Recalling the geometrical description of an arc (or a tube) in Fig. 1b, let us assume that S is its start point, $E(X_E, Y_E, Z_E)$ its end point, and $\overrightarrow{z_{j_1}}$ its tangent at the origin S. We consider that the robot sections lie in the same plane Π where $\Pi = (S, \overrightarrow{z_j}, \overrightarrow{SE})$. The output of this step is the homogeneous out-of-plane transformation ${}^{j-1}\mathbf{T}_j(\phi_j)$ defined in Eq. 1 such that:

$$ {}^{j-1}\mathbf{T}_j(\phi_j) = f(S, E) \quad \text{where} \quad \phi_j = atan2(Y_E, X_E) \tag{4} $$

The new frame at S is then $\left(\overrightarrow{x_j}, \overrightarrow{y_j}, \overrightarrow{z_j}\right)$ where $[\overrightarrow{u_j}, 1]^T = {}^{j-1}\mathbf{T}_j(\phi_j)[\overrightarrow{u_{j-1}}, 1]^T$ for $\overrightarrow{u} = \{\overrightarrow{x}, \overrightarrow{y}, \overrightarrow{z}\}$.

3.1.2 Planar Solution for a Single Arc

Once the plane Π is figured out, the process is performed in a planar way and thus reduces substantially the computation complexity. The desired point is projected onto the frame related to the plane Π such that its coordinates are ${}^{\Pi}X_E = \sqrt{X_E^2 + Y_E^2}$, ${}^{\Pi}Y_E = 0$, and ${}^{\Pi}Z_E = Z_E$. In all cases, for a tangent vector $\overrightarrow{z_j}$, the end-effector of an arc of a circle is described in its base frame $(S, \overrightarrow{x_j}, \overrightarrow{y_j}, \overrightarrow{z_j})$ (cf. Fig. 1b) by Eqs. 5 and 6.

$$ \kappa_j = \frac{2^{\Pi}X_E}{{}^{\Pi}X_E^2 + {}^{\Pi}Z_E^2}, \tag{5} $$

$$\ell_j = \begin{cases} \frac{1}{\kappa_j}\, acos\left(1 - \kappa_j\, {}^{\Pi}X_E\right) & \text{if } {}^{\Pi}Z_E > 0 \\ \frac{1}{\kappa_j}\left(2\pi - acos\left(1 - \kappa_j\, {}^{\Pi}X_E\right)\right) & \text{else.} \end{cases} \tag{6}$$

The homogeneous in-plane transformation matrix is then computed such as ${}^{j-1}\mathbf{T}_j$ in Eq. 1. This step is denoted $(\kappa_j, \ell_j, {}^{j-1}\mathbf{T}_j(\kappa_j, \ell_j)) = g(S, E, \overrightarrow{z_j})$.

3.1.3 Closed-Form Solution to the Translation Part

For the considered 2-tubes P2CuP2Cu, this single-arc inverse kinematic model g is computed for each arc separately, once the plane Π is defined. Both arcs meet at A which is the inflection point along the robot structure. Let us assure that for any $A \in \Pi$, closed-form solutions can be computed [14]. We compute the single arc solution following this order: (i) from the robot origin O to the first section end-effector A – assuming that O is also the world frame $(O, \overrightarrow{x_0}, \overrightarrow{y_0}, \overrightarrow{z_0})$ origin – then (ii) from A to the second section (and the robot) end-effector B:

$$\begin{cases} \left(\kappa_1, \ell_1, {}^{0}\mathbf{T}_1(\kappa_1, \ell_1)\right) = g(O, \overrightarrow{z_1}, A) & \text{for tube 1} \\ \left(\kappa_2, \ell_2, {}^{1}\mathbf{T}_2(\kappa_2, \ell_2)\right) = g(A, \overrightarrow{z_2}, {}^{2}B) & \text{for tube 2} \end{cases} \tag{7}$$

which can be expressed by the global closed-form function h:

$$\left(\kappa_1, \ell_1, \kappa_2, \ell_2, {}^{0}\mathbf{T}_1(\kappa_1, \ell_1), {}^{1}\mathbf{T}_2(\kappa_2, \ell_2)\right) = h(O, \overrightarrow{z_1}, A, B) \tag{8}$$

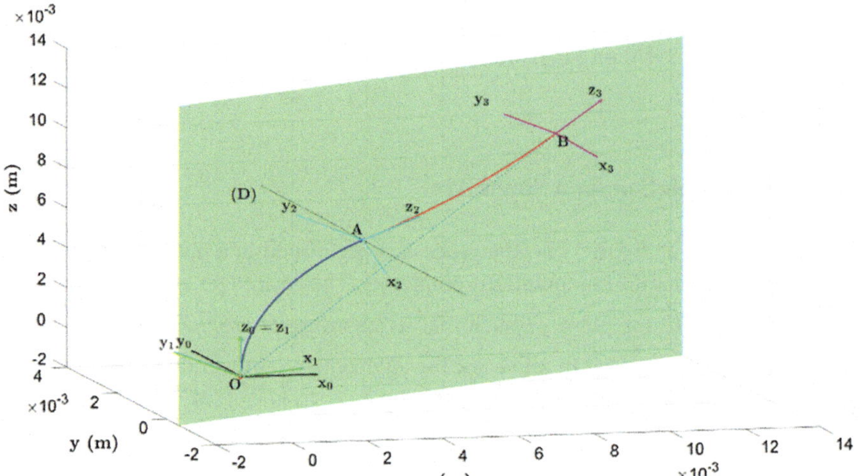

Fig. 2 Solving the closed-form inverse model for a desired point B after computing the plane Π (in *green*) with an arbitrary inflection point A. Section 1 plot is *blue* and Section 2 is *red*

Note that using $f(O, A)$, we define $F_1 = (O, \overrightarrow{x_1}, \overrightarrow{y_1}, \overrightarrow{z_1})$ the origin frame of the first arc and thus $\overrightarrow{z_1} = \overrightarrow{z_0}$. The output frames of the first and second arcs respectively are $F_2 = (A, \overrightarrow{x_2}, \overrightarrow{y_2}, \overrightarrow{z_2})$ and $F_3 = (B, \overrightarrow{x_3}, \overrightarrow{y_3}, \overrightarrow{z_3})$ computed using g function, and ${}^2B$ is the expression of B in the frame F_2 obtained by inverting ${}^0\mathbf{T}_1(\kappa_1, \ell_1)$. To check the validity of this solution, the results of Eq. 8 are fed through the forward model. It enables to draw the robot shape and pose. The end-effector position must match the initial desired point as described in Fig. 2 with orientation $\overrightarrow{z_3}$.

3.2 Monovariable Virtual Proportional Control

The closed-form model inversion presented above solves for the 3D position of the end-effector at B. In order to control the end-effector in-plane orientation, we use Algorithm 1. It is based on virtual proportional control of a single variable which is the in-plane orientation at B denoted $\overrightarrow{z_B}$ with a proportional gain λ. After initializing the inflection point A at an arbitrary position in the calculated plane Π (cf. Sect. 3.1.1), we solve the inverse kinematics for the position of point B through the function h in Eq. 8. The output orientation $\overrightarrow{z_3}$ at B is then compared to the desired orientation $\overrightarrow{z_B}$. At every step, the point A is slid on the perpendicular (D) to (OB) in the plane Π at A defined by its guiding vector $\overrightarrow{u_D}$.

Algorithm 1 Solving for position and in-plane orientation

Data: $O, B, \overrightarrow{z_B}$
Result: Arc variables $\chi = [\kappa_1\ \phi_1\ \kappa_2\ \phi_2\ \ell_1\ \ell_2]^T$
${}^0\mathbf{T}_2(\phi_2) = f(O, B)$;
Choose $\phi_1 = \phi_2$; $\{A \in \Pi\}$
while $\overrightarrow{z_3} \neq \overrightarrow{z_B}$ **do**
  $(\kappa_1, \ell_1, \kappa_2, \ell_2,\ {}^0\mathbf{T}_1(\kappa_1, \ell_1),\ {}^1\mathbf{T}_2(\kappa_2, \ell_2)) = h\,(O, \overrightarrow{z_1}, A, B)$;
  $\alpha = atan2\left(\overrightarrow{z_3}^T \overrightarrow{z_B},\ \overrightarrow{u_\Pi}\left(\overrightarrow{z_3} \times \overrightarrow{z_B}\right)\right)$;
  $\delta A = \lambda\alpha$; {Virtual proportional control}
  $A = A + \delta A\ \overrightarrow{u_D}$;
end while

Once the arc variables $\chi = [\kappa_1\ \phi_1\ \kappa_2\ \phi_2\ \ell_1\ \ell_2]^T$ are computed, the actuator configurations $\mathbf{q}$ are calculated by inverting Eqs. 2 and 3. For the brevity of this paper, such demonstrations are not detailed.

3.3 3T-1R Task Planning

Using the aforementioned kinematic inversion, examples of the path planning simulations of the P2CuP2Cu are presented hereby. For the intended medical applications, a sweeping scheme of the distal tube end-effector housing an optical imaging system

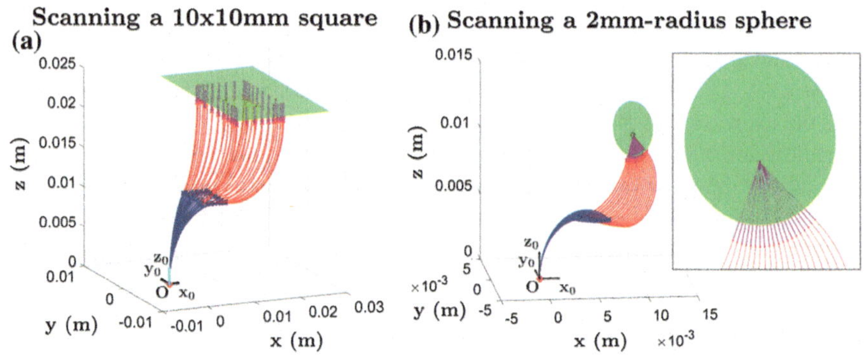

Fig. 3 For an arc length ratio $L_r = 1$, the robot poses during model inversion based path planning are shown for **a** a square scanning and **b** a spherical cap scanning schemes

(camera with fiber bundle, OCT probe, confocal microscope) is developed. More specifically, a square sweeping of $10 \times 10\,\text{mm}^2$ is validated by the simulations in Fig. 3a. One can note that the end-effector orientation is preserved (orthogonal to the examined tissue surface) which respects the constraints of a sweeping procedure. A circular path on a sphere of 2 mm radius is also performed (Fig. 3b) while keeping the orientation pointing to the sphere center as for a tissue examination with a sub-degree precision. These planning schemes were performed according to mechanical and fabrication constraints of a maximum arc lengths of 40 mm and maximum curvatures of $200\,\text{m}^{-1}$.

4 Conclusions

An alternative structure to the usual 3-tubes CTR was presented in this paper. This P2CuP2Cu is based on EAP soft micro-actuators and provides equivalent kinematic performances with only 2 tubes with a free lumen and a continuum shape approach. The major expected advantages of our robot are the compactness due to the embedded soft micro-actuation and the controllability especially for medical applications. Furthermore, the inverse kinematic model was analyzed in a geometrical approach and improved with an in-plane orientation control. This virtual control was validated by path planning schemes (namely tissue scanning) in simulations for 3T-1R tasks. Deriving the complete inverse kinematic model is a future challenge and will conduct to a full pose control.

Acknowledgements This work has been supported by the Labex ACTION project (contract ANR-11-LABX-0001-01) and by NEMRO project (contract ANR-14-CE17-0013).

References

1. Barrand, L., Chikhaoui, M.T., Cot, A., Rabenorosoa, K., Rougeot, P., Lakard, B., Lakard, S., Andreff, N.: Towards polypyrrole actuated flexible endomicroscope: synthesis. In: French Symposium on Emerging Technologies for Micro-Nanofabrication, Lyon, France (2015)
2. Burgner, J., Swaney, P.J., et al.: A bimanual teleoperated system for endonasal skull base surgery. In: IEEE/RSJ International Conference on Intelligent Robots and Systems, San Francisco, USA. pp. 2517–2523 (2011)
3. Burgner, J., Rucker, D., Gilbert, H., Swaney, P., Russell, P., Weaver, K., Webster, R.: A telerobotic system for transnasal surgery. IEEE/ASME Trans. Mechatron. **19**(3), 996–1006 (2014)
4. Burgner-Kahrs, J., Rucker, D., Choset, H.: Continuum robots for medical applications: a survey. IEEE Trans. Robot. **31**(6), 1261–1280 (2015)
5. Camarillo, D.B., Milne, C.F., Carlson, C.R., Zinn, M.R., Salisbury, J.K.: Mechanics modeling of tendon-driven continuum manipulators. IEEE Trans. Robot. **24**(6), 1262–1273 (2008)
6. Chikhaoui, M.T., Rabenorosoa, K., Andreff, N.: Kinematic modeling of an EAP actuated continuum robot for active micro-endoscopy. In: Lenarčič, J., Khatib, O. (eds.) Advances in Robot Kinematics, pp. 457–465. Springer, Berlin (2014)
7. Chikhaoui, M.T., Rabenorosoa, K., Andreff, N.: Towards clinical application of continuum active micro-endoscope robot based on EAP actuation. In: Surgetica, Chambéry, France (2014)
8. Chikhaoui, M.T., Rabenorosoa, K., Andreff, N.: Kinematics and performance analysis of a novel concentric tube robotic structure with embedded soft micro-actuation. Mech. Mach. Theory **104**, 234–254 (2016)
9. Cot, A., Chikhaoui, M.T., Rougeot, P., Rabenorosoa, K., Andreff, N.: Synthesis, encapsulation, and performance analysis of large deformation tri-layer polypyrrole actuator. In: IEEE International Conference on Advanced Intelligent Mechatronics, Banff, Canada. pp. 436–441 (2016)
10. Dupont, P.E., Lock, J., Itkowitz, B., Butler, E.: Design and control of concentric-tube robots. IEEE Trans. Robot. **26**(2), 209–225 (2010)
11. Gosline, A.H., Vasilyev, N.V., et al.: Metal MEMS tools for beating-heart tissue removal. In: IEEE International Conference on Robotics and Automation, Saint Paul, USA. pp. 1921–1926 (2012)
12. Gravagne, I.A., Rahn, C.D., Walker, I.D.: Large deflection dynamics and control for planar continuum robots. IEEE/ASME Trans. Mechatron. **8**(2), 299–307 (2003)
13. Hendrick, R.J., Herrell, S., Webster III, R.J.: A multi-arm hand-held robotic system for transurethral laser prostate surgery. In: IEEE International Conference on Robotics and Automation, Hong Kong, China. pp. 2850–2855 (2014)
14. Neppalli, S., Csencsits, M.A., Jones, B.A., Walker, I.D.: Closed-form inverse kinematics for continuum manipulators. Adv. Robot. **23**(15), 2077–2091 (2009)
15. Rucker, D.C., Croom, J.M., Webster III, R.J.: Aiming a surgical laser with an active cannula. J. Med. Devices **3**(2), 027506 (2009)
16. Webster III, R.J., Romano, J.M., Cowan, N.J.: Kinematics and calibration of active cannulas. In: IEEE International Conference on Robotics and Automation, Pasadena, USA. pp. 3888–3895 (2008)
17. Webster III, R.J., Romano, J.M., Cowan, N.J.: Mechanics of precurved-tube continuum robots. IEEE Trans. Robot. **25**(1), 67–78 (2009)
18. Xu, R., Asadian, A., Naidu, A., Patel, R.: Position control of concentric-tube continuum robots using a modified Jacobian-based approach. In: IEEE International Conference on Robotics and Automation, Karlsruhe, Germany. pp. 5813–5818 (2013)

Structural Synthesis of Hands for Grasping and Manipulation Tasks

Ali Tamimi, Alba Perez-Gracia and Martin Pucheta

Abstract In the kinematic synthesis of multi-fingered robotic hands for a specific task, the selection of the hand topology is an important step. Considerable research efforts have been directed to the structural synthesis of hand topologies for satisfying grasping and manipulation metrics such as mobility and force closure. In this work, we develop a structural synthesis, isomorphism-free enumeration method that combines the solvability for rigid-body guidance with the grasping and manipulation metrics, for general hands with a tree structure. An algorithmic implementation of the methodology is presented and illustrated with validation examples.

1 Introduction

Multi-fingered robotic hands are mechanical linkages where a common set of links spans a number of serial chains, designed for grasping and manipulation tasks. Traditionally, a robotic hand consists of a single link, or palm, spanning several subchains, which are the fingers. This definition can be extended to consider a common set of links and joints spanning the finger chains, possibly in several stages.

The structural or type synthesis of multi-fingered hands seeks to enumerate all possible topologies for a desired quality of the hand. Most of the previous work focuses on enumerating the topologies according to the mobility of the hand. Salisbury and Roth [4] studied the type synthesis of three-fingered hands with a single

A. Tamimi (✉) · A. Perez-Gracia
Department of Mechanical Engineering, Idaho State University, Pocatello, ID, USA
e-mail: tamiali@isu.edu

A. Perez-Gracia
e-mail: perealba@isu.edu

M. Pucheta
Centro de Investigación en Informática para la Ingeniería (CIII) and CONICET F.R. Córdoba, Universidad Tecnológica Nacional, Córdoba, Argentina
e-mail: mpucheta@frc.utn.edu.ar

J. Lenarčič and J.-P. Merlet (eds.), *Advances in Robot Kinematics 2016*,
Springer Proceedings in Advanced Robotics 4, DOI 10.1007/978-3-319-56802-7_7

palm and no wrist. They defined the degrees of freedom of the finger-object contact to synthesize all the topologies with full mobility and non-positive locked-joint mobility between palm and object. Based on this work, Lee and Tsai [2] undertook the structural synthesis of multi-fingered hands without wrist, with a single palm and identical type of finger contacts, with 3 to 7 fingers, to present an enumeration of *feasible* kinematic structures of mechanical hands.

Tischler, Samuel, and Hunt worked on the type synthesis for robotic hands with emphasis on the creation of a minimal-isomorphism list of kinematic chains [7]. They also considered a positive mobility between the ground and the grasped object and the connectivity between fingertips and the grasped object as selection criteria [8]. Their work imposes full-cycle mobility and restricts the results to full six-dof of mobility and point contact with friction for the finger-object contact. This contact is modelled as a spherical pair, realized as a 3R serial chain.

More recently, Özgür et al. [3] used the structural analogy between the palm-fingers-object system and a parallel robot consisting on a base-limbs-platform system (as was recognized before by [2, 8]). They adapted the procedures developed by Gogu [1] for parallel robot manipulators and worked on the structural synthesis of robotic hands for given values of dexterity, mobility, overconstraint, and redundancy.

All this previous work in structural synthesis of hands is focused on mobility and related metrics for grasping and manipulation of hands.

This paper presents a method for the structural synthesis of general hands (allowing multiple branchings of the tree topology) for grasping, mobility, and free motion of the fingertips. This method combines the checking for solvability for the rigid-body guidance dimensional synthesis problem [6] with the computation of a desired mobility and force closure for the hand-object system, for a given number of fingertips. The method generates an isomorphism-free list of structural solutions with a labelling approach which can be considered similar to [7].

2 Hand, Fingertip Contacts, and Hand-Object Representations

A multi-fingered hand is defined as a multi-body system with a common body - the wrist, which is a fundamental part of the hand manipulation- spanning several branches and ending in multiple end-effectors [5]. The kinematic chain of a multi-fingered hand has a tree topology that can be represented as rooted a tree graph [9], with the root vertex being fixed with respect to a reference system, see Fig. 1b.

A more general hand also has several palms arbitrarily branched and can be called a multi-fingered, multi-palm hand. A *palm* is an intermediate link whose degree is ternary or above. A *branch* of the hand is defined as the series of joints connecting the root node to one of the end-effectors, or fingertip. They are the main elements whose motion or contact with the environment is being defined by the task, see Fig. 1a. Hereafter, the *tree* of the hand will refer to the contracted tree of the hand obtained

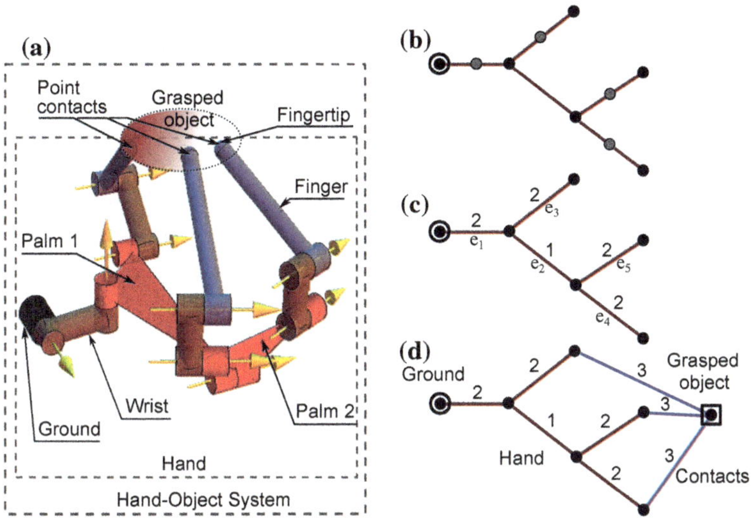

Fig. 1 A hand-object interaction and its graph representation. **a** A multi-fingered hand with 1-DOF revolute joints and 3 spherical fingertips. **b** The tree associated to the hand. Vertices filled in *black* are the root (*circled*), the palms, and the fingertip links; vertices filled in *grey* are binary links. **c** Contraction replaces the string of vertices with a unique edge labeled with its number of 1-DOF joints: 9 edges are contracted to 5 edges. **d** The contracted tree is connected by blue edges from each fingertip, each one labelled with the type (DOF) of contact, to the grasped object (*squared vertex*), to form the hand-object graph

by replacing the binary links between two higher order links by an edge labeled with their connectivity, which is equivalent to the number of 1-DOF joints between them; see for example Fig. 1c. Open hands with a hybrid topology can also be transformed into a contracted tree topology, adequate to perform its dimensional synthesis, by removing the internal loops [6].

Two arrays are defined for the tree topology with n vertices and $e = n - 1$ edges, which capture the incidence and adjacency properties as well as information of the edges. They are the *parent-pointer array* $\mathbf{p}$ and the *joint array* $\mathbf{j}$. A labelling of the graph edges from 1 to e is assumed for the entries of both arrays. The parent-pointer array implements the parent-pointer representation of the tree. The first edges incident at the root vertex take the value zero.

The relative motion allowed between each fingertip and the object can geometrically be classified as proposed by Salisbury and Roth [4]. The connectivity c of the grasped object relative to the fingertip, which is denoted as the degrees of freedom at the contact, can take any value from 0 (rigidly attached) to 6 (no contact). The Table 1 summarizes the description of the contacts [4, 8].

This work uses these contact types to extend the representation of the hand to the representation of the hand-object system. When an object is grasped, a loop is created in the graph of the hand-object system for any two fingers in contact with the object.

Table 1 Contact types between a fingertip and a grasped object

Degrees of freedom	Description
6	Free link (without contact)
5	Point contact without friction
4	Line contact without friction
3	Point contact with friction or plane contact without friction
2	Aea of contact with friction (*Soft finger* [4])
1	Plane contact with friction
0	Rigid attachment to object

An additional *fingertip array* **c** that contains the type of contact between the fingertip and the object. Figure 1d depicts the graph of the tree topology of the hand grasping an object with all fingertips in contact. In this case, the corresponding parent-pointer, joint, and fingertip arrays are respectively $\mathbf{p} = \{0, 1, 1, 2, 2\}$, $\mathbf{j} = \{2, 1, 2, 2, 2\}$, and $\mathbf{c} = \{3, 3, 3\}$.

For dimensional synthesis purposes, the general Chevychev-Grübler-Kutzbach mobility criteria for the hand-object system [4] is preferred to the more accurate methods developed by [1] and used in [3], because the information on the relative positions of the axes is not available; assuming a general position of the axes is appropriate in the general design problem where no geometric constraint on the unknown axes is prescribed. Once the degree-of-freedom of the fingertips are defined, the grasping and manipulation tasks for a body with a known shape can be defined to dimensionally size each of the feasible hand topologies found by the following algorithm.

3 Type Synthesis Algorithm for Free-Finger and Object-Contact Tasks

The goal is to find all hand topologies that can be paired with the task for dimensional synthesis, given a set of user-defined restrictions. User-defined inputs are the number of positions of the task m, the number of end-effectors or branches b, the range $[e_{\min}, e_{\max}]$ for the total number of edges of the graph e, the types of allowed fingertips **c** (from Table 1), and the desired mobility conditions.

The overall mobility M of the grasped object can be imposed as an input. In addition, a different mobility can be imposed at a given palm level, $M(T_{p_i})$, for in-palm manipulation or grasping, that can be different from M for certain sub-tasks of the task.

The output is the set of topologies that (i) meet the solvability criterion subject to these requirements, and (ii) meet the constraints related to the mobility.

3.1 Full-Tree Mobility Conditions

Given the tree of the hand T and its mobility M, any root-to-end-effector subgraph T_{sub} must satisfy a non-positive mobility M' if joints are locked and a mobility greater or equal than that of the overall tree.

$$\forall T_{\text{sub}} \in T \; : \quad \begin{cases} M'(T_{\text{sub}}) \leq 0 \\ M(T_{\text{sub}}) \geq M. \end{cases} \tag{1}$$

3.2 Variety

Tischler and Hunt [8] define the variety of a graph as the difference between its full mobility M and the minimum mobility of a subgraph containing a loop or set of loops, M_{min}, that is, $V = M - M_{min}$.

For the reduced and compacted tree graphs of the hand, all the loops contain the vertex corresponding to the grasped object. Imposing that the graphs have variety $V = 0$ ensures that the object has the desired degrees of freedom and that the locked-joints mobility is non-positive. This condition is imposed by identifying and checking the subgraphs created along the tree graph, starting at the root. Let the ternary or above vertices (palms) be labeled as p_i, and the subgraph starting at p_i in which all previous edges and vertices have been eliminated be T_{p_i}.

For a graph with variety $V = 0$, the following condition is imposed:

$$M(T_{p_i}) \geq M, \quad i = 1, \ldots, p, \tag{2}$$

where p is the total number of palms in the hand.

For task consisting of sub-tasks that require different mobility at different palm levels, the palms p_i are identified and the required mobility of the subtree starting at the edges spanning from the palm is checked for the candidate tree topologies,

$$\begin{aligned} M(T_{p_i}) &= M_{p_i}, \\ M'(T_{p_i}) &\leq 0. \end{aligned} \tag{3}$$

3.3 Solvability Conditions

The solvability criterion for the dimensional synthesis of a tree topology T passing through a number of m positions is the formula $m = s(T)$ proposed by Simo-Serra and Perez-Gracia [6], which also requires the analysis for each subtree T_{sub} of the graph. The tree is solvable iff

$$s(T_{\text{sub}}) \geq m \qquad \forall T_{\text{sub}} \in T \tag{4}$$

3.4 Algorithmic Implementation

The algorithm is divided in three main steps. In Step 1, the algorithm searches all possible topologies which satisfy user inputs. Then, Step 2 checks the solvability of candidate topologies and keeps only the topologies that are solvable. Finally, the mobility of the solvable candidates is computed in Step 3 and those topologies that satisfy the user inputs are presented as final answers. The method is described in Algorithm 1.

4 Results

The calculation for a binary hand with four fingertips is detailed in Table 2 for the overall mobility and in-palm mobility for different palms along the depth of the tree, removing first the wrist and then the depth-1 palm. For clarity, the solvability of this hand is calculated separately.

For comparison, the input used in the type synthesis example of Tischler and Hunt [8] is used here in the first example below. For the second example, we compare the output to the results of Salisbury and Roth [4] but using soft fingers instead of pointy fingers with friction. The number of positions for the synthesis is chosen so that the number of joints in the hand candidates is similar to those in the references used. The input values for both examples are shown in Table 3.

For the first example, the algorithm constructed 95 hand topologies and 10 of them were solvable. Out of those 10, only 3 topologies fulfilled the mobility requirements, that is, having $M = 6$ at the object with negative locked-joints mobility. The 3 topologies are shown in Table 4. Out of these topologies, one of them has a 1-dof wrist, which means that it has in-palm mobility equal to 5. The no-wristed hand obtained is the same that was obtained in the example from [8].

For the second example, 295 topologies are compatible with the rigid-body guidance task, out of which 78 are solvable. However only one topology, the one corresponding to three 4-dof fingers and no wrist, has the required mobility $M = 6$ without being constrained by any subgraph, and negative locked-joints mobility. This topology corresponds to the solution chosen in [4]. In this case several other topologies had the required overall mobility, but the additional constraint of having the same or higher in-palm mobility from any palm discarded those other topologies.

Algorithm 1 Type Synthesis Algorithm for Free-finger and Object-contact Tasks

(1) Find all the possible topologies.
Inputs: number of positions (m), number of branches (b), number of edges (e)
Outputs: Parent Pointer Array and Joints Array.

(1.1) Find parent pointer array (p). Parent pointer array must have length of e and b branches.

(1.2) Find joint array. For each parent pointer array in step 1.1, construct all possible joint arrays which meet the input criteria

(2) Solvability check. For each pair of parent pointer array and joint array found in step 1, calculates the number of positions for the exact kinematic synthesis. If the number of positions obtained for the kinematic task of all subtrees is greater or equal than the number of positions for the overall tree, the tree is solvable.

(2.1) Find all root to end effectors subgraphs. A graph with b branches has $2^b - 1$ subgraphs. Calculate m for all subgraphs and compare with m for the overall tree.

(2.2) Remove common edges. Common edges are the edges which are contained in all branches. In this step, an algorithm finds all common edges and removes them.

(2.3) Change root to one of end effectors. When the root of the graph is changed the value of the parent pointer array and joint array should be updated. The algorithm updates them in two steps. There is a path between the previous root and the new root.

- First, the parent-pointer value of the edges that are connected to this path is updated.
- Second, the parent-pointer value for the edges which are in the path is updated.
- Other edges which are not in the path or does not connect to the path do not need to be updated because the parents of them did not change.

(2.4) Iterate steps 2.1 to 2.3. This part will be stopped when only two end-effectors remain.

(3) Check Mobility. The output of the step 2 are the possible topologies. In this step the algorithm verifies that the mobility of the topology is equal to that defined as input when the grasping loops are created adding the fingertip contact array **c** to the graph.

(3.1) Remove unused part and calculate mobility. Since some part of rigid body may not participate in the grasping process, the algorithm removes them. For finding the used part, the algorithm finds all the edges that are in the branches from root to the end-effectors which contribute in grasping. The other edges are unused and the value of -1 is assigned to each corresponding element of parent pointer array and joint array. Then, calculate mobility for the resulting topology. If it equals to the user input, it is one of possible solutions.

(3.2) Find Mobility for subgraphs. Using the algorithm proposed in step 2.1, find all the root to end-effector subgraphs and calculate mobility (M) and locked joint mobility (M′) for them.

(3.3) Remove common edges (Palms). Using the algorithm proposed in step 2.2 remove palms.

(3.4) Calculate Mobility for the graph of part 3.3.
(3.5) Iterate step 3.3 and 3.4 until there is no common edge.

(3.6) Internal checks. If all the subgraphs fulfill the two following conditions, the topology is one of the solutions.

- $M'_{subgraph} \leq 0$
- $M_{subgraph} \geq M$

Table 2 Mobility calculations for a binary hand with four fingertips

Topology	Parameter	Symbol	Value
$\mathbf{p} = \{0, 1, 1, 2, 2, 3, 3\}$ $\mathbf{j} = \{2, 1, 1, 2, 4, 2, 4\}$	Number of task positions	m	9
	Number of branches (fingertips)	b	4
	Number of edges	e	7
	Type of fingertip contact	$\mathbf{c}$	{2, 2, 2, 2}
	Mobility	M	6
	Locked-joints mobility	M	−10
Subgraph 1 Remove Wrist $\mathbf{p} = \{0, 0, 1, 1, 2, 2\}$ $\mathbf{j} = \{1, 1, 2, 4, 2, 4\}$	Number of branches (fingertips)	b	4
	Number of edges	e	6
	Type of fingertip contact	$\mathbf{c}$	{2, 2, 2, 2}
	Mobility	M	4
	Locked-joints mobility	M	−10
Subgraph 2 Remove Palm 1 $\mathbf{p} = \{0, 0\}$ $\mathbf{j} = \{2, 4\}$	Number of branches (fingertips)	b	2
	Number of edges	e	2
	Type of fingertip contact	$\mathbf{c}$	{2, 2}
	Mobility	M	4
	Locked-joints mobility	M	−2

Table 3 Input values for the example

Parameter	Symbol	Example 1	Example 2
Number of task positions	m	5	9
Number of branches (fingertips)	b	3	3
Minimum and maximum number of edges	e	(2, 4)	(2, 5)
Type of fingertip contact	$\mathbf{c}$	{3, 3, 3}	{2, 2, 2}
Mobility	M	6	≥6

For the third case, we consider all topologies solvable for $m = 5$ positions, with $b = 3$ branches and $e = 4$ edges, and pointy fingertips with friction. This yields the parent-pointer array $p = \{0, 0, 1, 1\}$ shown in Table 5. It is required that the overall mobility of the tree is $M = 6$ with locked-joints immobilizing the object. In addition, it is only required to have $M_1 = 3$ for the 2-finger palm. The candidate topologies are shown in Table 5, with only the last topology fulfilling all conditions.

The results clearly show that the obtained hand topologies are general. Salisbury and Roth as well as Lee and Tsai procedures leads to hands with serial chain fingers

Table 4 Resulting topologies suited for the tasks of Examples 1 and 2

Example	Parent-pointer array	Joint array	Tree graph
Example 1	{0, 0, 0}	{3, 3, 3}	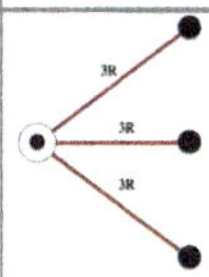
	{0, 0, 1, 1}	{1, 3, 2, 3}	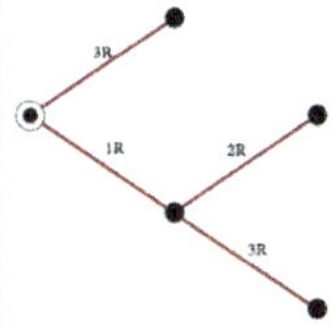
	{0, 1, 1, 1}	{1, 2, 3, 3}	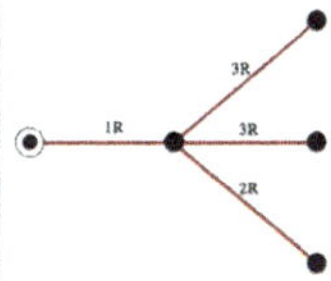
Example 2	{0, 0, 0}	{4, 4, 4}	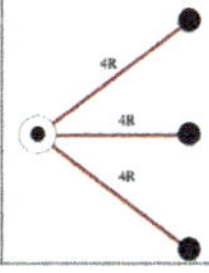

Table 5 Topology with different mobility at different palm levels

Parent-pointer array	Joint array	Mobility	Tree graph
{0, 0, 1, 1}	{1, 3, 2, 3}	$M = 6, M_1 = 5$	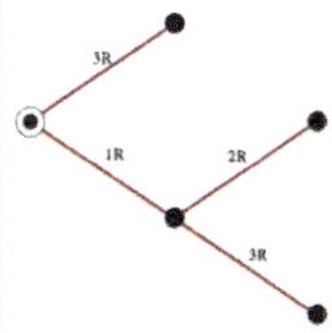
	{2, 3, 1, 3}	$M = 6, M_1 = 4$	
	{2, 3, 2, 2}	$M = 6, M_1 = 4$	
	{3, 3, 1, 2}	$M = 6, M_1 = 3$	

and a unique palm without wrist. Özgür methodology leads to serial and complex (chains with loops) parallel hand topologies analogous to parallel robots. Tischler et al. procedures have complex fingers with hybrid kinematic chains and produce topologies similar to the ones produced here for the case with a unique palm without wrist. Additionally, the tree topologies used here can be dimensioned through exact dimensional synthesis and when connected to the grasped object have serial, parallel, and hybrid topologies given more, or eventually new, design alternatives compared to those obtained in previous research.

The current algorithm also allows flexibility on where and when define the mobility. The current implementation imposes the same or higher mobility at each palm as that of the overall hand, but that can be modified to make some of the palms as grasping-only, for instance, while having different degrees of dexterity depending on the palm and fingers involved.

5 Conclusions

In this work, a structural synthesis procedure for general multi-fingered hands has been presented. The methodology considers the solvability of the hand for rigid-body guidance, and the mobility and locked-joints mobility when grasping an object, including the selection of the fingertips involved in the grasping and manipulation action. This procedure yields an isomorphism-free enumeration for compacted and reduced tree graphs. The presented examples show the adequacy of the methodology as a first step in the selection of a hand structure for a given general task that may include free-finger motion, grasping, and manipulation of the grasped objects.

Acknowledgements The first and second authors' work is supported by the National Science Foundation under Grant No. 1208385. The content is solely the authors' responsibility. The third author acknowledges to Argentinean projects, CONICET PIP 2011-01105, ANPCyT PICT 2013-2894, and UTN PID 3935.

References

1. Gogu, G.: Structural Synthesis of Parallel Robots-Part 1: Methodology. Solid Mechanics and Its Applications. Springer, Netherlands (2008)
2. Lee, J., Tsai, L.: Structural synthesis of multi-fingered hands. J. Mech. Design **124**, 272–276 (2002)
3. Ozgur, E., Gogu, G., Mezouar, Y.: Structural synthesis of dexterous hands. In: Intelligent Robots and Systems (IROS) Conference (2014)
4. Salisbury, J., Roth, B.: Kinematic and force analysis of articulated mechanical hands. J. Mech. Trans. Autom. Design **105**(1), 35–41 (1983)
5. Selig, J.M.: Geometric Fundamentals of Robotics. Monographs in Computer Science. Springer, Berlin (2004)
6. Simo-Serra, E., Perez-Gracia, A.: Kinematic synthesis using tree topologies. Mech. Mach. Theory **72 C**, 94–113 (2014)
7. Tischler, C., Samuel, A., Hunt, K.: Kinematic chains for robot hands - I. orderly number-synthesis. Mech. Mach. Theory **30**(8), 1193–1215 (1995)
8. Tischler, C., Samuel, A., Hunt, K.: Kinematic chains for robot hands - II. kinematic constraints, classification, connectivity and actuation. Mech. Mach. Theory **30**(8), 1217–1239 (1995)
9. Tsai, L.W.: Mechanism Design: Enumeration of Kinematic Structures According to Function. CRC Press, Boca Raton (2001)

Generalized Construction of Bundle-Folding Linkages

Shengnan Lu, Dimiter Zlatanov, Matteo Zoppi and Xilun Ding

Abstract A mechanism which is able to fold into a bundle is of particular interest: minimal size facilitates storage and transport. The paper presents a simple and general geometric method to design bundle-folding linkages based on one-degree-of-freedom spatial overconstrained loops. The so designed mechanism can be folded into a line bundle and deployed into a spatial shape. The geometric conditions, under which an overconstrained linkage can be folded into a bundle, are discussed. Case studies of bundle-folding designs are presented and validated using simulations.

1 Introduction

A deployable mechanism (DM) is capable of configuration change which dramatically alters its shape and size. This property enables many potential applications [5, 6, 25]. With good design, DMs can be folded into a bundle and deployed into different shapes: the compact folding facilitates storage and transport.

DMs are often constructed as networks of simple component mechanisms called deployable units (DUs). Typically, identical units (of one or several types) can be added to the assembly without limit resulting in arbitrarily large deployed structures.

This research has been supported by the National Science Foundation of China (51635002, 51605011, 51275015).

S. Lu · X. Ding
Beihang University, Beijing, China
e-mail: lvshengnan5@gmail.com

X. Ding
e-mail: xlding@buaa.edu.cn

D. Zlatanov (✉) · M. Zoppi
University of Genoa, Genoa, Italy
e-mail: zlatanov@dimec.unige.it

M. Zoppi
e-mail: zoppi@dimec.unige.it

J. Lenarčič and J.-P. Merlet (eds.), *Advances in Robot Kinematics 2016*,
Springer Proceedings in Advanced Robotics 4, DOI 10.1007/978-3-319-56802-7_8

A necessary condition for the compact folding of the network is that each DU can be reduced in size, ideally collapsed into a bundle. This is true for scissor-linkage elements, the most important subassembly used in DMs [15, 24]. Recently, the classic spatial overconstrained loops have been used as DUs. Pellegrino et al. studied a bundle-compacting form of the Bennett linkage with equal link lengths [17]. Similar research has also been done on the Myard [20] and Bricard linkages c [3, 4, 9, 17, 22]. In this study, we discuss the generalized construction of bundle folding mechanisms, focusing on 1-dof overconstrained hinged loops. Some general conditions and special cases are discussed and illustrated with examples and simulations.

Section 2 presents the geometric method for generating a bundle folding mechanism. Next, geometric conditions for a 1-dof loop with different numbers of revolute joints are analyzed. Case studies are performed on some typical overconstrained linkages; the obtained mechanisms have been simulated.

2 General Method for Obtaining Bundle Folding Mechanisms

In general, the geometric construction of R-jointed linkage with a given connectivity graph involves the determination of the spatial relationship among the hinge axes in each link. Beyond interference, the kinematics of the linkage is not affected by the geometric outlines of the links, yet these affect its physical appearance and utility.

The linkage is bundle-folding, if it has a configuration, in which the physical rigid links can be folded completely into a bundle without internal space gaps. We focus on single loops. A rigid bar which realizes a physical binary link should not be confused with the common normal used to geometrically represent the abstract link. The former can be any line segment with ends on the two joint axes sharing the link. The objective of the conceptual design of a maximally compact linkage is to find a line segment containing all rigid bars in some configuration.

Thus, a simple construction procedure can be proposed: choose a configuration and draw a line intersecting all joint axes. Then, take the segment of this line connecting the intersection points on any two adjacent R-axes as the physical rigid link. Thus, in the chosen configuration, the linkage will be compacted into a single line segment. Once the linkage moves to other configurations, (generally non-planar) polygons will be formed by these connecting link segments. Practically, the rigid links have finite cross section and so in the folded configuration, the physical shape of the mechanism will be a bundle. The mechanical design must ensure that this bundle is realized without gaps and interference.

3 Bundle Folding Conditions of 1-dof Overconstrained Mechanism

According to the Chebychev–Grübler–Kutzbach criterion, a spatial closed-loop linkage should have seven joints to be mobile. If realized with fewer hinges, the mechanism is called overconstrained. In the following, we will discuss the bundle folding conditions for loops with four, five, and six revolute joints.

3.1 4R Loops

Figure 1 [19] characterizes the Bennett linkage. Opposing links have the same length, a or b, and twist angle, α or β, respectively. All offsets are zero and $a \sin \beta = b \sin \alpha$ [1].

The linkage is mobile with one dof as in every configuration the four zero-pitch twists of the hinges are linearly dependent and span a three-system of screws. For general choices of the parameters a, b, α, β, and a general configuration, this is a general three system with one positive, one negative, and no zero principal pitches. So four revolute joint axes will be in the same regulus on a hyperboloid of one sheet (the zero-pitch quadric of the system). Hence, every line of the second regulus of the same hyperboloid intersects all four hinge axes, and can be used to construct a bundle-folding Bennett four-bar. (See [8, 19] for the properties of the general three-system and the one spanned by the Bennett axes.)

Special or degenerate cases occur when for special geometries or configurations. Thus, when the common normals of the links align, the two reguli, of the R joints and of their intersectors, both rule a hyperbolic paraboloid.

When $a = b \neq 0$, each pair of opposing revolute axes intersects at a point. The desired line can be the segment linking the two intersections, or any line passing through one of the points and lying in the plane of the two axes through the other. In this case the general three-system spanned by the joint twists has middle principal pitch equal to zero and the hyperboloid regulus of the four lines degenerates to a pair of intersecting planes.

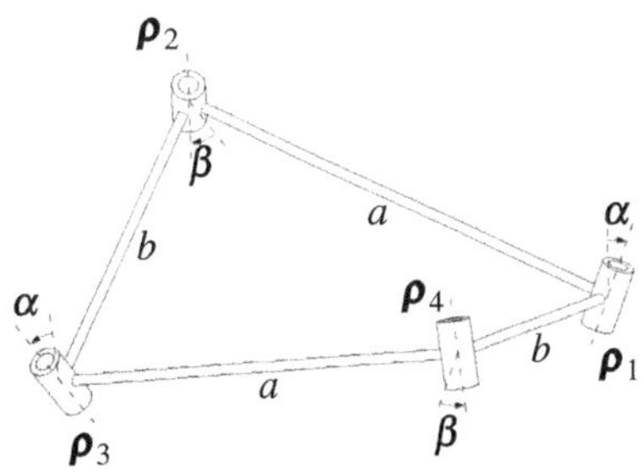

Fig. 1 The Bennett linkage and its geometric description [19]

When $\alpha = \beta = 0$, the mechanism becomes a planar parallelogram. In a general configuration no line intersects all axes (at finite points), but the four axes can become coplanar in a singularity of increased instantaneous mobility. Only in this configuration a bundle-generator line can be drawn.

Analogously, when $a = b = 0$, the mechanism is a collapsible spherical four-bar. At the configuration where all the axes are coplanar, there are (infinitely many) bundle generators.

3.2 5R Loops

In a 5R loop with one dof, the screw system spanned by the rotational twists is of dimension 4. Any line intersecting all five hinges is the axes of a zero-pitch screw in the reciprocal two-system. (Two zero pitch screws are reciprocal if and only if their axes are coplanar.) Therefore, the existence and multiplicity of a bundle generator depends on the type of the co-determined two- and four-systems. A two-system may contain zero, one, two, or infinitely many zero-pitch screws. The most general cases are of two or zero solutions.

It is important to distinguish the special case when the candidate generator is parallel rather than intersecting an axis. To construct a physical bundle-folding linkage the intersection point must be finite. In practice, the angle at the intersection is preferred to be close to $\pi/2$, although some useful solutions exist when the generator coincides with an axis.

An example of 5R-loop is the plane-symmetric Myard linkage, in Fig. 2. The zero-pitch joint screws and their axes are denoted $\boldsymbol{\rho}_i$ and $\ell(\boldsymbol{\rho}_i)$, $i = 1, \ldots, 5$, respectively. $\ell(\boldsymbol{\rho}_2)$ and $\ell(\boldsymbol{\rho}_5)$ always intersect, and so do $\ell(\boldsymbol{\rho}_3)$ and $\ell(\boldsymbol{\rho}_4)$. The two intersection points must be in the plane of symmetry, therefore, the line connecting the two points also intersects $\boldsymbol{\rho}_1$ (in some singular configurations, the two lines coincide). One more solution exists: the intersection of the plane defined by $\ell(\boldsymbol{\rho}_3)$ and $\ell(\boldsymbol{\rho}_4)$ and the one spanned by $\ell(\boldsymbol{\rho}_2)$ and $\ell(\boldsymbol{\rho}_5)$ intersects also with $\ell(\boldsymbol{\rho}_1)$, Fig. 2 [16, 19, 20, 23].

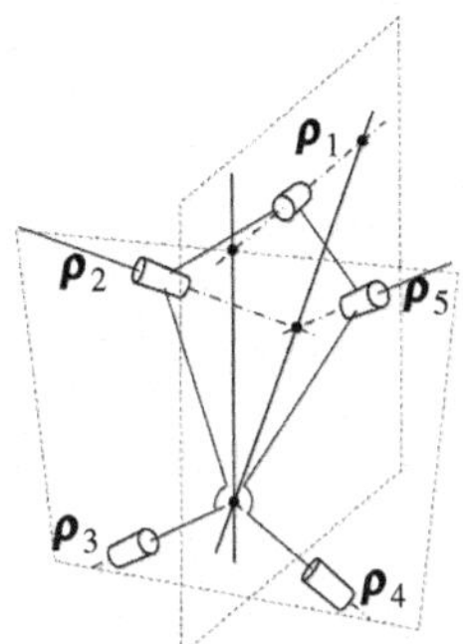

Fig. 2 Plane symmetric Myard linkage

3.3 6R Loops

In a 6R 1-dof loop, the joint rotations span a 5-system of twists. The reciprocal (wrench) system is defined by a unique screw. A (unique) bundle generator exists only if the pitch of this screw is zero. (In addition this line must not be parallel to any of the axes.) Let $\boldsymbol{\rho}_i = (\boldsymbol{\omega}_i, \mathbf{v}_i)$, $i = 1, \ldots, 5$, be five independent zero-pitch twists in the 5-system. A wrench, $\psi = (\mathbf{f}, \mathbf{m})$, with on the screw reciprocal to the 5-system satisfies the five homogeneous linear equations

$$\mathbf{f} \cdot \mathbf{v}_i + \mathbf{m} \cdot \boldsymbol{\omega}_i = 0 \tag{1}$$

This defines a one-system of screws whose pitch is zero only if $\mathbf{f} \cdot \mathbf{m} = 0$ for one (and every) solution. To avoid parallelism with a joint axis we need, for every i, either $\mathbf{f} \times \boldsymbol{\omega}_i \neq 0$ or $\mathbf{m}/|\mathbf{f}| \neq \mathbf{v}_i/|\boldsymbol{\omega}_i|$. In general, one cannot expect the additional conditions to be satisfied and therefore, for five lines in general position a sixth intersecting each of them does not exist.

However, the search for a bundle generator needs to succeed in only one configuration. If an IIM (increased instantaneous mobility) singularity exists, where no five of the lines are independent, finding a generator is easier. In particular, if the spanned system is three-dimensional, so is its reciprocal and then (usually) there will be infinitely many solutions, as explained above.

For instance, the systems of some Bricard linkages can degenerate in some configurations. Depending on the degree and type of the system, different numbers of solutions can be obtained. Examples are given in the following section.

4 Case Studies of Bundle Folding Loops

4.1 Bennett Linkage

We choose a Bennett linkage with parameters $a = 80, b = 128, \alpha = 30°, \beta = 51.13°$ as an example. The bundle is generated along the common intersecting line as shown in Fig. 3a. Here the folded configuration is the one where the joint axes have a common normal, which is the chosen bundle generators. There are infinitely many other common intersecting lines in this configuration, resulting in longer link segments. Bennett loops can be linked in a network forming intersting DMs. Bundle folding units are preferred; in particular the special case with $a = b$ [2, 12, 17, 18].

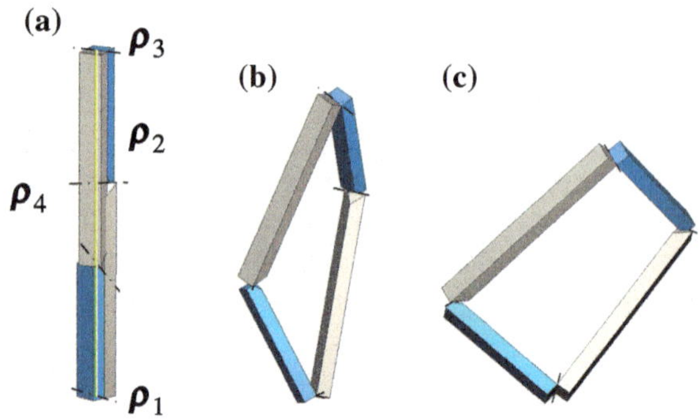

Fig. 3 Simulation of a bundle-folding Bennett linkage

4.2 *Type III Bricard Linkage*

Figure 4 shows the definition of a type III Bricard linkage in one of its two collapsed configurations. The edges AB, BC, AC', $A'B'$, $B'C'$, $A'C$ are the rotation axes of the mechanism. (For more details, see [7, 9].) In the example chosen, Figs. 4 and 5, the dimensions are: $r = 110, l_{OA} = l'_{OA} = 213.35, \angle AOA' = 27.93°, r = 110, R = 179.47$.

In the collapsed configurations (where all hinge axes are coplanar), the rank of the system spanned by the joint twists drops to three. Then, there are infinitely many possible bundle generators, each intersecting all the revolute axes. Indeed, any line in the plain which is not parallel to any of the six axes can be used. Obviously, none of these generators can be a common normal to the hinges.

The segment BB' is the folded bundle of the linkage. In the other collapsed configuration, the link segments form a square, Fig. 4. A 3D CAD model of the Bricard linkage bundle has been build. The cross-section of each bar is nearly a rectangle, slightly modified to avoid collisions during the motion. Simulation of the movement is illustrated in Fig. 5. Bricard linkages have also been used as deployable

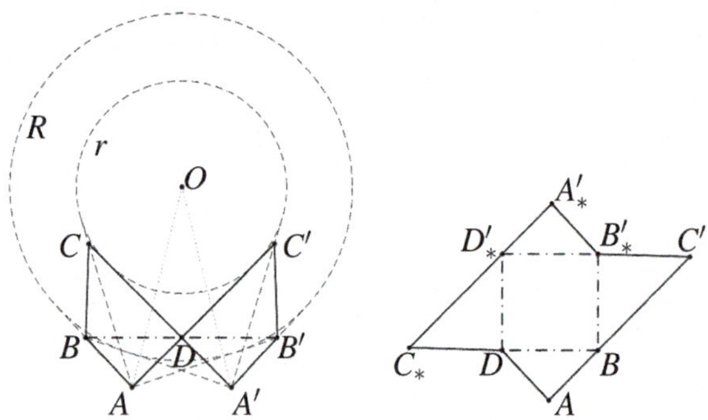

Fig. 4 Example of the type III Bricard linkage

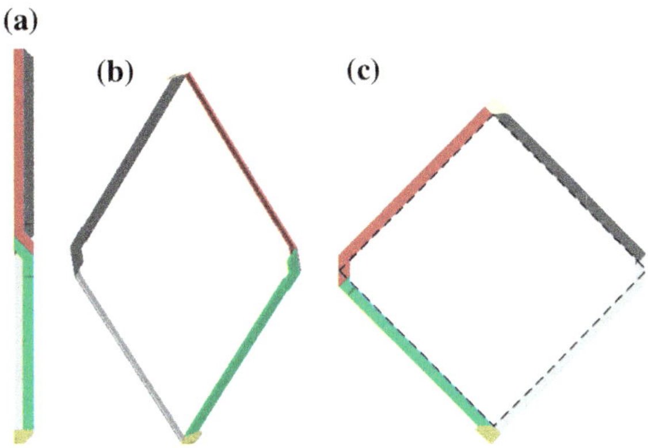

Fig. 5 Simulation of a Bricard linkage

units, usually bundle-folding [3, 4, 17, 22]. Type III Bricards can form indefinitely long chains [11] which can deploy and reconfigure in various ways [14].

4.3 Sarrus Linkage

The model and parameters of the Sarrus linkage [21] are shown in Fig. 6. The unit joint rotation twists are $\boldsymbol{\rho}_i$, $i = 1, \ldots, 6$. The adjacent axes of joints 1, 2 and 3 are parallel, and so are $\ell(\boldsymbol{\rho}_4)$, $\ell(\boldsymbol{\rho}_5)$, and $\ell(\boldsymbol{\rho}_6)$. The angle between the necessarily non-parallel directions of two groups of joint axes is α.

As discussed in Sect. 3.3, the desired common intersecting line (a zero-pitch reciprocal screw) does not always exist for a 6R loop. In a Sarrus linkage configuration where the joint twists span a five-system, there cannot be a pure force exerting no power on any hinge rotation. Indeed, if this were the case, then the reciprocal system would be of dimension at least two because there is always an infinite-pitch reciprocal screw (in the direction perpendicular to all axes). However, for many Sarrus linkages there are singular configurations with instantaneous mobility two (or three) and then a bundle-generating line can be found.

For example suppose that the two planar 3R serial subchains of the Sarrus can be maximally extended simultaneously. That is, there is a configuration where the parallel triples $\ell(\boldsymbol{\rho}_i), i = 1, 2, 3$ and $i = 4, 5, 6$, are coplanar in π_{123} and π_{456}, respectively.

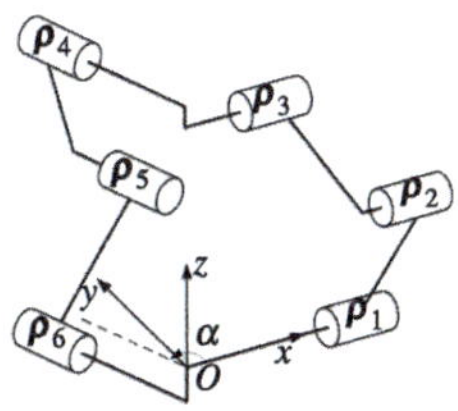

Fig. 6 The Sarrus linkage

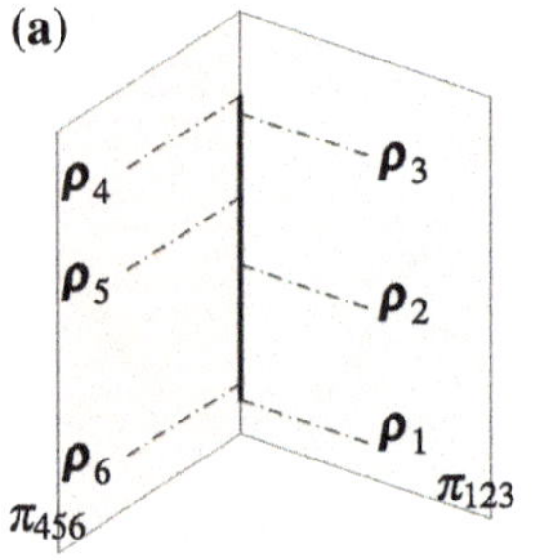

Fig. 7 Bundle-folding Sarrus linkage

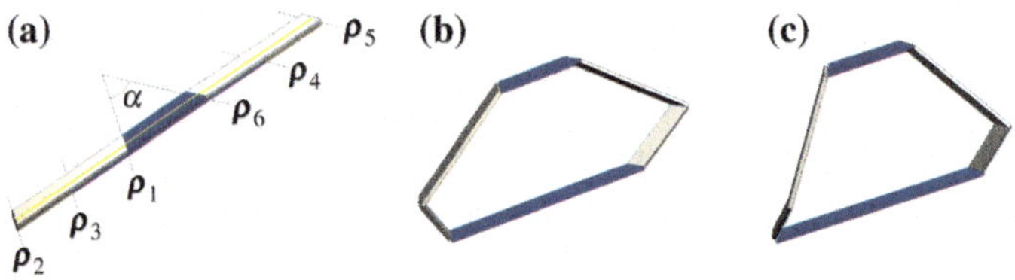

Fig. 8 Another bundle-folding Sarrus linkage

Then the intersection $\pi_{123} \cap \pi_{456}$ is the expected line. (The intersection line exists and is unique if the Sarrus is non-degenerate, allowing finite translation of link 3–4 with respect to 1–6.) An example mechanism is shown in Fig. 7.

Another example is when a Sarrus can be collapsed. Then, as with the type III Bricard, the joint twists span only a three-system, and any line on the plane not directed in either of the two joint-axis directions, can be the bundle generator, Fig. 8. Sarrus linkages can also be very useful when constructing deployable mechanisms: they can be used as equivalents of sliders in networks of scissor linkages [10, 13]. Bundle-folding Sarrus variants can be useful in such DM applications.

5 Conclusions and Future Work

A simple general geometric method for the conceptual design of bundle-folding realizations of a spatial 1-dof overconstrained loop is presented. The procedure involves the construction of a bundle-generating line, which intersects all hinge axes in a chosen configuration. Geometric conditions are given for the existence of such a generator for linkages composed of four, five, and six revolute joints. Case studies of bundle-folding designs of different 4R and 6R loops are presented and validated using simulations.

References

1. Bennett, G.: A new mechanism. Engineering **76**(12), 777–778 (1903)
2. Chen, Y.: Design of structural mechanisms. Ph.D. thesis, University of Oxford (2003)
3. Chen, Y., You, Z., Tarnai, T.: Threefold-symmetric Bricard linkages for deployable structures. Int. J. Solids Struct. **42**(8), 2287–2301 (2005)
4. Cui, J., Huang, H.L., Li, B., et al.: A novel surface deployable antenna structure based on special form of Bricard linkages. Advances in Reconfigurable Mechanisms and Robots I, pp. 783–792. Springer, Berlin (2012)
5. Durand, G., Sauvage, M., Bonnet, A., et al.: TALC: a new deployable concept for a 20-m far-infrared space telescope. In: SPIE Astronomical Telescopes+ Instrumentation, p. 91431A. International Society for Optics and Photonics (2014)
6. Escrig, F., Valcarcel, J.P., Sanchez, J.: Deployable cover on a swimming pool in Seville. Bull. Int. Assoc. Shell Spat. Struct. **37**(1), 39–70 (1996)
7. Goldberg, M.: Linkages polyhedral. Nat. Math. Mag. **16**(7), 323–332 (1942)
8. Hunt, K.H.: Kinematic Geometry of Mechanisms. Clarendon Press, Oxford (1990)
9. Lu, S., Zlatanov, D., Ding, X., Zoppi, M., Guest, S.: Folding type III Bricard linkages. In: Proceedings of the 14th IFToMM World Congress, pp. 455–462 (2015)
10. Lu, S.N., Zlatanov, D., Ding, X.L., Molfino, R., Zoppi, M.: Mechanisms with decoupled freedoms assembled from spatial deployable units. Advances in Robot Kinematics, pp. 517–525. Springer, Berlin (2014)
11. Lu, S.N., Zlatanov, D., Ding, X.L., Zoppi, M., Guest, S.D.: A network of type III Bricard linkages. In: ASME 2015 IDETC/CIE, 2015-47139 (2015)
12. Lu, S.N., Zlatanov, D., Ding, X.L.: Approximation of cylindrical surfaces with deployable Bennett network. In: ASME 2016 IDETC/CIE, 2016-59817 (2016)
13. Lu, S.N., Zlatanov, D., Ding, X.L., Molfino, R., Zoppi, M.: Novel deployable mechanisms with decoupled degrees-of-freedom. J. Mech. Robot. **8**(2), 021008 (2016)
14. Lu, S.N., Zlatanov, D., Ding, X.L., Zoppi, M., Guest, S.D.: Reconfigurable chains of bifurcating type III Bricard linkages. Advances in Reconfigurable Mechanisms and Robots II, pp. 3–14. Springer, Berlin (2016)
15. Maden, F., Korkmaz, K., Akgün, Y.: A review of planar scissor structural mechanisms: geometric principles and design methods. Archit. Sci. Rev. **54**(3), 246–257 (2011)
16. Myard, F.E.: Chaîne fermée à cinq couples rotoïdes, déformable au premier degré de liberté. Comptes Rendus Hebdomadaires des Séances de lAcadémie des Sciences **192**, 1352–1354 (1931)
17. Pellegrino, S., Green, C., Guest, S.D., et al.: SAR Advanced Deployable Structure. University of Cambridge, Department of Engineering (2000)
18. Perez, A., McCarthy, J.M.: Dimensional synthesis of Bennett linkages. J. Mech. Design **125**(1), 98–104 (2003)
19. Phillips, J.: Freedom in Machinery, vol. 1. Cambridge University Press, Cambridge (2007)
20. Qi, X.Z., Deng, Z.Q., Li, B., et al.: Design and optimization of large deployable mechanism constructed by Myard linkages. CEAS Sp. J. **5**(3–4), 147–155 (2013)
21. Sarrus, P.T.: Note sur la transformation des mouvements rectilignes alternatifs, en mouvements circulaires; et reciproquement. Academie des sciences, comtes rendus hebdomataires des sances **36**, 1036–1038 (1853)
22. Viquerat, A.D., Hutt, T., Guest, S.D.: A plane symmetric 6R foldable ring. Mech. Mach. Theory **63**, 73–88 (2013)
23. Waldron, K.: Overconstrained linkages. Environ. Plan. B Plan. Design **6**(4), 393–402 (1979)
24. Zhao, J.S., Chu, F.L., Feng, Z.J.: The mechanism theory and application of deployable structures based on SLE. Mech. Mach. Theory **44**(2), 324–335 (2009)
25. Zhao, J.S., Wang, J.Y., Chu, F.L., et al.: Mechanism synthesis of a foldable stair. J. Mech. Robot. **4**(1), 014502 (2012)

A Complete Analysis of Singularities of a Parallel Medical Robot

Josef Schadlbauer, Calin Vaida, Paul Tucan, Doina Pisla, Manfred Husty and Nicolae Plitea

Abstract This paper analyzes the singular poses of a 5-DOF parallel robot used for brachytherapy. In compliance with the latest safety protocols and requirements [3] the paper presents a new mathematical model using algebraic constraints and the Study parameterization of the Euclidian displacement group. Using algebraic methods combined with multidimension geometry proved to be efficient in the calculation of the kinematics of mechanisms and in the explanations of their behavior. The results obtained using this algebraic method were analyzed with respect to the data obtained from the experimental model of the robot by comparing theoretical computation results with the actual behavior of the robot. The analysis of the kinematics using these methods allows a complete description of working modes, singularities and robot behavior enabling a safe control throughout the medical task.

1 Introduction

Robotic architectures have been introduced in multiple medical fields to provide advanced tools for the doctors enhancing through their characteristics the medical act and ultimately the life quality of patients. Some of the most complex tasks are

J. Schadlbauer (✉) · M. Husty
Unit Geometry and CAD, University Innsbruck, Innsbruck, Austria
e-mail: josef.schadlbauer@uibk.ac.at

M. Husty
e-mail: manfred.husty@uibk.ac.at

C. Vaida · P. Tucan · D. Pisla · N. Plitea
Technical University of Cluj-Napoca, Cester, Romania
e-mail: calin.vaida@mep.utcluj.ro

P. Tucan
e-mail: paul.tucan@mep.utcluj.ro

D. Pisla
e-mail: doina.pisla@mep.utcluj.ro

N. Plitea
e-mail: nicolae.plitea@mep.utcluj.ro

J. Lenarčič and J.-P. Merlet (eds.), *Advances in Robot Kinematics 2016*,
Springer Proceedings in Advanced Robotics 4, DOI 10.1007/978-3-319-56802-7_9

those where the robot as a slave tool, controlled by the doctor from a master console, interacts intimately with the patient [4]. A new challenge refers to the development of new techniques for the curative and palliative treatment of malignant tumors. In the recent years a new technique, called brachytherapy (BT), has been developed, aiming to provide, local, targeted treatment of the tumors, by delivering specialized radioactive seeds directing into the tumor. The limited use of BT is caused by the seed placement accuracy required, which, especially for deeply located tumors is beyond human capability. Thus, as shown in [1, 5], robot assisted brachytherapy proves to be a necessity in order to: improve the accuracy of the needle placement and seed delivery; improve the consistency of the seed implant; avoid critical healthy areas; reduce radiation exposure. For a successful procedure the robot should introduce, needles with diameters varying from 0.6 mm up to 2 mm on distances up to 200 mm, following a linear trajectory, with a maximum positioning error of 1 mm [1]. This task requires both high accuracy and stiffness where parallel robots thrive. In full compliance with the latest challenges and requirements defined in [3] the CESTER team developed an innovative parallel robotic system, PARA-BRACHYROB, capable of targeting tumors located in the entire thoraco-abdominal area of the body under real-time CT monitoring [1, 2].

The prototype of PARA-BRACHYROB parallel robot has been built at CESTER and can be seen in Fig. 1. The study of this paper is the implementation of a new mathematical model that defines all the area where the robot becomes unstable, helping the controlling unit to avoid them, increasing the safety of the human patient and in the same time optimizing the command of the robotic system, enabling the manipulation in a singularity free workspace.

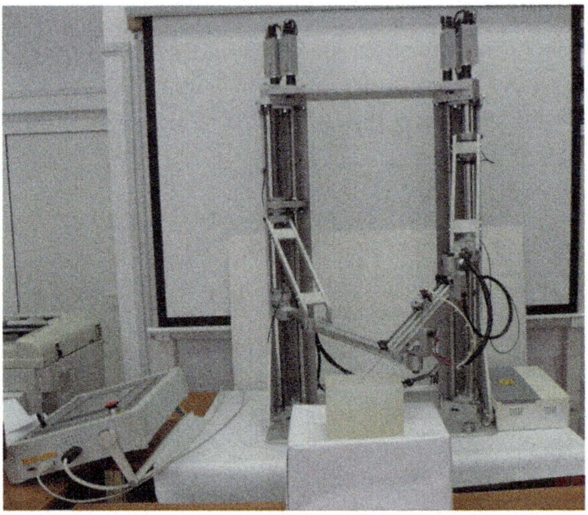

Fig. 1 PARA-BRACHYROB prototype

2 The Kinematics Using Study Parameters

PARA-BRACHYROB is a parallel robot with 2 modules with 3-DOF working in cylindrical coordinates, the first module having three actuated joints (two translations and one rotation), while the second has only two actuated joints (two translations). The two modules each with Cardan joints having the first axis parallel with the Z-axis, connect between them the needle-insertion module [5]. As described in detail in [5] the two modules of the robot have been decomposed into the PRPRR identical serial chains presented in Figs. 2 and 3. The matrices $\mathbf{L}_1$ resp. $\mathbf{L}_2$ characterize the Euclidian displacements for each chain, Equation 1 defining the direct kinematics computation. The terms of the matrix product are explained in the Table 1 [5].

$$\mathbf{L}_i = \mathbf{M}_{i1} \cdot \mathbf{T}_1 \cdot \mathbf{T}_2 \cdot \mathbf{T}_3 \cdot \mathbf{T}_4 \cdot \mathbf{M}_{i2} \cdot \mathbf{T}_5 \cdot \mathbf{M}_{i3}, \tag{1}$$

Summarizing the steps that lead to the direct kinematic description an ideal was generated for each kinematic chain. For the first chain the ideal κ ($\mathbf{L}_1$) is generated by four linear and one quadratic equation while the second ideal, κ ($\mathbf{L}_2$), is generated by ten quadratic equations. Taking into account the motion dependency between

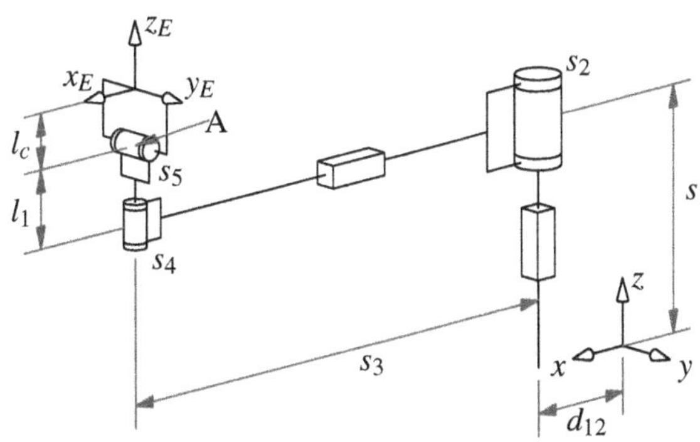

Fig. 2 Basic structure of the 2nd PRPRR chain

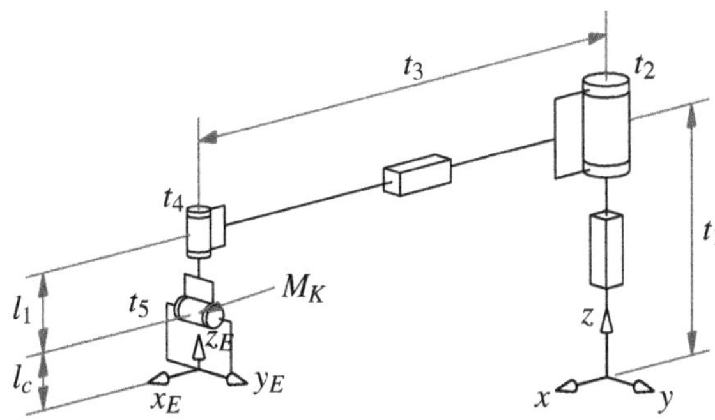

Fig. 3 Basic structure of the 1st PRPRR chain

Table 1 The geometric representation of the transformation matrices for $\mathbf{L}_1$ resp. $\mathbf{L}_2$

Matrix	Representation	Parameter	Type
$\mathbf{M}_{11}$	Basistransformation for leg 1 (identity matrix)	–	–
$\mathbf{M}_{21}$	Basistransformation for leg 2, translation along the x-axis	d_{12}	–
$\mathbf{T}_1$	Translation along z-axis	t_1 resp. s_1	Active
$\mathbf{T}_2$	Rotation around z-axis	t_2 resp. s_2	Active resp. passive
$\mathbf{T}_3$	Translation along x-axis	t_3 resp. s_3	Active
$\mathbf{T}_4$	Rotation around z-axis	t_4 resp. s_4	Passive
$\mathbf{M}_{12}$	Translation along z-axis (distance between A_1 C_1)	l_1	–
$\mathbf{M}_{22}$	Translation along z-axis (distance between A_2 C_2)	l_1	–
$\mathbf{T}_5$	Rotation around y-axis	t_5 resp. s_5	Passive
$\mathbf{M}_{13}$	Final translation along z-axis	l_c	–
$\mathbf{M}_{23}$	Final translation along z-axis	l_c	–

the two chains connected through a pair of Cardan joints, the direct kinematics solution is the union $\mathscr{I} = \mathscr{I}_1 \cup \mathscr{I}_2$, computed for every set of input parameters. The final result, which aims to determine a general Gröbner Basis, led to a result of a univariate polynomial of degree eight. The final polynomial was achieved in the most general case without defining any design parameter leading to a general, complete, description of the manipulator. The quadratic normalizing condition added in order to compute the basis doubles the number of solutions and it can be stated that the direct kinematics of this manipulator has four solutions. There are two solutions in the orientation module, consisting of the double cardan joints close to the end-effector coordinate system (x_E, y_E, z_E) (Figs. 2 and 3). These two solutions differ by a 180° rotation about the z_E axis. The two solutions of the positioning of the end-effector can be explained geometrically: when the input parameters s_1 and s_3 of the first chain are specified then the intersection point A of the two cardan axes is bound to move on a circle c_1 in a plane orthogonal to the z-axis of the base system in the height $s_3 + l_1$. When the input parameters t_1, t_2 and t_3 of the second system are specified, then the location of the intersection point M_K of both cardan axes of the second chain is defined. In the assembled robot the distance $2l_c$ between the points A and M_K is fixed, so the endpoint of a rod emanating from M_K with length $2l_c$ describes a sphere κ with center M_K and radius $2l_c$. The intersection of the sphere and the circle c_1 yields two possible locations for the point A (Fig. 4). A complete example of the solution of the direct kinematics can be found in [5].

3 Singularity Analysis

In this section a complete singularity analysis of the robot using the algebraic constraint equations from $\mathscr{I}_1$ and $\mathscr{I}_2$ will be presented. As it was mentioned in Sect. 2 the first ideal consists of five polynomials and the second on of ten polynomials. One

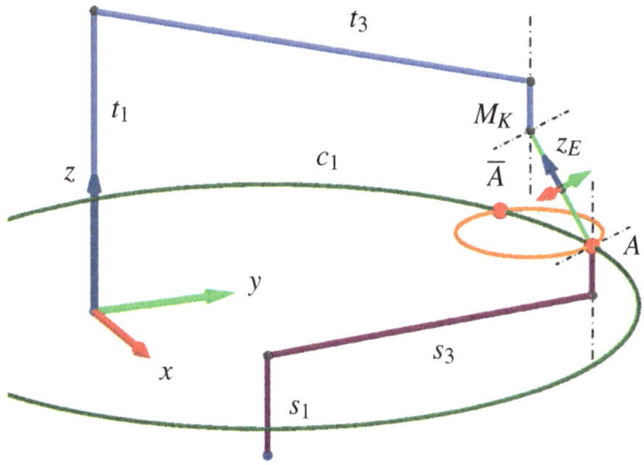

Fig. 4 Geometric interpretation of the direct kinematics

could rightfully question if this is a minimal description of the robot. Essentially it can be shown that the LIA algorithm comes up with a minimal set of polynomials that describe exactly the constraint variety of the robot.

In the following we are interested to compute all output singularities in the joint space. Having a single polynomial which describes all singular poses, allows its use as a parameter in the robot control, allowing the avoidance of any singular configuration, ensuring the safe robot behavior during the procedure. The number of 15 polynomial generates a serious difficulty in computing the Jacobian and the singularities. One could take subsets of eight polynomials respectively, differentiate with respect to the Study parameters (x_i, y_i), assemble the differentials into an 8×8 Jacobian matrix and compute the determinant. Unfortunately the resulting polynomials contain the Study parameters non linearly and an attempt to eliminate these parameters to obtain an equation in the input parameters solely is hopeless.

So another approach, based on the observation that in a singularity at least two solutions of the direct kinematics have to coincide, was followed. As mentioned in Sect. 2 the univariate polynomial of degree eight in one of the x_i Study parameters could be computed without specifying the input parameters (e.g. the univariate polynomial $P^8(x_3, t_1, t_2, t_3, s_1, s_3)$). It is easy to see that in a double solution the polynomial $DP = \frac{\partial P^8}{\partial x_3}$ and P^8 have to vanish. P^8 and DP can be taken to eliminate the last Study parameter. The result is a large polynomial which fortunately factors in eight terms:

$$Sing = P_1 \cdot P_2 \cdot P_3 \cdot P_4 \cdot P_5 \cdot P_6 \cdot P_7 \cdot P_8.$$

But this is not the end of the story, because the set of all double solutions in one Study parameter yields only a necessary condition for being a double solution of the direct kinematics. Sufficiency is only guaranteed when one can obtain double solutions in all Study parameters. This can be for example done by back substituting into the original system of constraint equations and computing the direct kinematics. The polynomials P_1, P_2, P_3, P_4 are simple enough to be displayed using the actual design parameters of PARA-BRACHYROB ($d_{12} = 615$ mm, $l_c = 85$ mm, $l_1 = 67$ mm):

$$P_1 : s_1 - t_1 + 304, \quad P_2 : s_1 - t_1 - 36, \quad P_3 : t_2^2 t_3 + 615\, t_2^2 - t_3 + 615$$
$$P_4 : t_2^3 t_3^2 + 1230\, t_2^2 + 378225\, t_2^2 + t_3^2 - 1230\, t_3 + 378225 \quad (2)$$

Substitution of the input parameters of P_1 and P_2 into the constraint equations reveals that they even lead to self motions of the manipulator. In both cases the vertical axes of the two cardan joints coincide and allow a full rotation of the end effector about its z_E axis. This motion was already detected in [5]. P_5, P_6, P_7 are not discussed because back substitution into the basis shows, that they do not lead to double solutions of the direct kinematic and hence, don't provide singular positions. The remaining cases will be discussed separately.

3.1 P_3

Here, one can solve for t_3 and obtains $t_3 = -\frac{615(t_2^2+1)}{t_2^2-1}$. Substituting t_3 into the set of original constraint equations and computing the Gröbner base with lexicographic term order reveals that the univariate polynomial is indeed squared and yields only four solutions, but one obtains one more polynomial in the base which is squared and this fact brings a total number of eight solutions. The conclusion is that the input parameter condition P_3 does not yield singular poses for the manipulator.

3.2 P_8

We start with this case because this case yields the general singularities which have to be avoided in the medical operation. P_8 is a polynomial of degree eight in the input parameters t_1, t_2, t_3, s_1, s_3. Figure 5 shows the singularity surface P_8 in the joint space t_1, t_2, t_3 for $s_1 = 200$ and $s_3 = 700$. This means, when four out of the five parameters are given, then up to eight values of the remaining input parameter will lead to a singular configuration of the end effector. These singularities have an obvious geometric interpretation which can be seen in Fig. 6. The intersection points of the sphere κ and the circle c_1 merge into one point ($A = \overline{A}$), or with other words: the circle touches the sphere, either from inside or outside. Computing the condition for this geometric situation yields also the polynomial P_8 and conforms the above statement. This polynomial codes all input configurations which yield the tangency configuration of both geometric objects and therefore yields singularities, an example is shown in Fig. 6. These poses have been tested on the experimental model and behaved as expected, generating singularities. They define the intersection of two working modes of the robot and shows that the robot cannot cross from one working mode into the other.

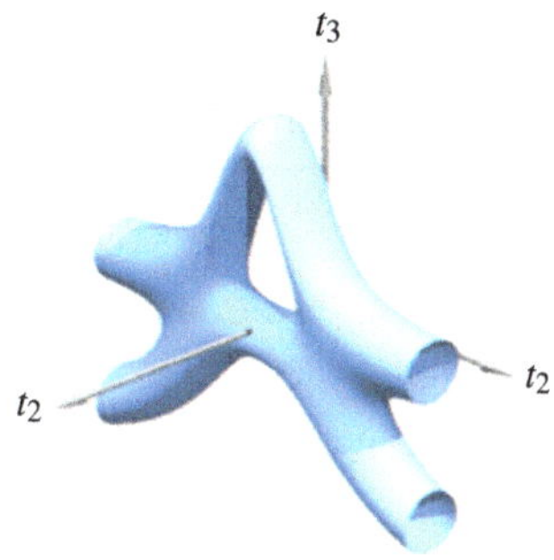

Fig. 5 Singularity set for $s_1 = 200$ and $s_3 = 700$

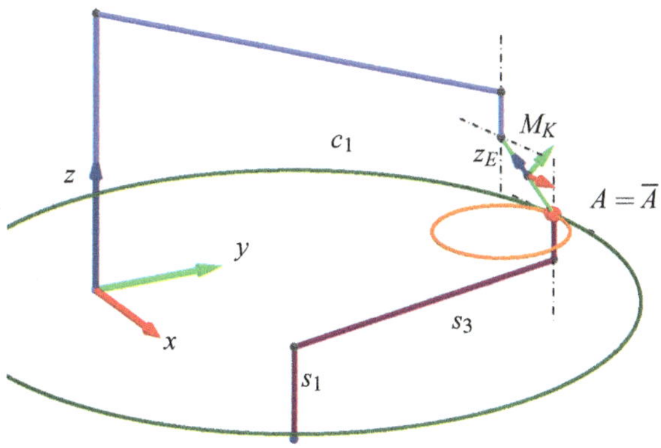

Fig. 6 Singularity condition according to P_8

3.3 P_4

Solving P_4 for t_2 yields $t_2 = \frac{I(t_3-615)}{t_3+615}$, $t_2 = -\frac{I(t_3-615)}{t_3+615}$, which is only real for $t_2 = 0$ and $t_3 = 615$. This set of parameters leads to an interesting geometric configuration of the manipulator, which can be seen in Fig. 7: the first rotation axis of the first chain and the first rotation axis of the cardan joint of the second chain coincide.

This means that the circle axis of c_1 passes through M_K and the intersection circle of κ in the plane of c_1 and c_1 are concentric. They have real intersections only when they are congruent. The condition for this is $s_3 = \sqrt{(s_1 - t_1 + 304)(s_1 - t_1 + 36)}$. But then point A together with the whole end effector system can freely rotate about the coinciding axes.

4 PARA-BRACHYROB Experimental Tests

The univariate polynomial that represents the complete description of the PARA-BRACHYROB singularities was implemented in the robot controller to enable real-time singularities avoidance. Even though the overall expression is big

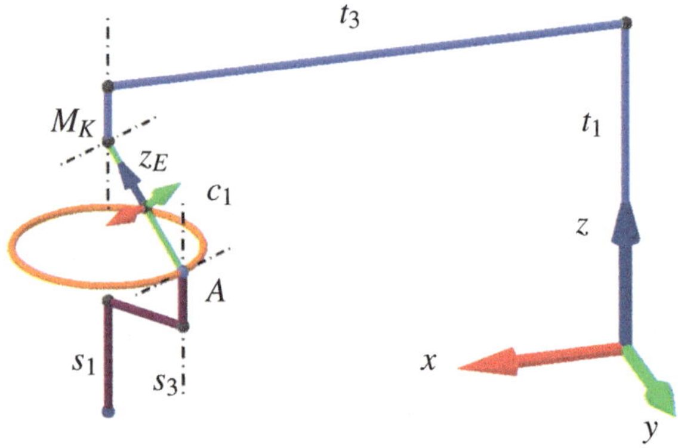

Fig. 7 Third self motion

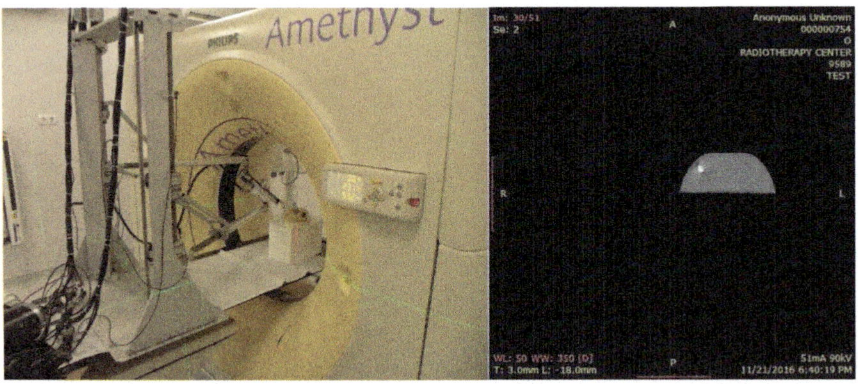

Fig. 8 PARA-BRACHYROB tests inside the CT

the computation time for the controller is very small without any negative effect on the robot motions. Having this safety condition satisfied, PARA-BRACHYROB underwent a series of tests in medical environment. With the support of an oncology clinic, Ametyst Cluj-Napoca, a first set of tests with the robot working inside the CT-Sim were performed. The robot setup is illustrated in Fig. 8.

With respect to the medical protocol [6, 7] of the brachytherapy a ballistic gel model was created, able to simulate the human tissue properties. Inside the model some bodies with a similar density with the hepatic tumors were inserted. During the entire procedure, the robotic system was controlled from a different room, the test being performed in a room containing a CT unit, known as a high emission of radiation source. Through its behavior during the entire procedure and based on the results obtained (after the brachytherapy procedure, the ballistic model was scanned using the CT) the parallel robotic system PARA-BRACHYROB has been validated as an efficient solution in the robotic assisted brachytherapy procedure.

5 Conclusions

Using algebraic constraint equations a complete singularity analysis of a medical robot designed for brachytherapy was performed. It turned out that this manipulator allows three kinds of self motions and all singularities can be found by input parameter sets that fulfill an eight degree polynomial. The zero set of this polynomial describes a degree eight hypersurface in the five dimensional joint space.

The polynomial expression was implemented in the control unit of the robot ensuring its motion in singularity free poses enabling a safe behavior during the medical procedure. The safe motion capability of the robot enables its use in a CT-Sim environment with real-time position control of the needle.

Acknowledgements This paper was supported by the Project no. 173/2012, code PN-II-PCCA-2011-3.2-0414, entitled "Robotic assisted brachytherapy, an innovative approach of inoperable cancers - CHANCE", financed by UEFISCDI, as well as the Austrian-Romanian WTZ project RO 06/2014 (UEFISCDI Code: 745/2014)

References

1. Pisla, D.: Application oriented desgin and simulation of an innovative parallel robot for brachytherapy. In: Proceedings of the ASME 2014 IDETC/CIE New-York, USA, 17–20 August 2014
2. Plitea, N.: Parallel robot for brachytherapy with two linkages for guiding a platform of cyl-u type. Patent: RO129698-A2 (2013)
3. Podder, T.: Aapm and gec-estro guidelines for image-guided robotic brachytherapy: report of task group 192. Med. Phys. **41**(10), 101501 (2014)
4. Schweikard, A., Ernst, F. (eds.): Medical Robotics. Springer, Berlin (2015). ISBN 978-3-319-22890-7
5. Vaida C.: Kinematic analysis of an innovative medical parallel robot using study parameters. In: 4th International Workshop on Medical and Service Robots, MESROB (2015)
6. Vaida.C., et al. (eds.): The Control System of a Parallel Robot for Brachytherapy. Springer, Berlin (2014). ISBN 978-3-319-09410-6
7. Yu Y.: Robotic system for prostate brachytherapy. Comput. Aided Surg. **12**(6) (2007)

Workspace Analysis of a 3-PS$\underline{P}$ Motion Platform

Luc Baron

Abstract This paper presents the workspace analysis of a 3-PS$\underline{P}$ motion platform to be used as a flight simulator. In this design, all passive joints are kept on the base plate rather than under the moving platform, thus ensuring a secure and easy to construct configuration. With a proper description of the constrained rotation of the platform, the inverse kinematics can be solved in an analytical form. The resulting workspace shows varying tilting amplitudes with respect to the vertical displacement, the azimuth and the actuator strokes.

1 Introduction

For flight simulation applications, parallel manipulators are first-choice mechanisms to provide the 6-degrees-of-freedom (dof) mobility of the motion platform. In general, these manipulators are particularly worthy of note because they have a high carrying capacity, as well as a lower workspace volume, more singularity problems and an increased complexity when solving the direct kinematic problem compared to serial manipulators of equivalent size [1]. However, since a full 6-dof is not required for every type of flight simulation training, the use of lower mobility mechanisms may often fulfill the same training objectives at a much lower cost. In particular, 3-dof motion platforms providing a vertical displacement and pitch and roll rotations, without either the lateral and longitudinal displacement or the yaw rotation (see Fig. 1), are the most important mobilities [2] required from the point of view of flight simulation. Like 3-PSP platforms, these mechanisms have many advantages in terms of simplicity of construction/control and reduced cost, although they also present some problems due to the coupled orientation and position of the platform. Most of the research works published in the literature on these platforms have been done on the 3-$\underline{P}$SP variant [3–7], i.e., where the actuated prismatic joint is located on the base, unlike our 3-PS$\underline{P}$ variant [8], where the actuated prismatic joint are located on the platform. In our design, all passive joints (possibly harmful) are located at the

L. Baron (✉)
École Polytechnique, Montréal, Canada
e-mail: Luc.Baron@polymtl.ca

J. Lenarčič and J.-P. Merlet (eds.), *Advances in Robot Kinematics 2016*,
Springer Proceedings in Advanced Robotics 4, DOI 10.1007/978-3-319-56802-7_10

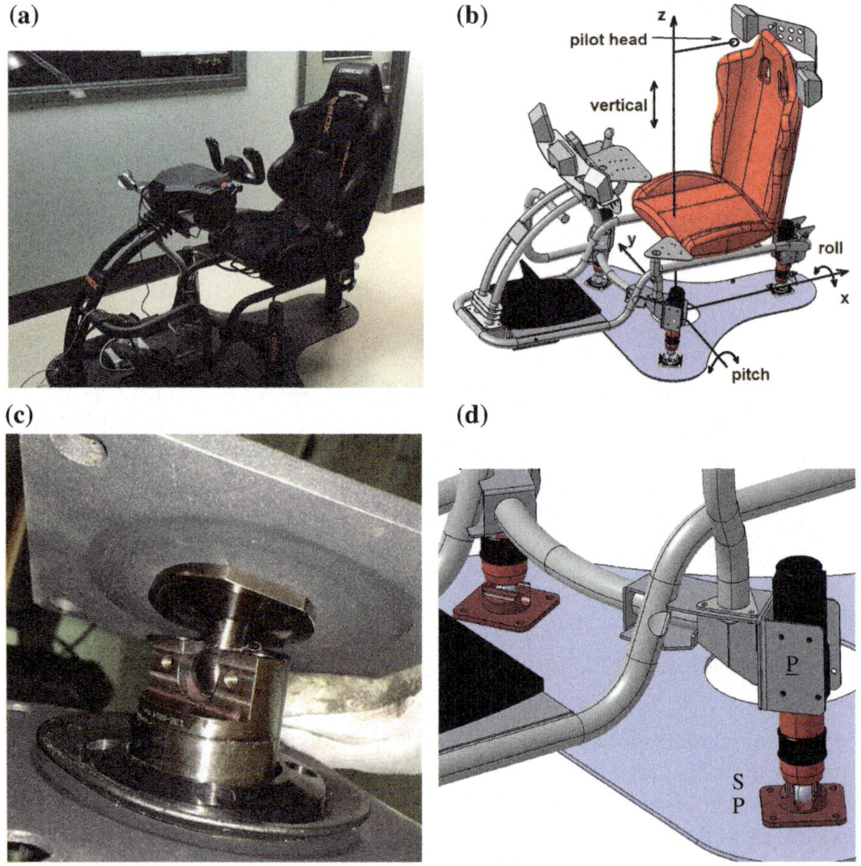

Fig. 1 The 3-PSP̲ motion platform: **a** actual prototype; **b** CAD model; **c** actual passive PS joints; and **d** modeled passive PS joints

attachment of the leg tip with the base plate through a passive combined PS joints. As shown in Fig. 1c, d, the actual implementation is realized with a sphere (attached to the leg tip) sliding and rotating into a partially open horizontal hole (attached to the base plate). It is noteworthy that here we develop a kinematic model of general geometry, while in our previous study [8] we assumed an equilateral triangle and not addressing the influence of the actuator stroke.

2 Platform Mobility

The 3-PSP̲ mechanism provides a constrained spatial displacement of the platform, i.e., a generic point of it moves in 3-dimensional space, while the platform rotates around three axes, but only 3 out of 6 parameters are independent. Consequently, the platform has only 3-dof relative to the base. The constrained rotation of the platform can be described with only two independent coordinates together with its vertical displacement. Thus, the orientation of the platform is not arbitrary and can be described with the tilt and torsion angles [9] such that

$$\mathbf{R}_{T\&T}(\phi, \theta, \sigma) = \mathbf{R}_Z(\phi)\mathbf{R}_Y(\theta)\mathbf{R}_Z(\sigma - \phi) \tag{1}$$

where the tilt angle θ is a rotation around axis a located in the original xy-plane, while the orientation of the latter is given by the azimuth angle ϕ. As shown in Fig. 2b, the torsion angle σ is about z^*-axis. It has been pointedout [9] that 3-PSP mechanisms have a zero torsion angle, i.e., $\sigma = 0$, and are hence,

$$\begin{aligned} \mathbf{R}(\phi, \theta) &= \mathbf{R}_Z(\phi)\mathbf{R}_Y(\theta)\mathbf{R}_Z(-\phi) \\ &= \begin{bmatrix} \cos^2\phi\cos\theta + \sin^2\phi & \sin\phi\cos\phi(\cos\theta - 1) & \cos\phi\sin\theta \\ \sin\phi\cos\phi(\cos\theta - 1) & \sin^2\phi\cos\theta + \cos^2\phi & \sin\phi\sin\theta \\ -\sin\theta\cos\phi & -\sin\theta\sin\phi & \cos\theta \end{bmatrix} \end{aligned} \tag{2}$$

Equation (2) is nothing else then the Euler ZYZ angles with the third angle being constrained to $-\phi$. This representation of the constrained rotation of the platform will allow a compact formulation of the kinematic model.

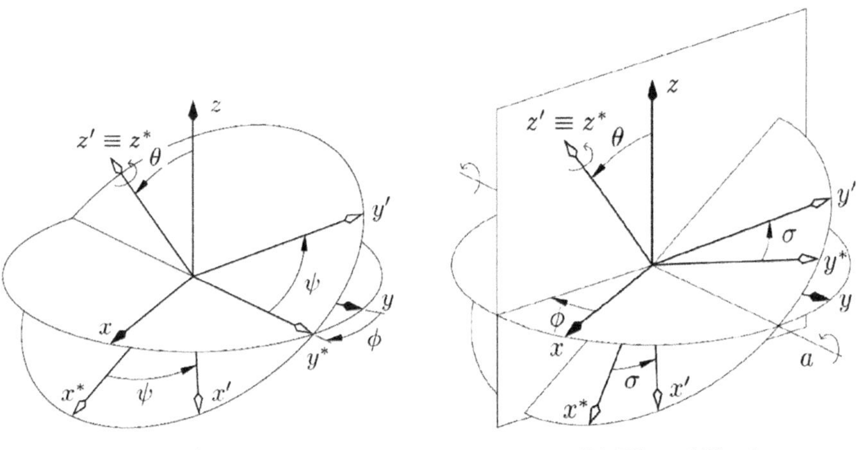

(a) EulerZYZ (b) Tilt and Torsion

Fig. 2 Orientation parameters: **a** Euler ZYZ angles (rotation of ϕ around z, rotation of θ around y^* and a rotation of ψ around z^*); **b** the tilt and torsion [9] (tilting of θ around a being oriented in xy-plan at ϕ; then torsion of σ around z^*)

3 Kinematic Model

Let us attach frame $\mathscr{A}$ to the fixed base A and frame $\mathscr{B}$ to the moving body B, its z-axis parallel to the three actuated joints axes, as shown in Fig. 3. Frames $\mathscr{A}$ and $\mathscr{B}$ are coincident when all actuated prismatic joints are fully retracted. Finally, a prime is used to denote vectors expressed in the moving frame $\mathscr{B}$, while every other vectors are by default expressed in frame $\mathscr{A}$. Table 1 gives the geometry of the actual prototype.

3.1 Position Analysis

The position vector of point B_i is expressed in $\mathscr{B}$ as

$$\mathbf{b}_i' = [r_i \cos\alpha_i \quad r \sin\alpha_i \quad 0]^T, \tag{3}$$

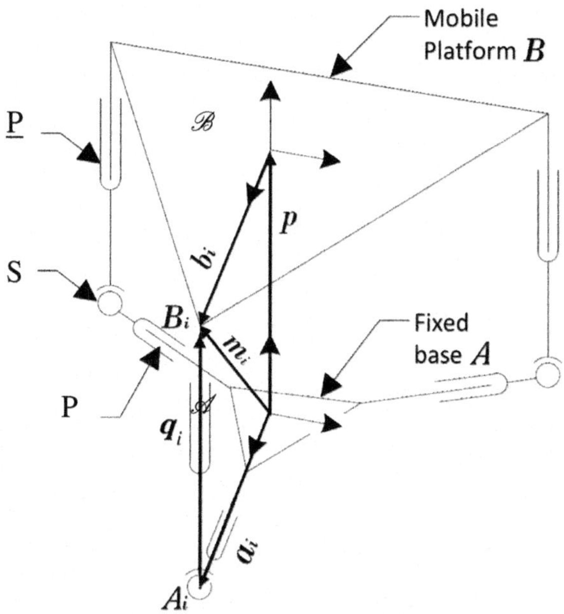

Fig. 3 The kinematic loop of a leg

Table 1 Geometry of the 3-PS$\underline{\text{P}}$ prototype

i	1	2	3	unit
$\{q_i\}_{min}$	0	0	0	mm
$\{q_i\}_{max}$	100	100	100	mm
r_i	475	400	400	mm

with α_i being defined as

$$\alpha_i \equiv 2(i-1)\pi/3, \quad i = 1, 2, 3. \tag{4}$$

In the base frame $\mathscr{A}$, the unit vectors along the three passive prismatic joints can be written as

$$\mathbf{n}_i = [\cos\alpha_i \quad \sin\alpha_i \quad 0]^T, \tag{5}$$

or alternatively, in the moving frame $\mathscr{B}$, as

$$\mathbf{n}'_i = \mathbf{R}^T\mathbf{n}_i. \tag{6}$$

The position vector of the origin of $\mathscr{B}$ relative to the origin of $\mathscr{A}$ expressed in the moving frame $\mathscr{B}$ is given as

$$\mathbf{p}' = [x' \quad y' \quad z']^T, \tag{7}$$

and alternatively, in the base frame $\mathscr{A}$ as

$$\mathbf{p} = \mathbf{R}\mathbf{p}' = [x \quad y \quad z]^T. \tag{8}$$

The position vector of point B_i relative to the origin of $\mathscr{A}$, but expressed in $\mathscr{B}$, is given as

$$\mathbf{m}'_i = \mathbf{p}' + \mathbf{b}'_i. \tag{9}$$

Alternatively, the same position vector can also be obtained as

$$\mathbf{m}'_i = \mathbf{a}'_i + \mathbf{q}'_i = a_i\mathbf{n}'_i + q_i\mathbf{k}'_i. \tag{10}$$

Since the closure equation requires vectors $\mathbf{m}'_i$, $\mathbf{n}'_i$ and $\mathbf{k}'_i$ of each individual leg i to be coplanar, they must satisfy the following equation

$$\det\left[\mathbf{m}'_i \quad \mathbf{n}'_i \quad \mathbf{k}'_i\right] = 0, \tag{11}$$

where $\mathbf{k}'_i = [0 \quad 0 \quad 1]^T$. Using Eq. (11), we can algebraically obtain the solution of the *inverse kinematic problem*, i.e., the actuated joint position q_i as

$$\begin{aligned} q_1 &= \frac{z - r_1\sin\theta\cos\phi}{\cos\theta} \\ q_2 &= \frac{2z + r_2\sin\theta\cos\phi - \sqrt{3}r_2\sin\theta\sin\phi}{2\cos\theta} \\ q_3 &= \frac{2z + r_3\sin\theta\cos\phi + \sqrt{3}r_3\sin\theta\sin\phi}{2\cos\theta} \end{aligned} \tag{12}$$

from a given position and orientation of the platform, i.e., z, θ, ϕ. It is worth noting that z is an independent coordinate that can be chosen freely, while θ and ϕ are the specified orientation of the platform. An important feature of this mechanism is the parasitic displacement in x and y as

$$x = \frac{2r_{12}(r_{11} - \sqrt{3}r_{21})r_1 + r_{11}(r_{12} - 3r_{21} - \sqrt{3}(r_{11} + r_{22}))r_2}{2\sqrt{3}(r_{11}r_{22} - r_{12}r_{21})} \tag{13}$$

$$y = \frac{-2r_{12}(r_{12} - \sqrt{3}r_{22})r_1 + -r_{12}(r_{12} - 3r_{21} - \sqrt{3}(r_{11} + r_{22}))r_2}{2\sqrt{3}(r_{11}r_{22} - r_{12}r_{21})}, \tag{14}$$

which depend on the platform geometry r_i and orientation θ and ϕ through r_{ij}.

3.2 Velocity Analysis

On differentiating Eq. (12), it is easy to discover the mechanism's general velocity relationship, i.e.,

$$\mathbf{A}\dot{\pi} = \mathbf{B}\dot{\mathbf{q}} \tag{15}$$

where $\dot{\pi}$, namely the *twist* of the platform, and $\dot{\mathbf{q}}$, namely the actuated joint velocities, are defined as

$$\dot{\pi} \equiv [\dot{z} \quad \dot{\theta} \quad \dot{\phi}]^T, \quad \dot{\mathbf{q}} \equiv [\dot{q}_1 \quad \dot{q}_2 \quad \dot{q}_3]^T. \tag{16}$$

Matrices **A** and **B** are the so-called parallel and serial Jacobian matrices given as

$$\mathbf{A}(z, \theta, \phi) = \begin{bmatrix} a_{11} & a_{12} & a_{13} \\ a_{21} & a_{22} & a_{23} \\ a_{31} & a_{32} & a_{33} \end{bmatrix}, \quad \mathbf{B} = \mathbf{1}_{3\times3}, \tag{17}$$

with

$$
\begin{aligned}
a_{11} &= \frac{1}{\cos\theta}, \\
a_{12} &= \frac{q_1 \sin\theta - r_1 \cos\theta \cos\phi}{\cos\theta}, \\
a_{13} &= \frac{r_1 \sin\theta \sin\phi}{\cos\theta} \\
a_{21} &= \frac{1}{\cos\theta}, \\
a_{22} &= \frac{2q_2 \sin\theta + r_2 \cos\theta \cos\phi - \sqrt{3} r_2 \cos\theta \sin\phi}{2\cos\theta}, \\
a_{23} &= \frac{-(\sqrt{3} r_2 \sin\theta \cos\phi + r_2 \sin\theta \sin\phi)}{2\cos\theta}, \\
a_{31} &= \frac{1}{\cos\theta}, \\
a_{32} &= \frac{2q_3 \sin\theta + r_3 \cos\theta \cos\phi + \sqrt{3} r_3 \cos\theta \sin\phi}{2\cos\theta}, \\
a_{33} &= \frac{\sqrt{3} r_3 \sin\theta \cos\phi - r_3 \sin\theta \sin\phi}{2\cos\theta}.
\end{aligned}
\tag{18}
$$

Matrix **A** degenerates when $\theta = \pi/2$. At this tilting angle, the legs become of infinite length, and do not produce any motion of the platform. Obviously, the mechanism is not able to reach such a high value of θ. In fact, the prototype is able to reach a maximum tilt angle of approximately $\pi/6$.

4 Workspace

The workspace of the 3-PSP motion platform is determined by varying the following three variables z, ϕ and θ, and computing the corresponding joint positions $\{q_i\}_1^3$ with Eq. (12). The first loop varies z from 0 to $\{q_i\}_{max}$. The second loop varies ϕ from $-\pi$ to $+\pi$, and the third loop varies θ from 0 to the angles for which one of the q_i reaches its minimum or maximum. In order to study the displacement the pilot's head with the platform motion, let us define the position of the pilot head in frame $\mathscr{B}$ as $\mathbf{h}' = [100\ 0\ 1200]^T$ mm and compute the corresponding head position in frame $\mathscr{A}$ as $\mathbf{h}$. Figures 4 and 5 show a set of contours limiting the workspace of constant z' values of $\{0, 10, 20, \ldots, \{q_i\}_{max}\}$ mm. Apparently, the amplitude of head displacement and orientation (corresponding to the pitch and roll angles of the aircraft) are both varying with z'. All the azimuth ϕ are reachable with different amplitude of tilt angle. Increasing the actuator strokes from 100 to 300 mm allows to increase the maximum tilting angle from 8° to approximatively 25°.

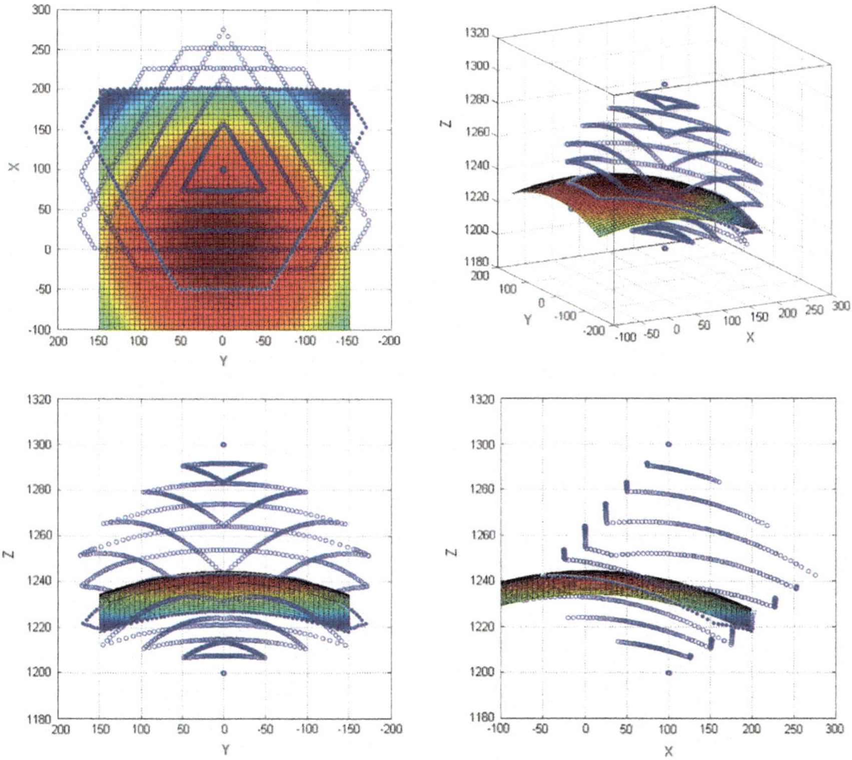

Fig. 4 Workspace of the 3-PS<u>P</u> of Table 1 with $\mathbf{h}' = [100\ 0\ 1200]^T$ mm

5 Conclusions

The 3-PSP̲ variant, i.e., with the three actuated prismatic joints rigidly attached to the platform, is secure for the pilot and easy to construct. It is particularly well-suited for the flight simulation applications. The inverse kinematic model can be solved in analytical form with a proper description of the constrained rotation of the platform. Although every azimuth are obtained, the amplitude of tilting is varying with both the azimuth and the vertical displacement. The tilting amplitude also greatly dependes on the actuator strokes.

Acknowledgements The author would like to acknowledge the financial support provided by NSERC (National Sciences and Engineering Research Council of Canada) under grant OGPIN-203618 and the CAE R. Fraser Elliott Modeling and Simulation Laboratory. Thanks also go to Master of Engineering student Jaime Andres Pacheco for developing programs enabling the validation of the kinematic model and the control of the motion platform.

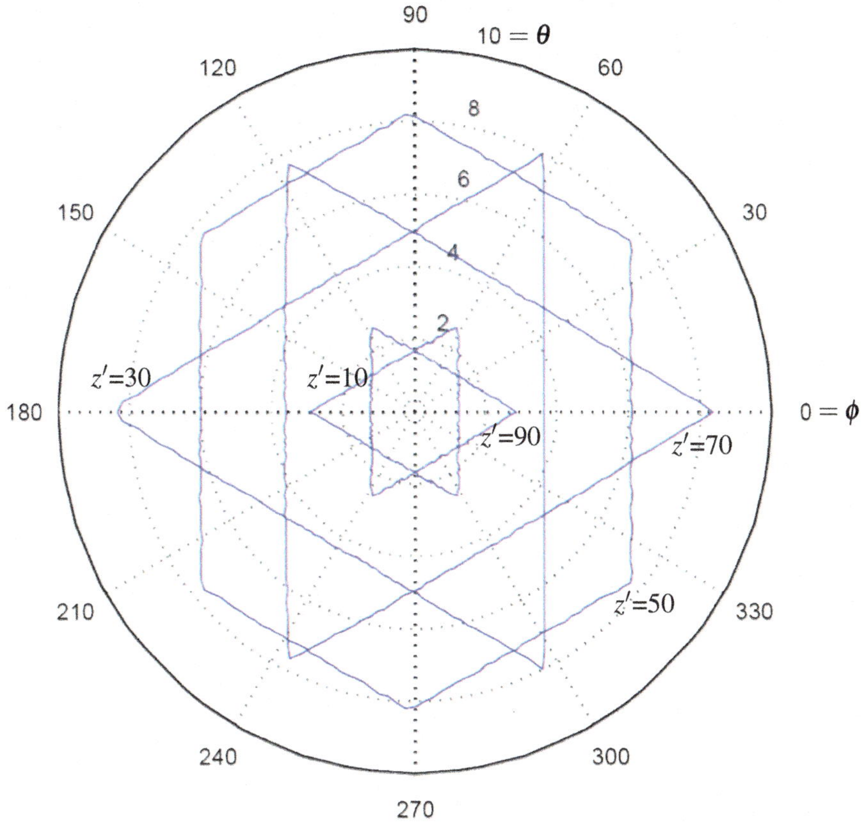

Fig. 5 Orientation workspace of the prototype for different $z' = \{10, 30, 50, 70, 90\}$ mm

References

1. Vermette, Y., Baron, L.: Planar 3 degrees of freedom parallel manipulator with an articulated platform featuring a planetary gearbox. In: Proceedings. of the 13th World Congress on the Theory of Machines and Mechanisms, IFToMM, June 19–25, Guanajuato, Mexico (2011)
2. Pouliot, N., Gosselin, C., Nahon, N.: Motion simulation capabilities of three-degree-of-freedom flight simulators. AIAA J. Aircr. **35**(3), 9–17 (1998)
3. Hao, Q., Wang, L., Guan, L., Liu, X.-J.: Dynamic analysis of a novel 3-PSP 3-DOF parallel manipulator. In: Proceeding of the International Conference on Reconfigurable Mechanisms and Robots, ASME/IFToMM, pp. 309–314 (2009)
4. Chen, C., Huang, Y., Cheng, X. and Guo, S.: Study and analysis of the 3-PRS parallel mechanism. In: IEEE International Conference on Mechatronics and Automation, pp. 1515–1520 (2009)
5. Nia, P.M., Tootoonchi, A.A.: Dynamics modeling of 3-PSP parallel manipulator using natural orthogonal complement method (NOC). In: 17th Annual Conference on Mechanical Engineering, University of Tehran, ISME-2009-3028 (2009)
6. Di Gregorio, R., Parenti-Castelli, V.: Position analysis in analytical form of the 3-PSP mechanism. ASME J. Mech. Design **123**, 51–57 (2001)

7. Rezaei, A., Akbarzadeh, A., Aahmoodi, P., Akbarzadeh, M.: Position, Jacobian and workspace analysis of a 3-PSP spatial parallel manipulator. Robot. Comput.-Integr. Manuf. **29**, 158–173 (2013)
8. Baron, L.: Kinematic analysis of a 3-PSP spatial motion platform. In: ASME 2015 International Design Engineering Technical Conferences, IDETC/CIE 2015, August 2–5, Boston, USA (2015)
9. Bonev, I.: Geometric analysis of parallel mechanisms. Ph.D. Thesis, Laval University, Quebec, Quebec, Canada, November, 174 p. (2002)
10. Gosselin, C., Angeles, J.: Singularity analysis of closed-loop kinematic chains. IEEE Trans. Robot. Autom. **6**(3), 281–290 (1990)

Posture Optimization of a Functionally Redundant Parallel Robot

David Corinaldi, Jorge Angeles and Massimo Callegari

Abstract The use of parallel-kinematics machines (PKM) for manufacturing operations is attractive because of the high accuracy they can ensure. These robots might perform a task that requires less degrees of freedom than those offered by the robot. This is the case of a robot facing a functional redundancy, which can be exploited to further increase the accuracy of the task, e.g. upon minimizing the condition number of the Jacobian matrix. A practical case study of a spherical manipulator performing a pointing task are reported, to show how posture-optimization can be used as a redundancy-resolution means for functionally redundant PKMs. The kinematics of the machine and the orientation of the pointing task is used to build, respectively, the objective function and the constraint equations. Sequential Quadratic Programming is conducted to solve the nonlinear constrained optimization problem and to find the end-effector pose corresponding to the robot posture of minimum condition number for every direction of a given pointing path. Lastly, the constrained problem is rewritten as one of unconstrained optimization of one objective function in one design variable.

1 Introduction and Case Study

The improvement of task performance using robotic manipulators is a recurrent challenge in robotics research. The need of a robot with high accuracy often arises in industry when manufacturing tasks have to be performed without giving up the flexibility provided by a manipulator. This feature often drives the choice of the manipulator, namely the most generic type: robots aimed at displacing the end-effector (EE) with six degrees of freedom (dof). Due to their wide range of applications, six-dof

D. Corinaldi (✉) · M. Callegari
Polytechnic University of Marche, Ancona, Italy
e-mail: d.corinaldi@univpm.it

M. Callegari
e-mail: m.callegari@univpm.it

J. Angeles
McGill University, Montreal, Canada
e-mail: angeles@cim.mcgill.ca

J. Lenarčič and J.-P. Merlet (eds.), *Advances in Robot Kinematics 2016*,
Springer Proceedings in Advanced Robotics 4, DOI 10.1007/978-3-319-56802-7_11

robots can manipulate an axially symmetric tool, thereby freeing a rotation around its axis of symmetry. This class of tasks is characterized by five-dof, of particular interest to industry, since machining, arc-welding [5] and deburring operations all fall into this category. A six-dof robot that performs a five-dof task is said to be functionally redundant [1]: more dof are available than needed; therefore, these robots can be exploited to accomplish a secondary task. More specifically, the robot considered is *functionally redundant* because it has an operational-space dimension (reachable Cartesian space of the EE) greater than its operational task-space dimension (Cartesian space of the task). Functional redundancy can be used to increase the accuracy of the manipulator above what is currently available, as reported by Léger and Angeles for serial robots, or to improve dynamic performance for high-speed task [6]. The secondary objective in the Léger's paper being to avoid singularities, the local dexterity index used was the condition number, which quantifies the error amplification between joint and EE relative errors [7]: by lowering the condition number value, the propagation of joint errors to the EE pose is reduced, thus increasing accuracy. In this work the authors report a new step in the same direction and hence, apply the same concept to parallel manipulators, known to offer many advantages over their serial counterparts, like a lightweight structure and a high stiffness, which are ideal for the task at hand: they are the best candidates for tasks that require high accuracy.

One of the issues in the analysis of parallel manipulators with six-dof is the complexity of their kinematics, which can affect adversely in the path-planning. Often what is done in conventional machining operations is to decompose the full-mobility operations into elementary sub-tasks, to be performed by separate machines with lower mobility [4]. The authors envisaged the architecture of a mechatronic system with six-dof, i.e., two parallel robots cooperating while performing a five-dof assembly task, as shown in Fig. 1. The kinematics of both machines is based upon the 3-CPU topology, but the joints are differently assembled so as to obtain: one

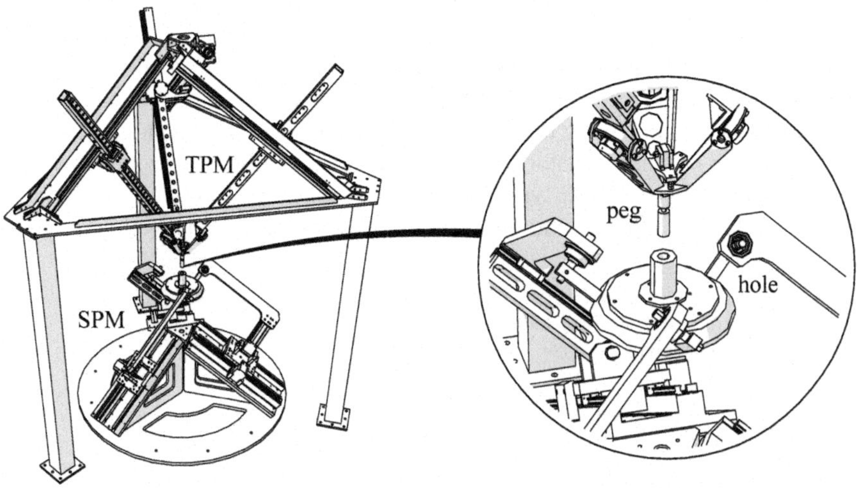

Fig. 1 System architecture of an assembly cell based on two cooperating 3-dof parallel robots

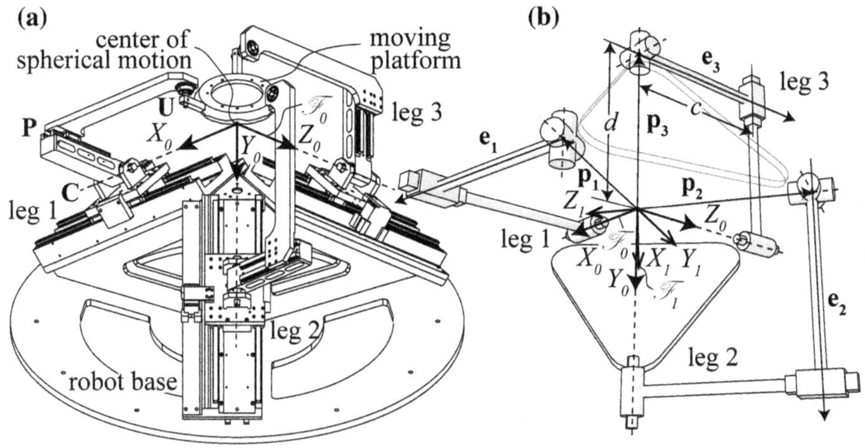

Fig. 2 Virtual model **a** of the spherical 3-CPU manipulator Sphe.I.Ro. Reference frames and geometrical parameters (**b**)

translating parallel machine (TPM) with one mechanism and one spherical parallel machine (SPM) with the other. The operation that the two PKMs have to perform is a peg-in-hole assembly task with axisymmetric tools, a classical five-dof assembly task. The SPM on the bottom, called Sphe.I.Ro., shown in Fig. 2a, holds the piece with the hole, while the TPM on the top translates the cylindrical peg. The accuracy of the two manipulators is often decisive for the assembly to be successful. Hence, functional redundancy, pertaining to the SPM, is exploited to orient the axis of the hole along the directions given by a pointing-path coming from the assembly strategy. The manipulator, which is still able to rotate around such directions, will attain an optimal posture that minimizes the condition number of the SPM.

2 Jacobian Matrices and Their Condition Number

The objective function of the optimization problem is the square of the condition number of the Jacobian matrix of Sphe.I.Ro. The Jacobian is obtained by means of the theory of screws; its formulation for the 3-CPU PKM architecture was reported in [4]. The moving-platform (MP) twist $\mathbf{t} = \left[\boldsymbol{\omega}^T \ \mathbf{v}^T\right]^T$ can be expressed as a linear combination of the velocities that each joint in the kinematic chain provides. The expression is simplified by multiplication of the screws reciprocal to all passive joint screws of each leg: the screw directions $\mathbf{e}_1$, $\mathbf{e}_2$ and $\mathbf{e}_3$ are indicated in Fig. 2b. By doing so, the influence on end-effector velocity of non actuated joint rates is eliminated. In the specific case of a pure rotational tripod robot, the kinematic relations follow:

$$\underbrace{\left[\mathbf{n}_1\ \mathbf{n}_2\ \mathbf{n}_3\right]^T}_{\mathbf{J}} \underbrace{\left[\omega_1\ \omega_2\ \omega_3\right]^T}_{\boldsymbol{\omega}} = \underbrace{\left[\dot{q}_1\ \dot{q}_2\ \dot{q}_3\right]^T}_{\dot{\mathbf{q}}} \tag{1}$$

where $\mathbf{n}_i$ is the moment of the direction $\mathbf{e}_i$ with respect to the $\mathscr{F}_0$-origin and $\dot{q}_i$ is the velocity of the actuated joint of the of ith leg. Since the inverse Jacobian matrix is constant and equal to the identity matrix, the study focuses only on the direct Jacobian $\mathbf{J}$. To get an analytic function of the condition number, the Frobenius norm of the 3×3 Jacobian $\mathbf{J}$ is calculated. This is the only matrix norm that is infinitely many times differentiable with respect to its arguments. This choice allows a straightforward minimum search by virtue of its smoothness properties. Based on the Frobenius norm of $\mathbf{J}$, the square of the condition number $\kappa_F^2(\mathbf{J})$ becomes

$$\kappa_F^2(\mathbf{J}) = \|\mathbf{J}\|_F^2 \|\mathbf{J}^{-1}\|_F^2 = \frac{1}{n^2}\mathrm{tr}\left(\mathbf{J}\mathbf{J}^T\right)\mathrm{tr}\left(\mathbf{J}^{-1}\mathbf{J}^{-T}\right) \tag{2}$$

where $\|\mathbf{J}\|$ is obtained as the positive square root of $\|\mathbf{J}\|_F^2 = \mathrm{tr}(\mathbf{J}\mathbf{J}^T)/3$. Each trace of this expression is expanded, to express it as a function of the row vectors of the Jacobian matrix, namely, the first trace is equal to the summation of the 2-norm square of the $\mathbf{n}_i$ vectors. Regarding the second factor of Eq. 2 $\mathrm{tr}(\mathbf{J}^{-1}\mathbf{J}^{-T})$, the inverse of $\mathbf{J}^T$ can be expressed in terms of its columns explicitly, without introducing components, if the concept of *reciprocal bases* is recalled [2]. Hence, Eq. 2 becomes

$$\kappa_F^2(\mathbf{J}) = \frac{\left(\|\mathbf{n}_1\|^2 + \|\mathbf{n}_2\|^2 + \|\mathbf{n}_3\|^2\right)\left(\|\mathbf{n}_2 \times \mathbf{n}_3\|^2 + \|\mathbf{n}_3 \times \mathbf{n}_1\|^2 + \|\mathbf{n}_1 \times \mathbf{n}_2\|^2\right)}{9(\mathbf{n}_1 \times \mathbf{n}_2 \cdot \mathbf{n}_3)^2} \tag{3}$$

which can be written as a function of the parameters describing the orientation through the rotation matrix since to an EE pose corresponds a single posture of Sphe.I.Ro, as reported by Carbonari [4]. To this end the $\mathbf{n}_i$ vectors are expressed in terms of the rotation matrix $\mathbf{Q}$, represented in the reference system $\mathscr{F}_0$ that maps a vector from the mobile frame $\mathscr{F}_1$ to the fixed frame $\mathscr{F}_0$. This is possible since the unit vectors of the reciprocal screws $\mathbf{e}_i$ represent the axes of a reference frame fixed to the ground, while the directions of the chosen vectors $\mathbf{p}_i$ represent an orthonormal triad of directions of a reference frame fixed to the mobile platform. The three vectors $\mathbf{n}_i$ can then be written as

$$\begin{aligned}
\mathbf{n}_1 &= \mathbf{p}_1 \times \mathbf{e}_1 = -d\mathbf{j}_1 \times \mathbf{i}_0 = \mathbf{i}_0 \times d\mathbf{j}_1 \Rightarrow [\mathbf{n}_1]_0 = d[\mathbf{i}_0]_0 \times [\mathbf{Q}]_0[\mathbf{j}_1]_1 \\
\mathbf{n}_2 &= \mathbf{p}_2 \times \mathbf{e}_2 = -d\mathbf{k}_1 \times \mathbf{j}_0 = \mathbf{j}_0 \times d\mathbf{k}_1 \Rightarrow [\mathbf{n}_2]_0 = d[\mathbf{j}_0]_0 \times [\mathbf{Q}]_0[\mathbf{k}_1]_1 \\
\mathbf{n}_3 &= \mathbf{p}_3 \times \mathbf{e}_3 = -d\mathbf{i}_1 \times \mathbf{k}_0 = \mathbf{k}_0 \times d\mathbf{i}_1 \Rightarrow [\mathbf{n}_3]_0 = d[\mathbf{k}_0]_0 \times [\mathbf{Q}]_0[\mathbf{i}_1]_1
\end{aligned} \tag{4}$$

Hence, $\mathbf{p}$ is the vector connecting the origin of the reference frame $\mathscr{F}_0$ to a point of the axis of the screw. Among the various vectors, we choose the one from the center of the spherical manipulator to the intersection of the two universal joint axes.

3 Formulation of the Optimization Problem

In trying to resolve the functional redundancy of Sphe.I.Ro. performing a two-dof task, the problem can be formulated as one of solving a constrained optimization problem, the constraints being imposed by the pointing specification, while the objective function is $\kappa_F^2(\mathbf{J})$. Then the problem takes the form:

$$f(\mathbf{x}) \equiv \kappa_F^2(\mathbf{J}) \rightarrow \min_{\mathbf{x}}, \quad \text{s.t. } \mathbf{h}(\mathbf{x}) = \mathbf{0} \tag{5}$$

where $\mathbf{h}$ is a vector of constraints and $\mathbf{x}$ the array of unknowns. The constrained problem can be formulated by means of points lying on a sphere with unit radius. Choosing this approach, a rotation about the hole axis that passes through the center of the spherical manipulator is the functional redundancy. The constraint is then the coincidence of the vertical unit vector $\mathbf{e}_v$, taken as reference vector and mapped by the rotation matrix, and the actual unit vector of the hole direction $\mathbf{e}_h$, both expressed in the same reference frame, i.e., the fixed frame $\mathscr{F}_0$.

$$\mathbf{h}(\mathbf{x}) = \mathbf{Q}\mathbf{e}_v - \mathbf{e}_h = \mathbf{0} \quad \Rightarrow \quad \mathbf{h}(\mathbf{x}) = [\mathbf{Q}(\mathbf{x})]_0[\mathbf{e}_v]_0 - [\mathbf{e}_h]_0 = \mathbf{0} \tag{6}$$

According to the two reference systems chosen, $[\mathbf{e}_v]_0 \stackrel{\triangle}{=} [\mathbf{e}_h]_1 = -\sqrt{3}/3\begin{bmatrix}1 & 1 & 1\end{bmatrix}^T$.

So far the optimization problem is written in terms of the rotation matrix $\mathbf{Q}$ that needs a proper parametrization to describe the orientation of the mobile platform: among the various representations that describe the rotation, we choose the one based on the rotation invariants, and in particular, the Euler-Rodrigues Parameters (ERPs), i.e., the four scalars $\mathbf{r}$ and r_0 [1]:

$$\mathbf{Q} = (r_0^2 - \mathbf{r}^T\mathbf{r})\mathbf{1} + 2\mathbf{r}\mathbf{r}^T + 2r_0\mathbf{R}, \quad \mathbf{R} = \text{CPM}(\mathbf{r}) \tag{7}$$

where $\text{CPM}(\mathbf{r})$ denotes the cross-product matrix[1] of $\mathbf{r} \equiv \mathbf{e}\sin(\phi/2)$. This representation is more robust than the others because it does not entail any singularity; therefore, smoother trajectories can be obtained. The array of unknowns becomes $\mathbf{x} = \begin{bmatrix}\mathbf{r}^T & r_0\end{bmatrix}^T$, which are not independent, for they obey the constraint $\|\mathbf{r}\|^2 + r_0^2 = 1$. A non linear system of four algebraic equations in four unknowns is thus obtained, that seems to leave no room for optimization. Actually, the system conceals a non-linear dependency between the variables [2], since the 4×4 gradient of the system of equations is singular. As a matter of fact, the 4th equation of the system naturally comes out from the Euclidean norm of the two sides of Eq. 6 by introducing Eq. 7 and imposing that $\mathbf{e}_v$ and $\mathbf{e}_h$ be unit vectors. In order to provide the system with a straightforward physical interpretation, a projection of the Eq. 6 is operated along known directions as shown below:

[1]That is, $\text{CPM}(\mathbf{r}) = \partial(\mathbf{r} \times \mathbf{v})/\partial\mathbf{v}, \ \forall\mathbf{v}, \ \mathbf{x} \in \mathbb{R}^3$.

$$\mathbf{r}^T\mathbf{e}_v = \mathbf{r}^T\mathbf{e}_h; \quad \mathbf{r}^T\mathbf{E}_v\mathbf{e}_h = 2r_0(\mathbf{r}^T\mathbf{r} - (\mathbf{r}^T\mathbf{e}_v)^2); \quad \mathbf{e}_v^T\mathbf{e}_h = 2(r_0^2 + (\mathbf{r}^T\mathbf{e}_v)^2) - 1 \quad (8)$$

From the first scalar equation the angle between the vector $\mathbf{r}$ and the two unit vectors $\mathbf{e}_v$ and $\mathbf{e}_h$ of the vertical and actual hole axes must be equal: this vector is bound to lie on the bisecting plane defined by to the two unit vectors $\mathbf{e}_v$ and $\mathbf{e}_h$. Vector $\mathbf{r}$ can sweep this plane upon rotating around the origin of the frames. Within this set of vectors $\mathbf{r}$ that brings $\mathbf{e}_v$ to overlay with $\mathbf{e}_p$, the solution of the problem is the one that optimally orients the EE in the $\mathbf{e}_h$ pointing direction.

The constrained optimization can be rewritten as an unconstrained problem by decomposing the $\mathbf{Q}$ matrix into two rotation matrices, namely, $\mathbf{Q} = \mathbf{Q}_1\mathbf{Q}_2$. For the sake of simplicity, the two factors are described here in terms of their linear invariants $\mathbf{Q}_i(\mathbf{e}_i, \vartheta_i)$, even though ERPs where used for practical implementation. The first rotation is determined by imposing the coincidence of the two given vectors: among the infinite number of rotation matrices, the geodetic one that minimizes the angle of rotation ϑ_1 is chosen [8]

$$\mathbf{Q}_1 = \mathbf{Q}_1\left(\frac{\mathbf{e}_v \times \mathbf{e}_h}{\|\mathbf{e}_v \times \mathbf{e}_h\|}, \operatorname{atan2}\left(\|\mathbf{e}_v \times \mathbf{e}_h\|, \mathbf{e}_v^T\mathbf{e}_h\right)\right) \quad (9)$$

The matrix $\mathbf{Q}_2(\mathbf{e}_h, \vartheta_2)$ describes a rotation around the unit vector $\mathbf{e}_h$ of an angle ϑ_2. Then, the whole rotation $\mathbf{Q}$ can be used to compute the square of the condition number κ_F^2 as a function of ϑ_2, which represents the only unknown. Such manipulations allow the writing down of the problem as an unconstrained optimization of the objective function κ_F^2 in the single design variable ϑ_2.

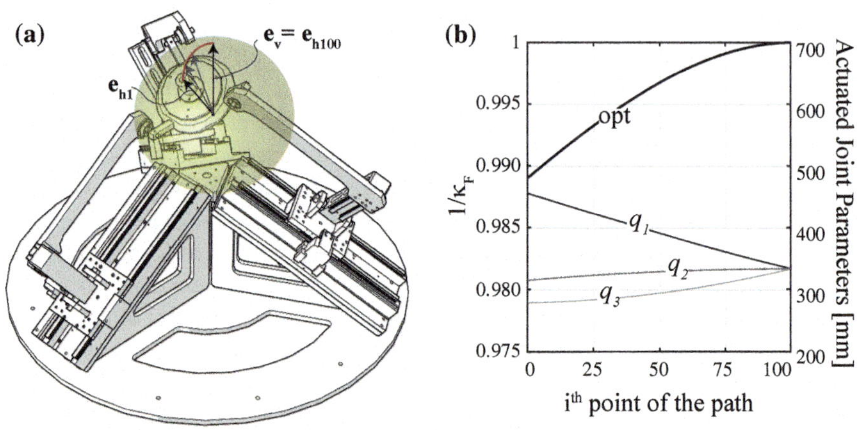

Fig. 3 Sphe.I.Ro under trajectory-following: **a** at the initial posture; and **b** evolution of $1/\kappa_f$ (opt) along the trajectory, with the time-histories (q_i) of the joint-coordinates, for $i = 1, 2, 3$

4 Implementation and Results

Sphe.I.Ro is to perform a pointing task while keeping the Frobenius condition number of its Jacobian matrix at a minimum. In order to prove the effectiveness of the procedure, an arbitrary array of ERPs $\left[\mathbf{r}^T\ r_0\right]^T = \left[0.140\ 0.210\ 0.280\ 0.926\right]^T$ is chosen to reproduce a geometric path for the EE. In practical cases instead, such path is given by the task. The finite rotation between the two frames is taken as the pointing path of the hole axis, represented by $\mathbf{e}_{hi}$: a total of 100 path points on the unit sphere are used to describe the axis path, keeping fixed the axis of rotation while decreasing the angle of rotation φ by $\Delta\varphi$, as shown in Fig. 3a. For each prescribed direction of the hole axis an optimization problem is solved in order to find the EE orientation that minimizes the Jacobian condition number. To solve the optimization problem of a nonlinear function with nonlinear constraint equations we use the Sequential Quadratic Programming (SQP) algorithm, while the gradient of the objective function is estimated using finite differences at every step. After having found the first point of the optimum joint trajectory, the remainder of the path follows in a similar way, using the previous trajectory point as an initial guess. The next posture is, consequently, the closest minimizer at the current posture. To verify the results of the SQP method, the condition number is evaluated at each iteration upon varying the angle of rotation ϑ_2 about the axis $\mathbf{e}_h$ of the unconstrained problem: the plots in Fig. 4a, b shows the evaluation in a rich sample of argument values for a full rotation of 2π, i.e. the middle and final point of the path. In Fig. 4b the reciprocal of the condition number indicates the passing through an isotropic configuration and through a singular one; moreover, the plot rightfully appears as 2π −periodic. From the first results of the optimization problem we conclude that the implementation of Eq. 6 together with $\|\mathbf{r}\|^2 + r_0^2 = 1$, leads to an excessive number of iterations, often without reaching a minimum. The exclusion of the unit norm equation from the system makes the dof of the functional redundancy explicit, thereby leading to

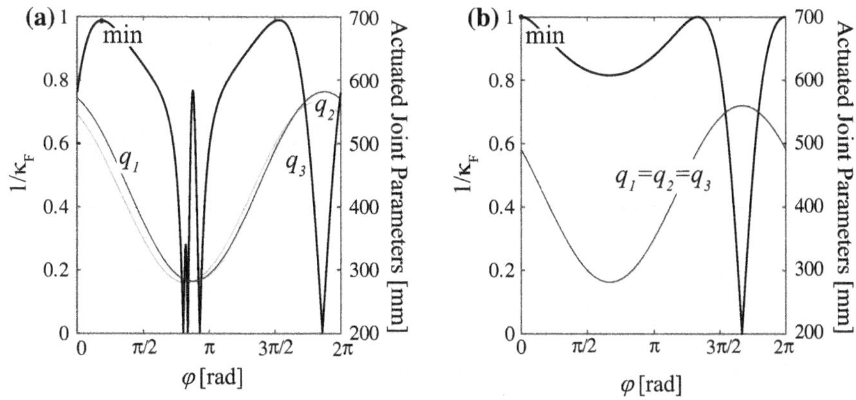

Fig. 4 Reciprocal of the condition number and joint values **a** time-histories for the middle (50th); and **b** final (100th) trajectory point

a faster convergence. The comparison of the history of the condition number along the path for the optimization problem, Fig. 3b, versus the unconstrained function evaluation, shows the same results; hence, the solver is capable of finding the local minimum. In Figs. 3 and 4 the histories of the actuated joints $q_i(t)$ are also plotted, for the respective condition number obtained using the inverse kinematics relationships $\mathbf{q}_{1,i} = \left[c - dQ_{1,2},\ c - dQ_{2,3},\ c - dQ_{3,1}\right]^T$, where c is the constant length between the prismatic pair direction and the universal-joint center along the cylindrical-joint axis, as per Fig. 2b, while $Q_{i,j}$ represents the (i, j) entry of the rotation matrix. The pointing-path was verified looking for potential problems such as joint-limit violation and singular postures. No such problems were found for this particular path. For every practical purpose, this result shows that the optimization scheme led to a constant condition number, at a remarkably low level. It is noteworthy that the optimization problem avoids singularities only if at least one non-singular posture is available for a particular pointing direction; since this is not ensured for every pointing direction, it can happen that the solution of the problem is a pose either singular or close to a singular configuration. In any case, the proposed procedure is not aimed at seeking a singularity-free trajectory, but at optimizing a path that should be generated a priori and verified for singularity avoidance. In particular, for each path it should be guaranteed that the condition number is lower than a safe threshold. The pointing path discussed here is not defined in the time domain; for this reason it does not represent a specific trajectory: a motion planner able to perform such verification and to provide a time history for the subsequent configurations is still a matter of study, that is to be further investigated.

In conclusion, the paper shows how posture-optimization can be used as a redundancy-resolution means for functionally redundant PKMs. The constrained optimization problem here defined is solved with the SQP algorithm. The problem is recast as one of unconstrained optimization, whose objective function is evaluated over a rich sample of argument values to verify the SQP results by inspection.

References

1. Angeles, J.: Fundamentals of Robotic Mechanical Systems, vol. 2. Springer, Berlin (2002)
2. Brand, L.: Advanced Calculus: an Introduction to Classical Analysis. Courier Corporation, New York (2013)
3. Corinaldi, D., Callegari, M., Palpacelli, M.C., Palmieri, G., Carbonari, L.: Dynamic optimization of pointing trajectories exploiting the redundancy of parallel wrists. ASME 2017 Int. Des. Eng. Tech. Conf. & Comp. And Inf. In Eng. Conf., August 6–9, Clevelana, USA (2017). Accepted
4. Carbonari, L., Callegari, M., Palmieri, G., Palpacelli, M.C.: Analysis of kinematics and reconfigurability of a spherical parallel manipulator. IEEE Trans. Robot. **30**(6), 1541–1547 (2014)
5. Huo, L., Baron, L.: Kinematic inversion of functionally-redundant serial manipulators: application to arc-welding. Trans. Can. Soc. Mech. Eng. **29**(4), 679–690 (2005)
6. Léger, J., Angeles, J.: Off-line programming of six-axis robots for optimum five-dimensional tasks. Mech. Mach. Theory **100**, 155–169 (2016)
7. Merlet, J.P.: Jacobian, manipulability, condition number, and accuracy of parallel robots. J. Mech. Des. **128**(1), 199–206 (2006)
8. Taubin, G.: Graphics math & code editor: 3d rotations (2011)

Analysis Methods for the 3-$\underline{R}$RR with Uncertainties in the Design Parameters

Joshua K. Pickard, Juan A. Carretero and Jean-Pierre Merlet

Abstract Accounting for uncertainties in the design variables of a parallel manipulator is important for a reliable analysis of a mechanism. The design of the 3-$\underline{R}$RR planar parallel manipulator is modelled with uncertainties. Interval analysis techniques are utilised to solve for the reachable workspace and the collision-free workspace. It is necessary to ensure that a fabricated design can achieve some desired criteria. Here, we consider generating a desired set of wrenches at the end-effector. The wrench capabilities under uncertainties are verified throughout the collision-free workspace. The results describe the set of poses which are guaranteed to be collision free and satisfy the desired wrench capabilities given the uncertainties in the specified design.

1 Introduction

A useful concept in parallel manipulator analysis is to evaluate certain properties of a mechanism when the design geometries are not precisely known, but can be determined with a tolerance. A desired geometry can be fabricated to within a certain set of tolerances. In order to ensure that the desired properties will be present within the fabricated mechanism, the fabrication tolerances must be taken into account when analysing the mechanism. These properties can include, but are not limited to, identifying collision-free regions [2], singularity-free regions [2, 8], and wrench capability analysis [3, 9].

Self-collisions are typically prevalent throughout the reachable workspace of parallel manipulators (especially for the 3-$\underline{R}$RR mechanism). To ensure that the fabricated mechanism avoids self-collisions, the components of the design (e.g., the

J.K. Pickard (✉) · J.A. Carretero
University of New Brunswick, Fredericton, Canada
e-mail: Joshua.Pickard@unb.ca

J.A. Carretero
e-mail: Juan.Carretero@unb.ca

J.-P. Merlet
INRIA Sophia-Antipolis, Sophia-Antipolis, France
e-mail: Jean-Pierre.Merlet@inria.fr

J. Lenarčič and J.-P. Merlet (eds.), *Advances in Robot Kinematics 2016*,
Springer Proceedings in Advanced Robotics 4, DOI 10.1007/978-3-319-56802-7_12

actuator and mounting hardware, joints, links, platform, end-effector), and the arrangement of these components, should be accounted for when analysing the mechanism. Several techniques have been proposed to deal with self-collisions. Ketchel and Larochelle [5] exploit the geometry of right circular cylindrical objects to facilitate the detection of collisions from three dimensional motions [5]. Merlet and Daney [6] apply interval analysis techniques to determine the minimum distance between cylindrical members to detect self-collisions in the Stewart-Gough platform. These works cannot be directly applied here as the incorporation of uncertainties does not allow for an exact description of the limb, but rather gives an overestimated description. It is simple enough to ensure that two members do not collide by verifying that the distance between the members is always greater than some minimum allowable distance, however ensuring that the members always collide is a difficult problem that should be addressed. The problem of detecting collisions under uncertainties via interval analysis is considered here.

Interval analysis is a mathematical framework which allows for computation using interval quantities, such that an interval variable $[x]$ denotes the natural extension of the closed interval $[x] = [\underline{x}, \overline{x}] = \{x \mid x \in \mathbb{R},\ \underline{x} \leq x \leq \overline{x}\}$. Through the use of interval analysis techniques, a variable's tolerances can be accounted for during all computations. The evaluation of a function $f(x)$ over an interval $[x]$ yields the interval solution $[f]$ of the function. This function $[f]$ is called the *inclusion function* for $f([x])$, such that $f([x]) = \{f(x) \mid x \in [x]\} \subseteq [f]$. The converse inclusion does not hold in general, and $[f]$ overestimates $f([x])$. This overestimation is a result of the well known *wrapping effect* and *dependency problem* in interval analysis [4]. Several techniques are commonly applied to manage interval overestimation: interval contracting techniques, branch and bound methods, and alternative representations of the inclusion functions (e.g., centred, mixed centred, Taylor) [4, 7].

In Sect. 2, interval analysis is incorporated into the design specifications of the 3-RRR parallel manipulator. This allows uncertainties in the mechanism to be accounted for during analysis. Then, in Sect. 3, the inverse kinematics for the mechanism are solved using interval analysis and both self-collisions and joint limitations are accounted for. Section 4 introduces the description of a desired task in terms of workspace and wrench requirements. The wrench capabilities of the mechanism are evaluated and verified against the wrench requirements of the task.

2 Design Specifications

The design of a 3-RRR will be described in terms of a set of design parameters, denoted $\mathscr{D}$. Illustrated in Fig. 1, these design parameters include link lengths (r_i and l_i), moving platform and fixed base geometries (described by $\mathbf{d}_i$ and $\mathbf{a}_i$, respectively), and actuator torque capabilities (τ_i). In addition to the discrete values for the design parameters, interval analysis provides the ability to easily model tolerances on each of the variables. It is therefore necessary to denote a design with tolerances as $[\mathscr{D}]$. The length of a link may normally be discretely represented as l_i, whereas the inclusion of

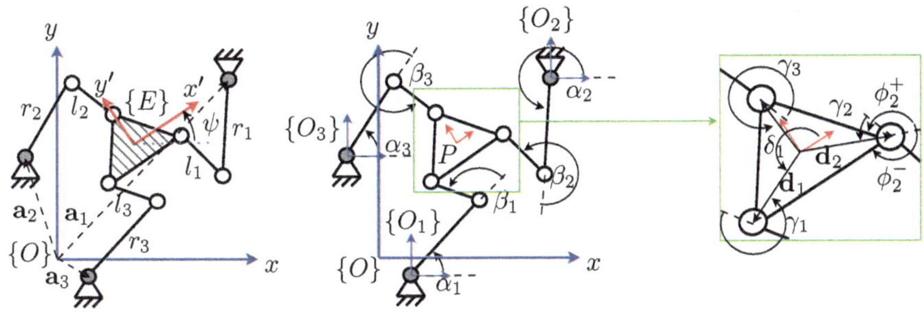

Fig. 1 The 3-RRR architecture with 3-DOF (x, y, and ψ) (adapted from [9])

a manufacturing tolerance ρ allows for the length to be represented as $[l_i] = l_i \pm \rho$. Also, consider that the location of each actuator is not exact; thus, it is useful to add tolerances ($[\mathbf{a}_i]$). Lastly, it may be necessary to include positioning and orientation tolerance on the desired pose ($[\mathbf{p}]$). The remainder of this work proposes and demonstrates algorithms which are developed to analyse mechanisms with interval design parameters, such that each design $\mathscr{D} \in [\mathscr{D}]$ and each pose $\mathbf{p} \in [\mathbf{p}]$ are accounted for.

3 Reachability and Collisions

Oetomo et al. [8] formulated a set of constraint equations using the inverse and direct kinematics equations of a 3-RRR (for brevity these equations are not repeated here). Reachability of the 3-RRR for a given pose interval $[\mathbf{p}]$ is determined by applying interval constraint propagation and numerical constraint satisfaction techniques to the set of constraint equations for the manipulator in order to solve the inverse kinematics. A solution to the inverse kinematics yields consistent interval domains for the joint variables: $\cos([\alpha_i])$, $\sin([\alpha_i])$, $\cos([\beta_i])$, $\sin([\beta_i])$, $\cos([\gamma_i])$, $\sin([\gamma_i])$ for $i = 1 \ldots 3$, such that each of the joint variables is consistent with the domain $[-1, 1]$. This domain can be narrowed to limit the allowable range of motion of a joint. Elbow configurations can be specified by selecting the appropriate domain for $\sin([\beta_i])$. The allowable range of motion in the elbow joint is specified by selecting the appropriate domain for $\cos([\beta_i])$. The allowable range of motion between the distal link and platform is specified by adding restrictions on the allowable domains of $\cos([\gamma_i])$ and $\sin([\gamma_i])$. To account for a non-symmetric platform design, let ϕ_i^+ and ϕ_i^- describe the edges of the platform relative to the platform attachment point of limb i (see Fig. 1). The restrictions are described by a union of constraints as follows

$$\begin{pmatrix} \cos([\gamma_i]) \subset [\cos(\phi_i^-), 1] \\ \sin([\gamma_i]) \subset [0, 1] \end{pmatrix} \bigcup \begin{pmatrix} \cos([\gamma_i]) \subset [\cos(\phi_i^+), 1] \\ \sin([\gamma_i]) \subset [-1, 0] \end{pmatrix} \tag{1}$$

An 'inside-outside' classification test solves for the set of pose intervals which are completely inside the reachable workspace ($\mathscr{P}_{RW}$)[1] and the complementary set of pose intervals which are completely outside the $\mathscr{P}_{RW}$. A pose interval $[\mathbf{p}]$ satisfies the inside test when the set of interval joint variable solutions satisfies all of the domain constraints. Alternatively, a pose interval $[\mathbf{p}]$ satisfies the outside test if any of the joint variable solutions falls in the complement of a domain constraint. When a pose interval fails to be classified as inside or outside, the pose interval $[\mathbf{p}]$ is bisected using an interval bisection routine. For example, a largest-first bisection would bisect $[\mathbf{p}]$ along the dimension with the largest width, resulting in two new pose intervals $[\mathbf{p}_1]$ and $[\mathbf{p}_2]$, which can then be tested. This takes place until every unclassified pose interval is below some desired width threshold, denoted ϵ.

An algorithm for detecting collisions between limbs and platform will now be proposed. As a result of the uncertainties, it is necessary to determine if there is always a collision or when there is never a collision for all $\mathbf{p} \in [\mathbf{p}]$. The proposed collision algorithm consists of a *partial collision test* and a *full collision test*. The partial collision test first determines if a collision is possible. The full collision test is then applied to determine if a collision will always occur. The design of a planar mechanism may consist of several non-colliding layers which aid in enlarging the collision-free workspace at the end-effector. The methods presented here can be applied to designs consisting of many layers. It is not necessary to check for collisions between every member during each pose test; instead, it is useful to track which members are guaranteed not to collide and pass this information along in the bisection. Only the members which have the possibility of colliding are rechecked.

Let $[\mathbf{b}_i]$ describe the location of the elbow joint in limb i. A line segment, denoted $\overline{\mathbf{a}_i\mathbf{b}_i}$, is used to represent the proximal[2] link for limb i. Similarly, a line segment, denoted $\overline{\mathbf{b}_i\mathbf{c}_i}$, is used to represent the distal[3] link for limb i. The ith limb can be described by the union of segments: $L_i = \overline{\mathbf{a}_i\mathbf{b}_i} \cup \overline{\mathbf{b}_i\mathbf{c}_i}$ (see Fig. 2). The width of link k on limb i is represented by w_{ik}. Two links will collide when they are in the same layer and the distance (dist) between the links is less than or equal to the sum of the half-widths of the corresponding links. For convenience, L_{ik} will denote the kth link on limb i and L_{jh} will denote the hth link on limb j. Assuming that two limbs i and j are in the same layer, a collision will always occur when

$$\forall \mathbf{a}_i \in [\mathbf{a}_i],\ \forall \mathbf{a}_j \in [\mathbf{a}_j],\ \forall \mathbf{b}_i \in [\mathbf{b}_i],\ \forall \mathbf{b}_j \in [\mathbf{b}_j],\ \forall \mathbf{c}_i \in [\mathbf{c}_i],\ \forall \mathbf{c}_j \in [\mathbf{c}_j],$$
$$\text{dist}(L_{ik}, L_{jh}) \leq \left(\frac{w_{ik}}{2} + \frac{w_{jh}}{2}\right), \text{ for } k = 1, 2,\ h = 1, 2 \quad (2)$$

such that each pair of links is checked for collisions. Alternatively, k and h in Eq. (2) may be modified to only consider links in the same layer.

It is also necessary to account for collisions between the links and the platform. The distal link is ignored since it shares a joint with the platform and restrictions on

[1]The symbol $\mathscr{P}$ denotes a set of poses and the subscript refers to the relevant workspace.

[2]The link closest to the base with length r_i.

[3]The link attached to the moving platform with length l_i.

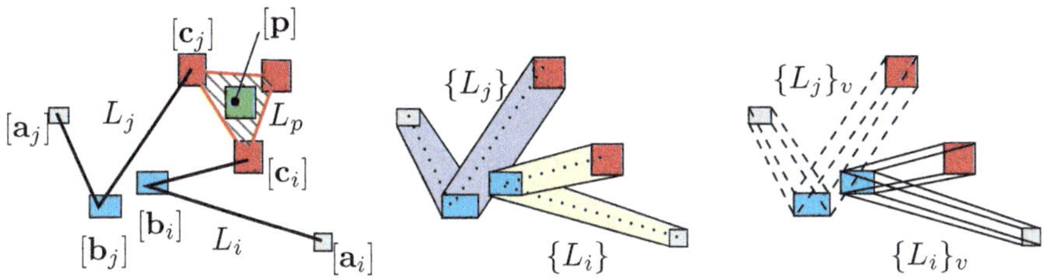

Fig. 2 Limb collisions

γ_i can be used to prevent collisions. The following method can be extended for limbs consisting of two or more links. Let L_p denote the set of line segments corresponding to the platform: $L_p = \overline{\mathbf{c}_1\mathbf{c}_2} \cup \overline{\mathbf{c}_2\mathbf{c}_3} \cup \overline{\mathbf{c}_3\mathbf{c}_1}$. The edges of the platform are located at a distance of w_p from L_p. Assuming that the elbow joint $[\mathbf{b}_i]$ does not intersect the platform, a collision will always occur between limb i and the platform when

$$\forall \mathbf{a}_i \in [\mathbf{a}_i],\ \forall \mathbf{b}_i \in [\mathbf{b}_i],\ \forall \mathbf{c}_1 \in [\mathbf{c}_1],\ \forall \mathbf{c}_2 \in [\mathbf{c}_2],\ \forall \mathbf{c}_3 \in [\mathbf{c}_3], \\ \text{dist}(L_{ik}, L_p) \leq \left(\frac{w_{ik}}{2} + w_p\right),\ \text{for } k = 1. \tag{3}$$

3.1 *Partial Collision Test*

Applying a convex hull routine (conv), the set of all line segments for the proximal link ($\{L_{i1}\}$), distal link ($\{L_{i2}\}$), and platform ($\{L_p\}$) are represented as

$$\{L_{i1}\} = \text{conv}([\mathbf{a}_i], [\mathbf{b}_i]) \tag{4}$$

$$\{L_{i2}\} = \text{conv}([\mathbf{b}_i], [\mathbf{c}_i]) \tag{5}$$

$$\{L_p\} = \text{conv}([\mathbf{c}_1], [\mathbf{c}_2], [\mathbf{c}_3]) \tag{6}$$

Let $\{L_i\}$ be the union of $\{L_{i1}\}$ and $\{L_{i2}\}$ (see Fig. 2). The distance between each pair of links $\{L_{ik}\}$ and $\{L_{jh}\}$ is computed and a collision will not occur between limbs i and j if

$$\text{dist}(\{L_{ik}\}, \{L_{jh}\}) > \left(\frac{w_{ik}}{2} + \frac{w_{jh}}{2}\right),\ \text{for } k = 1, 2,\ h = 1, 2 \tag{7}$$

If the elbow joint $[\mathbf{b}_i]$ does not intersect $\{L_p\}$, a collision between the proximal link on the ith limb and the platform will not occur if

$$\text{dist}(\{L_{ik}\}, \{L_p\}) > \left(\frac{w_{ik}}{2} + w_p\right),\ \text{for } k = 1 \tag{8}$$

The possibility of a collision exists when Eq. 7 or 8 is not satisfied.

3.2 Full Collision Test

To solve the full collision test, the use of a vertex representation of $\{L_i\}$ and $\{L_p\}$ is proposed. The vertex representations, denoted as $\{L_i\}_v$ and $\{L_p\}_v$ respectively, are finite in terms of line segments and are computed by selecting a vertex from the first joint and connecting this to each consecutive corresponding vertex (see Fig. 2). Since the mechanism is planar, $\{L_i\}_v$ and $\{L_p\}_v$ will each have four elements, corresponding to each of the vertices.

A collision will always occur between limbs i and j when

$$\forall L_i \in \{L_i\}_v,\ \forall L_j \in \{L_j\}_v,\ \text{dist}(L_{ik}, L_{jh}) \leq \left(\frac{w_{ik}}{2} + \frac{w_{jh}}{2}\right),\ \text{for}\, k = 1, 2,\ h = 1, 2 \tag{9}$$

A collision will always occur between the proximal link on the ith limb and the platform when

$$\forall L_i \in \{L_i\}_v,\ L_p \in \{L_p\}_v,\ \text{dist}(L_{ik}, L_p) \leq \left(\frac{w_{ik}}{2} + w_p\right),\ \text{for}\ k = 1 \tag{10}$$

If the full collision test fails, then only a partial collision can be guaranteed and the pose interval $[\mathbf{p}]$ must be bisected for additional testing.

3.3 Applying the Reachability and Collision Algorithms

The 3-RRR planar parallel manipulator with design $[\mathscr{D}]$, described by the following interval design parameters, is analysed. All of the links of the manipulator have a width of $w = 0.0150\,\text{m}$, while the platform has $w_p = 0.0075\,\text{m}$. A constant orientation is selected with a platform orientation specified as $[\psi] = [0.0]$ rad. A resolution of $\epsilon = 0.001$ m is used for all workspaces.

$$\mathbf{a}_1 = \begin{pmatrix} [0.0997, 0.0998] \\ [0.1077, 0.1078] \\ [0.0] \end{pmatrix} \text{m},\ \mathbf{a}_2 = \begin{pmatrix} [-0.2419, -0.2418] \\ [0.05427, 0.05428] \\ [0.0] \end{pmatrix} \text{m},\ \mathbf{a}_3 = \begin{pmatrix} [0.07265, 0.07266] \\ [-0.2370, -0.2369] \\ [0.0] \end{pmatrix} \text{m};$$

$$\mathbf{d}_1 = \begin{pmatrix} [0.0498, 0.0502] \\ [-0.0001, 0.0001] \\ [0.0] \end{pmatrix} \text{m},\ \mathbf{d}_2 = \begin{pmatrix} [-0.0251, -0.0249] \\ [0.0432, 0.0434] \\ [0.0] \end{pmatrix} \text{m},\ \mathbf{d}_3 = \begin{pmatrix} [-0.0251, -0.0249] \\ [-0.0434, -0.0432] \\ [0.0] \end{pmatrix} \text{m};$$

$$r_1 = [0.1648, 0.1650]\text{m},\quad r_2 = [0.1350, 0.1352]\text{m},\quad r_3 = [0.1479, 0.1481]\text{m};$$

$$l_1 = [0.2411, 0.2413]\text{m},\quad l_2 = [0.2955, 0.2957]\text{m},\quad l_3 = [0.2895, 0.2897]\text{m};$$

$$\tau_1 = [-10, 10]\text{Nm},\quad \tau_1 = [-10, 10]\text{Nm},\quad \tau_1 = [-10, 10]\text{Nm};$$

The constant orientation workspace neglecting collisions is provided in Fig. 3a. The set of poses with the *inside* classification are guaranteed to be reachable for the design $[\mathscr{D}]$. Alternatively, the set of poses with the *outside* classification are guaranteed to be unreachable. The full $\mathscr{P}_{RW}$ can be computed by accounting for

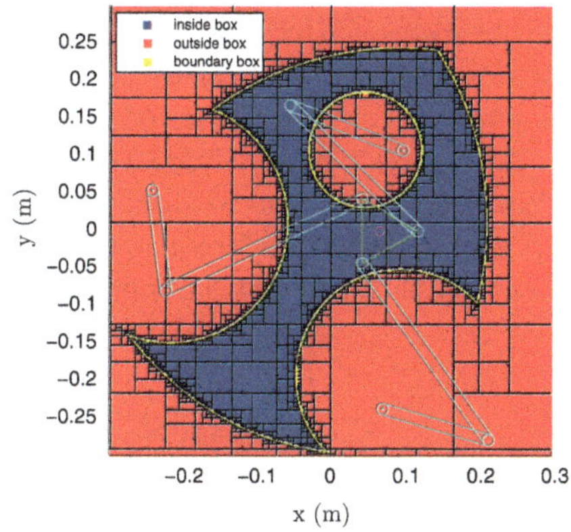

(a) Neglecting self-collisions.

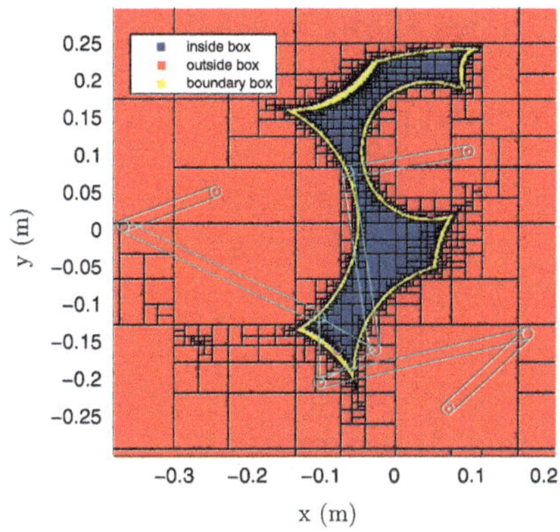

(b) Accounting for self-collisions with all links in layer 1.

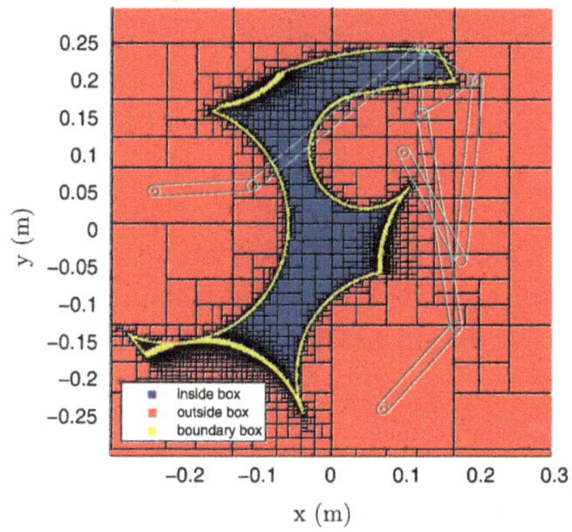

(c) Accounting for self-collisions with the proximal links in layer 1, the distal links in layer 2, and the platform in layer 3.

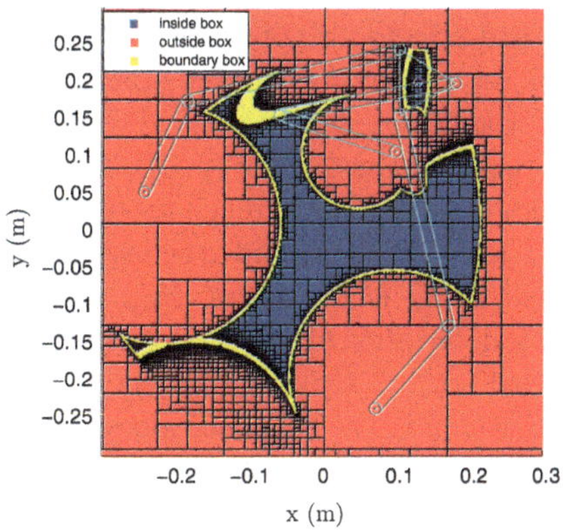

(d) Accounting for self-collisions with the proximal links and platform in layer 1 and the distal links in layer 2.

Fig. 3 Constant orientation workspaces for the 3-RRR planar parallel manipulator

every platform orientation. This can be done by simply allowing $[\psi] = [-\pi, \pi]$ at the start of the solving routine.

A design with all of the links on layer 1 is considered in Fig. 3b. The joints limitations are set as $\alpha_i \in [0°, 360°]$, $\beta_i \in [0°, 165°]$, and $\gamma_i \in [-135°, 135°]$ for $i = 1, \ldots, 3$. Collisions between the proximal links, distal links and the platform are accounted for.

A design with the proximal links in layer 1, the distal links in layer 2, and the platform in layer 3 is considered in Fig. 3c. Distal-distal and proximal-proximal collisions are accounted for. The joints limitations are $\alpha_i \in [0°, 360°]$, $\beta_i \in [0°, 180°]$, and $\gamma_i \in [0°, 360°]$ for $i = 1, \ldots, 3$.

A design with the proximal links and platform in layer 1 and the distal links in layer 2 is considered in Fig. 3d. Note also that the elbow configurations have been changed. Distal-distal, proximal-proximal and proximal-platform collisions are accounted for. The joints limitations are $\alpha_i \in [0°, 360°]$, $\beta_1 \in [180°, 360°]$, $\beta_2 \in [180°, 360°]$, $\beta_3 \in [0°, 180°]$ and $\gamma_i \in [0°, 360°]$ for $i = 1, \ldots, 3$. It can be noted that two disconnected components of the workspace are computed corresponding to different assemblies of the mechanism. A single assembly can be considered by adding restrictions to γ_1.

4 Task Requirements

Given a design $[\mathscr{D}]$ containing tolerances, it is important to understand the capabilities of a mechanism in generating wrenches throughout $\mathscr{P}_{RW}$. The term *wrench capability* ($\mathscr{F}$) denotes the complete set of wrenches that a mechanism can generate at its end-effector in a given pose. Bouchard et al. [1] considered a discrete representation for a mechanism's design and pose and presented an algorithm for an exact polytopic representation of $\mathscr{F}$. The selected design, $\mathscr{D}$, and pose, $\mathbf{p}$, both affect the resulting wrenches, i.e., $\mathscr{F}(\mathbf{p}, \mathscr{D})$. The *minimum allowable wrench capability* ($\mathscr{F}_{\text{min}}$) defines the minimum wrench set required for a specific task. The set of poses with $\mathscr{F}_{\text{min}} \subseteq \mathscr{F}(\mathbf{p}, \mathscr{D})$ are said to be inside the mechanism's *Wrench Workspace* ($\mathscr{P}_{WW}$). That is, $\mathscr{P}_{WW}$ is a subset of $\mathscr{P}_{RW}$ which can be used by the desired task.

The description of a task used in this work will contain: (1) the set of pose intervals, denoted $\mathscr{P}_{\text{task}} = \{[\mathbf{p}_1], \ldots, [\mathbf{p}_k]\}$ ($\mathscr{P}_{\text{task}}$ may also be set equal to $\mathscr{P}_{RW}$), and (2) the $\mathscr{F}_{\text{min}}$ required to be generated along the trajectory. $\mathscr{P}_{WW}$ defined in terms of interval variables is

$$\mathscr{P}_{WW} = \{[\mathbf{p}] \mid [\mathbf{p}] \in \mathscr{P}_{\text{task}},\ \forall\, \mathbf{p} \in [\mathbf{p}],\ \forall\, \mathscr{D} \in [\mathscr{D}],\ \mathscr{F}_{\text{min}} \subseteq \mathscr{F}(\mathbf{p}, \mathscr{D})\}. \tag{11}$$

4.1 Verifying Task Requirements

Interval analysis derived tests can be used to verify the wrench capabilities of a mechanism. Gouttefarde et al. [3] proposed an inside-outside test for the verification of the wrench capabilities of cable-driven parallel manipulators. Their inside test makes use of a strong feasibility theorem proposed by Rohn [11] which relies on the assumption that the solution of an interval linear system of equations is nonnegative. This restriction makes the tests of Gouttefarde et al. suitable for cable-driven architectures which must maintain nonnegative cable tensions. Pickard and Carretero [10] proposed an alternative formulation of the inside test which makes use of the strong solvability theorem proposed by Rohn [11] which removes the nonnegative assumption. The formulation in [10] removes the restriction on the actuator limits and their tests can be applied to other parallel manipulator architectures, such as the 3-$\underline{\text{R}}$RR.

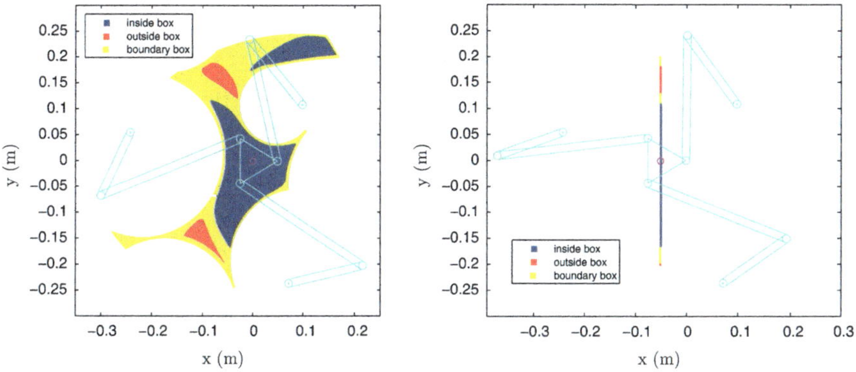

(a) Wrench capability verification throughout the constant orientation workspace.

(b) Wrench capability verification along a trajectory.

Fig. 4 Wrench capability verification for the 3-RRR planar parallel manipulator

A task may be specified in terms of wrench requirements at the end-effector. For example, the manipulator is used to perform a task which requires $\mathscr{F} = ([f_x], [f_y], [m_z])^T = ([-30, 30]\,\text{N}, [-30, 30]\,\text{N}, [0]\,\text{N})^T$. The interval analysis tests described by Pickard and Carretero [10] can be utilized to verify that the manipulator's wrench capabilities exceed the task's wrench requirements. The associated wrench workspace considering the task's requirements and self-collisions is provided in Fig. 4a with a resolution of $\epsilon = 0.001$ m. Here, $\mathscr{P}_{\text{task}}$ is set equal to $\mathscr{P}_{RW}$, and the design of the mechanism is the same as in Fig. 3c. As a result of the tolerances on the design variables and the inherent overestimation with interval analysis, there are thick regions of unclassified (*boundary*) poses; that is, these poses cannot be classified as *inside* or *outside* with the wrench workspace tests. Therefore, a trajectory for the task must remain within the region of *inside* poses to guarantee that the desired wrench set can be generated for the design $[\mathscr{D}]$.

If a desired trajectory and wrench set are known, the pose intervals containing the trajectory are used to populate $\mathscr{P}_{\text{task}}$. Depending on the resolution of the sensors used by the manipulator and the control techniques applied, the manipulator may also have a pose error. The same $\mathscr{F}$ is selected as before, and a linear trajectory is to be completed. A positioning error of ± 0.1 mm is assumed for the manipulator. The associated wrench workspace considering the task's requirements and self-collisions is provided in Fig. 4b with a resolution of $\epsilon = 0.0001$ m. The benefit of considering only the pose intervals containing the trajectory is that the problem can be solved much more quickly. This is important when multiple designs are evaluated, as is required in design optimization.

5 Conclusions

Techniques for analysis of the 3-RRR planar parallel manipulator with tolerances in the design parameters have been presented. These techniques are designed to account for self-collisions and have been demonstrated by evaluating the reachable and wrench workspaces of a given design with reasonable tolerances on each variable. The techniques presented here may easily be extended to other planar mechanisms. In fact, with some additional consideration given to collisions between links and the platform, these techniques may also be extended to spatial mechanisms. An interesting application of these techniques is for the design synthesis of a mechanism which accounts for tolerances in the design parameters.

Acknowledgements The authors would like to thank the Natural Sciences and Engineering Research Council of Canada (NSERC), Mitacs, and Inria for their funding of this research.

References

1. Bouchard, S., Gosselin, C.M.: Workspace optimization of a very large cable-driven parallel mechanism for a radiotelescope application. In: Proceedings of the 2007 ASME DETC, Las Vegas, Nevada, USA, 4–7 Sept 2007
2. Caro, S., Chablat, D., Goldsztejn, A., Ishii, D., Jermann, C.: A branch and prune algorithm for the computation of generalized aspects of parallel robots. Artif. Intell. **211**, 34–50 (2014)
3. Gouttefarde, M., Daney, D., Merlet, J.: Interval-analysis-based determination of the wrench-feasible workspace of parallel cable-driven robots. IEEE Trans. Robot. **27**(1), 1–13 (2011)
4. Jaulin, L., Kieffer, M., Didrit, O., Walter, E.: Applied Interval Analysis. Springer, Berlin (2001)
5. Ketchel, J.S., Larochelle, P.M.: Self-collision detection in spatial closed chains. J. Mech. Design **130**(9), 92305-1–92305-9 (2008)
6. Merlet, J.P., Daney, D.: Legs interference checking of parallel robots over a given workspace or trajectory. In: Proceedings 2006 IEEE ICRA, pp. 757–762, Orlando, FL, 15–19 May (2006)
7. Moore, R., Kearfott, R., Cloud, M.: Introduction to Interval Analysis. Society for Industrial and Applied Mathematics, Philadelphia (2009)
8. Oetomo, D., Daney, D., Shirinsadeh, B., Merlet, J.: An interval-based method for workspace analysis of planar flexure-jointed mechanism. J. Mech. Design **131**(1), 011014-1–011014-11 (2008)
9. Pickard, J., Carretero, J.: Design optimisation of the 3-RRR planar parallel manipulator via wrench capability analysis. In: Proceedings of 14th World Congress, Taipei, Taiwan, 25–30 Oct (2015)
10. Pickard, J., Carretero, J.: An interval method for wrench workspace determination of parallel manipulator architectures. In: Proceedings of 2015 CCToMM M^3, Ottawa, Ontario, Canada, 28–29 May 2015
11. Rohn, J.: Solvability of systems of interval linear equations and inequalities. Linear Optimization Problems with Inexact Data, pp. 35–77. Springer, Berlin (2006)

A Study on Simplified Dynamic Modeling Approaches of Delta Parallel Robots

Jan Brinker, Philipp Ingenlath and Burkhard Corves

Abstract This contribution presents a study on simplified dynamic modeling approaches of the Delta parallel robot. Complete and simplified dynamic modeling approaches are reviewed and compared in respect to their computation times. Also, the dependency of the accuracy on the mass distribution of the distal link is analyzed in detail and assessed based on a single industry-relevant pick-and-place trajectory as well as randomly generated Lissajous curves for (more) general validity.

1 Introduction

In the 1980s, Reymond Clavel (professor at EPFL École Polytechnique Fédérale de Lausanne) invented a parallel robot known as Delta robot [4]. A Delta robot consists of three symmetric kinematic chains of the type RRPaR, RUU or $R(SS)_2$ (where R: revolute joint, U: universal joint, S: spherical joint, Pa: parallelogram). The three spatial four-bar parallelograms, each attached distally to one of the rotationally actuated links, restrain completely the orientation of the mobile platform which remains with three purely translational degrees of freedom. With this, the rods within the parallelogram only need to transmit axial forces allowing for light-weight materials and thus, very low inertia compared to serial articulated robots. Research in the general fields of kinematics, dynamics, control, singular configurations, workspace, calibration, and mechanical design of Delta robots has been conducted extensively during the last decades. A comprehensive overview about the historical, academic, and industrial development of such mechanisms is presented by the authors of this contribution [1].

Dynamic modeling is generally used to predict the desired actuator torque in order to dimension parts and develop efficient control schemes. Against this background, the dichotomy between simulation accuracy and computation time poses high challenges to the development of robot dynamics models. The most common analytical

J. Brinker (✉) · P. Ingenlath · B. Corves
Department of Mechanism Theory and Dynamics of Machines, RWTH Aachen University, Aachen, Germany
e-mail: brinker@igm.rwth-aachen.de

J. Lenarčič and J.-P. Merlet (eds.), *Advances in Robot Kinematics 2016*,
Springer Proceedings in Advanced Robotics 4, DOI 10.1007/978-3-319-56802-7_13

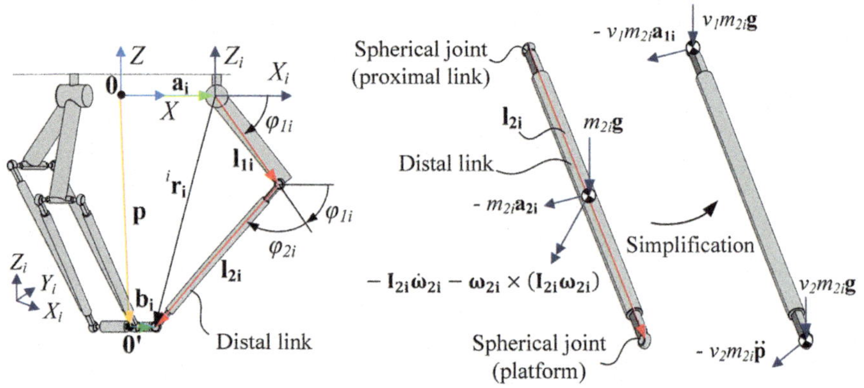

Fig. 1 Geometric relations and notations for the basic structure (*left*) and for the complete and simplified modeling approaches of the distal link (*right*)

approaches for dynamic modeling are: the Principle of Virtual Work, the Newton–Euler Formulation, and the Lagrangian Formulation [3]. Besides basic assumptions of complete non-simplified models (e.g., frictional effects are neglected, links are modeled as rigid cylinders, etc.), the description of the dynamic behavior of a system and its components can be further simplified to increase the computational efficiency for control purposes. With regard to Delta robots, in these models usually the rotational inertias of the light-weight distal links are neglected and their masses are distributed to the corresponding joints (see Fig. 1). Due to the light-weight design of the connecting rods, the significance of this simplification in respect to the accuracy of the model is expected to be fairly low. However, neglecting the distal link's rotational inertia and thus, changing the direction of the resultant dynamic forces may cause significant errors (i.e., deviations in actuator torques) of up to 10% depending on the system parameters [9]. Current market developments demand adaptations of materials and designs leading to increased inertias and component weights. This may influence the accuracy of the models even more and thus, motivates this study.

On the one hand, adaptations of materials may be required due to hygiene issues within the food processing industry (i.e., the most relevant field of application of Delta robots). Here, machines need to adhere to strict regulations. This also concerns carbon fiber components as commonly used for the links of a Delta robot.

On the other hand, adaptations of designs can be observed as a consequence of adding serial mechanisms to the original purely parallel Delta architecture in order to obtain orientation capabilities. These industrial concepts consist of the basic Delta structure and a serial wrist which is mounted on the platform and usually driven by three motors. These motors are fixed on the frame (e.g., FANUC M-1), attached to the platform or to the distal links (e.g., FANUC M-3). Hence, modifications of the basic structure including reinforcements of the distal links are required.

In this context, this contribution is concerned with two key aspects. First, complete and simplified dynamic modeling approaches are reviewed and compared in respect to their computation times. Second, the dependency of the accuracy on the mass distribution of the distal link is analyzed and assessed.

2 Problem Statement

As mentioned before, for simplified modeling approaches, the rotational inertias of the light-weight distal links are usually neglected and their masses are distributed to the corresponding joints. To determine the ratio of mass distribution, the mass distribution factors ν_1 and ν_2 are introduced (where $\nu_1,\nu_2 \in [0, 1]$ and $\nu_1 + \nu_2 = 1$ with subscripts 1 and 2 denoting the proximal and distal connecting points, see Fig. 1 right). Analyses on the determination of these factors are performed first by Codourey [5]. Computing the actuation torques of a specific system and a given trajectory, it was found that best results are obtained for $\nu_1 = 2/3$, i.e., two-thirds of the mass are allocated to the tip of the proximal link. This distribution is also applied in, e.g., [6, 7]. Stamper [13] presents a Delta-based structure with revolute joints only. Here, the masses of the connecting rods are distributed evenly to the tip of the proximal joint and the platform joint, respectively ($\nu_1 = \nu_2 = 1/2$). Contrary to the Delta variant investigated in this study, the intermediate links of this particular manipulator connecting the proximal link to the parallelogram rotate and thus, the related inertial effects are considered crucial. However, the comparison of the accuracies of complete to simplified dynamic modeling approaches in [12] solely considers an even mass distribution based on a single test trajectory. Also, it is not distinguished between the deviations resulting from neglecting the motions of the transmission bars and the deviations resulting from the simplification that the mass is distributed to the joints. The present paper considers varying mass distribution factors and randomly generated Lissajous curves imposed as trajectories for (more) general validity. Contributions with even distributions are found in, e.g., [11, 14].

The dissemination of the first type of distribution (i.e., $\nu_1 = 2/3$ and $\nu_2 = 1/3$) is due to the fact that the simplified consideration of the distal link corresponds to the complete model under certain assumptions. These are that: (1) the velocity of the joint connecting the proximal and distal link, i.e., the position of the distributed mass $\nu_1 m_{2i}$, is assumed to be zero, (2) the distal links are modeled as rigid cylinders with a moment of inertia of $I = 1/3ml^2$, and (3) static force deviations among the models are neglected. For the second type of distribution (i.e., $\nu_1 = \nu_2 = 1/2$) the resultant gravitational forces of the simplified and complete consideration correspond. It is thus, rather suitable for slow (quasi-static) applications since the effects of the dynamic forces on the actuation torques are not taken into account adequately (cf. Sect. 5). The following analyses are based on the kinematic and mass parameters, test trajectory, and detailed specifications as introduced in [2].

3 Complete and Simplified Dynamic Modeling Approaches

All three dynamic modeling approaches can be used for complete and simplified dynamic analyses. For the sake of brevity and for further analyses (cf. Sect. 5), solely the energy-based Lagrangian approach is reviewed briefly. Comprehensive analyses are conducted in [2]. The desired actuator torques can be computed by the Lagrange's equations of the first kind. Thus, the kinetic and potential energies (T and Π) need to be derived for each component of link i in order to obtain the Lagrangian equation:

$$L = T - \Pi \tag{1}$$

The actuation torques can then be derived by:

$$\tau_i = \frac{d}{dt}\left(\frac{\partial L}{\partial \dot{\varphi}_{1i}}\right) - \frac{\partial L}{\partial \varphi_{1i}} - \lambda_i \frac{\partial \Gamma_i}{\partial \varphi_{1i}} \tag{2}$$

with $p_x, p_y, p_z, \varphi_{11}, \varphi_{12}, \varphi_{13}$ as generalized coordinates (see Fig. 1), the constraint equations Γ_i, and the multipliers λ_i. The kinetic and potential energies are given by the sum of energies of each component (i.e., proximal links, distal links and platform). The following analyses solely consider the distal links. Thus, the energies of the other components are not introduced. For the complete approach (cf. [9]), the kinetic energies of the distal links are:

$$T_{2i} = \frac{1}{2} m_{2i} \left(\dot{\mathbf{p}}^T \mathbf{v}_{1i} + \frac{1}{3} \left(\dot{\mathbf{p}} - \mathbf{v}_{1i}\right)^{\mathbf{T}} \left(\dot{\mathbf{p}} - \mathbf{v}_{1i}\right)\right) \tag{3}$$

where $\dot{\mathbf{p}}$ and $\mathbf{v}_{1i}$ denote the platform velocity and the velocity at the tip of the proximal link, respectively. Their potential energies can be stated as:

$$\Pi_{2i} = \frac{1}{2} m_{2i} g \left(p_z - l_{1i} \sin \varphi_{1i}\right) \tag{4}$$

whereby the zero point lies within the origin of the coordinate system $\mathbf{0}$.

In simplified models (indicated with *), the rotational inertias of the light-weight distal links are neglected. Thus, the energy equations can be simplified to:

$$^{*}T_{2i} = \frac{1}{2} m_{2i} \left(\nu_1 \cdot l_{1i}^2 \dot{\varphi}_{1i}^2 + \nu_2 \cdot \left(\dot{p}_x^2 + \dot{p}_y^2 + \dot{p}_z^2\right)\right) \tag{5}$$

$$^{*}\Pi_{2i} = m_{2i} g \left(-\nu_1 \cdot l_{1i} \sin \varphi_{1i} + \nu_2 \cdot p_z\right) \tag{6}$$

assuming identical distribution factors for each link. Models based on the Principle of Virtual Work and Newton–Euler can be formulated and simplified accordingly.

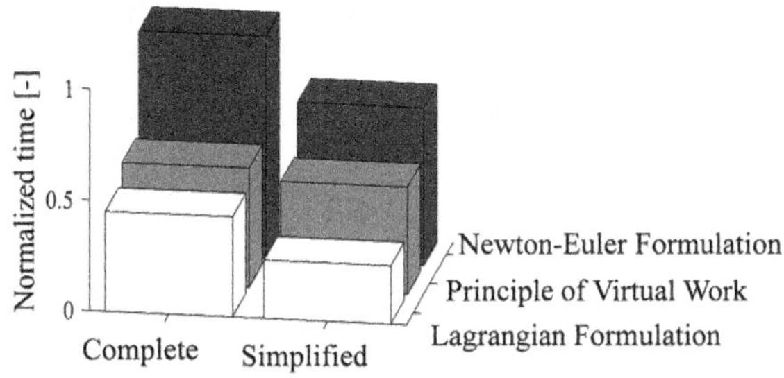

Fig. 2 Comparison of the computation times of complete and simplified approaches

4 Analysis of Computation Times

The three approaches for dynamic modeling are theoretically equivalent, but their computational intensity may vary. A comparative study of complete models shows that for the basic structure fastest processing times can be reached applying the Lagrangian Formulation [2]. Assume a normalized reference time of 1 as a base for the longest computation time (i.e., the time required to compute the complete model based on Newton–Euler), the processing times of simplified and complete models can be compared as displayed in the following Fig. 2.

It can be seen that independent of the underlying model shortest processing times are achieved applying the Lagrangian approach. The relative saving of time between complete and simplified models is 42, 10, and 29% for the Newton–Euler Formulation, the Principle of Virtual Work, and the Lagrangian Formulation, respectively. Referring to the simplified Newton–Euler Formulation, the application of the Principle of Virtual Work reduces the computation time by 32%. Applying the simplified Lagrangian approach, relative saving of time is 63% compared to the Newton–Euler Formulation.

To sum up, the minimum computation time is 74% less than the reference time and obtained using the simplified Lagrangian approach. It should be noted that the efficiency of a model not only depends on the analyzed mechanical structure, but also on the computational scheme and the program structure (e.g., number and kind of operations).

5 Assessment of Accuracy

This section provides analyses in order to show that the deviations of torques between complete and simplified models are related to the mass distribution factor. Moreover, the impact of optimizing the mass distribution factor in respect to the reduction of the torque deviations is assessed.

From (2) it can be seen that the calculation of actuator torques can be performed by an energy-based approach. Thus, to analyze the torque deviations and their

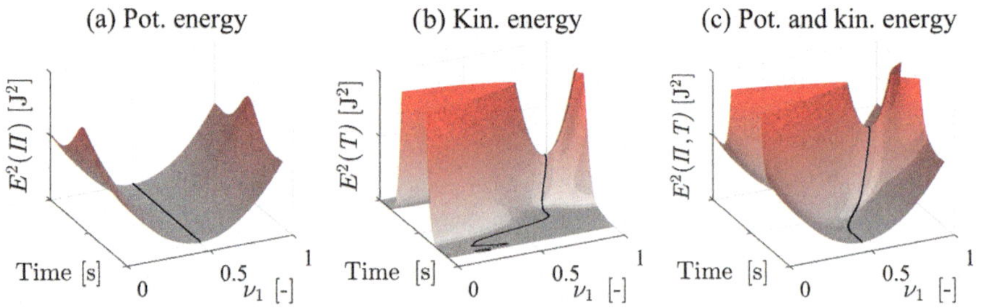

Fig. 3 Comparison of the potential (**a**), kinetic (**b**), and superposition of both (**c**) energy deviations

origins, the energy deviations of the distal links are used. To increase the weighting of relatively high deviations, the sum of squares of the energy deviations is considered which gives:

$$E^2(\Pi) = \sum_i \left(\Pi_{2i} - {}^*\Pi_{2i}\right)^2 \tag{7}$$

$$E^2(T) = \sum_i \left(T_{2i} - {}^*T_{2i}\right)^2 \tag{8}$$

$$E^2(\Pi, T) = \sum_i \left(\Pi_{2i} - {}^*\Pi_{2i}\right)^2 + \left(T_{2i} - {}^*T_{2i}\right)^2 \tag{9}$$

referring to the potential (7), kinetic (8), and superposition of both (9) energy deviations.

Provided that $\nu_2 = 1 - \nu_1$, these energy deviations can be visualized as a function of time (resulting from the imposed trajectory) and the mass distribution factor ν_1. The results are displayed in Fig. 3 with gray and red colors denoting low and high deviations, respectively. Minimal deviations are indicated by the black line.

Exclusively considering the deviations of the potential energy (Fig. 3a), it can be seen that independently from the trajectory (denoted by Time [s]) minimal deviations are achieved for $\nu_1 = 0.5$. The reason for this is that for any position the resultant of the gravitational forces of the distributed masses within the simplified model corresponds to the gravitational force of the complete model (also cf. (4) and (6)).

Figure 3b shows the deviations of the kinetic energy. The imposed trajectory represents a standard pick-and-place cycle with maximum velocity reached half way between pick and place positions. Inherently, velocities are minimal at these positions. Thus, it can be found that deviations are maximal and minimal at maximum and minimum speed, respectively. More importantly, minimal deviations vary along the distribution factor (cf. black line in Fig. 3b). Deviations are maximal for poorly chosen mass distribution factors (e.g., $\nu_1 = 0$ or $\nu_1 = 1$). However, the optimal distribution factor (referring to minimal deviations along the trajectory) cannot be identified a priori. At this point it becomes evident that simply choosing a factor for

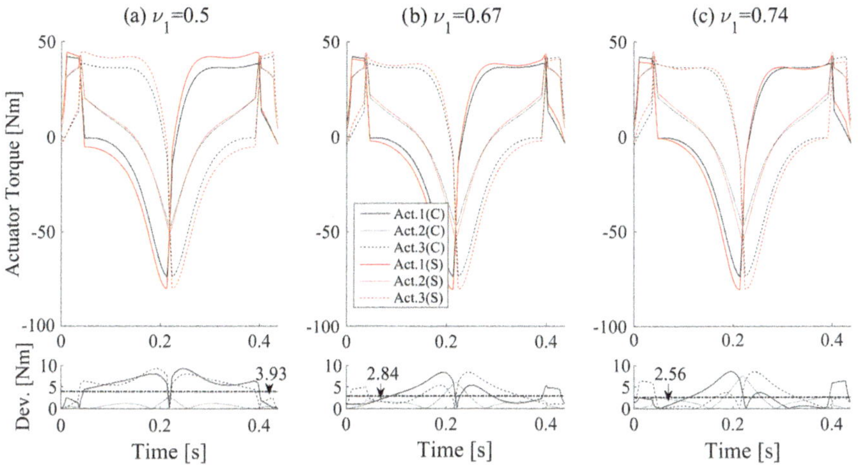

Fig. 4 Comparison of the deviations of actuator torques

distribution may not be ideal. Figure 3c depicts the superposition of the deviations of both potential and kinetic energies (cf. (9)).

The consideration of the energy deviations helps to understand the influences of the imposed trajectory and the chosen distribution factors on the accuracy of the simplified modeling approaches. However, the primary motivation of the dynamics modeling is the accurate computation of the actuator torques. Therefore, the optimal mass distribution factor is identified by minimizing the torque deviations along the trajectory. For that, the average deviation of all three actuators is taken into account. This value in turn is averaged along the trajectory and finally gives a target value for the optimization. Accordingly, minimum torque deviations are obtained for a mass distribution factor of $\nu_1 = 0.74$. For the overall system, the upper part of Fig. 4 shows the torque curves of the three actuators as modeled with a complete (C, black) and simplified (S, red) approach.

The results for the commonly chosen factors $\nu_1 = 0.5$ (Fig. 4a) and $\nu_1 = 0.67$ (Fig. 4b) are compared with the results applying the optimal distribution factor, i.e., $\nu_1 = 0.74$ (Fig. 4c). The lower part of Fig. 4 displays the absolute torque deviations for each actuator. The horizontal dash-dot line refers to the averaged torque deviation as used as target value for the optimization.

As outlined previously, the factor $\nu_1 = 0.5$ is most suitable for very slow (quasi-static) applications (for example, in surgical robotics [8]). Thus, for the fast application in hand, the average torque deviation of 3.9 N m is highest for $\nu_1 = 0.5$. Applying the distribution factor of $\nu_1 = 0.67$ closely approximates the outcome of a complete modeling approach with an average torque deviation of 2.8 N m. Further improvements can be achieved by an optimized value of $\nu_1 = 0.74$. In this way, the average torque deviation can be reduced to 2.5 N m.

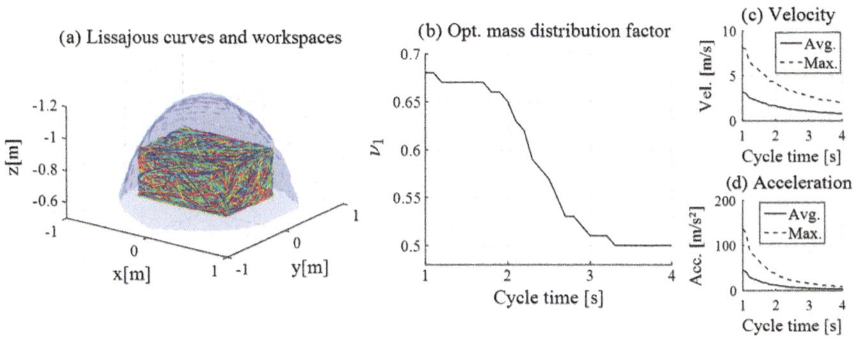

Fig. 5 Analyses of the mass distribution factor based on Lissajous curves

In order to obtain general validity, the analyses are extended taking into account randomly generated Lissajous curves within a prescribed workspace as similarly proposed in [10]. Lissajous curves are infinitely differentiable periodic functions given by:

$$\begin{aligned} x(t) &= A_1 \sin(a_1 s(t) + \delta_1) \\ y(t) &= A_2 \sin(a_2 s(t) + \delta_2) \\ z(t) &= A_3 \sin(a_3 s(t) + \delta_3) \end{aligned} \tag{10}$$

where the constants $A_1, A_2, A_3, a_1, a_2, a_3, \delta_1, \delta_2$, and δ_3 are chosen randomly and where the path profiles are specified based on a fifth order polynomial $s(t)$. Figure 5a depicts 200 open Lissajous curves covering a cuboid prescribed workspace of $1100 \cdot 800 \cdot 300\ \text{mm}^3$ within the reachable workspace (illustrated in Fig. 5a in blue). The acceleration and velocity profiles for a randomly generated curve depend on the imposed cycle time. For each curve of the set of 200 Lissajous curves the cycle time is varied between 1 and 4 s (with a step size of 0.1 s) which give an overall number of 6200 trajectories. Each trajectory is analyzed in respect of the optimal mass distribution factor deploying optimization algorithms. The results of the optimization are shown in Fig. 5b. For each evaluated cycle time, the average and maximum velocity and acceleration, respectively, for each of the curves are averaged over the set of 200 Lissajous curves (see Fig. 5c, d).

The results demonstrate that, irrespective of the imposed path, the optimum of the mass distribution factor depends on the application speed. In contrast to the outcome of the analyses based on a single trajectory, the optimum distribution factor solely exceeds the identified limit values (i.e., $\nu_1 = 0.5$ and $\nu_1 = 0.67$) for very short cycle times (less than 1.2 s).

6 Conclusion

In this study complete and simplified approaches for dynamic analyses of a Delta robot were reviewed and compared in respect to their computation times and accuracies. It was found that compared to the complete Newton–Euler approach computation times can be reduced by 74% applying a simplified Lagrangian approach. Analyses of the energy deviations of the distal link revealed that mass distribution factors are highly dependent on the imposed trajectory. For quasi-static motions an optimum mass distribution factor of $\nu_l = 1/2$ was identified. The deviations of kinetic energies showed that simply choosing a factor for mass distribution may not be ideal. Applying an optimized mass distribution factor of $\nu_l = 0.74$, the average deviations of the actuator torque were reduced by 9.7% compared to the commonly chosen factor of $\nu_l = 0.67$. Analyses taking into account randomly generated Lissajous curves prove a more general validity of the achievements.

The results should be taken as an example to show that a careful evaluation of the mass distribution factor may help to further improve the accuracy of the simplified model. The impact needs to be assessed in the overall context and confronted with the influence of frictional effects and the accuracy of the complete model itself. A long-term goal of future investigations is the derivation of a single parameter which can be used to find an optimal mass distribution factor related to a given system and handling task.

References

1. Brinker, J., Corves, B.: A survey on parallel robots with Delta-like architecture. In: Proceedings of the 14th World Congress in Mechanism and Machine Science. Taipei, Taiwan (2015)
2. Brinker, J., Corves, B., Wahle, M.: A comparative study of inverse dynamics based on Clavel's Delta robot. In: Proceedings of the 14th World Congress in Mechanism and Machine Science. Taipei, Taiwan (2015)
3. Briot, S., Khalil, W.: Dynamics of parallel robots from rigid bodies to flexible elements. Mechanisms and Machine Science, vol. 35. Springer, Switzerland (2015)
4. Clavel, R.: Device for the movement and positioning of an element in space. US Patent 4976582, 1990
5. Codourey, A.: Contribution á la commande des robots rapides et prècis: application au robot Delta á entraînement direct. Ph.D. thesis, EPFL, Lausanne (1991)
6. Codourey, A.: Dynamic modeling of parallel robots for computed-torque control implementation. Int. J. Robot. Res. **17**(12), 1325–1336 (1998)
7. Codourey, A., Burdet, E.: A body-oriented method for finding a linear form of the dynamic equation of fully parallel robots. In: Proceedings of the IEEE International Conference on Robotics and Automation (ICRA), pp. 1612–1618. Albuquerque (1997)
8. Courteille, E., Deblaise, D., Maurine, P.: Design optimization of a Delta-like parallel robot through global stiffness performance evaluation. In: Proceedings of the IEEE/RSJ International Conference on Intelligent Robots and Systems (IROS), pp. 5159–5166. St. Louis (2009)
9. Miller, K., Clavel, R.: The Lagrange-based model of Delta-4 robot dynamics. Robotersysteme **8**, 49–54 (1992)
10. Nefzi, M.: Analysis and optimisation of 4UPS - 1UPU parallel robots. Ph.D. thesis, RWTH Aachen University, Aachen (2010)

11. Park, S., Kim, H., Song, C., Kim, K.: Dynamics modeling of a Delta-type parallel robot. In: Proceedings of the International Symposium on Robotics (ISR), pp. 1–5. Seoul (2013)
12. Staicu, S.: Recursive modelling in dynamics of Delta parallel robot. Robotica **27**(2), 199–207 (2009)
13. Stamper, R.E.: A three degree of freedom parallel manipulator with only translational degrees of freedom. Ph.D. thesis, University of Maryland, Maryland (1997)
14. Tsai, L.W.: Robot Analysis: the Mechanics of Serial and Parallel Manipulators. Wiley, New York (1999)

Hidden Cusps

Michel Coste, Philippe Wenger and Damien Chablat

Abstract This paper investigates a situation pointed out in a recent paper, in which a non-singular change of assembly mode of a planar 2-RPR-PR parallel manipulator was realized by encircling a point of multiplicity 4. It is shown that this situation is, in fact, a non-generic one and gives rise to cusps under a small perturbation. Furthermore, we show that, for a large class of singularities of multiplicity 4, there are only two types of stable singularities occurring in a small perturbation: these two types are given by the complex square mapping and the quarto mapping. Incidentally, this paper confirms the fact that, generically, a local non-singular change of solution must be accomplished by encircling a cusp point.

1 Introduction

The non-singular change of assembly mode in parallel manipulators, first observed by C. Innocenti and V. Parenti-Castelli [1], is often associated with the presence of cusps and the non-singular change of assembly mode is realized by turning around a cusp point, or a cuspidal edge of the singularity surface (see for instance [2–5]). It has also been reported that non-singular change of assembly modes can be realized by following an "alpha curve" (i.e. a fold curve intersecting itself transversally) [6, 7], and that the presence of cusps is not necessary for the existence of non-singular assembly mode changes [8].

Authors partially supported by ANR-14-CE34-0008-01 Kapamat.

M. Coste (✉)
IRMAR, Université de Rennes 1 – CNRS, Rennes, France
e-mail: michel.coste@univ-rennes1.fr

P. Wenger · D. Chablat
LS2N, CNRS – École Centrale de Nantes, Nantes, France
e-mail: philippe.wenger@ls2n.fr

D. Chablat
e-mail: damien.chablat@ls2n.fr

J. Lenarčič and J.-P. Merlet (eds.), *Advances in Robot Kinematics 2016*,
Springer Proceedings in Advanced Robotics 4, DOI 10.1007/978-3-319-56802-7_14

A recent paper [9] exhibits an example of a 2-dof parallel manipulator with an isolated singularity of multiplicity 4 of the inverse kinematics mapping, such that circling around the image of this singularity in the joint space results in a non-singular assembly mode change; moreover, after a second loop around the singularity, one is back in the same assembly mode. There is no cusp in the picture, but we intend to explain in the present paper that actually the cusps are hidden. Precisely, the singularity of multiplicity 4 is not a stable singularity, which means that it disappears under a small perturbation of the geometry of the manipulator, giving rise to three cusp points; in the joint space, the isolated singularity is perturbed into a deltoid curve with three cusps. Hence, circling around the singularity of multiplicity 4 was actually circling around 3 degenerate cusps.

H. Whitney [10] has shown that the only stable singularities of mappings between spaces of dimension 2 are folds and cusps. Any other higher order singularity becomes a combination of folds and cusps after perturbation, which amounts to say that these higher order singularities are degenerations of folds and cusps. We shall show that the case study of the perturbation of the 2-dof manipulator actually describes two main cases of singularities of multiplicity 4 (complex square and quarto mappings) leading to two different perturbations (the former with three cusps, the latter with one cusp).

2 A Case Study

2.1 2RP̲R-PR with Higher Order Singularities

We begin with the example given in [9]. It is a 2RP̲R-PR planar manipulator with architecture described in Fig. 1.

The output coordinates are the angle φ and the y-coordinate of the revolute joint B which is constrained to move on the vertical axis. The input coordinates are the square ℓ_1^2 and ℓ_2^2 of the lengths of the legs A_1B_1 and A_2B_2. The equations for the inverse kinematic mapping are

$$\ell_i^2 = y^2 - 2\,b_i\,y\,\sin(\varphi) + a_i^2 - 2\,a_i\,b_i\,\cos(\varphi) + b_i^2 \quad \text{for } i = 1, 2 \tag{1}$$

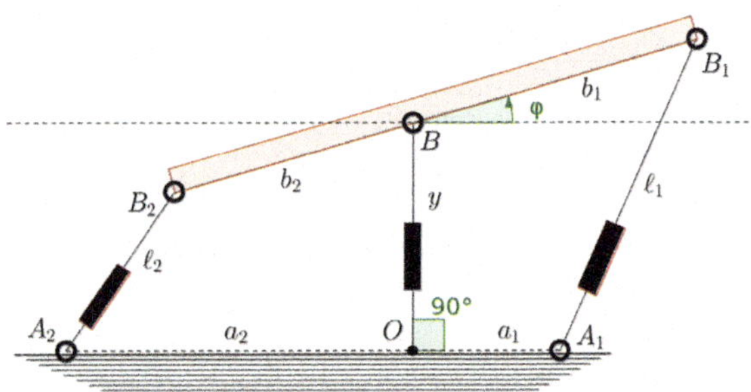

Fig. 1 Architecture of 2RP̲R-PR

The Jacobian matrix Jac of the inverse kinematic mapping and its Jacobian determinant J (up to a factor 4) are:

$$\mathrm{Jac} = 2 \begin{pmatrix} -b_1\, y\, \cos(\varphi) + a_1\, b_1\, \sin(\varphi) & y - b_1\, \sin(\varphi) \\ -b_2\, y\, \cos(\varphi) + a_2\, b_2\, \sin(\varphi) & y - b_2\, \sin(\varphi) \end{pmatrix}$$
$$J = (b_1 + b_2)\, \cos(\varphi)\, y^2 + (a_1\, b_1 - a_2\, b_2)\, \sin(\varphi)\, y - (a_1 + a_2)\, b_1\, b_2\, \sin(\varphi)^2\,. \tag{2}$$

The possible cusp points and higher order singularities may be detected by adding to $J = 0$ the equations

$$\mathrm{Jac} \begin{pmatrix} -J_y \\ J_\varphi \end{pmatrix} = \begin{pmatrix} 0 \\ 0 \end{pmatrix}, \tag{3}$$

where J_φ and J_y denote the partial derivatives of J with respect to φ and y. These equations express that the curve of singular points in the workspace either has a singularity or has a tangent vector in the kernel of the Jacobian matrix. We observe that $(\varphi, y) = (0 \bmod \pi, 0)$ satisfy Jac = the zero matrix (and hence also $J = 0$ and Eq. (3) hold); these singularities are not cusps, but higher order singularities.

2.2 An Example

We compute the singularities in the workspace and in the joint space for an example with $a_1 = 3, a_2 = 7, b_1 = 6, b_2 = 5$, which is the same as the one considered in [9]. In this case we can check that the only real solutions of $J = 0$ and the Eq. (3) are $(\varphi, y) = (0 \bmod \pi, 0)$.

For $(\varphi, y) = (0, 0)$, we have $\ell_1 = 3$, $\ell_2 = 2$. Developing the equations for the inverse kinematic mapping and for J in a neighbourhood of $(0, 0)$ we get

$$\ell_1^2 - 9 = y^2 + 12\, y\, \varphi + 18\, \varphi^2 + \text{h.o.t.}\,, \quad \ell_2^2 - 4 = y^2 - 10\, y\varphi + 35\, \varphi^2 + \text{h.o.t.}$$
$$J = 11\, y^2 - 17\, y\, \varphi - 300\, \varphi^2 + \text{h.o.t.}\,, \tag{4}$$

where h.o.t. stands for "higher order terms". This shows that the singularity is of multiplicity 4, and that the point $(\varphi = 0, y = 0)$ is a node of the curve of singularities (the discriminant $\Delta = (-17)^2 - 4 \times 11 \times (-300)$ of the quadratic part of the development of J at $(0, 0)$ is positive).

For $(\varphi, y) = (\pi, 0)$, we have $\ell_1 = 9$, $\ell_2 = 12$. Developing $\ell_1^2 - 81$, $\ell_2^2 - 144$ and J in a neighbourhood of $(\pi, 0)$, with $\psi = \phi - \pi$, we get

$$\ell_1^2 - 81 = y^2 - 12 y\psi - 18\psi^2 + \text{h.o.t.}\,, \quad \ell_2^2 - 144 = y^2 + 10 y\psi - 35\psi^2 + \text{h.o.t.}$$
$$J = -11\, y^2 + 17\, y\, \psi - 300\, \psi^2 + \text{h.o.t.}\,. \tag{5}$$

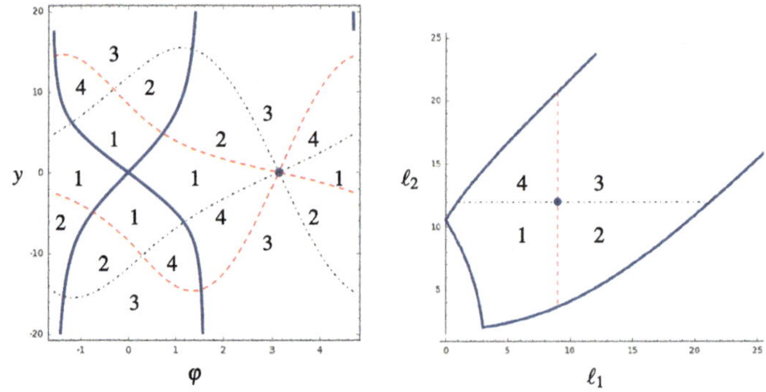

(a) Workspace (b) Joint space

Fig. 2 Workspace and joint space of the manipulator

This shows that the singularity is also of multiplicity 4, and that the point ($\varphi = \pi$, $y = 0$) is an isolated double point of the curve of singularities (the discriminant $\Delta = 17^2 - 4 \times (-11) \times (-300)$ of the quadratic part of the development of J at $(\pi, 0)$ is negative).

Figure 2a represents the workspace of the manipulator; it must be understood that the right side ($\varphi = 3\pi/2$) has to be identified with the left side ($\varphi = -\pi/2$). The singularity curve is represented in thick blue; one can see the node at $\varphi = 0$, $y = 0$ and the isolated double point at $\varphi = \pi$, $y = 0$. The dash-dot black curve is the level curve $\ell_1 = a_1 + b_1 = 9$ and the dashed red curve is the level curve $\ell_2 = a_2 + b_2 = 12$. The numbers in the zones delimited by these curves indicate the corresponding images by the inverse kinematic mapping in the joint space.

The joint space is represented in Fig. 2b (the same figure appears in [9]). One can see the image of the singularity curve in blue. Inside the domain delimited by this curve, the direct kinematic problem has four solutions, except at the image ($\ell_1 = 9$, $\ell_2 = 12$) of the isolated singularity point where there is one solution of multiplicity 4 and two other solutions. Above each point of the image singularity curve, there are two double solutions, except at the point $\ell_1 = 3$, $\ell_2 = 2$ (image of the node) where there is one solution of multiplicity 4. The zones numbered 1, 2, 3, 4 are the images of the zones with the corresponding numbers in the workspace.

It can be seen that circling around the isolated singularity point in the joint space, following the numbering 1-2-3-4-1, yields a non-singular assembly mode change leading from a zone numbered 1 in the workspace touching the isolated singularity point to the other one. A second loop makes one return to the initial assembly mode. This phenomenon cannot be faithfully represented in a 3-dimensional reduced configuration space: one cannot have a ramp turning around the singular configuration and returning to the start level after two turns without an artificial self-intersection.

One can see in this example a non-singular assembly mode change by circling around a singularity which is not a cusp. The hidden cusps are revealed by slightly perturbing the geometry of the manipulator.

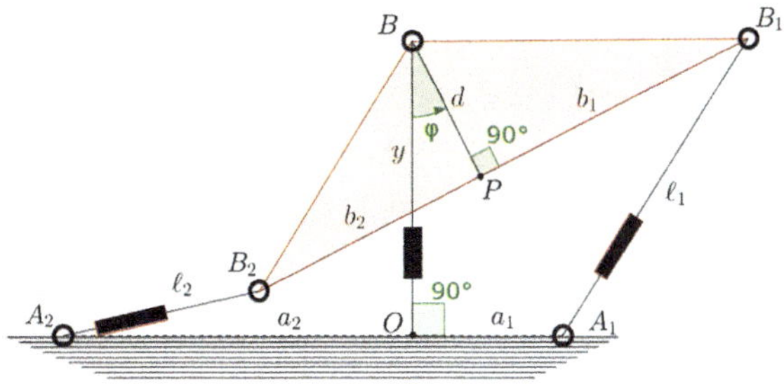

Fig. 3 The modified manipulator

2.3 *Revealing the Hidden Cusps*

The manipulator is modified so that the revolute joint B on the platform is no longer on the line B_1B_2, but at a distance d from this line: see Fig. 3.

We compute the example with $a_1 = 3$, $a_2 = 7$, $b_1 = 6$, $b_2 = 5$ and $d = 3$. The equations for the inverse kinematic mapping are now

$$\begin{aligned} \ell_1^2 &= y^2 - 6y(\cos(\varphi) - 2\sin(\varphi)) - 36\cos(\varphi) - 18\sin(\varphi) + 54 \\ \ell_2^2 &= y^2 - 2y(3\cos(\varphi) + 5\sin(\varphi)) - 70\cos(\varphi) + 42\sin(\varphi) + 83\ , \end{aligned} \tag{6}$$

and the Jacobian determinant is, up to a constant factor,

$$\begin{aligned} J = 11\,y^2\,\cos(\varphi) - y\,(30\,\cos(\varphi) + 17\,\sin(\varphi) + 33) \\ + 390\,\cos(\varphi)^2 - 30\,\cos(\varphi)\,\sin(\varphi) - 300 \end{aligned} \tag{7}$$

We can detect the cusps or higher order singularities by solving the system formed by $J = 0$ and Eq. (3). We get four real solutions, which are

$$\begin{aligned} (\varphi \simeq -0.0023,\ y \simeq 2.9069),\ (\varphi \simeq 2.6492,\ y \simeq -2.2190), \\ (\varphi \simeq -2.7368,\ y \simeq -1.2968),\ (\varphi \simeq 3.0855,\ y \simeq 2.6935) \end{aligned} \tag{8}$$

It can be checked that all four points are actually cusp points.

The curve of singularities in the workspace is represented in thick blue in Fig. 4a. Note that it retains the overall features of the original singularity curve in Fig. 2a, except for the node at $(0, 0)$ which is simplified in two non-intersecting branches and the isolated double point at $(\pi, 0)$ which has evolved into an oval. The characteristic curves in the workspace (defined in [11]) are represented in green. One can recognize the cusp points as the points of tangency of the characteristic curves with the curve of singularities: there is one cusp point on one branch of the simplification of the node, and three cusp points on the oval obtained by perturbing the isolated double point of the singularity curve.

Figure 4b represents the joint space. One can see the four cusps, three on the central deltoid (image of the oval) and one on a branch of the outer curve (the two

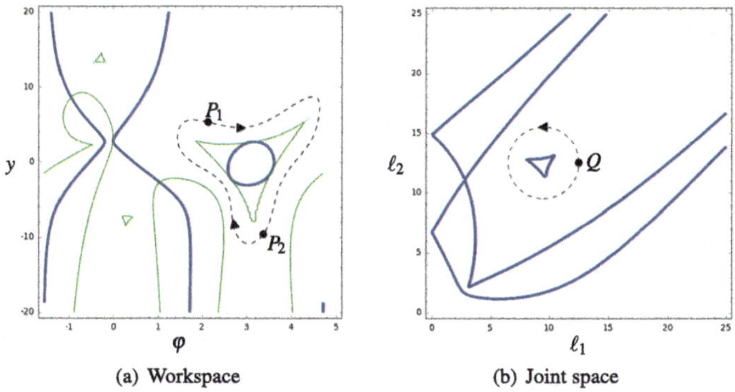

Fig. 4 Workspace and joint space of the modified manipulator

branches have also a crossing point). There are six solutions to the direct kinematic problem inside the deltoid, and 4, 2 or 0 solutions as one proceeds towards the outer regions.

Note that the connected zone in the workspace encircling the large green deltoid is not a uniqueness domain [11]: it is a 2-sheeted covering of the zone around the deltoid in the joint space. This latter zone is not simply connected, so we cannot deduce that a connected component of its preimage is a uniqueness domain.

The picture of the joint space shows that circling around the isolated singularity in the joint space was actually circling around three degenerate cusps. The dashed circle from Q to Q around the deltoid in the joint space lifts to the dashed trajectory from P_1 to P_2 in the workspace; a second turn on the circle completes the circuit from P_2 back to P_1.

3 The General Mathematical Picture: Unfolding of a Singularity of Multiplicity 4 of a Mapping of Surfaces

We explain here how the observations made for the modified 2RPR-PR fit into a general mathematical framework. We begin by recalling two examples described in [12].

3.1 Complex Square Mapping and Its Unfolding

The first example is given by $f : (x, y) \longmapsto (u = x^2 - y^2,\ v = 2xy)$ which is the complex square function $z \mapsto z^2$, written in real and imaginary parts; this shows that every point in $\mathbb{R}^2$ is the image by f of two points in $\mathbb{R}^2$, except the origin which is the image of the origin only. The Jacobian determinant of f is, up to a constant factor, $x^2 + y^2$. The only singularity of f is at the origin, and this singularity has multiplicity 4 (the dimension of the quotient algebra $\mathbb{R}[x, y]/(x^2 - y^2, 2xy)$).

Now we perturb the mapping f to

$$\tilde{f} : (x, y) \longmapsto (u = x^2 - y^2 + 4ax,\ v = 2xy + 4by)\ . \tag{9}$$

The Jacobian determinant of $\tilde{f}$ becomes, up to a constant factor, $(x + a + b)^2 + y^2 - (a - b)^2$. If $b \neq a$, the set of singular points of $\tilde{f}$ is the circle with centre $(-a - b, 0)$ and radius $|a - b|$. There are three cusp points on this circle, and the image curve in the (u, v) plane is a deltoid with three cusps. A point inside the deltoid has four preimages, outside two. Circling around the deltoid permutes the two preimages (as circling around the origin does for the complex square root).

Figure 5a shows the situation at the source (coordinates (x, y)) and at the target (coordinates (u, v)), in the case $a = 1$, $b = -1$. The blue curves are the curves of singularities. The green curve at the source is the characteristic curve; the cusps points are the points where the blue and green curves are tangent.

3.2 *Quarto Mapping and Its Unfolding*

The second example is the mapping $g : (x, y) \longmapsto (u = x^2,\ v = y^2)$; this mapping is named "quarto" because it folds the plane (x, y) onto the first quadrant of the plane (u, v), which is covered by four sheets. The Jacobian determinant of g is, up to a constant factor, xy. The set of singular points of g is the union of the two axes.

Let us perturb the mapping g to

$$\tilde{g} : (x, y) \longmapsto (u = x^2 + 2ay,\ v = y^2 + 2bx)\ . \tag{10}$$

The Jacobian determinant becomes, up to a constant factor, $xy - ab$. This is the equation of an equilateral hyperbola, if $ab \neq 0$. Its image by $\tilde{g}$ is a curve in the (u, v) plane with two branches, one of which has a cusp; inside the cusp, each point has four preimages by $\tilde{g}$, between the branches two, and zero elsewhere.

Figure 5b represents the situation at the source and at the target in the case $a = b = 1$ in the same way as for the preceding example. One can see the cusp point at the source.

3.3 *General Case*

The two examples above are actually the complete list of the stable singularities that can be obtained by perturbing a singularity of multiplicity 4 where the 2×2 Jacobian matrix is the zero matrix. These are the elliptic (complex square case) and hyperbolic (quarto case) Σ^2 singularities which are studied in [13], Part I §3. The notation Σ^2

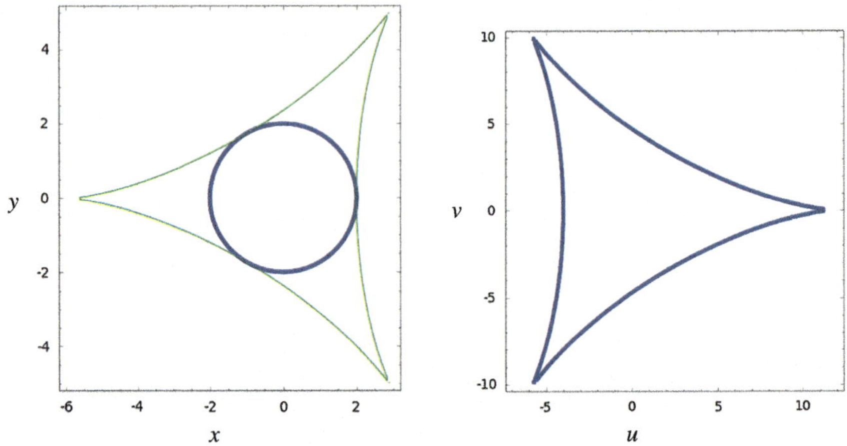

(a) Perturbation of complex square

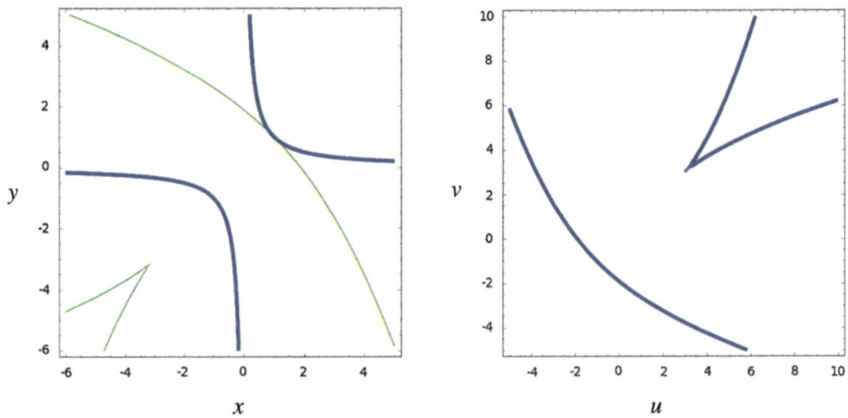

(b) Perturbation of quarto

Fig. 5 Perturbations of the complex square mapping and of the quarto mapping

means that the Jacobian matrix has corank 2, i.e. is the zero matrix in dimension 2, and in this case multiplicity 4 is equivalent to the fact that the discriminant Δ of the quadratic part of the Taylor expansion of the Jacobian determinant at the singularity is nonzero. The elliptic case corresponds to $\Delta < 0$ and the hyperbolic case to $\Delta > 0$.

We can now return to the example of the 2R$\underline{\text{P}}$R-PR. We have $\Delta > 0$ at the singularity $(\varphi, y) = (0, 0)$ (see (4)): we are here in the case "quarto mapping" and we can clearly see the relevant parts of Fig. 4 corresponding to Fig. 5b. We have $\Delta < 0$ at the singularity $(\varphi, y) = (\pi, 0)$ (see (5)): we are now in the case "complex square mapping" and we can compare the relevant parts of Fig. 4 with Fig. 5a.

4 Conclusion

We have shown that the singularities of multiplicity 4 that appear in the study of the kinematics of the 2RPR-PR are not generic and give rise to cusps under a small perturbation. We have also shown that these singularities belong to a family of singularities which splits in two cases according to the sign of the discriminant of the quadratic part of the Jacobian determinant: the "complex square mapping" case and the "quarto mapping" case which are well known in the theory of singularities of differentiable mappings. In the first case, the singularity is isolated and circling around it in the joint space results in an exchange of two solutions to the direct kinematic problem. A small perturbation to stable singularities gives three cusp points which were in some sense "hidden" in the singularity of multiplicity 4, and so one can argue that this example does not invalidate the rule that, generically, local non-singular assembly mode changes arise by circling around cusps.

We have limited our study to the 2-dof case. In a future work, we shall examine the perturbation of the second example in [9], which is interesting because it gives a fully parallel generic 3-R$\underline{P}$R manipulator with properties similar to the ones we have seen for the constrained 2R$\underline{P}$R-PR. We shall also discuss in more details characteristic surfaces and uniqueness domains [11].

References

1. Innocenti, C., Parenti-Castelli, V.: Singularity-free evolution from one configuration to another in serial and fully-parallel manipulators. J. Mech. Design **120**, 73–79 (1998)
2. McAree, P.R., Daniel, R.W.: An explanation of never-special assembly changing motions for 3–3 parallel manipulators. Int. J. Robot. Res. **18**(6), 556–574 (1999)
3. Zein, M., Wenger, P., Chablat, D.: Singular curves in the joint space and cusp points of 3-RPR parallel manipulators. Robotica **25**(6), 717–724 (2007)
4. Caro, S., Wenger, P., Chablat, D.: Non-singular assembly mode changing trajectories of a 6-DOF parallel robot. In: Proceedings of the ASME 2012 International Design Engineering Technical Conferences and Computers and Information in Engineering Conference IDETC/CIE 2012, pp. 1245–1254 (2012)
5. Husty, M., Schadlbauer, J., Caro, S., Wenger, P.: The 3-RPS manipulator can have nonsingular assembly-mode changes. Computational Kinematics, Mechanisms and Machine Science, pp. 339–348. Springer, Berlin (2014)
6. Bamberger, H., Wolf, A., Shoham, M.: Assembly mode changing in parallel mechanisms. IEEE Trans. Robot. **24**(4), 765–772 (2008)
7. Macho, E., Altuzarra, O., Pinto, C., Hernandez, A.: Transitions between multiple solutions of the direct kinematic problem. Advances in Robot Kinematics: Analysis and Design, pp. 301–310. Springer, Berlin (2008)
8. Coste, M., Chablat, D., Wenger, P.: Nonsingular change of assembly mode without any cusp. Advances in Robot Kinematics, pp. 105–112. Springer, Berlin (2014)
9. Peidro, A., Marin, J.M., Gil, A., Reinoso, O.: Performing nonsingular transitions between assembly modes in analytic parallel manipulators by enclosing quadruple solutions. J. Mech. Design **137**(12), 122302–122302-11 (2015)
10. Whitney, H.: On singularities of mappings of Euclidean spaces. I. Mappings of the plane into the plane. Ann. Math. **62**, 374–410 (1955)

11. Wenger, P., Chablat, D.: Uniqueness domains in the workspace of parallel manipulators. In: I.F.A.C-Symposium on Robot Control (SYROCO97), pp. 431–436 (1997)
12. Callahan, J.: Singularities and plane maps. Am. Math. Mon. **81**, 211–240 (1974)
13. Arnold, V.I., Gusein-Zade, S.M., Varchenko, A.N.: Singularities of Differentiable Maps, vol. 1. Birkhäuser, Basel (1985)

Some Mobile Overconstrained Parallel Mechanisms

J.M. Selig

Abstract The Griffis–Duffy platform is an example of an overconstrained parallel mechanism. Although it has 6 SS legs joining its platform to its base it is still mobile. In this work similar structures are found but with different types of legs. The key to finding these structures is a pair of theorems concerning 3 degree-of-freedom mechanisms subjected to a translation or a half-turn. Although these results are not new concise statements and proofs are given. These constructions are then applied to parallel mechanisms consisting of 3 RPS legs and 3UPU legs. Some details of the rigid-body motions that the platform of these mechanisms can execute are found. This is facilitated by the observations that rigid displacements permitted by an RPS leg are the displacements which constrain a point to a fixed plane, while the displacements of a UPU leg constrain a line to be coplanar to a fixed line.

1 Introduction

There has been much interest in overconstrained, single loop mechanisms such as the Bennett, Goldberg and various Bricard mechanisms. With interest turning to parallel mechanisms workers have also begun to look at over constrained parallel mechanisms. These are sometimes described as mechanisms which are architecturally singular. A key example of such a mechanism was the Griffis–Duffy platform as explained by Husty and Karger, [2]. Here these ideas are extended to platforms with other types of legs, in particular RPS and UPU. First we consider a pair of constructions which guarantee the mobility of the mechanisms.

J.M. Selig (✉)
School of Engineering, London South Bank University, London SE1 0AA, UK
e-mail: seligjm@lsbu.ac.uk

J. Lenarčič and J.-P. Merlet (eds.), *Advances in Robot Kinematics 2016*,
Springer Proceedings in Advanced Robotics 4, DOI 10.1007/978-3-319-56802-7_15

2 Line-Symmetry and Translations

The results considered here are well known in general terms but giving formal statements of the results clarifies the underlying geometry. In both cases the symmetries discussed confine the rigid motion to the intersection of a 5-dimensional projective space (a 5-plane) with the Study quadric.

We consider arbitrary mechanisms. These are to be thought of as systems of links and joints but we don't specify their arrangement so the mechanism could be a serial chain, a single loop or a parallel mechanism, for example. All we require is that one link is fixed and called the base link. We will concentrate our attention on another link in the mechanism and will refer to this as the coupler or platform of the mechanism. If the coupler has 3-degrees-of-freedom then the possible displacements of the coupler, relative to the base link, can be specified using three parameters.

Lemma 1 *Let $\mathscr{M}$ be an arbitrary mechanism having a coupler with 3 degrees-of-freedom. Duplicate the mechanism $\mathscr{M}$ and subject the new one to a fixed translation. The translation must include all links and joints including the base link. After the translation the translated base-link is again fixed. Rigidly join the coupler bars of the two mechanisms to form a combined coupler. This combined coupler bar will be able to move and will, in general, follow a 1 degree-of-freedom Schönflies motion.*

Proof Assume $g(\mu_1, \mu_2, \mu_3)$ is the dual quaternion representing the three parameter motion performed by $\mathscr{M}$. After a translation t, the shifted mechanism will be able to perform the motion, $tg(\mu_1, \mu_2, \mu_3)t^-$, where t^- is the dual quaternion conjugate of t. When the couplers are joined together, any motion performed must satisfy,

$$g(\mu_1, \mu_2, \mu_3) = tg(\mu_1, \mu_2, \mu_3)t^-.$$

This relation will have solutions for all displacements $g(\mu_1, \mu_2, \mu_3)$ that commute with t. The set of all elements in the group which commute with a translation consist of the subgroup of all translations and all rotations about axes parallel to t. The centraliser of a translation is a Schönflies group. In the Study quadric a Schönflies group is the intersection of the Study quadric with a 5-plane. Intersecting with the 3-dimensional set of displacements $g(\mu_1, \mu_2, \mu_3)$ generally gives a 1-dimensional set, necessarily lying in the Schönflies subgroup. □

Only the direction of the translation is important, any translation in the same direction will give the same Schönflies group. For the parallel mechanisms below this means that the same motion can be generated by a machine with an arbitrary number of legs.

The second result is probably even more well known, the statement and simple proof are still instructive.

Lemma 2 *Let $\mathscr{M}$ be an arbitrary mechanism that has a coupler with 3 degrees-of-freedom. Again, duplicate the mechanism $\mathscr{M}$ but now subject the new one to a half-turn about a line ℓ_0. This time the base-link of the new mechanism is rigidly* *fixed to the coupler link of the original and the coupler link of the new machine is fixed to the base. The coupler bar of the new combined mechanism will generally follow a 1 degree-of-freedom line-symmetric motion.*

Proof After the half-turn the motion of the coupler will be $\ell_0 g(\mu_1, \mu_2, \mu_3)\ell_0^-$ but the motion of the base with respect to the coupler will be, $\ell_0 g^-(\mu_1, \mu_2, \mu_3)\ell_0^-$. After connecting the mechanism as specified the motion of the combined coupler will satisfy,

$$g(\mu_1, \mu_2, \mu_3) = \ell_0 g^-(\mu_1, \mu_2, \mu_3)\ell_0^-.$$

This can be rearranged to produce,

$$g(\mu_1, \mu_2, \mu_3)\ell_0^- + \ell_0 g^-(\mu_1, \mu_2, \mu_3) = 0,$$

since $\ell_0^- = -\ell_0$. In [6] it was shown that this equation characterises line-symmetric motions, moreover, line-symmetric motions were shown to lie in the intersection of the Study quadric with a 5-plane. □

3 RPS Legs

The legs considered here are each composed of a revolute, prismatic and a final spherical joint, see Fig. 1. Keeping the base of the leg fixed and moving the platform attached to the spherical joint gives a set of possible rigid-body displacements allowed by the leg. In all these displacements the centre of the spherical joint remains in contact with a fixed plane normal to the axis of the first revolute joint. Clearly, the set of displacements allowed by such a leg coincides with the point-plane constraint varieties discussed in [4] for example. These point-plane constraint varieties can be thought of as the intersection of the Study quadric in $\mathbb{P}^7$ with another quadric hypersurface.

Now the construction of Lemma 1 is applied to a parallel mechanism consisting of three general RPS legs, see Fig. 2. Assume the axis of the Schönflies motion is the z-axis so the rotation matrix and translation vector can be written

Fig. 1 An RPS leg

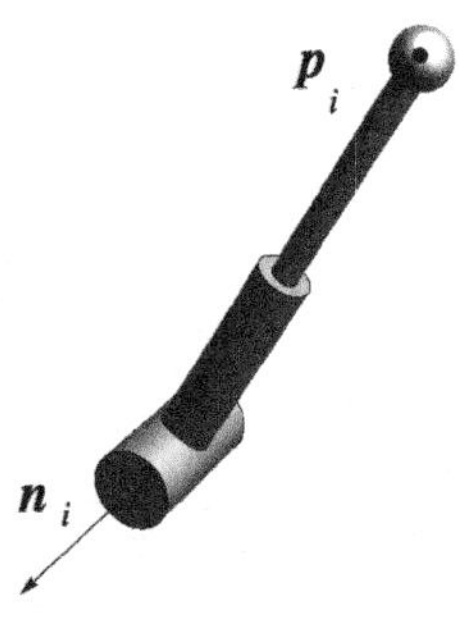

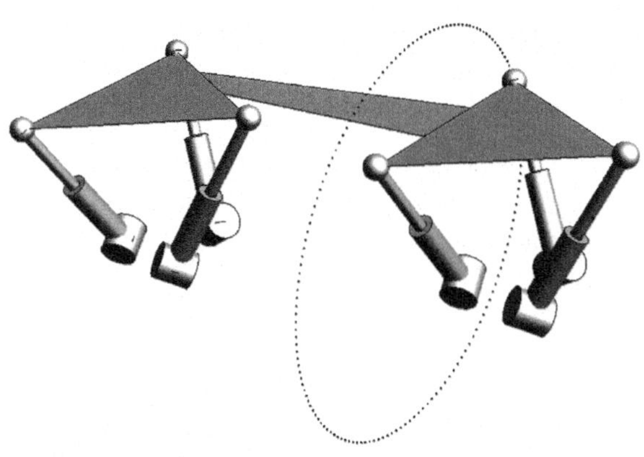

Fig. 2 A mobile 6RPS parallel mechanism constructed by translating a 3RPS linkage and joining the coupler bars. The *dots* represent the elliptical path of the centre of one of the spherical joints

$$R = \begin{pmatrix} \cos\phi & -\sin\phi & 0 \\ \sin\phi & \cos\phi & 0 \\ 0 & 0 & 1 \end{pmatrix} \quad \text{and} \quad \mathbf{t} = \begin{pmatrix} t_x \\ t_y \\ t_z \end{pmatrix}.$$

The three point-plane constraints can be written

$$(\mathbf{n}_i^T,\ -d_i)\begin{pmatrix} R & \mathbf{t} \\ 0 & 1 \end{pmatrix}\begin{pmatrix} \mathbf{p}_i \\ 1 \end{pmatrix} = \mathbf{n}_i^T(R\mathbf{p}_i + \mathbf{t}) - d_i = 0, \qquad i = 1,\ 2,\ 3, \tag{1}$$

where $\mathbf{n}_i$ is the unit normal to the plane, d_i the perpendicular distance from the plane to the origin and $\mathbf{p}_i$ the position vector of the point at the centre of the spherical joint. These equations can be written in matrix form as

$$N\mathbf{t} = \boldsymbol{\delta}, \qquad \text{where} \qquad N = \begin{pmatrix} \mathbf{n}_1^T \\ \mathbf{n}_2^T \\ \mathbf{n}_3^T \end{pmatrix} \qquad \text{and} \qquad \boldsymbol{\delta} = \begin{pmatrix} d_1 - \mathbf{n}_1^T R\mathbf{p}_1 \\ d_2 - \mathbf{n}_2^T R\mathbf{p}_2 \\ d_3 - \mathbf{n}_3^T R\mathbf{p}_3 \end{pmatrix}. \tag{2}$$

Assuming that the points $\mathbf{p}_i$ lie on their respective planes at the start of the motion, when $R(0) = I_3$, the row of $\boldsymbol{\delta}$ can be written $\mathbf{n}_i^T(I_3 - R)\mathbf{p}_i$. Elements of $I_3 - R$ can be written in terms of the sine and cosine of the rotation angle ϕ as R can only be rotations about the z-axis. The matrix N can be inverted symbolically,

$$N^{-1} = \frac{1}{\mathbf{n}_1 \cdot (\mathbf{n}_2 \times \mathbf{n}_3)}\Big(\mathbf{n}_2 \times \mathbf{n}_3 \ \Big|\ \mathbf{n}_3 \times \mathbf{n}_1 \ \Big|\ \mathbf{n}_1 \times \mathbf{n}_2\Big).$$

So $\mathbf{t} = N^{-1}\boldsymbol{\delta} = \boldsymbol{\alpha}(1 - \cos\phi) + \boldsymbol{\beta}\sin\phi$, where

$$\boldsymbol{\alpha} = \frac{1}{\mathbf{n}_1 \cdot (\mathbf{n}_2 \times \mathbf{n}_3)}\Big((n_{1x}p_{1x} + n_{1y}p_{1y})\mathbf{n}_2 \times \mathbf{n}_3 + \\ (n_{2x}p_{2x} + n_{2y}p_{2y})\mathbf{n}_3 \times \mathbf{n}_1 + (n_{3x}p_{3x} + n_{3y}p_{3y})\mathbf{n}_1 \times \mathbf{n}_2\Big)$$

and

$$\boldsymbol{\beta} = \frac{1}{\mathbf{n}_1 \cdot (\mathbf{n}_2 \times \mathbf{n}_3)}\Big((n_{1x}p_{1y} - n_{1y}p_{1x})\mathbf{n}_2 \times \mathbf{n}_3 + (n_{2x}p_{2y} - n_{2y}p_{2x})\mathbf{n}_3 \times \mathbf{n}_1 + (n_{3x}p_{3y} - n_{3y}p_{3x})\mathbf{n}_1 \times \mathbf{n}_2\Big)$$

Using the familiar tan-half-angle substitutions, $\cos\phi = (1-t^2)/(1+t^2)$ and $\sin\phi = 2t/(1+t^2)$, it can be seen that the motion of a general point in the platform will be a conic curve. Hence this is a Darboux motion, see [1, Chap. IX, Sect. 3]. The fact that three point-plane constraints restricted to a Schönflies motion produces a Darboux motion is well known, see for example [7]. Notice however, that in [3] it was shown that a parallel mechanism with 3 RPS legs could perform a vertical Darboux motion, a particular Darboux motion that is also line-symmetric. The above shows that the mechanism can perform a general Darboux motion and indicates how to construct such a mechanism.

Next, the construction from Lemma 2 is applied to three RPS legs. As above there are three point-plane constraints given in Eq. (1). Since a line-symmetric motion consists of successive half-turns about the generators of a ruled surface, the rigid-body displacements can be given by the exponential of a line

$$\begin{pmatrix} R & \mathbf{t} \\ 0 & 1 \end{pmatrix} = e^{\pi L}, \quad \text{where} \quad L = \begin{pmatrix} 0 & -P_{03} & P_{02} & P_{23} \\ P_{03} & 0 & -P_{01} & P_{31} \\ -P_{02} & P_{01} & 0 & P_{12} \\ 0 & 0 & 0 & 0 \end{pmatrix}.$$

Here P_{ij} are the Plücker coordinates of the line L. The explicit dependence of L on time has been suppressed for brevity. Using the Rodrigues formula, the exponential can be written:

$$e^{\pi L} = I_4 + 2L^2,$$

since $L^3 = -L$ and assuming $P_{01}^2 + P_{02}^2 + P_{03}^2 = 1$. Writing,

$$\Omega = \begin{pmatrix} 0 & -P_{03} & P_{02} \\ P_{03} & 0 & -P_{01} \\ -P_{02} & P_{01} & 0 \end{pmatrix}, \quad \boldsymbol{\omega} = \begin{pmatrix} P_{01} \\ P_{02} \\ P_{03} \end{pmatrix}, \quad \text{and} \quad \mathbf{v} = \begin{pmatrix} P_{23} \\ P_{31} \\ P_{12} \end{pmatrix},$$

the rotation matrix is $R = I + 2\Omega^2$ and the translation vector $\mathbf{t} = 2\boldsymbol{\omega} \times \mathbf{v}$, where $\boldsymbol{\omega}$ and $\mathbf{v}$ consist of the Plücker coordinates of a line and so satisfy $\boldsymbol{\omega} \cdot \mathbf{v} = 0$.

Assume that the motion passes through the identity element of the group and that the points lie on their respective planes in this position. Then a line-symmetric motion is given by reflecting the three points $\mathbf{p}_1$, $\mathbf{p}_2$ and $\mathbf{p}_3$ in the initial line of the ruled surface L_0, and then reflecting in the successive lines of the surface, so that,

$$\begin{pmatrix} R & \mathbf{t} \\ 0 & 1 \end{pmatrix} = (I_4 + 2L^2)(I_4 + 2L_0^2).$$

To be definite assume that the initial line L_0 is the z-axis. Let $\mathbf{p}_i' = (I_3 + 2\Omega_0^2)\mathbf{p}_i$, where Ω_0 is the direction of L_0. The equations for the three point-plane constraints become, $\mathbf{n_i}^T(I_3 + 2\Omega^2)\mathbf{p}_i' + 2\mathbf{n_i}^T(\boldsymbol{\omega} \times \mathbf{v}) - d_i = 0$. Since $d_i = \mathbf{n_i}^T(I_3 + 2\Omega_0^2)\mathbf{p}_i'$, we get, $\mathbf{n_i}^T(\Omega^2 - \Omega_0^2)\mathbf{p}_i' + \mathbf{n_i}^T(\boldsymbol{\omega} \times \mathbf{v}) = 0$, for $i = 1, 2, 3$. These equations can be made homogeneous by multiplying the Ω_0^2 term by the square of the norm of the vector $\boldsymbol{\omega}$, denoted $|\boldsymbol{\omega}|^2$. This results in three homogeneous equations,

$$\mathbf{n_i}^T(\Omega^2 - |\boldsymbol{\omega}|^2\Omega_0^2)\mathbf{p}_i' + \mathbf{n_i}^T(\boldsymbol{\omega} \times \mathbf{v}) = 0, \qquad i = 1, 2, 3. \tag{3}$$

Including the equation for the Klein quadric $\boldsymbol{\omega} \cdot \mathbf{v} = 0$, gives 4 homogeneous quadratic equations for the ruled surface generating the line-symmetric motion. The intersection of these quadrics is not a complete intersection as they clearly vanish on the 2-dimensional plane of "lines at infinity" $\boldsymbol{\omega} = \mathbf{0}$.

The equation given in (3) can be written in the same matrix vector form as in (2) but with

$$\mathbf{t} = \boldsymbol{\omega} \times \mathbf{v} \qquad \text{and} \qquad \boldsymbol{\delta} = \begin{pmatrix} \mathbf{n}_1^T(\Omega^2 - |\boldsymbol{\omega}|^2\Omega_0^2)\mathbf{p}_1' \\ \mathbf{n}_2^T(\Omega^2 - |\boldsymbol{\omega}|^2\Omega_0^2)\mathbf{p}_2' \\ \mathbf{n}_3^T(\Omega^2 - |\boldsymbol{\omega}|^2\Omega_0^2)\mathbf{p}_3' \end{pmatrix}.$$

The variables $\mathbf{v}$ can be eliminated by multiplying the equation $N\mathbf{t} = \boldsymbol{\delta}$, by the inverse or adjugate of N and then taking the scalar product with $\boldsymbol{\omega}$. The result is

$$\begin{aligned}\boldsymbol{\omega} \cdot (\mathbf{n}_2 \times \mathbf{n}_3)\mathbf{n}_1^T(\Omega^2 - |\boldsymbol{\omega}|^2\Omega_0^2)\mathbf{p}_1' + \boldsymbol{\omega} \cdot (\mathbf{n}_3 \times \mathbf{n}_1)\mathbf{n}_2^T(\Omega^2 - |\boldsymbol{\omega}|^2\Omega_0^2)\mathbf{p}_2' + \\ \boldsymbol{\omega} \cdot (\mathbf{n}_1 \times \mathbf{n}_2)\mathbf{n}_3^T(\Omega^2 - |\boldsymbol{\omega}|^2\Omega_0^2)\mathbf{p}_3' = 0.\end{aligned}$$

This equation determines a plane cubic curve in the variables P_{01}, P_{02} and P_{03}. This is the base surface of the motion's direction cone or spherical indicatrix. Generally a plane cubic curve is rational or elliptic (has genus 0 or 1) depending on whether or not it has a singularity. Computing with a few random examples shows that the curve can be non-singular. Hence, in general, the curve is elliptic. However, there may be particular examples where the cubic acquires a singularity and hence becomes rational. Eliminating the moment vector $\mathbf{v}$ can be seen as a linear projection with centre of projection given by the 2-plane of lines "at infinity" $\boldsymbol{\omega} = \mathbf{0}$. To recover the moments of the generator lines multiply the equation $N\mathbf{t} = \boldsymbol{\delta}$, by the inverse of N and then take the vector product with $\boldsymbol{\omega}$ to get,

$$\begin{aligned}|\boldsymbol{\omega}|^2\mathbf{v} = \frac{1}{\mathbf{n}_1 \cdot (\mathbf{n}_2 \times \mathbf{n}_3)}\big(&\boldsymbol{\omega} \times (\mathbf{n}_2 \times \mathbf{n}_3)\mathbf{n}_1^T(\Omega^2 - |\boldsymbol{\omega}|^2\Omega_0^2)\mathbf{p}_1' + \\ &\boldsymbol{\omega} \times (\mathbf{n}_3 \times \mathbf{n}_1)\mathbf{n}_2^T(\Omega^2 - |\boldsymbol{\omega}|^2\Omega_0^2)\mathbf{p}_2' + \boldsymbol{\omega} \times (\mathbf{n}_1 \times \mathbf{n}_2)\mathbf{n}_3^T(\Omega^2 - |\boldsymbol{\omega}|^2\Omega_0^2)\mathbf{p}_3'\big).\end{aligned}$$

The triple product $\boldsymbol{\omega} \times (\boldsymbol{\omega} \times \mathbf{v})$, has been simplified using the equation for the Klein quadric. This give a rational cubic map from the plane with homogeneous coordinates $\boldsymbol{\omega} = (P_{01} : P_{02} : P_{03})$ to the Klein quadric. The image of the cubic spherical indicatrix will be a degree 9 ruled surface meeting the center of the linear projection, the space of lines at infinity, with multiplicity 6.

4 6UPU Legs

Next we study parallel mechanisms composed of 6 UPU legs with either line symmetry or a translational symmetry. We require the axes of the first and last revolute joints of the two U joints to be coplanar, see Fig. 3. This is not the most general configuration for such a leg but it is a commonly used design. The key observation is that the rigid displacement allowed by such a UPU leg will maintain the coplanarity of these lines. In [5] the problem of finding the set of rigid displacements which move a line in such a way that it remains in a linear line complex was studied. The set of lines meeting or parallel to a fixed line form a special linear line complex so the displacements of a UPU leg are a special case of the quadratic constraint found in [5]. The displacements achievable by the leg lie on the intersection of the Study quadric with another quadric hypersurface in $\mathbb{P}^7$. There are several serial kinematic chains with the same property, the UPU, PSP and RRPC chains for example. The remarks below thus apply to any of these chains.

It is convenient here to use the adjoint representation of $SE(3)$. Consider a pair of lines given in Plücker coordinates as,

$$\ell_a = \begin{pmatrix} \boldsymbol{\omega}_a \\ \mathbf{v}_a \end{pmatrix}, \quad \text{and} \quad \ell_b = \begin{pmatrix} \boldsymbol{\omega}_b \\ \mathbf{v}_b \end{pmatrix}.$$

Two lines are coplanar if and only if they are reciprocal, this can be represented as,

$$(\boldsymbol{\omega}_a^T, \mathbf{v}_a^T) \begin{pmatrix} 0 & I_3 \\ I_3 & 0 \end{pmatrix} \begin{pmatrix} \boldsymbol{\omega}_b \\ \mathbf{v}_b \end{pmatrix} = 0,$$

where I_3 is the 3×3 identity matrix. Rigid displacements moving ℓ_b so that it remains coplanar to ℓ_a thus satisfy

Fig. 3 A UPU leg

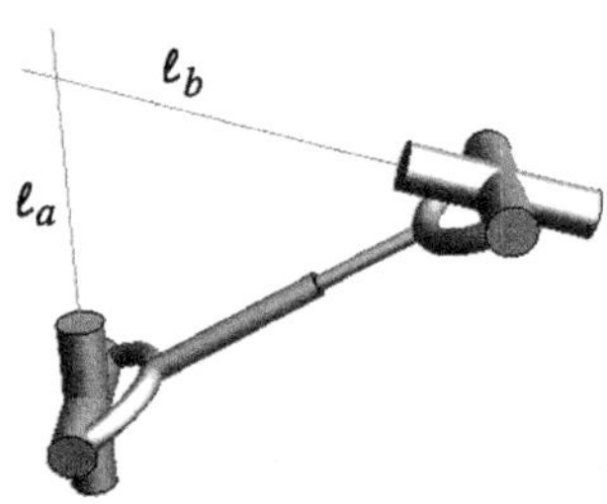

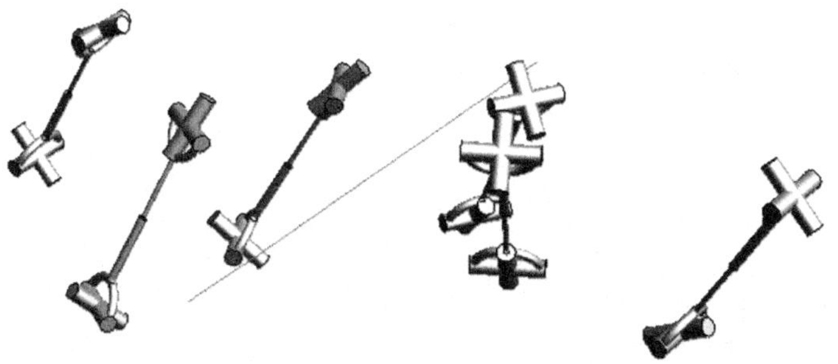

Fig. 4 A mobile 6UPU parallel mechanism constructed by reflecting three legs in the line shown. For clarity, the base and platform are not shown, the base connects the lower R joint of each leg and the platform joins the uppermost R joints

$$(\boldsymbol{\omega}_a^T, \mathbf{v}_a^T)\begin{pmatrix} 0 & I_3 \\ I_3 & 0 \end{pmatrix}\begin{pmatrix} R & 0 \\ TR & R \end{pmatrix}\begin{pmatrix} \boldsymbol{\omega}_b \\ \mathbf{v}_b \end{pmatrix} = 0,$$

where R is a rotation as above and T is the translation vector written as a 3×3 anti-symmetric matrix. Expanding the equation above produces

$$\boldsymbol{\omega}_a^T T R \boldsymbol{\omega}_b + \boldsymbol{\omega}_a^T R \mathbf{v}_b + \mathbf{v}_a^T R \boldsymbol{\omega}_b = 0.$$

Three such legs yield an equation of the form $N\mathbf{t} = \boldsymbol{\delta}$ again, now with

$$N = \begin{pmatrix} \left(\boldsymbol{\omega}_{a1} \times (R\boldsymbol{\omega}_{b1})\right)^T \\ \left(\boldsymbol{\omega}_{a2} \times (R\boldsymbol{\omega}_{b2})\right)^T \\ \left(\boldsymbol{\omega}_{a3} \times (R\boldsymbol{\omega}_{b3})\right)^T \end{pmatrix} \quad \text{and} \quad \boldsymbol{\delta} = \begin{pmatrix} \boldsymbol{\omega}_{a1}^T R\mathbf{v}_{b1} + \mathbf{v}_{a1}^T R\boldsymbol{\omega}_{b1} \\ \boldsymbol{\omega}_{a2}^T R\mathbf{v}_{b2} + \mathbf{v}_{a2}^T R\boldsymbol{\omega}_{b2} \\ \boldsymbol{\omega}_{a3}^T R\mathbf{v}_{b3} + \mathbf{v}_{a3}^T R\boldsymbol{\omega}_{b3} \end{pmatrix}.$$

Consider the 6UPU that can perform a Schönflies motion. Parameterising the rotation by tan-half-angles, as in Sect. 3, we can solve for the translation $\mathbf{t}$. This gives a degree 6 solution in the tan-half angle. The trajectories of general points on the platform will follow rational curves of degree 6.

Finally consider the line symmetric 6UPU, see Fig. 4. The spherical indicatrix of the base surface of the line-symmetric motion will be a planar curve of degree 7. From the genus-degree formula the maximum genus of such a curve is $\frac{1}{2}(7-1)(7-2) = 15$.

5 Conclusion

Although much is known about motions that constrain points to planes and to spheres, motions in which lines remain coplanar seems less well studied.

References

1. Bottema, O., Roth, B.: Theoretical Kinematics. Dover Publications, New York (1990)
2. Husty, M.L., Karger, A.: Self-motions of Griffis–Duffy type parallel manipulators. In: Proceedings of the IEEE International Conference on Robotics and Automation, ICRA'00, San Francisco, CA, vol. 1, pp. 7–12 (2000)
3. Nurahmi, L., Schadlbauer, J., Husty, M., Wenger, P., Caro, S.: Motion capabilities of the 3-RPS cube parallel manipulator. In: Lenarčič, J., Khatib, O. (eds.) Advances in Robot Kinematics, pp. 527–535. Springer, Switzerland (2014)
4. Selig, J.M.: On the geometry of point-plane constraints on rigid-body displacements. Acta Appl. Math. **116**(2), 133–155 (2011)
5. Selig, J.M.: On the geometry of the homogeneous representation for the group of proper rigid-body displacements. Rom. J. Tech. Sci. Appl. Mech. **58**(1–2), 27–50 (2013). (Special issue on New Trends in Advanced Robotics)
6. Selig, J.M., Husty, M.: Half-turns and line symmetric motions. Mech. Mach. Theory **46**(2), 156–167 (2011)
7. Zsombor-Murray, P.J., Gfrerrer, A.: A unified approach to direct kinematics of some reduced motion parallel manipulators. ASME J. Mech. Robot. **2**(2), 021006 (2010). (10 pages)

On the Line-Symmetry of Self-motions of Linear Pentapods

Georg Nawratil

Abstract We show that all self-motions of pentapods with linear platform of Type 1 and Type 2 can be generated by line-symmetric motions. Thus this paper closes a gap between the more than 100 year old works of Duporcq and Borel and the extensive study of line-symmetric motions done by Krames in the 1930s. As a consequence we also get a new solution set for the Borel Bricard problem. Moreover we discuss the reality of self-motions and give a sufficient condition for the design of linear pentapods of Type 1 and Type 2, which have a self-motion free workspace.

1 Introduction

The geometry of a linear pentapod is given by the five base anchor points M_i in the fixed system Σ_0 and by the five collinear platform anchor points m_i in the moving system Σ (for $i = 1, \ldots, 5$). Each pair $(\mathsf{M}_i, \mathsf{m}_i)$ of corresponding anchor points is connected by a SPS-leg, where only the prismatic joint is active.

If the geometry of the linear pentapod is given as well as the lengths R_i of the five pairwise distinct legs, it has generically mobility 1. This degree of freedom corresponds to the rotational motion about the carrier line p of the five platform anchor points. As this rotation is irrelevant for applications with axial symmetry (e.g. 5-axis milling, laser or water-jet engraving/cutting, spot-welding, spray-based painting, etc.), these manipulators are of great practical interest. Nevertheless configurations should be avoided where the linear pentapod gains an additional uncontrollable mobility, which is referred as self-motion.

G. Nawratil (✉)
Institute of Discrete Mathematics and Geometry, Vienna University of Technology, Vienna, Austria
e-mail: nawratil@geometrie.tuwien.ac.at

J. Lenarčič and J.-P. Merlet (eds.), *Advances in Robot Kinematics 2016*,
Springer Proceedings in Advanced Robotics 4, DOI 10.1007/978-3-319-56802-7_16

1.1 Review on Self-motions of Linear Pentapods

The self-motions of linear pentapods represent interesting solutions to a problem posed 1904 by the French Academy of Science for the *Prix Vaillant*, which is also known as Borel-Bricard problem (cf. [2, 3]). This still unsolved kinematic challenge reads as follows: *"Determine and study all displacements of a rigid body in which distinct points of the body move on spherical paths."*

For the special case of five collinear points the Borel-Bricard problem was studied by Darboux [5, p. 222], Mannheim [6, p. 180ff] and Duporcq [7] (see also Bricard [3, Chap. III]). A contemporary and accurate reexamination of these old results, which also takes the coincidence of platform anchor points into account, was done in [1] yielding a full classification of linear pentapods with self-motions.

Beside the architecturally singular linear pentapods [1, Corollary 1] and some trivial cases with pure rotational self-motions [1, Designs α, β, γ] or pure translational ones [1, Theorem 1] there only remain the following three designs:

Under a self-motion each point of the line p has a spherical (or planar) trajectory. The locus of the corresponding sphere centers is a cubic space curve P, where the mapping from p to P is named σ. P intersects the ideal plane in one real point W and two conjugate complex ideal points, where the latter ones are the cyclic points I and J of a plane orthogonal to the direction of W. P is therefore a so-called straight cubic circle. The following subcases can be distinguished:

- P is irreducible:
  - σ maps the ideal point U of p to W (Type 5 according to [1]).
  - σ maps U to a finite point of P (Type 1 according to [1]).
- P splits up into a circle and a line, which is orthogonal to the carrier plane of the circle and intersects the circle in a point Q. Moreover σ maps U to a point on the circle different from Q (Type 2 according to [1]).

1.2 Basics on Line-Symmetric Motions

Krames (e.g. [4, 10]) studied special one-parametric motions (*Symmetrische Schrotung* in German), which are obtained by reflecting the moving system Σ in the generators of a ruled surface of the fixed system Σ_0, which is the so called *basic surface*. These so-called *line-symmetric motions* were also studied by Bottema and Roth [8, Sect. 7 of Chap. 9], who gave an intuitive algebraic characterization in terms of Study parameters $(e_0 : e_1 : e_2 : e_3 : f_0 : f_1 : f_2 : f_3)$, which are shortly repeated next.

All real points of the Study parameter space P^7 (7-dimensional projective space), which are located on the so-called Study quadric $\Psi : \sum_{i=0}^{3} e_i f_i = 0$, correspond to an Euclidean displacement with exception of the 3-dimensional subspace $e_0 = e_1 = e_2 = e_3 = 0$, as its points cannot fulfill the condition $N \neq 0$ with $N := e_0^2 + e_1^2 +$

$e_2^2 + e_3^2$. The translation vector $\mathbf{s} := (s_1, s_2, s_3)^T$ and the rotation matrix $\mathbf{R}$ of the corresponding Euclidean displacement $\mathbf{m}_i \mapsto \mathbf{R}\mathbf{m}_i + \mathbf{s}$ are given for $N = 1$ by:

$$s_1 = -2(e_0 f_1 - e_1 f_0 + e_2 f_3 - e_3 f_2), \quad s_2 = -2(e_0 f_2 - e_2 f_0 + e_3 f_1 - e_1 f_3),$$
$$s_3 = -2(e_0 f_3 - e_3 f_0 + e_1 f_2 - e_2 f_1),$$
$$\mathbf{R} = \begin{pmatrix} r_{11} & r_{12} & r_{13} \\ r_{21} & r_{22} & r_{23} \\ r_{31} & r_{32} & r_{33} \end{pmatrix} = \begin{pmatrix} e_0^2 + e_1^2 - e_2^2 - e_3^2 & 2(e_1 e_2 - e_0 e_3) & 2(e_1 e_3 + e_0 e_2) \\ 2(e_1 e_2 + e_0 e_3) & e_0^2 - e_1^2 + e_2^2 - e_3^2 & 2(e_2 e_3 - e_0 e_1) \\ 2(e_1 e_3 - e_0 e_2) & 2(e_2 e_3 + e_0 e_1) & e_0^2 - e_1^2 - e_2^2 + e_3^2 \end{pmatrix}.$$

There always exists a moving frame (in dependence of a given fixed frame) in a way that $e_0 = f_0 = 0$ holds for a line-symmetric motion. Then $(e_1 : e_2 : e_3 : f_1 : f_2 : f_3)$ are the Plücker coordinates (according to the convention used in [8]) of the generators of the basic surface with respect to the fixed frame.

1.3 Line-Symmetric Self-motions of Linear Pentapods

It is well known (cf. [7, Sect. 15], [3, Sect. 12]) that the self-motions of Type 5 are obtained by restricting the Borel-Bricard motions[1] (also known as BB-I motions) to a line. Note that Krames gave a detailed discussion of this special case in [4, Sect. 5], where he also pointed out the line-symmetry of BB-I motions.

Beside these BB-I motions, there also exist line-symmetric motions (so-called BB-II motions), where every point of a hyperboloid carrying two reguli of lines has a spherical path. It is known (cf. [9, p. 24] and [10, p. 188]) that the corresponding sphere centers of lines, belonging to one regulus,[2] constitute irreducible straight cubic circles, which imply examples of Type 1 self-motions. It should be noted that there also exist degenerated cases where the hyperboloid splits up into the union two orthogonal planes, which contain examples of Type 2 self-motions.

A simple count of free parameters shows that not all self-motions of Type 1 (5-parametric set[3] of motions where all points of a line have spherical paths) can be generated by BB-II motions (which produce only a 4-parametric set[4]). The same argumentation holds for Type 2 self-motions and the mentioned degenerated case.

As a consequence the question arise whether all self-motions of linear pentapods of Type 1 and Type 2 can be generated by line-symmetric motions. If this is the case

---

[1]These are the only non-trivial motions where every point of the moving space has a spherical trajectory (cf. [3, Chap. VI]).

[2]The corresponding sphere centers of lines belonging to the other regulus are again located on lines (cf. [9, p. 24]), which imply linear pentapods with an architecturally singular design.

[3]With respect to the notation introduced in Sect. 2 these five parameters are C, a_r, a_c, a_4 and p_5 or R_1 (cf. Eq. (7)) by canceling the factor of similarity by setting $A = 1$.

[4]These are the parameters a, c, g, k used in [9, Sect. 2.3].

we can apply a construction proposed by Krames [4, p. 416], which is discussed in Sect. 4, yielding new solutions to the Borel-Bricard problem.

Finally it should be noted that a detailed review on line-symmetric motions with spherical trajectories is given in [11, Sect. 1].

2 On the Line-Symmetry of Type 1 and Type 2 Self-motions

For our calculations we do not select arbitrary pairs $(\mathsf{m}_i, \mathsf{M}_i)$ of p and P, which are in correspondence with respect to σ ($\Leftrightarrow \sigma(\mathsf{m}_i) = \mathsf{M}_i$), but choose the following special ones:

M_4 equals W, M_2 coincides with I and M_3 with J. The corresponding platform anchor points are denoted by m_4, m_2 and m_3, respectively. As M_i are ideal points the corresponding points m_i are not running on spheres but in planes orthogonal to the direction of M_i. Therefore these three point pairs imply three so-called Darboux conditions Ω_i for $i = 2, 3, 4$. Moreover we denote U as m_5 and its corresponding finite point under σ by M_5. This point pair describes a so-called Mannheim condition Π_5 (which is the inverse of a Darboux condition). The pentapod is completed by a sphere condition Λ_1 of any pair of corresponding finite points m_1 and M_1.

In [1] we have chosen the fixed frame $\mathscr{F}_0$ in a way that M_1 equals its origin and M_4 coincides with the ideal point of the z-axis. Moreover we located the moving frame $\mathscr{F}$ in a way that p coincides with the x-axis, where m_1 equals its origin.

For the study at hand it is advantageous to select a different set of fixed and moving frames $\mathscr{F}_0'$ and $\mathscr{F}'$, respectively:

- As M_2 and M_3 coincides with the cyclic points, we can assume without loss of generality (w.l.o.g.) that M_5 is located in the xz-plane (as a rotation about the z-axis does not change the coordinates of $\mathsf{M}_1, \ldots, \mathsf{M}_4$). Moreover we want to apply a translation in a way that M_5 is in the origin of the new fixed frame $\mathscr{F}_0'$. Summed up the coordinates with respect to $\mathscr{F}_0'$ read as:

$$\mathbf{M}_5 = (0, 0, 0), \quad \mathbf{M}_1 = (A, 0, C) \quad \text{with} \quad A \neq 0 \tag{1}$$

  as $A = 0$ implies a contradiction to the properties of P for Type 1 and Type 2 pentapods given in Sect. 1.1. Moreover, M_2, M_3 and M_4 are the ideal points in direction $(1, i, 0)^T$, $(1, -i, 0)^T$ and $(0, 0, 1)^T$, respectively.
- With respect to $\mathscr{F}_0'$ the location of p is undefined, but the coordinates $\mathbf{m}_i$ of m_i can be parametrized as follows for $i = 1, \ldots, 4$:

$$\mathbf{m}_i = \mathbf{n} + (a_i - a_r)\mathbf{d} \quad \text{with} \quad a_1 = 0, \; a_2 = a_r + i a_c, \; a_3 = a_r - i a_c \tag{2}$$

  where $a_r, a_c \in \mathbb{R}$ and $a_c \neq 0$ holds. m_5 is the ideal point in direction of the unit-vector $\mathbf{d} = (d_1, d_2, d_3)^T$, which obtains the rational homogeneous parametrization of the unit-sphere, i.e.

$$d_1 = \frac{2h_0h_1}{h_0^2+h_1^2+h_2^2}, \quad d_2 = \frac{2h_0h_2}{h_0^2+h_1^2+h_2^2}, \quad d_3 = \frac{h_1^2+h_2^2-h_0^2}{h_0^2+h_1^2+h_2^2}. \tag{3}$$

Now we are looking for the point $\mathbf{n} = (n_1, n_2, n_3)^T$ and the direction $(h_0 : h_1 : h_2)$ in a way that for the self-motion of the pentapod $e_0 = f_0 = 0$ holds. We can discuss Type 1 and Type 2 at the same time, just having in mind that $a_4 \neq 0 \neq C$ has to hold for Type 1 and $a_4 = 0 = C$ for Type 2 (according to [1]).

By setting $\mathbf{r}_i := (r_{i1}, r_{i2}, r_{i3})^T$ for $i = 1, 2, 3$ the Darboux and Mannheim constraints with respect to $\mathscr{F}_0'$ and $\mathscr{F}'$ can be written as:

$$\Omega_2 : (s_1 + \mathbf{r}_1\mathbf{m}_2) - i(s_2 + \mathbf{r}_2\mathbf{m}_2) - p_2N = 0, \qquad \Omega_4 : (s_3 + \mathbf{r}_3\mathbf{m}_4) - p_4N = 0, \tag{4}$$

$$\Omega_3 : (s_1 + \mathbf{r}_1\mathbf{m}_3) + i(s_2 + \mathbf{r}_2\mathbf{m}_3) - p_3N = 0, \qquad \Pi_5 : (\mathbf{Rd})(\mathbf{s} + \mathbf{Rp}_5)N^{-1} = 0, \tag{5}$$

with $\mathbf{p}_5 = \mathbf{n} + (p_5 - a_r)\mathbf{d}$, which is the coordinate vector of the intersection point of the Mannheim plane and p with respect to $\mathscr{F}'$. Moreover $(p_j, 0, 0)^T$ for $j = 2, 3$ (resp. $(0, 0, p_4)^T$) are the coordinates of the intersection point of the Darboux plane and the x-axis (resp. z-axis) of $\mathscr{F}_0'$.

Remark 1 As from the Mannheim constraint Π_5 of Eq. (5) the factor N cancels out, all four constraints $\Omega_2, \Omega_3, \Omega_4, \Pi_5$ are homogeneous quadratic in the Study parameters and especially linear in $f_0, \ldots, f_3$. ◇

According to [1, Theorems 13 and 14] the leg-parameters $p_2, \ldots, p_5, R_1$ have to fulfill the following necessary and sufficient conditions for the self-mobility (over $\mathbb{C}$) of a linear pentapod of Type 1 and Type 2, respectively:

$$p_2 = \frac{Aa_3v}{(a_3-a_4)^2}, \qquad p_3 = \frac{Aa_2v}{(a_2-a_4)^2}, \qquad p_4 = -\frac{Ca_4v}{(a_2-a_4)(a_3-a_4)}, \tag{6}$$

$$(a_2 - a_4)^2(a_3 - a_4)^2\left[2wp_5 - vR_1^2 - (2w - va_4)a_4\right] + vw^2(A^2 + C^2) = 0, \tag{7}$$

with $v := a_2 + a_3 - 2a_4$ and $w := a_2a_3 - a_4^2$. Therefore if we set p_2, p_3, p_4 as given in Eq. (6) then only one condition in p_5 and R_1 remains in Eq. (7). Therefore these pentapods have a 1-dimensional set of self-motions.

Theorem 1 *Each self-motion of a linear pentapod of Type 1 and Type 2 can be generated by a 1-dimensional set of line-symmetric motions. For the special case* $p_5 = a_4 = a_r$ *this set is even 2-dimensional.*

Proof W.l.o.g. we can set $e_0 = 0$ as any two directions $\mathbf{d}$ of p can be transformed into each other by a half-turn about their enclosed bisecting line. Note that this line is not uniquely determined if and only if the two directions are antipodal.

W.l.o.g. we can solve $\Psi, \Omega_2, \Omega_3, \Omega_4$ for f_0, f_1, f_2, f_3 and plug the obtained expressions into Π_5, which yields in the numerator a homogeneous quartic polynomial $G[1563]$ in e_1, e_2, e_3, where the number in the brackets gives the number of terms. Moreover the numerator of the obtained expression for f_0 is denoted by $F[600]$, which is a homogeneous cubic polynomial in e_1, e_2, e_3.

General Case ($v \neq 0$)**:** The condition $G = 0$ already expresses the self-motion as G equals Λ_1 if we solve Eq. (7) for R_1. Moreover $F = 0$ has to hold if the self-motion of the line p can be generated by a line-symmetric motion. As for any solution $(e_1 : e_2 : e_3)$ of $F = 0$ also $G = 0$ has to hold, G has to split into F and a homogeneous linear factor L in e_1, e_2, e_3.

Now $L = 0$ cannot correspond to a self-motion of the linear pentapod, but has to arise from the ambiguity in representing a direction of p mentioned at the beginning of the proof. This can be argued indirectly as follows:

Assumed $L = 0$ implies a self-motion, then it has to be a Schönflies motion (with a certain direction v of the rotation axis) due to $e_0 = 0$. As under such a motion the angle enclosed by v and p remains constant[5] the ideal point U of p has to be mapped by σ to the ideal point V of v. This implies that V has to coincide with W, which can only be the case for pentapods of Type 5; a contradiction.

Therefore there has to exist a pose of p during the self-motion, where it is oppositely oriented with respect to the fixed frame and moving frame, respectively. As a consequence we can set $L = d_1e_1 + d_2e_2 + d_3e_3$ which yields the ansatz $\Delta : \lambda LF - G = 0$. The resulting set of four equations arising from the coefficients of $e_1^3e_2$, $e_1^3e_3$, $e_1e_3^3$ and $e_2e_3^3$ of Δ has the unique solution:

$$n_1 = a_c d_2, \quad n_2 = -a_c d_1, \quad n_3 = (a_r - a_4)d_3, \quad \lambda = 2(h_0^2 + h_1^2 + h_2^2). \tag{8}$$

Now Δ splits up into $(e_1^2 + e_2^2 + e_3^2)^2(h_0^2 + h_1^2 + h_2^2)H[177]$, where H is homogeneous of degree 4 in h_0, h_1, h_2. For more details on $H = 0$ please see Remark 3, which is given right after this proof.

Remark 2 Note that all self-motions of the general case can be parametrized as the resultant of G and the normalizing condition $N - 1$ with respect to e_i yields a polynomial, which is only quadratic in e_j for pairwise distinct $i, j \in \{1, 2\}$. ◇

Special Case ($v = 0$)**:** If $v = 0$ holds, we cannot solve Eq. (7) for R_1. The conditions $v = 0$ and Eq. (7) imply $p_5 = a_4 = a_r$. Now G is fulfilled identically and the self-motion is given by $\Lambda_1 = 0$, which is of degree 4 in e_1, e_2, e_3. Moreover for this special case $F = 0$ already holds for $\mathbf{n}$ given in Eq. (8). Therefore any direction $(h_0 : h_1 : h_2)$ for p can be chosen in order to fix the line-symmetric motion. □

Remark 3 $H = 0$ represents a planar quartic curve, which can be verified to be entirely circular. Moreover $H = 0$ can be solved linearly for p_5. The corresponding graph is illustrated in Fig. 1.

If we reparametrize the $h_0h_1h_2$-plane in terms of homogenized polar coordinates by:

[5]This angle condition can be seen as the limit of the sphere condition (cf. [12, Sect. 4.1]).

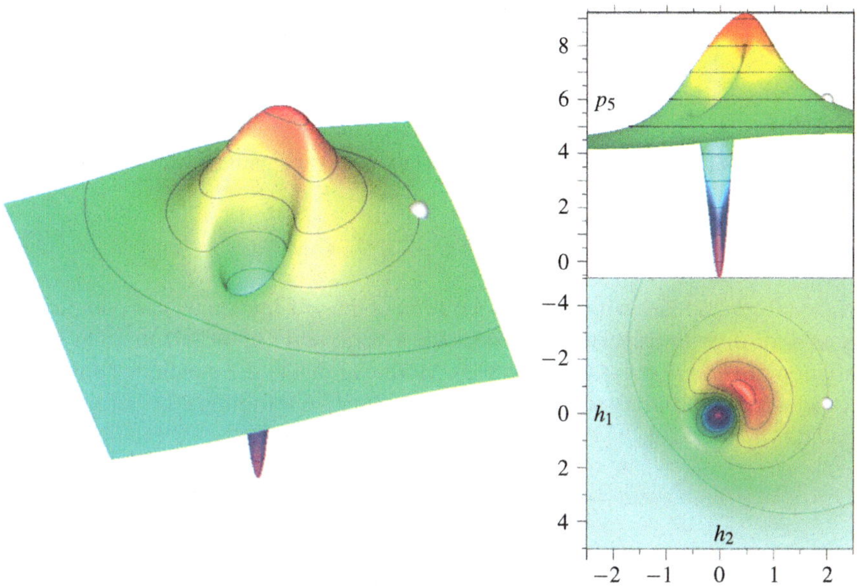

Fig. 1 For a type 1 pentapod with self-motion given by the parameters $a_4 = 2$, $A = -1$, $C = -5$, $a_r = 7$ and $a_c = 4$, the graph of p_5 in dependency of h_1 and h_2 with $h_0 = 1$ is displayed in the axonometric view on the *left* and in the front resp. *top* view on the *right* side. The highlighted point at height 6 corresponds to the values $h_1 = -\frac{489262}{226525} + \frac{488}{226525}\sqrt{675091}$ and $h_2 = \frac{535336}{226525} + \frac{446}{226525}\sqrt{675091}$

$$h_0 = (\tau_1^2 + \tau_0^2)\rho_0, \qquad h_1 = (\tau_1^2 - \tau_0^2)\rho_1, \qquad h_2 = 2\tau_0\tau_1\rho_1, \tag{9}$$

where $(\tau_0, \tau_1) \neq (0, 0) \neq (\rho_0, \rho_1)$ and $\tau_0, \tau_1, \rho_0, \rho_1 \in \mathbb{R}$ hold, then H factors into $(\tau_0^2 + \tau_1^2)^3(H_2\tau_1^2 + H_1\tau_0\tau_1 + H_0\tau_0^2)$ with

$$\begin{aligned}
H_1 &= 8\rho_0\rho_1 A(a_4 - a_r)(\rho_1^2 + \rho_0^2)(a_r^2 - a_4^2 + a_c^2)a_c, \\
H_0 - H_2 &= 8\rho_0\rho_1 A(a_4 - a_r)(\rho_1^2 + \rho_0^2)[a_r(a_r - a_4)^2 + a_c^2(a_r - 2a_4)], \\
H_0 + H_2 &= 2\left[(a_r - a_4)^2 + a_c^2\right][2a_4(\rho_1^4 - \rho_0^4)(a_4 - a_r)C \\
&\quad + \left((a_r - a_4)^2 + a_c^2\right)\left((\rho_0^4 + \rho_1^4)(a_4 - p_5) + 2\rho_0^2\rho_1^2(2a_r - a_4 - p_5)\right)].
\end{aligned} \tag{10}$$

Therefore this equation can be solved quadratically for the homogeneous parameter $\tau_0 : \tau_1$. Note that the value p_5 is fixed during a self-motion. ◇

3 On the Reality of Type 1 and Type 2 Self-motions

A similar computation to [1, Example 1] shows that for any real point $\mathsf{p}_t \in \mathsf{p}$ with $t \in \mathbb{R}$ and coordinate vector $\mathbf{p}_t = \mathbf{n} + (t - a_r)\mathbf{d}$ with respect to $\mathscr{F}'$ the corresponding real point $\mathsf{P}_t \in \mathsf{P}$ has the following coordinate vector $\mathbf{P}_t$ with respect to $\mathscr{F}_0'$:

$$\mathbf{P}_t = \left(\frac{A(a_r^2+a_c^2-ta_r)}{(t-a_r)^2+a_c^2}, -\frac{Aa_ct}{(t-a_r)^2+a_c^2}, \frac{Ca_4}{a_4-t} \right)^T. \quad (11)$$

As $L = 0$ corresponds with one configuration of the self-motion we can compute the locus $\mathscr{E}_t$ of p_t with respect to $\mathscr{F}_0'$ under the 1-parametric set of self-motions by the variation of $(h_0 : h_1 : h_2)$ within $L = 0$. Moreover due to the mentioned ambiguity we can select an arbitrary solution $(e_0 : e_1 : e_2)$ for $L = 0$ fulfilling the normalization condition $N = 1$; e.g.:

$$e_1 = \frac{h_2}{\sqrt{h_1^2+h_2^2}}, \quad e_2 = -\frac{h_1}{\sqrt{h_1^2+h_2^2}} \quad \text{and} \quad e_3 = 0. \quad (12)$$

Now the computation of $\mathbf{R}\mathbf{p}_t + \mathbf{s}$ yields a rational quadratic parametrization of $\mathscr{E}_t$ in dependency of $(h_0 : h_1 : h_2)$.

Note that this approach also includes the special case $(v = 0)$ as there always exists a value for R_1^2 (in dependency of $(h_0 : h_1 : h_2)$) in a way that $\Lambda_1 = 0$ holds.

For $t \neq a_4$ all $\mathscr{E}_t$ are ellipsoids of rotation (see Fig. 2a), which have the same center point C and axis of rotation c. In detail, C is the point of the straight cubic circle (11) for the value $t = c$ with $c := \frac{a_4^2-a_c^2-a_r^2}{2(a_4-a_r)}$ (for $a_4 = a_r$ we get $c = \infty$ thus $\mathsf{p}_\infty = \mathsf{U} = \mathsf{m}_5$ holds, which implies $\mathsf{C} = \mathsf{M}_5$) and c is parallel to the z-axis of $\mathscr{F}_0'$. Moreover the vertices on c have distance $|a_4 - t|$ from C and the squared radius of the equator circle equals $(a_r - t)^2 + a_c^2$. Note that for $a_4 \neq a_r$ the only sphere within the described set of ellipsoids is $\mathscr{E}_c$. For $a_4 = a_r$ no such sphere exists.

$\mathscr{E}_{a_4}$ is a circular disc in the Darboux plane $z = p_4$ (w.r.t. $\mathscr{F}_0'$) centered in C.

Remark 4 The existence of these ellipsoids was already known to Duporcq [7, Sect. 9], who used them to show that the spherical trajectories are algebraic curves of degree 4 (intersection curve of $\mathscr{E}_t$ and the sphere Φ_t centered in P_t illustrated in Fig. 2b). ◇

Based on this geometric property, recovered by line-symmetric motions, we can formulate the condition for the self-motion to be real as follows:

- $w \neq 0$: We can reduce the problem to a planar one by intersecting the plane spanned by $\mathsf{P}_0 = \mathsf{M}_1$ and c with $\mathscr{E}_0$ and the sphere with radius R_1 centered in P_0. Now there exists an interval $I_0 =]I_-, I_+[$ such that for $R_1 \in I_0$ the two resulting conics have at least two distinct real intersection points. It is well known (e.g. [14]) that the computation of the limits I_- and I_+ of the reality interval I_0 leads across an algebraic problem of degree 4 (explicitly solvable). Thus for a real self-motion we have to choose $R_1 \in I_0$ and solve Eq. (7) for p_5.

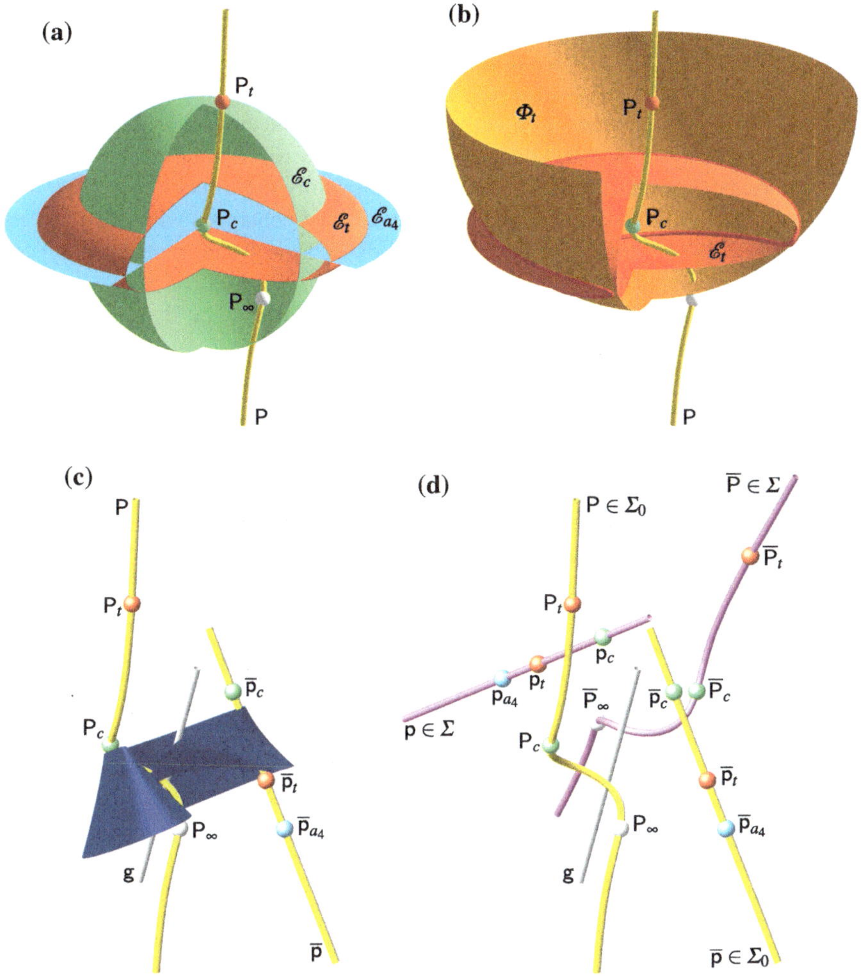

Fig. 2 Type 1 pentapod with self-motion given by $a_4 = 2$, $A = -1$, $C = -5$, $a_r = 7$ and $a_c = 4$. **a** The loci $\mathscr{E}_{a4}$, $\mathscr{E}_c$ and $\mathscr{E}_t$ with $t = \frac{69}{20}$ are sliced (along the not drawn axis of rotation c) in order to visualize their positioning with respect to the cubic P on which the points $\mathsf{P}_\infty = \sigma(\mathsf{U})$, $\mathsf{P}_c = \mathsf{C}$ and P_t are highlighted. Note that $\mathsf{P}_{a4} = \mathsf{W}$ is the real ideal point of P. **b** By setting $p_5 = 6$ a one-parametric self-motion μ is fixed. The trajectory of p_t under μ is illustrated as the intersection curve of $\mathscr{E}_t$ and the sphere Φ_t centered in P_t. **c** A strip of the basic surface of μ is illustrated for the value highlighted in Fig. 1. In addition P and $\overline{\mathsf{p}}$ are visualized, where the latter denotes the pose of p such that its half-turns about the generators of the basic surface yield the self-motion μ. **d** Krames's construction is illustrated with respect to the generator g of the basic surface: As P_{a4} (resp. $\overline{\mathsf{p}}_\infty$) is the real ideal point of P (resp. $\overline{\mathsf{p}}$), the trajectory of p_{a4} (resp. $\overline{\mathsf{P}}_\infty$) under μ is planar. The (Mannheim) plane $\in \Sigma$, which contains the point P_∞ (resp. $\overline{\mathsf{p}}_{a4}$) and is orthogonal to the direction of the real ideal point p_∞ (resp. $\overline{\mathsf{P}}_{a4}$) of p (resp. $\overline{\mathsf{P}}$) in the displayed pose, slides through the point P_∞ (resp. $\overline{\mathsf{p}}_{a4}$) during the complete motion μ

- $w = 0$: Now P_0 coincides with C and the interval collapses to the single value $R_1 = |a_4|$, which can be seen from Eq. (7). Moreover p_5 can be chosen arbitrarily.

These considerations also show that any pentapod of Type 1 and 2 has real self-motions if the leg-parameters are chosen properly. Note that this is e.g. not the case for some designs of Type 5 pentapods described in [1, Sect. 6], where it was also proven that pentapods with self-motions have a quartically solvable direct kinematics. It is possible to use this advantage (closed form solution) of pentapods with self-motions without any risk,[6] by designing linear pentapods of Type 1 and Type 2, which are guaranteed free of self-motions within their workspace.

A sufficient condition for that is that (at least) for one of the five legs $\mathsf{p}_t\mathsf{P}_t$ of the pentapod the corresponding reality interval I_t is disjoint with the interval of the maximal and minimal leg length implied by the mechanical realization. This condition for a self-motion free workspace gets especially simple if $\mathsf{p}_c\mathsf{P}_c$ is this leg.

Remark 5 Due to limitation of pages, we refer for detailed examples to the paper's corresponding arXiv version [13], which also show that for the general case ($v \neq 0$) the basic surface is of degree 5 (see Fig. 2c) and that a general point has a trajectory of degree 6 under the corresponding line-symmetric motion.[7] Note that the latter also holds for a general point of the cubic $\overline{\mathsf{P}}$ explained in the next section. ◇

4 Conclusion and Open Problem

Krames [4, p. 416] outlined the following construction (see Fig. 2d): Assume that p is in an arbitrary pose of the self-motion μ with respect to P, where g denotes the generator of the basic surface, which corresponds to this pose. Moreover $\overline{\mathsf{p}}$ and $\overline{\mathsf{P}}$ are obtained by the reflexion of p and P, respectively, with respect to g, where $\overline{\mathsf{p}}$ belongs to the fixed system Σ_0 and $\overline{\mathsf{P}}$ to the moving system Σ. Then under the self-motion μ also the points of $\overline{\mathsf{P}}$ are located on spheres with centers on the line $\overline{\mathsf{p}}$.

We can apply this construction for each line-symmetric motion of Theorem 1, which yields new solutions for the Borel Bricard problem, with the exception of one special case where $\mathsf{W} \in \overline{\mathsf{p}}$ holds (i.e. $h_1 = h_2 = 0$ or $h_0 = 0$), which was already given by Borel in [2, Case Fa4]. Moreover for this case Borel noted that beside p and $\overline{\mathsf{P}}$ only two imaginary planar cubic curves ($\in$ isotropic planes through p) run on spheres. The example of [13] shows that this also holds true for the general case.

Thus the problem remains to determine all line-symmetric motions of Theorem 1 where additional real points (beside those of p and $\overline{\mathsf{P}}$) run on spheres. Until now the only known examples with this property are the BB-II motions (cf. Sect. 1.3).

Acknowledgements This research is funded by Grant No. P 24927-N25 of the Austrian Science Fund FWF within the project "Stewart Gough platforms with self-motions".

---

[6] A self-motion is dangerous as it is uncontrollable and thus a hazard to man and machine.

[7] Note that all basic surfaces and trajectories can be parametrized due to Remark 2.

References

1. Nawratil, G., Schicho, J.: Self-motions of pentapods with linear platform. In: Robotica (in press). doi:10.1017/S0263574715000843. arXiv:1407.6126
2. Borel, E.: Mémoire sur les déplacements à trajectoires sphériques. Mémoire présenteés par divers savants à l'Académie des Sciences de l'Institut National de France **33**(1), 1–128 (1908)
3. Bricard, R.: Mémoire sur les déplacements à trajectoires sphériques. Journal de École Polytechnique(2) **11**, 1–96 (1906)
4. Krames, J.: Zur Bricardschen Bewegung, deren sämtliche Bahnkurven auf Kugeln liegen. Monatsheft für Mathematik und Physik **45**, 407–417 (1937)
5. Koenigs, G.: Leçons de Cinématique (avec notes par G. Darboux). Paris (1897)
6. Mannheim, A.: Principes et Développements de Géométrie Cinématique. Paris (1894)
7. Duporcq, E.: Sur le déplacement le plus général d'une droite dont tous les points décrivent des trajectoires sphériques. Journal de mathématiques pures et appliquées **5**(4), 121–136 (1898)
8. Bottema, O., Roth, B.: Theoretical Kinematics. North-Holland Publishing Company, Amsterdam (1979)
9. Hartmann, D.: Singular stewart-gough platforms. Master Thesis, Department of Mechanical Engineering, McGill University, Montreal, Canada (1995)
10. Krames, J.: Die Borel-Bricard-Bewegung mit punktweise gekoppelten orthogonalen Hyperboloiden. Monatsheft für Mathematik und Physik **46**, 172–195 (1937)
11. Gallet, M., Nawratil, G., Schicho, J., Selig, J.M.: Mobile Icosapods. Advances in Applied Mathematics (in press). doi:10.1016/j.aam.2016.12.002. arXiv:1603.07304
12. Nawratil, G.: Correcting Duporcq's theorem. Mech. Mach. Theory **73**, 282–295 (2014)
13. Nawratil, G.: On the Line-symmetry of Self-motions of Linear Pentapods. [arXiv:1510.03567]
14. Chernova, N., Wijewickrema, S.: Algorithms for projecting points onto conics. J. Comput. Appl. Math. **251**, 8–21 (2013)

On Some Notable Singularities of 3-$R\underline{P}R$ and 3-$\underline{R}RR$ PPRMs

Khaled Assad Arrouk, Belhassen Chedli Bouzgarrou and Grigore Gogu

Abstract This paper highlights the existence of some notable type-II singular configurations for certain planar parallel robotic manipulators (PPRMs). These singularities are characterized by intrinsic geometric conditions of alignment or coincidence between geometric entities of the fixed base, the mobile platform and the limbs. Thus, in the general case of the 3-$R\underline{P}R$ manipulator, a set of 6 such singular configurations can be identified for each orientation of the mobile platform. Moreover, for 6 particular orientations of the mobile platform, a set of positions of the end-effector, defined by two concurrent lines, can be identified as notable degenerated singularity curves. On another side, in the general case of the 3-$\underline{R}RR$ manipulator, a set of 24 curves in the 3-dimensional operational space (x, y, β) can be identified as singular poses. All these singularities are easy to determine by means of simple geometric graphical constructions. In this paper, we try to exploit the existence of such particular singularities for kinematic analysis and design of PPRMs. For instance, we can construct the singularity surface of the 3-$R\underline{P}R$ manipulator by using a pure graphical approach and without any need of algebraic or analytic formulations.

1 Introduction

Despite the apparently simple architecture of parallel planar robotic manipulators (PPRMs), the characterization of their singularities remains a challenging problem in robotic research field. In [5], the singularities of PRMs have been classified according to three categories: the serial or type-I singularities, parallel or type-II singularities, and the combined or type-III singularities (combination of types I and II). Constraint

---

K.A. Arrouk (✉) · B.C. Bouzgarrou · G. Gogu
Pascal Institute, UMR6602 UBP/CNRS/SIGMA-Clermont, BP10448,
F63000 Clermont-Ferrand, France
e-mail: khaled.arrouk@sigma-clermont.fr

B.C. Bouzgarrou
e-mail: belhassen-chedli.bouzgarrou@sigma-clermont.fr

G. Gogu
e-mail: grigore.gogu@sigma-clermont.fr

J. Lenarčič and J.-P. Merlet (eds.), *Advances in Robot Kinematics 2016*,
Springer Proceedings in Advanced Robotics 4, DOI 10.1007/978-3-319-56802-7_17

singularities have been introduced the first time by Zlatanov as phenomenon occurring in parallel mechanisms with reduced freedoms when the screw system, formed by the constraint wrenches in all legs, loses rank [9]. More recently, parallel mechanism singularities have been characterized in terms of structural parameters [1, 4]. Three types of singularities have been defined: constraint singularities, redundant singularities and constraint-redundant singularities [1]. Parallel singularities depend on the choice of the passive and active joints in the mechanism. In a parallel singularity, the actuators cannot control the motions of the mobile platform MPF.

The kinematic mapping method has been used to discuss all singularities of the 3-$R\underline{P}R$ manipulator in a uniform way, showing that all singularities are on a degree four surface in the kinematic image space [6]. By fixing one of the actuated joints and analyzing the configuration-space as a surface in a three-dimensional space, Zein et al. have determined all possible non-singular assembly-mode changing by using loop trajectories encircling a cusp points of the 3-$R\underline{P}R$ manipulator [8]. The present paper falls within the context of previous researches in which we have presented a new unified CAD-based graphical approach, quite apart from the existing methods in the literature, for determining and representing the main kinematic properties of PPRMs such as the 3D total workspace, the parallel singularity surface, and the assembly modes associated with the direct kinematic problem solutions [2]. It has been shown that the graphical superimposing of the 3D total workspace, the singular surface and the direct kinematic problem solutions enables a rapid and accurate identification of all singularity-free regions (aspects), as 3D solids, and generating singularity-free trajectories between different assembly modes [3].

In this work, we highlight the existence of particular parallel singular configurations of certain PPRMs, characterized by intrinsic geometric conditions of alignment or coincidence between geometric entities of the fixed base, the mobile platform and the limbs. Therefore, they are easy to determine by means of geometric graphical constructions. These particular singularities are mainly useful in kinematic analysis and design of PPRMS. In the case of the 3-$R\underline{P}R$ manipulator, they are useful for graphically determining the singularity curves associates with all singular configurations for a given orientation angle of the MPF. For these reasons, we consider these particular singular configurations as remarkable or notable singularities.

In this work, we highlight the existence of particular parallel singular configurations of certain PPRMs, characterized by intrinsic geometric conditions of alignment or coincidence between geometric entities of the fixed base, the mobile platform and the limbs. Therefore, they are easy to determine by mean of geometric graphical constructions. That's why they can be considered as remarkable singularities. In this paper, we try to emphasize the use of such particular poses for kinematic analysis and design of PPRMs.

The paper is organized as follows. In Sect. 2, notable singularities of 3-$R\underline{P}R$ and 3-$\underline{R}RR$ manipulators are introduced. The geometric graphical operations used for their determination are presented. As an immediate exploitation of such a result, a new geometric-based graphical construction of singularity surface of 3-$R\underline{P}R$ manipulator is illustrated in Sect. 3.

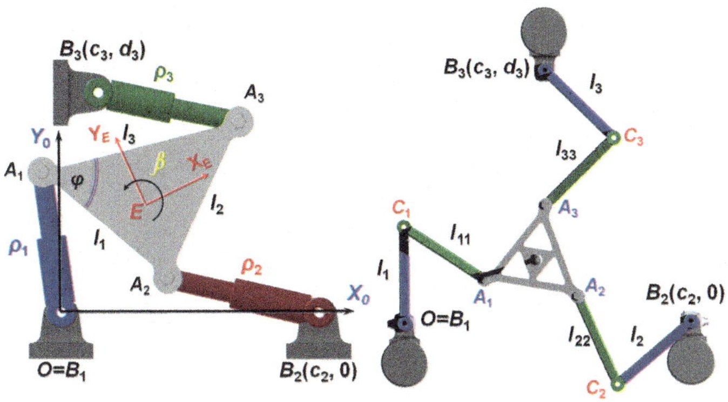

Fig. 1 Parameterizing of 3-$R\underline{P}R$ and 3-$\underline{R}RR$ manipulators

2 Notable Singularities of 3-$R\underline{P}R$ and 3-$\underline{R}RR$ PPRMs

In this section we consider two types of PPRMs: 3-$R\underline{P}R$ and 3-$\underline{R}RR$ manipulators (Fig. 1). Even though they have been extensively studied in the literature, some notable singular configurations of these manipulators deserve to be considered. These singularities are characterized by intrinsic geometric conditions of alignment or coincidence between geometric entities of the fixed base, the mobile platform and the limbs. For the 3-$R\underline{P}R$ manipulator, these singularities are formed by a finite set of points for each fixed orientation of the MPF. Whereas, a set of curves in the operational space (x, y, β) are identified as notable singular poses for the 3-$\underline{R}RR$ manipulator.

2.1 *Notable Singular Points of the 3-$R\underline{P}R$ Manipulator*

The 3-$R\underline{P}R$ manipulator is composed of three identical kinematic limbs connecting the fixed base (FB) $B_1B_2B_3$ to the mobile platform $A_1A_2A_3$. Each limb has two passive revolute joints (on the extremities of each limb), and one actuated prismatic joint relating the first and the second links. The directed distance between B_i and A_i along the direction of the prismatic joint direction is ρ_i $(i = 1, 2, 3)$. A parametrization of the mechanism is given in Fig. 1. It is well known that the 3-$R\underline{P}R$ manipulator is in a parallel singular configuration when the lines passing through its passive revolute joint centers in each limb intersect in one point or are parallel [7].

For a given orientation of the MPF, the first subset of notable singularities corresponds to the configurations for which two passive revolute joints of a limb i are superposed: $A_i = B_i$. These configurations verify the condition $\rho_i = 0$ for a given

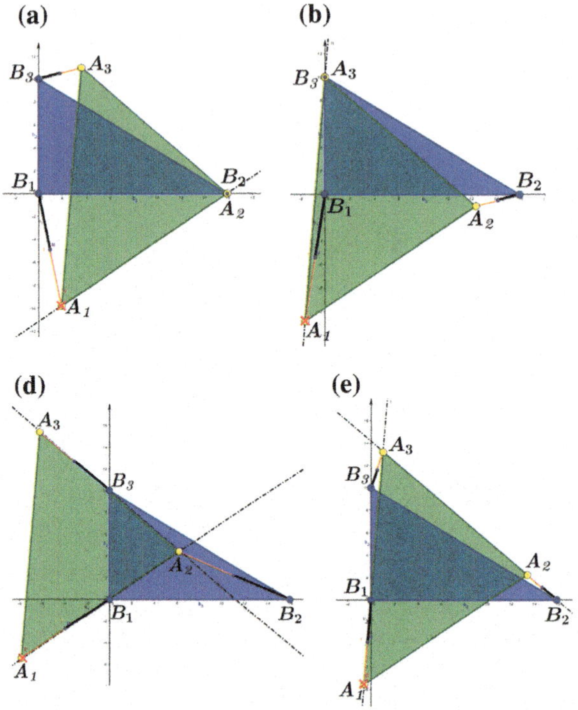

Fig. 2 Notable singular points of the 3-$R\underline{P}R$ manipulator

limb i ($i \in \{1, 2, 3\}$). Therefore, this subset is formed by 3 points which are the positions in the (x, y) plane of the end-effector characteristic point (EECP) associated with each of these configurations. They are shown in Fig. 2a, b and c, where A_1 is considered as the EECP.

Likewise, for a given orientation of the MPF, the second subset of notable singularities corresponds to the configurations for which two limbs, B_iA_i and B_jA_j, are aligned, respectively, with two edges of the mobile platform, A_iA_k and A_jA_k ($i, j, k \in \{1, 2, 3\}$). In this case, the three limbs are concurrent in A_k, which correspond to a type-II singular configurations. These 3 configurations can be obtained by superposing the vertex A_k of the mobile platform with the intersection point of two lines: the line parallel to the edge A_iA_k and passing through B_i, and the line parallel to the edge A_jA_k and passing through B_j. This second subset is also formed by 3 points which are the positions in the (x, y) plane of the EECP associated with each of these configurations. They are shown in Fig. 2d, e and f, where A_1 is considered as the EECP.

2.2 Notable Degenerate Singularity Curves of the 3-$R\underline{P}R$ Manipulator

In addition to the set of six singular points that can be determined for a given orientation of the MPF, notable singular lines can be identified for particular orientations of the MPF. Indeed, if the orientation of the MPF is such that its edge $A_i A_j$ is aligned with the edge $B_i B_j$ of the FB, then the two links $B_i A_i$ and $B_j A_j$ are also aligned with the edge $B_i B_j$. In this case all the links of the manipulator are concurrent in one point belonging to the line $B_i B_j$, which corresponds to a type-II singular configuration. The set of positions described by the EECP verifying this condition is a line parallel to the edge $B_i B_j$ and passing by the EECP when the edge $A_i A_j$ of the MPF is placed wherever on the line $B_i B_j$. This set of positions is the first notable singular line. It is illustrated in Fig. 3 where $i = 2$, $j = 3$, $k = 1$ and A_1 is the EECP.

Since a singularity curve, for a given orientation of the 3-$R\underline{P}R$ manipulator, is a conic in the general case, the existence of a singularity line must correspond to a degeneration of the conic. Thus, a second line necessarily exists. Even if this line is less evident then the first one, it is easy to be determined by using the singular points we have introduced in Sect. 2.1. Then, it is sufficient to use two notable singular points, which are not on the first singularity line, to define the second notable singularity line. As shown in Fig. 3, the second line passes through the singular points B_1 and A_1 when the limbs B_2A_2 and B_3A_3 are respectively aligned with the MPF edges A_2A_1 and A_3A_1.

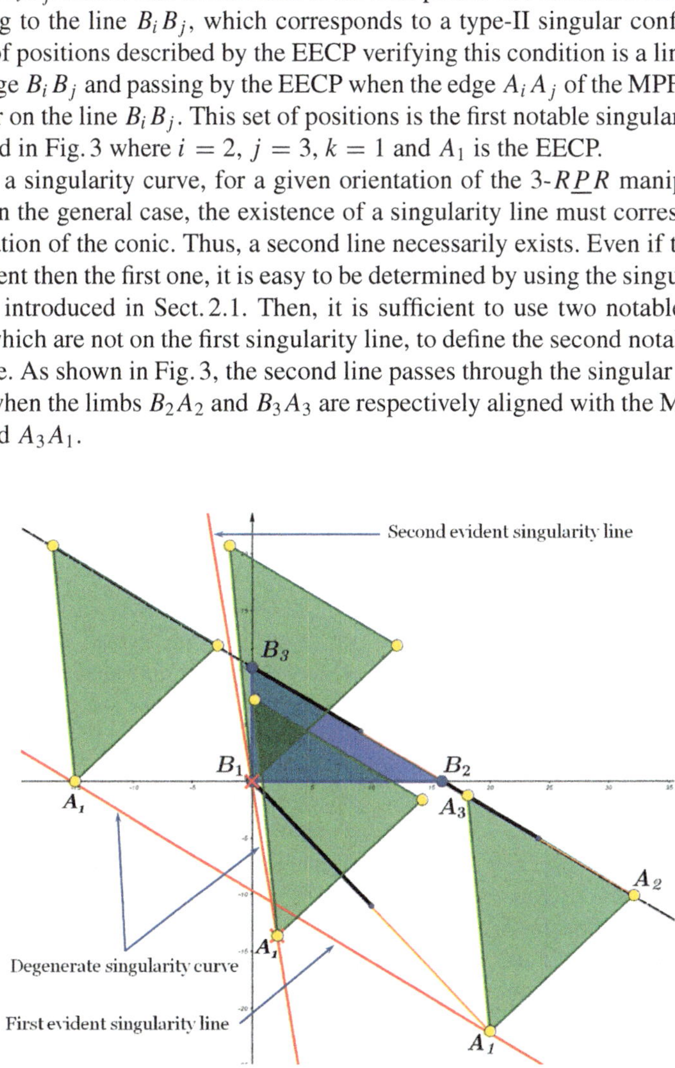

Fig. 3 Degenerate singularity curve of the 3-$R\underline{P}R$ manipulator

There are two possible orientations, at 180 deg of rotation from each other, to have aligned the edges $B_i B_j$ and $A_i A_j$ with $i, j \in 1, 2, 3$. Therefore, it can be stated, in the general case, that the singularity curves associated with different MPF orientations of the 3-$R\underline{P}R$ meet at least 6 cases of degeneration from an hyperbola into two concurrent lines.

2.3 Notable Singularities of the 3-$\underline{R}RR$ Manipulator

The 3-$\underline{R}RR$ manipulator is composed of three limbs connecting the FB $B_1B_2B_3$ to the MPF $A_1A_2A_3$. Each limb is composed by two links and three revolute joints. The actuated revolute joint relates the first link to the FB. The first link of a limb i $(i = 1, 2, 3)$ is defined by the segment B_iC_i and the distal link by the segment C_iA_i. A parametrization of the mechanism is given in Fig. 1.

In this paragraph, we show that we can also identify a set of notable type-II singular configurations of the 3-$\underline{R}RR$ manipulator that we can graphically determine. Indeed, this mechanism is in a parallel singularity when its distal links are concurrent in one single point. Among these configurations, we can identify a first category of notable singularities for which two distal links are aligned with two distinct edges of the MPF. If these links are C_iA_i and C_jA_j, attached to the vertices A_i and A_j of the MPF, the considered edges are respectively A_iA_k and A_jA_k. In this case, these links are necessary concurrent in the triangle vertex A_k, which belongs to the remaining third link C_kA_k.Therefore, all distal links are concurrent in the point A_k, which corresponds to a parallel singularity of the manipulator. Figure 4a, b and c illustrate these singularities through three possibilities associated with three kinematic loops of the 3-$\underline{R}RR$ manipulator.

The second category of notable singularities of the 3-$\underline{R}RR$ manipulator corresponds to the configurations for which two distal links are aligned with an edge of the mobile platform. If these links are C_iA_i and C_jA_j, the considered edge is A_iA_j. In this case, these two links are aligned and concurrent in the same point with the remaining distal link C_kA_k. In this case, we also obtain a type-II singular configuration. Figure 4d, e and f illustrate these singularities through three possibilities associated with three kinematic loops of the 3-$\underline{R}RR$ manipulator.

A geometric graphical approach can be used in order to determine the set of configurations associated with these two categories of singularities. For the first category, the set of points described by the EECP when the singularity condition is maintained as kinematic constraints, i.e. maintaining aligned the links C_iA_i and C_jA_j respectively with the MPF edges A_iA_k and A_jA_k, is a coupler curve in the (x, y) plane. This coupler curve is obtained by considering the four-bar mechanism $B_iC_iC_jB_j$, formed by links B_iC_i and B_jC_j and having as a coupler part the triangle $C_iC_jA_k$. The considered EECP for the four-bar mechanism is the same as in the 3-$\underline{R}RR$ manipulator.

For the second category of these singularities, the set of points described by the EECP when maintaining aligned the links C_iA_i and C_jA_j the MPF edge A_iA_j is

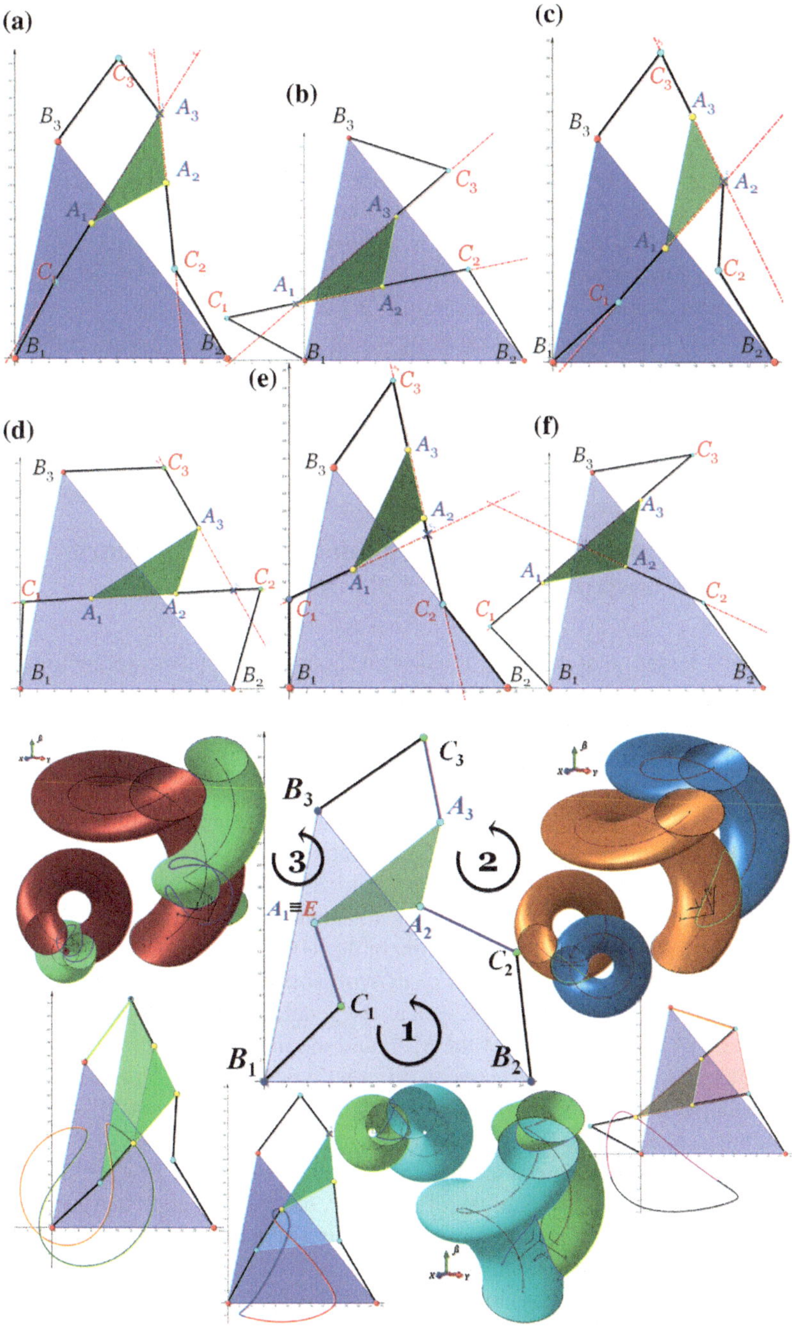

Fig. 4 Notable singular configurations of the 3-$\underline{R}RR$ manipulator and their representation in the operational space

also a coupler curve in the (x, y) plane. This curve, called singular coupler curve, is obtained by considering the four-bar mechanism $B_iC_iC_jB_j$, formed by links B_iC_i and B_jC_j and as a coupler part the triangle $C_iC_jA_k$ with the same EECP as in the 3-$\underline{R}RR$ manipulator. Figure 4d, e and f illustrate these singularities.

In previous works [3], we have shown that the domain reached by the MPF belonging to a 2 dof limb isolated from the rest of the mechanism can be represented in the 3-dimensional operational space (x, y, β) by a helical surface. The domain reached by the coupler part of a four-bar mechanism in the 3-dimensional operational space results from the intersections of the two domains associated with its two limbs. Thereafter, this domain is the intersection of two helical surfaces, which gives a 3-dimensional curve. The projection of this curve on the (x, y) plane is nothing but a singular coupler curve. These results are illustrated in Fig. 4 for the first category of 3-$\underline{R}RR$ notable singularities we have defined. The determination of the second category of 3-$\underline{R}RR$ notable singularities can be performed similarly.

3 Exploitation of the Notable Singularities for Singularity Surface Construction of the 3-$R\underline{P}R$ Manipulator

For a given orientation of the MPF, it has been shown in [7] that the geometrical loci of the EECP in the operational workspace, associated with parallel singular configurations of the 3-$R\underline{P}R$ manipulator, form a quadratic conic curve (i.e., a hyperbola, a parabola or an ellipse), unless there is an architectural singularity. This latter can be easily predicted at the earlier stage of robot design [7]. In this section, we propose to exploit this property as well as the existence of the notable singularities. Indeed, one can perfectly determine the conic curve associated with all singular configurations for a given orientation angle of the MPF by using only 5 points among the 6 singular points introduced in Sect. 2.1. Figure 5a illustrates the graphical determination of the conics associated with different orientation of the MPF. This determination is purely graphic and has been implemented in *GeoGebra* software. The construction procedure of the notable singular points as well as the generation of the associated conic is parametrized by the orientation of the MPF and the mechanism design parameters. It can be noticed the transitions between different types of conics: from an ellipse to a hyperbola passing by a parabola. All these graphical constructions didn't require any analytical or algebraic formulation by the user.

The automatic generation of all the singularity curves, in the (x, y) plane, associated with the different orientations of the MPF, can also be exploited to generate the singularity surface of the manipulator in the 3-dimensional operational space (x, y, β). Singularity curves can be generated in a 3D CAD environment, based on geometric determination of notable singular points, for different orientations of the MPF. Since then, this can be performed in purely graphic approach by generating these conics in a 3D CAD environment, at different altitudes corresponding the orientation of MPF with enough small step. Thereafter, these curves can be fitted by

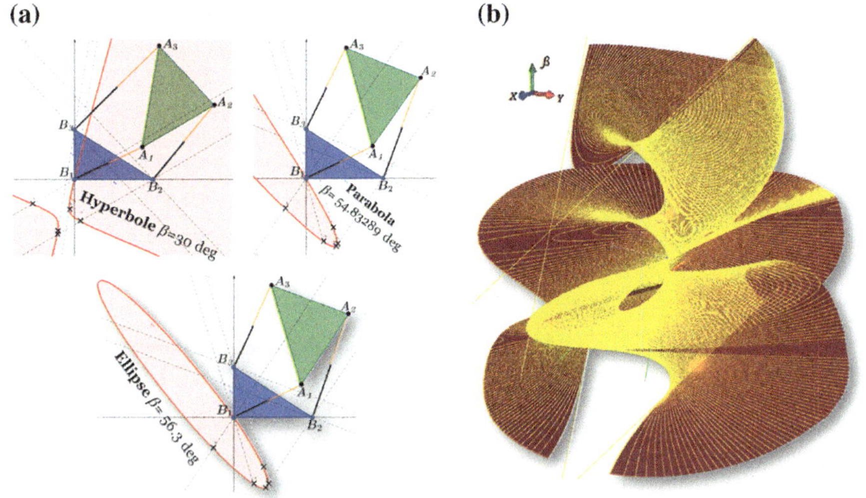

Fig. 5 Graphical determination, based on notable singular points, of singularity curves **a** and surface **b** of the 3-$R\underline{P}R$ manipulator

a surface, such as NURBS in *CATIA®* software, with quite a good precision. This surface represents the singularity loci of the 3-$R\underline{P}R$ manipulator in the operational space as illustrated in Fig. 5b.

4 Conclusion

In this paper notable type-II singular configurations of certain PPRMs have been highlighted. These singularities are easily determined by using geometric graphical constructions. It has been shown for the 3-$R\underline{P}R$ manipulator that the identification of 6 singular poses for each orientation of the MPF allows the construction of the singularity surface in the 3-dimensional operational space by using a pure geometric graphical approach and with no need of any analytical or algebraic formulations. This result has undoubtedly practical advantages in mechanism design and analysis. Moreover, 6 degenerated singularity curves, in the general case, can be determined immediately by simple geometric considerations. For the 3-$\underline{R}RR$ manipulator, the identified singular poses form a set of 24 curves in the operational space. Each curve is obtained by the intersection of two helical surfaces associated with the reachable domains of each limb of a four-bar mechanism. The projections of these singular curves on the (x, y) plane are coupler curves. Further exploitation of these notable singularities can be addressed in future developments for optimal mechanism synthesis. The geometric parameters of the 3-$R\underline{P}R$ manipulator, for a prescribed singularity curve corresponding to a given orientations of the MPF, can be determined so as the working zone of the manipulator can be kept far from this curve.

Acknowledgements This work has been sponsored by the French government research program "Investissements d'Avenir" through the IMobS3 Laboratory of Excellence (ANR-10-LABX-16-01).

References

1. Aimedee, F., Gogu, G., Dai, J.S., Bouzgarrou, C., Bouton, N.: Redundant singularities versus constraint singularities in parallel mechanisms. Proc. Inst. Mech. Eng. Part C J. Mech. Eng. Sci. (2015)
2. Arrouk, K., Bouzgarrou, B., Gogu, G.: Cad-based unified graphical methodology for solving the main problems related to geometric and kinematic analysis of planar parallel robotic manipulators. Robot. Comput. Integr. Manufact. **37**, 302–321 (2016)
3. Arrouk, K., Bouzgarrou, B.C., GOGU, G.: On the resolution of forward kinematic problem using cad graphical techniques: application on planar parallel robotic manipulators. In: Flores, P., Fernado Viadeiro, F. (eds.) Proceedings of the 5th European Conference on Mechanism Science, EUCOMES 2014. New Trends in Mechanism and Machine Science, vol. 24, pp. 211–219. Springer, Guimaraes (2014)
4. Gogu, G.: Structural Synthesis of Parallel Robots: Part 1: Methodology (2008)
5. Gosselin, C., Angeles, J.: Singularity analysis of closed-loop kinematic chains. IEEE Trans. Robot. Autom. **6**(3), 281–290 (1990)
6. Husty, M.L., Gosselin, C.: On the singularity surface of planar 3-rpr parallel mechanisms. Mech. Based Design Struct. Mach. **36**, 411–425 (2008)
7. Sefrioui, J., Gosselin, C.: On the quadratic nature of the singularity curves of planar three degree of freedom parallel manipulators. Mech. Mach. Theory **30**(4), 533–551 (1995)
8. Zein, M., Wenger, P., Chablat, D.: Non-singular assembly mode changing motions for 3rpr parallel manipulators. Mech. Mach. Theory **43**, 480–490 (2008)
9. Zlatanov, D., Bonev, I., Gosselin, C.: Constraint singularities of parallel mechanisms. In: Proceedings of the IEEE International Conference on Robotics and Automation. Washington, DC (2002)

Minimized-Torque-Oriented Design of Parallel Modular Mechanism for Humanoid Waist

Mouna Souissi, Vincent Hugel, Samir Garbaya and John Nassour

Abstract This article focuses on the design and integration of a parallel modular mechanism inside the waist of a human-sized biped robot to enable tilting motion of the torso. The mechanism for each tier is adapted from the parallel 2-degree-of-freedom tilting part of an existing 3-rotation flight simulator structure. The main contribution of this work is the design of a minimized-torque-oriented optimization process that takes into account the upper mass load to be supported by the mechanism, the constrained volume of the waist, a minimal dexterity threshold, and the tilting range required. The design process aims to determine the relative size and position of the different parts of the mechanism. The objective consists of minimizing the actuator average torque over the entire tilt range, and to evaluate how much torque reduction this parallel mechanism can bring compared with the use of a serial mechanism. Up to three modules can be stacked inside the waist to limit the actuator torques and to reach the required tilting range for sitting and bending movements.

1 Introduction

This work was developed in the framework of the ROMEO project that aims at building an innovative humanoid robot capable of assisting humans at home and equipped with an actuated waist. The mobility of the waist is useful to achieve human-like bending-down and sitting-down movements.

The trunk of most existing humanoid robots is equipped with one up to four serial degrees of freedom (DOF). Asimo [2], Hubo [8], or Lola [12] have a single yaw joint

M. Souissi
HEI, Lille, France

V. Hugel (✉)
Université de Toulon, Toulon, France
e-mail: vincent.hugel@univ-tln.fr

S. Garbaya
ENSAM, Paris, France

J. Nassour
Technische Universität Chemnitz, Chemnitz, Germany

J. Lenarčič and J.-P. Merlet (eds.), *Advances in Robot Kinematics 2016*,
Springer Proceedings in Advanced Robotics 4, DOI 10.1007/978-3-319-56802-7_18

in the trunk. Additional pitch and roll joints are used to execute specific whole-body movements or to design human-like locomotion gaits [18].

Bio-inspired prototypes named *musculoskeletal* humanoids [15, 19] were designed to imitate the human spine, using wire cables to drive replica of the vertebrae and inter-vertebral discs. Other bio-inspired prototypes without wire systems feature flexible properties thanks to the use of silent block as intervertebral disc [1] or the use of hydraulic pistons associated with springs [20].

Liang et al. [11] used rigid parallel mechanisms to model a waist-trunk system with a 6-DOF a Gough/Stewart platform and a 3-rotary-DOF mechanism. 3-DOF orientation manipulators were also designed as parallel wrists, e.g. wrists equipped with a central mast and three U$\underline{\text{P}}$S legs [17], spherical wrists with converging actuator axes [4], wrists with converging passive joint axes [24], or wrists including cylindrical joints [9].

Parallel rigid architectures have the advantages of higher rigidity and high load capacity [10] in comparison with serial-based architectures. The waist mechanism proposed in this paper is inspired by the parallel part of the Sabrié's flight simulator mechanism [21] which is composed of two asymmetric legs and a central mast with a total of eight passive joints. Together with the adaptation of this mechanism to the waist of a human-scale humanoid, this work presents a design optimization process that aims to minimize the actuator torques required over the entire tilting range. The process takes into account the mass load to be supported, the available volume, and a given dexterity threshold. This approach is different from traditional design methods that aim to minimize the condition number, maximize the dexterity [16], the global conditioning index [13], some combined index [23], the workspace [14], or aim to fit some prescribed workspace with accuracy requirements [5, 6].

Section 2 describes the parallel mechanism. Section 3 is devoted to the statics analysis, and Sect. 4 with the design optimization. Results are presented in Sect. 5 and discussed in Sect. 6.

2 Parallel Mechanism

The parallel module consists of a base platform CA_3B_3 and a payload moving platform OA_1B_1 linked by a fixed central rod CO and two legs arranged at 90[deg] in the initial position (Fig. 1). The central rod joins the top platform through a Universal joint. The leg $A_1A_2A_3$ is a planar UR$\underline{\text{R}}$ mechanism. The leg $B_1B_2B_3$ is a 3D US$\underline{\text{R}}$ mechanism. The revolute joint at A_3, resp. B_3, are to be actuated to drive the system and produce the pitch, resp. roll, motion. The kinematic analysis leads to an actual mobility of 2 for this mechanism.

The mechanism for the trunk DOF has to be located in the waist, which matches the biological observation in humans that the vertebrae located in the lumbar part are the most mobile ones [7]. It is required to support a total weight estimated at 15 [kg] that includes the masses of the trunk, the head and the arms. Preliminary studies showed that a forward tilting range of 30[deg] of the thorax is sufficient to execute

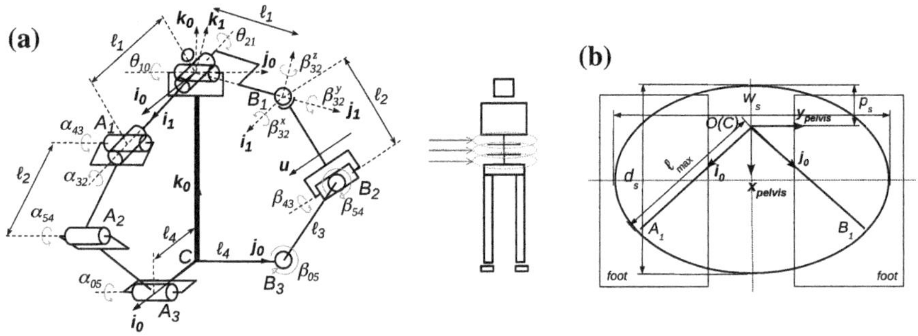

Fig. 1 A. Perspective view of one parallel mechanism module after pitch and roll rotations, resp. about fixed axis $\boldsymbol{j}_0$, and about axis $\boldsymbol{i}_1$. B. *Top* view of waist and footprints. Available volume: height $h_s = 0.30\,[\mathrm{m}]$, width, $w_s = 0.25\,[\mathrm{m}]$ and depth $d_s = 0.20\,[\mathrm{m}]$

flexion motions such as sitting down on a chair [22]. Given the constraint of total height and limitation of miniaturization, it is possible to stack three modules inside the ROMEO's waist, with a tilting range of up to 10[deg] each.

3 Statics Analysis

The objective is to express the torques exerted by the rotary actuators as a function of the geometric parameters of the mechanism. These torques are denoted by τ_a for the 2D leg and τ_b for the 3D leg. The masses of the parallel mechanism are neglected with respect to the mass M to be supported by the top platform. The mass M is assumed to be concentrated on a point G located at a height h_G above O along the normal $\boldsymbol{k}_1$ to the top platform.

The equations that govern the equilibrium of the 2D leg, the 3D leg, and the top platform lead to:

$$\begin{bmatrix}\tau_a\\ \tau_b\end{bmatrix} = \boldsymbol{J}^T \begin{bmatrix}\tau_a^g\\ \tau_b^g\end{bmatrix},$$

$$\begin{bmatrix}\dot{\theta}_{10}\\ \dot{\theta}_{21}\end{bmatrix} = \boldsymbol{J} \begin{bmatrix}\dot{\alpha}_{05}\\ \dot{\beta}_{05}\end{bmatrix},$$

$$\boldsymbol{J} = \begin{bmatrix} r_a & 0\\ -r_a r_c & r_b\end{bmatrix}$$

$$r_a = \frac{[\boldsymbol{A_3A_2}, \boldsymbol{u}_{A_2A_1}, \boldsymbol{j}_0]}{[\boldsymbol{OA_1}, \boldsymbol{u}_{A_2A_1}, \boldsymbol{j}_0]}$$

$$r_b = \frac{[\boldsymbol{B_3B_2}, \boldsymbol{u}_{B_2B_1}, \boldsymbol{i}_0]}{[\boldsymbol{OB_1}, \boldsymbol{u}_{B_2B_1}, \boldsymbol{i}_1]}$$

$$r_c = \frac{[\boldsymbol{j}_1, \boldsymbol{u}_{B_2B_1}, \boldsymbol{j}_0]}{\boldsymbol{k}_1.\boldsymbol{u}_{B_2B_1}}$$

$$\tau_a^g = -M\left[\boldsymbol{OG}, \boldsymbol{g}, \boldsymbol{j}_0\right]$$
$$\tau_b^g = -M\left[\boldsymbol{OG}, \boldsymbol{g}, \boldsymbol{i}_1\right]$$

where $\boldsymbol{J}$ is the Jacobian matrix of forward kinematics. τ_a^g and τ_b^g can be considered as the torques to be produced by the legs at O about axes $\boldsymbol{i}_1$ and $\boldsymbol{j}_0$ respectively, to compensate the moment due to the weight of mass M. These torques would be the active torques in the case of the serial mechanism constituted of the central rod only and where both rotary joints of the U-joint (θ_{10} and θ_{21}) at the top would be motorized.

4 Optimization Process

The aim of the optimization process consists of determining the adequate lengths of the segments that compose the kinematic chains, namely ℓ_1, ℓ_2, ℓ_3 and ℓ_4 (Fig. 1), in order to minimize the actuators' average torque over the entire tilting range. The maximal tilting angle of the torso is the same in all directions, which means that the robot can bend forward, backward and sideways, right and left, until reaching this maximal tilting angle, named θ^{max}. The four parameters to be optimized are normalized with respect to the height of the mechanism h_v, i.e. $(\ell_i^* = \ell_i / h_v)_{i\in\{1,2,3,4\}}$.

The mechanical structure is arranged to form a V-shape in top view (Fig. 1b), which means that legs are placed at ±45[deg] with respect to the longitudinal axis. This enables to benefit from a maximal lever arm inside the available volume. The central rod is placed at the rear of the lumbar part like in the human spine. The corresponding offset p_s from the external boundary of the lumbar region is set to $d_s/4$, but this value can be adjusted. The center of mass of the thorax, the head and the arms in the rest position must be at the vertical of the central rod. The maximal length for ℓ_1 is calculated taking into account the volume constraints ($\ell_{max} = 0.15\,[\mathrm{m}]$).

Active torques τ_a and τ_b depend on the tilting angles, but also on the lengths ℓ_i of the mechanism. The optimal values $(\overline{\ell_i^*})_{i\in\{1,2,3,4\}}$ are defined such that:

$$\Gamma(\overline{\ell_i^*}) = \min_{\ell_i^*} \Gamma(\ell_i^*)$$

where Γ is the cost function of the optimization process:

$$\Gamma(\ell_i^*) = -K \int_{\varphi=0}^{2\pi} \int_{\theta=0}^{\theta^{max}(\varphi)} (exp[-\tau_a^2(\varphi,\theta,\ell_i^*)] + exp[-\tau_b^2(\varphi,\theta,\ell_i^*)]) \sin\theta d\theta d\varphi$$

where K is a scaling factor, with $-2K < \Gamma(\ell_i^*) \leq 0$.

The cost function allows selecting the parameter values that involve reduced average torques over the entire tilting range of the torso. In the case of high torques over the tilting range, the cost function will tend to 0. When torques decrease over the tilting range the value of the cost function also decreases.

The following constraints are taken into account in the optimization process:

1. The upper bounds of the normalized parameters are defined taking into account the available volume in the lumbar part of the trunk. We set:

$$\ell^*_{min} \leq \ell^*_1, \ell^*_4 \leq \ell^*_{max},$$
$$\ell^*_{min} \leq \ell^*_2, \ell^*_3 \leq 1.5\ \ell^*_{max}$$

   with $\ell^*_{max} = \ell_{max}/h_v$ and $\ell^*_{min} = \ell^*_{max}/10$.

   The height of the lumbar part is limited to 0.30 [m]. Since we plan to incorporate 3 tiers for the mechanism, the height h_v of one module is set to 0.1 [m], and $\ell^*_{max} = 0.15/0.1 = 1.5$.
2. Three additional constraints for each leg are necessary to ensure the feasibility of the mechanism, i.e. the legs must connect the bottom platform to the top platform:

$$z^*_{A_3} < z^*_{A_1}$$
$$z^*_{B_3} < z^*_{B_1} \tag{1}$$
$$|\ell^*_2 - \ell^*_3| < \max_{\theta,\varphi}(A_1A^*_3) < \ell^*_2 + \ell^*_3$$
$$|\ell^*_2 - \ell^*_3| < \max_{\theta,\varphi}(B_1B^*_3) < \ell^*_2 + \ell^*_3 \tag{2}$$

3. A supplementary constraint on dexterity is also used to ensure a minimal dexterity of the mechanism.

$$dext(\varphi = \pm 45[\text{deg}], \pm\theta^{max}(\varphi)) > dext_{min} \tag{3}$$

   The dexterity of the mechanism is calculated as follows:

$$dext = \frac{\sigma_{min}}{\sigma_{max}} = \sqrt{\frac{p - \sqrt{q}}{p + \sqrt{q}}} \tag{4}$$

   where σ_{min} and σ_{max} are the square roots of resp. the minimal and maximal singular values of $\boldsymbol{J}$, i.e. the eigenvalues of $\boldsymbol{J}^T\boldsymbol{J}$, and $p = (1 + r_c^2).r_a^2 + r_b^2$, $q = p^2 - 4.r_a^2 r_b^2$. Here we assume that the dexterity is at the minimum when the mechanism tilts in the direction of either leg, i.e. at ±45[deg] frontward or backward. This constraint was introduced to prevent the mechanical legs from reaching singularities whereby points A_1, A_2 and A_3, or points B_1, B_2 and B_3 get aligned. In these configurations, forces on the end effector do not get transmitted to the actuators any more.

The optimization process is carried out using matlab *fmincon* function. The *fmincon* function uses a sequential quadratic programming method (SQP) that calculates an estimate of the Hessian of the Lagrangian at each iteration, then generates a QP subproblem that is solved using the *active set* strategy [3].

The initial values of the four parameters ($\ell_1^*, \ell_2^*, \ell_3^*, \ell_4^*$) are set using a preliminary grid exploration, to start the search for solutions from specific regions that are promising in terms of minimal cost function, and to avoid local solutions.

5 Results

Table 1 presents the optimization results obtained with a single module for a uniform inclination range of 10, 15, 20, 25 and 30[deg]. It gives the optimal values for the lengths (ℓ_i^*), the final value of the cost function, the average torque $< \tau >$, the maximal torque τ_{max}, the ratios of the average torque and the maximal torque over the maximal gravity torque $\tau^g = Mg\, h_G \sin \theta^{max}$, and the dexterity. The values in the first part of the table were obtained without dexterity constraint, and the values in the second part were obtained with a minimal dexterity constraint of 0.8 (inequality 3). Regarding lengths, ℓ_1 is always set to the upper bound, this allows obtaining a maximal lever arm.

The results also show that length ℓ_3 must be smaller than ℓ_2, which enables to reduce the lever arm about A_3 from A_1A_2 and therefore decrease ratios r_a and r_b.

With no dexterity constraint, ℓ_3 is increased and ℓ_2 remains constant as the inclination rises. In the case with minimal dexterity, ℓ_3 is much more increased than in the no-dexterity-constraint case, and ℓ_2 decreases. There is also a tendency to place A_3, resp. B_3 closest to the center – ℓ_4 set to minimal value –, whatever the inclination range. We can explain this fact as a consequence of the increase of ℓ_3 in order to keep the lever arm about O from A_1A_2 as large as possible. Without the lower bound

Table 1 Results from the optimization process. $h_v = 0.1$ [m]. $h_G = 0.1$ [m]. $M = 15$ [kg]. ℓ_1^* is always kept to the upper bound of 1.50 by the process. Underlined values refer to active constraints

Max. tilt	ℓ_2^*	ℓ_3^*	ℓ_4^*	Γ	$< \tau > / \tau_{max}$	$\frac{< \tau >}{\tau^g}$	$\frac{\tau_{max}}{\tau^g}$	Dexterity
[deg]	[0.15, 2.25]		[0.15, 1.5]		[Nm]/[Nm]			[0, 1]
No dexterity constraint								
10	1.45	0.18	0.44	−1870	0.14/0.22	0.06	0.09	[0.02, 1.00]
15	**1.44**	**0.27**	**0.45**	**−3894**	**0.31 / 0.50**	**0.08**	**0.13**	**[0.02, 1.00]**
20	**1.46**	**0.35**	**0.42**	**−5824**	**0.52 / 0.88**	**0.10**	**0.17**	**[0.02, 1.00]**
25	1.44	0.44	0.43	−6913	0.79/1.38	0.13	0.23	[0.02, 1.00]
30	1.42	0.53	0.44	−7175	1.12/2.02	0.15	0.27	[0.01, 1.00]
With minimal dexterity constraint of 0.8 (Ineq. (3))								
10	1.60	0.26	0.21	−1769	0.28 / 0.50	0.11	0.20	[0.80, 1.00]
15	**1.59**	**0.40**	**0.15**	**−3101**	**0.61 / 1.09**	**0.16**	**0.29**	**[0.80, 1.00]**
20	**1.49**	**0.59**	**0.15**	**−3490**	**1.11/2.00**	**0.22**	**0.40**	**[0.80, 1.00]**
25	1.28	0.89	0.15	−3221	1.90 / 3.47	0.31	0.57	[0.80, 1.00]
30	1.06	1.23	0.15	−2947	3.02 / 5.57	0.41	0.76	[0.75, 1.00]

constraint of ℓ_4, there would be no such decrease of ℓ_2 for bending angles above 15[deg]. However this lower bound was set due to technological considerations of integration. The case of 30[deg] bending is specific. The lowest dexterity is 0.75 because it was not possible to find a solution that could combine a dexterity above 0.8 over the entire range and a torque value always below the maximal gravity torque.

Figure 2 shows the results related to a maximal bending of 20[deg]. The plottings on the left-hand side show the torques issued from the parallel mechanism and the

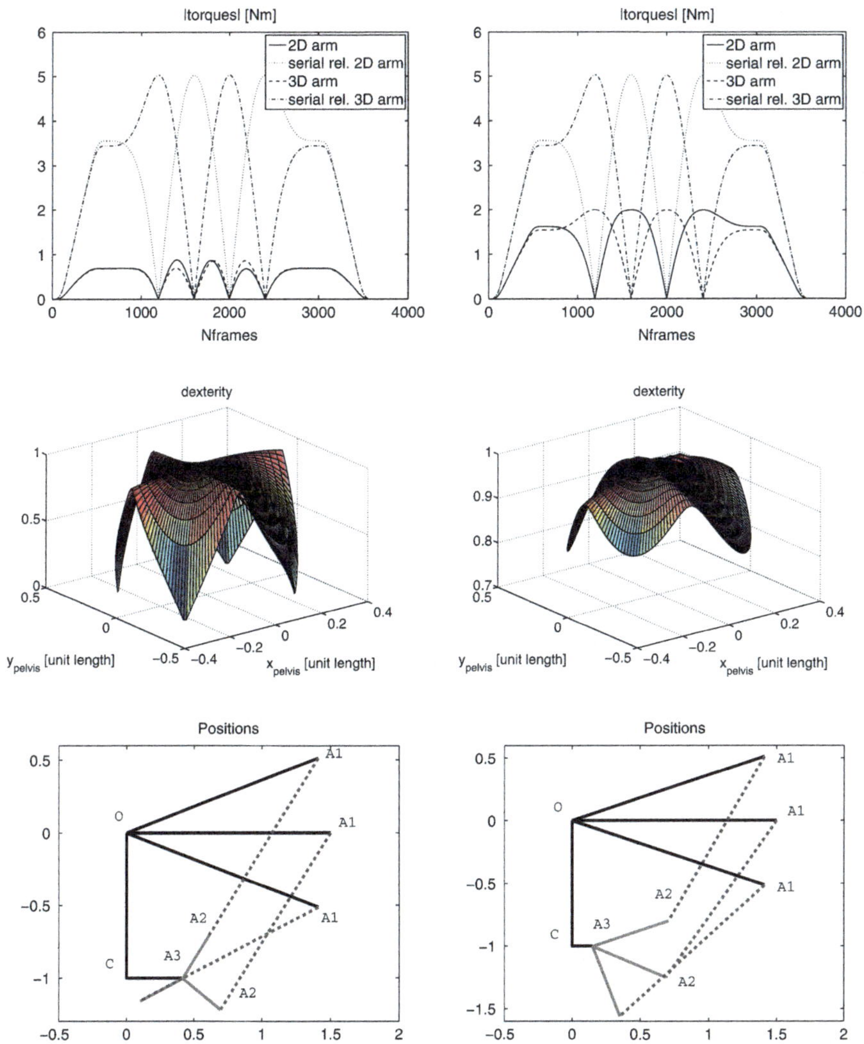

Fig. 2 Optimization results with $\theta^{max} = 20$[deg]. 1st row: comparison of torques (serial and parallel) over circular trajectory - 2nd row: dexterity - 3rd row: length configuration of leg $A_1A_2A_3$. 1st column: no dexterity constraint - 2nd column: minimal dexterity of 0.8

torques issued from the virtual serial mechanism where the top U-joint at O would be actuated; here the mass M tilts frontward to θ^{max} first, then makes a 360[deg] turn before getting back to the straight initial position. The ratio of maximal torque between the parallel mechanism and the serial mechanism is 17% with no dexterity constraint, and 40% with the dexterity constraint, which is a significant reduction. Obviously the dexterity constraint is met at the expense of torque reduction.

However the single module obtained with no dexterity constraint leads to a solution where the legs reach a singularity at $\varphi = \pm 45$[deg] when bending is maximal. For each of these 4 configurations, the dexterity falls to 0.02 (Fig. 2, 1st colum-2nd row), and one of the legs is either completely stretched or flexed (Fig. 2, 1st colum-3rd row).

The results obtained with a single module show that it is preferable to avoid the use of the parallel mechanism for bending above 15[deg]. Actually the larger the bending angle, the less the reduction ratio between maximal torque and gravity torque. This ratio is equal to 0.29 for 15[deg] bending with a 0.8 minimal dexterity threshold.

6 Discussion

The optimization process proposed aims at minimizing the actuated torques in the working volume of the mechanism. Torque minimization does not prevent the process from selecting a solution that includes singular configurations because torque tends to 0 in the vicinity of singularities. It could be possible to avoid singularities by adding geometric constraints similar to (1)–(2) using margin coefficients, which is specific. The solution of adding a dexterity constraint proposed here is better because it allows controlling the dexterity over the entire working volume, and it avoids the singular configurations.

In order to increase the bending angle of the torso it is possible to stack 3 parallel modules to build up a vertebral column. The tilt angle of each module can be set to one third of the bending angle of the thorax. However motors that drive the active joints need to be sized with increased power and nominal torque from the bottom to the top. Compared with the serial mechanism equipped with actuated θ_{10} and θ_{21} joints, the parallel mechanism allows reducing the average torque ratio and the maximal torque ratio to respectively 0.16 and 0.29 for each module tilted by 15[deg]. This result shows that the use of a parallel mechanism can save actuator weight by using motors with tree times less nominal torque than in the case of serial actuation.

The actuated torques and the dexterity were computed analytically using the duality between statics and kinematics of the mechanism. The maximal tilting angle of 30[deg] was identified for quasi-static bending and sitting down movements. In the case of walking, a revised cost function must be built up to take into account horizontal accelerations the waist mechanism must resist. The optimization process should help determine the maximum tilting angle, which can be used for trunk oscillations to enhance the dynamic balance of the walk.

7 Conclusion

This paper proposed an approach that consists of adapting a flight simulator parallel mechanism for the design of a three-tier waist mechanism that can execute pitch and roll motion. A minimized-torque-oriented optimization procedure was proposed to determine the adequate lengths of both lateral legs' segments of each module. A minimal dexterity threshold was introduced to avoid singularities and ensure the controllability of the mechanism. The study carried out shows that the maximal bending angle per module must be limited to enable enough dexterity in the direction of the legs, and to have acceptable nominal torques for the motors of the actuated joints. Compared with the actuated serial mechanism of the top U-joint, the use of the parallel mechanism allows reducing the nominal torques of the actuators by one third and consequently saving motor weight inside the humanoid trunk.

References

1. Cibert, C., Hugel, V.: Compliant intervertebral mechanism for humanoid backbone: Kinematic modeling and optimization. Mech. Mach. Theory. **66**, 32–55 (2013)
2. Dariush, B., Gienger, M., Jian, B., Goerick, C., Fujimura, K.: Whole body humanoid control from human motion descriptors. In: IEEE International Conference on Robotics and Automation, pp. 2677–2684 (2008)
3. Gill, P., Murray, W., Wright, M.: Practical Optimization. Academic Press, Dublin (1981)
4. Gosselin, C.: Kinematics analysis optimization and programming of parallel robotic manipulators. Ph.D. thesis, Mc Gill University (1988)
5. Hao, F., Merlet, J.P.: Multi-criteria optimal design of parallel manipulators based on interval analysis. Mech. Mach. Theory **40**, 151–157 (2005)
6. Hay, A., Snyman, J.A.: Methodologies for the optimal design of parallel manipulators. Int. J. Numer. Methods Eng. **59**(1), 131–152 (2003)
7. Kapandji, I.A.: The Physiology of the Joints: The Trunk and the Vertebral Column, vol. 3 (2007)
8. Kim, J., Park, I., Lee, J., Kim, M., Cho, B., Oh, J.: System design and dynamic walking of humanoid robot KHR-2. In: IEEE/RSJ International Conference on Robotics and Automation, pp. 1431–1436 (2005)
9. Kong, X., Gosselin, C.: Type synthesis of three-degree-of-freedom spherical parallel manipulators. Int. J. Robot. Res. **23**, 237–245 (2004)
10. Lenarčič, J., Bajd, T., Stanišić, M.: Robot mechanisms. In: International Series on Intelligent Systems, Control and Automation–Science and Engineering, vol. 60. Springer, Berlin (2013)
11. Liang, C., Ceccarelli, M.: Design and simulation a waist-trunk system for a humanoid robot. Mech. Mach. Theory **53**, 50–65 (2012)
12. Lohmeier, S., Buschmann, T., Ulbrich, H., Pfeiffer, F.: Modular joint design for performance enhanced humanoid robot lola. In: IEEE International Conference on Robotics and Automation, pp. 88–93 (2006)
13. Lopes, A.M., Pires, E.S., Barbosa, M.R.: Design of a parallel robotic manipulator using evolutionary computing. Int. J. Adv. Robot. Syst. 9–27 (2012)
14. Lou, Y., Liu, G., Chen, N., Li, Z.: Optimal design of parallel manipulators for maximum effective regular workspace. In: IEEE/RSJ International Conference on Intelligent Robotics and System, pp. 795–800 (2005)

15. Marques, H.G., Jäntsch, M., Wittmeier, S., Holland, O., Alessandro, C., Diamond, A., Lungarella, M., Knight, R.: Eccel: the first of a series of anthropomimetic musculoskelal upper torsos. In: International Conference on Humanoid Robotics, pp. 391–396 (2010)
16. Mehta, V., Dasgupta, B.: A general approach for optimal kinematic design of 6-DOF parallel manipulators. In: Sãdhanã, Indian Academy of Sciences, vol. 36. Springer, Berlin (2011)
17. Merlet, J.P.: Parallel Robots, Solid Mechanics and its Applications, 2nd edn., vol. 128. Springer, Berlin (2006)
18. Ogura, Y., Aikawa, H., Shimomura, K., Kondo, H., Morishima, A.: Development of a new humanoid robot wabian-2. In: IEEE International Conference on Robotics and Automation, pp. 76–81 (2006)
19. Osada, M., Izawa, T., Urata, J., Nakanishi, Y., Okada, K., Inaba, M.: Approach of planar muscle suitable for musculoskeletal humanoids, especially for their body trunk with spine having multiple vertebral. In: IEEE-RAS International Conference on Humanoid Robotics, pp. 358–363 (2011)
20. Roos, L., Guenter, F., Guignard, A., Billard, A.G.: Design of a biomimetic spine for the humanoid robot robota. In: International Conference on Biomedical Robotics and Biomechatronics, pp. 329–334 (2006)
21. Sabrié, E.: Analyse d'un mécanisme de simulation de vol sphérique et son contrôle en temps réel. Master's thesis, Faculté des sciences et de génie universitaire, Laval, Québec (2004)
22. Souissi, M., Hugel, V., Blazevic, P.: Influence of the number of humanoid vertebral column pitch joints in flexion movements. In: International Conference on Automation, Robotics and Application, pp. 277–282 (2011)
23. Stamper, R.E., Tsai, L.W., Walsh, G.C.: Optimization of a three-dof translational platform for well-conditioned workspace. In: IEEE International Conference on Robotics and Automation, pp. 3250–3255 (1997)
24. Zlatanov, D., Bonev, I.A., Gosselin, C.M.: Constraint singularities of parallel mechanisms. In: IEEE International Conference on Robotics and Automation, vol. 1, pp. 496–502 (2002)

Kinematic Analysis of the Delthaptic, a New 6-DOF Haptic Device

Margot Vulliez, Said Zeghloul and Oussama Khatib

Abstract The need for multi-purpose haptic devices is extensively emerging. This paper presents a novel 6-DOF versatile haptic device, the Delthaptic, with a large and singularity-free workspace. The originality of the device comes from its structure combining two Delta robots to obtain rotational and translational movements. The advantage of this type of paired parallel robots is to form 6-DOF manipulators by connecting well-known parallel structures. The paper focuses especially on the kinematic analysis of the Delthaptic. A method to evaluate the kinematics of paired parallel robots is described. The approach allows evaluating the full robot kinematics by considering each parallel manipulator separately. Then, the behavior of the device along its workspace is highlighted by a singularity analysis.

1 Introduction

The increasing need for kinesthetic and tactile senses in varied activity areas has led to manifold developments of dedicated haptic devices. But the design of an optimal and versatile haptic interface remains a major research concern. Only a few multi-purpose 6-DOF interfaces can be found in the literature. Ones of the most advanced solutions are the Sigma.7 of Force Dimension [12], and its previous model the Omega.7. These 7-DOF haptic devices allow a large workspace based on a hybrid structure composed of a parallel Delta mechanism, a serial wrist, and an active grasping extension. Other hybrid interfaces have been developped as multi-purpose haptic devices such as the Delta Haptic Device [5], the Falcon of Novint Technologies, or the Delta-R [1]. As inertia represents a critical issue for the transparency of haptic devices, parallel

M. Vulliez (✉) · S. Zeghloul
PPRIME Institute CNRS, University of Poitiers, ENSMA, UPR 3346, Poitiers, France
e-mail: margot.vulliez@univ-poitiers.fr

S. Zeghloul
e-mail: said.zeghloul@univ-poitiers.fr

O. Khatib
A.I. Lab, Stanford University, Stanford, CA 94305, USA
e-mail: ok@cs.stanford.edu

J. Lenarčič and J.-P. Merlet (eds.), *Advances in Robot Kinematics 2016*,
Springer Proceedings in Advanced Robotics 4, DOI 10.1007/978-3-319-56802-7_19

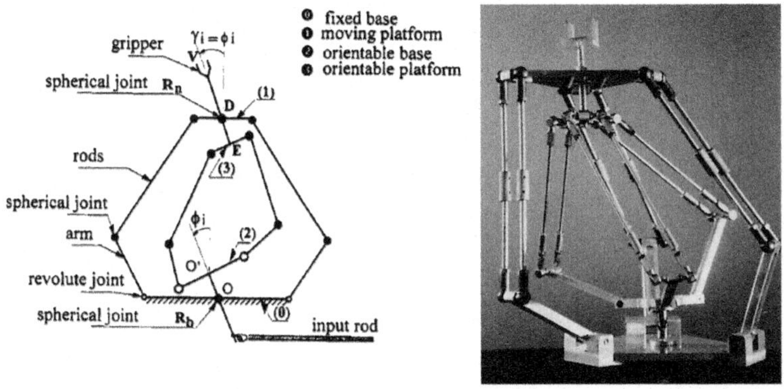

Fig. 1 2-Delta parallel robot of Lallemand et al. [6]

mechanisms can be preferred to serial structures. Some authors propose 6-DOF fully parallel mechanisms such as Yoon et al. [13] through pantograph linkages, Lee et al. [8] with a double-chain leg structure, or the desktop application of Gosselin et al. [4]. However the major drawbacks of parallel manipulators are the proximity of singular points and their limited workspace, particularly for rotational motions as studied by Merlet [10]. The challenge for the Delthaptic, proposed in this paper, is to constitute a low-inertia and high-stiffness versatile haptic interface with a large and singularity-free workspace.

A new parallel robot family, characterized by the pairing of two independant parallel manipulators, is introduced by Lallemand et al. [6] with the 2-Delta. This 6-DOF robot, in Fig. 1, is made of two Delta robots [3] whose moving platforms are connected to the end-effector. The external Delta controls the translational movements when the internal Delta generates the rotational movements of the end-effector. The main advantage of this new family is to build a 6-DOF manipulator by connecting simple and well-known parallel structures. Unfortunately the 2-Delta rotational workspace is reduced due to the limitation of the spherical joint motion range between the moving platform and the end-effector.

The proposed 6-DOF haptic device is part of this paired parallel robot family. The Delthaptic is a structure coupling two Delta robots to obtain rotational and translational movements. In that way the Delthaptic conserves the benefits of parallel structures while providing a large workspace with respect to the human-being rotations. This paper highlights the kinematic analysis of this new type of paired parallel robots that is essential to valid their operation as haptic interface and to understand their behavior toward singularity.

The paper is organized as follows. In Sect. 2 the design and the prescribed workspace of the Delthaptic are introduced. Section 3 details the calculation of the kinematic model of the system. Then, the singularity analysis along the Delthaptic workspace is carried out in Sect. 4.

2 Principle of the Delthaptic Device

2.1 *Prescribed Workspace*

The Delthaptic is designed to constitute a versatile haptic device with 6 active DOFs. It aims at ensuring a large singularity free workspace suitable for various applications. To fulfill this goal the prescribed workspace has to be defined based on the human hand mobilities. The Fig. 2 shows the hand rotations. Bending-extension and pronation-supination movements have an amplitude of $\pm 90°$ when ulnar and radial deviations admit around $\pm 45°$ as angular displacement. For the sake of simplicity the prescribed rotational workspace shall be assimilated to a half-sphere around the handle axis. This wide rotational workspace represents a challenge for the design of the Delthaptic that traditional parallel manipulators can not achieve.

The size of the prescribed translational workspace remains a subjective decision. The workspaces of some 6-DOF multi-purpose haptic devices are shown in Table 1. According to this short review the prescribed translational workspace is chosen within a cube of $200 \times 200 \times 200$ mm.

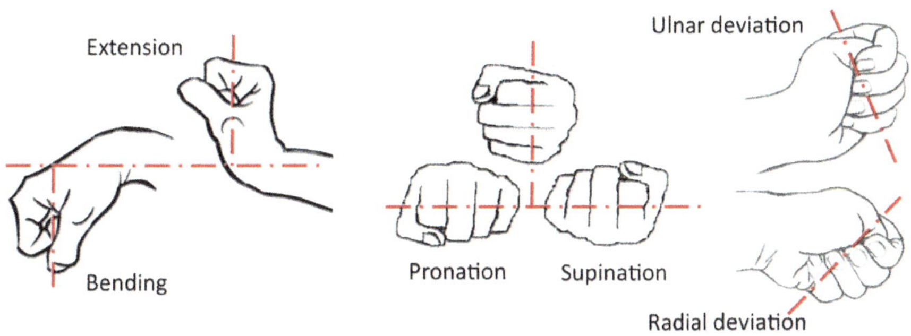

Fig. 2 Human hand rotational workspace

Table 1 Workspaces of different 6-DOF haptic devices

Workspace	Translations (mm)	Rotations
Sigma.7, Force dimension	$\Phi 190 \times 130$	$235 \times 140 \times 200°$
Omega.7, Force dimension	$\Phi 160 \times 110$	$240 \times 140 \times 180°$
Delta-R [1]	$\Phi 500 \times 200$	$\pm 80 \pm 80 \pm 80°$
6-DOF Desktop haptic device [4]	Sphere $\Phi 150$	$\pm 45 \pm 45 \pm 45°$
6-DOF haptic device [13]	$186 \times 214 \times 203$	$\pm 45 \pm 45 \pm 45°$

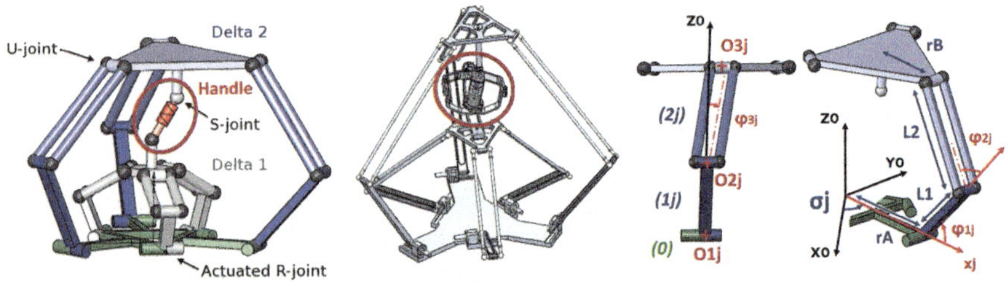

(a) Delthaptic's schematic (b) Delthaptic CAD design (c) Delta robot parameterization

Fig. 3 Delthaptic kinematical diagram and parameterization

2.2 *Mechanical Design*

To provide an efficient transparency a versatile 6-DOF haptic interface must meet many requirements such as having a large and singularity-free workspace, a low inertia, a high stiffness, minimal joint clearance and friction. The proposed structure is part of the paired parallel robot family. Since the interface is composed by two parallel manipulators, it ensures high dynamics, high stiffness and high precision. Thanks to the paired structure it provides the wide prescribed workspace. Then, it is designed to avoid singularity issues along the operational workspace. In the current literature, only serial or hybrid devices are able to allow a such full workspace. However these mechanisms have the disadvantages of serial structures.

The mechanism, showed in Fig. 3, consists of two coupled Delta robots connected to the same fixed base. Their two mobile platforms are linked to the both extremities of the handle. In that way the displacements of the two Delta robots can generate the (X, Y, Z) translational movements and the (ϕ, θ) tilt movements of the handle. The self-rotation (ψ) is obtained through a ball screw system located inside the handle. Rotational movements are described by using the tilt-and-torsion modified Euler angles presented by Bonev et al. [2] which are shown in Fig. 4. Each Delta robot i is actuated by three motors driving respectively the angles $q_i = [\varphi_{11,i}\varphi_{12,i}\varphi_{13,i}]^T$.

The zoom on the detailed CAD design of the handle in Fig. 4 shows the achievement of the links between the handle and the platforms. These links are especially important for ensuring the transparency of the device. Indeed the handle must transfer all the user movements to both Delta robots without any restriction and be able to transmit the forces of haptic feedback. To cope with its limited range of motion, the spherical joint (S-joint) between the handle and the platform of the Delta 2 is replaced by a pivot linkage and a modified universal joint (U-joint) with an axis offset d. A ball screw with a thread of $p = 20$ mm is selected to get a full reversibility of the self-rotation and be able to transmit the torques. The lower U-joint is achieved through a 2-DOF gimbal mechanism. It allows to reduce the moments of inertia of the handle and the size of the Delta 2.

The unknown vector I of design parameters for the Delthaptic, defined in the Figs. 3 and 4, is $I = [L_{1,1}, L_{1,2}, L_{2,1}, L_{2,2}, r_1, r_2, a_1, a_2, H]$, where $L_{1,i}$ and $L_{2,i}$ are

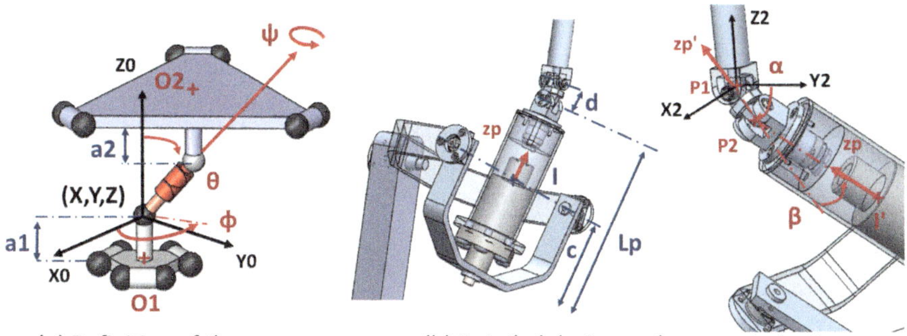

(a) Definition of the handle orientation

(b) Detailed design and parameterization of the handle

Fig. 4 Delthaptic handle design and parameterization

the leg lengths, $r_i = r_{B,i} - r_{A,i}$ is the difference between the both platform radius, a_i is the distance between the mobile platform and the handle joint and H represents the height of the center of the prescribed workspace from the base platform. An optimal solution for these parameters is chosen from the resulting Pareto front of a multi-objective optimization process. The optimization problem is solved thanks to penalty method. Two penalty components are considered: one to ensure the accessibility to the Prescribed Workspace (PW) and the other to avoid singularity as well as change of assembly and working modes. The first fitness function is chosen to minimize the distance between the Delthaptic workspace and the PW and the second one to maximize the distance to singularities, assimilated to the dexterity of the both Delta robots. The optimization process is carried out on the discretized prescribed workspace. Each point of the discretized half-sphere of tilt motion is checked for the two extreme self-rotations and to respect the translational workspace.

The additional parameters of the handle such as $(c, d, \sigma_{1,1}, \sigma_{1,2})$ are fixed to be easily handled by an operator. $L_p = L_{ini} + \psi \times \dfrac{p}{2\pi}$ is the handle lenght that depends on the self-rotation through the ball screw tranformation for the proposed design. The geometric analysis of the Delta has been widely studied in the literature. More details about its modeling and dimensional synthesis can be found in [7].

3 Kinematic Analysis

3.1 Kinematic Model of the Delta Robot

The kinematic model of the Delta robot is defined in this subsection. The platform velocity of the Delta i is considered as the sum of the contributions of each bar. The following equation is obtained for the leg j.

$$\mathbf{v_i} = \omega_{1j/0} \times \mathbf{O_{1j}O_{2j}} + \omega_{2j/0} \times \mathbf{O_{2j}O_{3j}} \\ [\dot{X}_i] \cdot \mathbf{O_{2j}O_{3j}} = [\dot{q_{j,i}}] \cdot (\mathbf{O_{1j}O_{2j}} \times \mathbf{O_{2j}O_{3j}}) \tag{1}$$

Then, the combination of the Eq. 1 applied to the three legs leads to the expression of the Delta robot kinematics. In the Eq. 2, J_i is the Delta i Jacobian matrix, $[\dot{X}_i]$ the mobile platform velocity vector and $[\dot{q}_i]$ the actuated angular velocities.

$$A_i[\dot{X}_i] = B_i[\dot{q}_i] \ \ with \ \ J_i = A_i^{-1} B_i \tag{2}$$

This Delta robot kinematic model is fully detailed by Pierrot et al. [11]. An overview of singularity can be found in [9].

3.2 Kinematic Model of the Delthaptic

The kinematic analysis of the Delthaptic is essential to ensure its proper functioning. The proposed method to evaluate the kinematics of a paired robot is to consider separately the kinematic model of each parallel manipulator composing the robot. Then, the complete robot model is obtained by writing the relationships between the end-effector and the different manipulators.

Kinematic analysis of this type of paired parallel robots is illustrated through the following modeling of the Delthaptic. The kinematic model of the Delta robot is first reminded in Sect. 3.1. Then the velocity of the handle $\mathbf{V_{I',h/0}}$ is expressed with respect to the kinematics of the joints with the two Delta mobile platforms. The link between the handle and the Delta 2 mobile platform leads to Eq. 3. The relationship between the handle and the Delta 1 through the gimbal mechanism is expressed by Eq. 4, where h represents the handle and s the screw.

$$\mathbf{V_{I',h/0}} = \mathbf{V_{P_2,h/0}} + \omega_{\mathbf{h/0}} \times \mathbf{P_2I'} = \mathbf{v_2} + \dot{\alpha}\mathbf{x_{1,2}} \times \mathbf{P_1P_2} + \omega_{\mathbf{h/0}} \times \mathbf{P_2I'} \tag{3}$$

$$\mathbf{V_{I',h/0}} = \mathbf{V_{I',s/0}} + \frac{p}{2\pi}\omega_{\mathbf{h/s}} = \mathbf{v_1} + \omega_{\mathbf{s/0}} \times \mathbf{II'} + \frac{p}{2\pi}\omega_{\mathbf{h/s}} \tag{4}$$

The handle angular velocity is deduced from these equations.

$$\mathbf{IP_2} \times \omega_{\mathbf{h/0}} = \dot{L_p} \cdot \mathbf{z_p} + \mathbf{v_1} - (\mathbf{v_2} + \dot{\alpha}\mathbf{x_{1,2}} \times \mathbf{P_1P_2}) = \mathbf{V_0} \\ \omega_{\mathbf{h/0}} = \omega_{\mathbf{h/0}}^{\mathbf{0}} + \frac{2\pi}{p}\dot{L}_p\mathbf{z_p} \ \ with \ \ \omega_{\mathbf{h/0}}^{\mathbf{0}} = \frac{\mathbf{V_0} \times \mathbf{IP_2}}{\| \mathbf{IP_2} \|^2} \tag{5}$$

The translational velocity is given by:

$$\mathbf{V_{I',h/0}} = \mathbf{v_1} + \omega_{\mathbf{s/0}} \times (L_p - L_{ini}) \cdot \mathbf{z_p} + \dot{Lp} \cdot \mathbf{z_p} \\ with \ \ \dot{Lp} = (\mathbf{v_2} + \dot{\alpha}\mathbf{x_{1,2}} \times \mathbf{P_1P_2} - \mathbf{v_1}) \cdot \mathbf{z_p} \tag{6}$$

Then, the global kinematic model of the Delthaptic can be written by integrating the two individual Delta robot kinematics.

$$\begin{bmatrix} \dot{X} \\ \dot{Y} \\ \dot{Z} \\ \omega_x \\ \omega_y \\ \omega_z \end{bmatrix} = J \times \begin{bmatrix} \dot{\varphi_{11,1}} \\ \dot{\varphi_{12,1}} \\ \dot{\varphi_{13,1}} \\ \dot{\varphi_{11,2}} \\ \dot{\varphi_{12,2}} \\ \dot{\varphi_{13,2}} \end{bmatrix} = \left| \begin{matrix} \begin{bmatrix} \mathbf{J_{x1}} \\ \mathbf{J_{y1}} \\ \mathbf{J_{z1}} \end{bmatrix} \mathbf{v_1} + \begin{bmatrix} \mathbf{J_{x2}} \\ \mathbf{J_{y2}} \\ \mathbf{J_{z2}} \end{bmatrix} \mathbf{v_2} \Bigg) \; linear \\ \begin{bmatrix} \mathbf{J_{\omega x1}} \\ \mathbf{J_{\omega y1}} \\ \mathbf{J_{\omega z1}} \end{bmatrix} \mathbf{v_1} + \begin{bmatrix} \mathbf{J_{\omega x2}} \\ \mathbf{J_{\omega y2}} \\ \mathbf{J_{\omega z2}} \end{bmatrix} \mathbf{v_2} \Bigg) \; angular \end{matrix} \right. \tag{7}$$

$$J = \begin{bmatrix} J_t \\ J_r \end{bmatrix} = \underbrace{\begin{bmatrix} \begin{bmatrix} \mathbf{J_{x1}} \\ \mathbf{J_{y1}} \\ \mathbf{J_{z1}} \end{bmatrix} & \begin{bmatrix} \mathbf{J_{x2}} \\ \mathbf{J_{y2}} \\ \mathbf{J_{z2}} \end{bmatrix} \\ \begin{bmatrix} \mathbf{J_{\omega x1}} \\ \mathbf{J_{\omega y1}} \\ \mathbf{J_{\omega z1}} \end{bmatrix} & \begin{bmatrix} \mathbf{J_{\omega x2}} \\ \mathbf{J_{\omega y2}} \\ \mathbf{J_{\omega z2}} \end{bmatrix} \end{bmatrix}}_{J_c} \times \begin{bmatrix} A_1^{-1}B_1 & zeros(3,3) \\ zeros(3,3) & A_2^{-1}B_2 \end{bmatrix}$$

This method allows to simply express the paired robot kinematic model as a function of the kinematic model of each Delta robot and a linkage matrix J_c.

The actuator torques τ required to provide the force feedback F are then evaluated through the static model as follows.

$$\tau = J^T F \tag{8}$$

4 Singularity Analysis

The system kinematics is particularly important to understand the behavior of the interface toward singularity. Singular configurations can be highlighted by evaluating the Jacobian matrix determinant along the workspace. The proposed approach of the kinematic model in Eq. 7 allows to isolate the effects of the singularities of each parallel manipulator.

$$\mid det(J) \mid = \mid det(J_c) \times det(J_1) \times det(J_2) \mid \tag{9}$$

The singular configurations are reached when $det(J) = 0$. Equation 9 demonstrates that the Delthaptic singularities are those of each Delta i, when $det(J_i) = 0$, and coupling singularities if $det(J_c) = 0$. There is any coupling singularity for the presented mechanism.

The Jacobian matrix determinant is plotted in Fig. 5 over the constant orientation workspace at $[\phi, \theta, \psi] = [0, 0, 0]$.

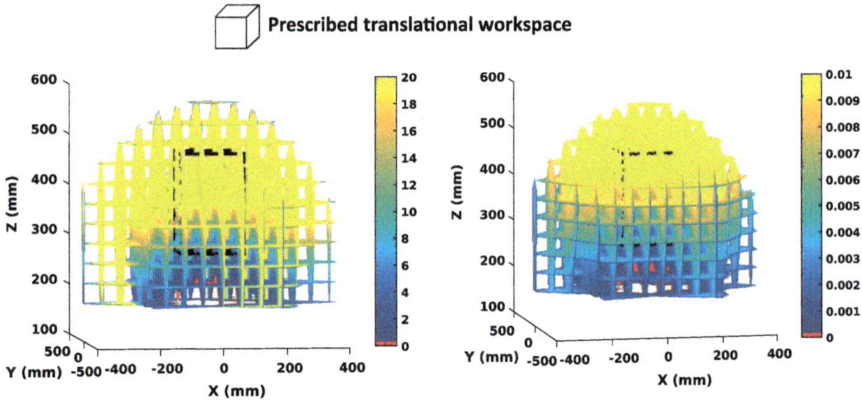

(a) Delthaptic singularity: |det(J)| *(b) Delta 1 serial singularity: |det(B1)|*

Fig. 5 Singularity analysis along the constant orientation workspace at $[\phi, \theta, \psi] = [0, 0, 0]$

The Fig. 5 emphasizes the conclusions lead by the Eq. 9 about the singularity analysis of paired robots. The Delthaptic admits the singular configurations of its both Delta robots and particularly, for this design, the serial singularity of the Delta 1 when $det(B_1) = 0$. Due to the multi-objective optimization of the design parameters, the device is designed to have no singularity in the prescribed workspace. This is verified for the constant orientation $[\phi, \theta, \psi] = [0, 0, 0]$ in the Fig. 5 along the prescribed translational workspace.

5 Conclusion

The Delthaptic, a novel 6-DOF versatile haptic device, is proposed in this paper. This paired parallel robot combines two Delta robots to obtain rotational and translational movements. In that way the structure conserves the benefits of parallel structures (low inertia, high stiffness, high precision) while providing a large workspace with full human-based rotations. A method to evaluate the kinematics of this type of paired parallel robots is described. The approach allows considering the full robot kinematics by considering each parallel manipulator separately. Then, the behavior of the Delthaptic along its workspace is validated by a singularity analysis.

References

1. Arata, J., Kondo, H., Ikedo, N., Fujimoto, H.: Haptic device using a newly developed redundant parallel mechanism. IEEE Trans. Robot. **27**(2), 201–214 (2011). doi:10.1109/TRO.2010.2098272
2. Bonev, I., Zlatanov, D., Gosselin, C.: Advantages of the modified euler angles in the design and control of pkms. In: 2002 International Conference of Parallel Kinematic Machines, pp. 171–188 (2002). doi:10.1109/WHC.2005.33
3. Clavel, R.: Delta, a fast robot with parallel geometry. In: 18th International Symposium on Industrial Robots, Lausanne, 1988, pp. 91–100 (1988)
4. Gosselin, F., Martins, J., Bidard, C., Andriot, C., Brisset, J.: Design of a new parallel haptic device for desktop applications. In: 2005 Eurohaptics Conference and Symposium on Haptic Interfaces for Virtual Environment and Teleoperator Systems, 2005. World Haptics, pp. 189–194 (2005). doi:10.1109/WHC.2005.33
5. Grange, S., Conti, F., Rouiller, P., Helmer, P., Baur, C.: The delta haptic device. Technical report (2001)
6. Lallemand, J., Goudali, A., Zeghloul, S.: The 6-dof 2-delta parallel robot. Robotica **15**(04), 407–416 (1997). doi:10.1017/S0263574797000507
7. Laribi, M., Romdhane, L., Zeghloul, S.: Analysis and dimensional synthesis of the delta robot for a prescribed workspace. Mech. Mach. Theory **42**(7), 859–870 (2007). doi:10.1016/j.mechmachtheory.2006.06.012
8. Lee, J., Eom, K.: Design of a new 6-dof parallel haptic device. In: IEEE International Conference on Robotics and Automation, ICRA 2001, vol. 1, pp. 886–891 (2001). doi:10.1109/ROBOT.2001.932662
9. López, M., Castillo, E., García, G., Bashir, A.: Delta robot: inverse, direct, and intermediate jacobians. Proc. Inst. Mech. Eng. Part C J. Mech. Eng. Sci. **220**(1), 103–109 (2006). doi:10.1243/095440606X78263
10. Merlet, J.P.: Determination of the orientation workspace of parallel manipulators. J. Intell. Robot. Syst. **13**(2), 143–160 (1995). doi:10.1007/BF01254849
11. Pierrot, F., Reynaud, C., Fournier, A.: Delta: a simple and efficient parallel robot. Robotica **8**(02), 105–109 (1990). doi:10.1017/S0263574700007669
12. Tobergte, A., Helmer, P., Hagn, U., Rouiller, P., Thielmann, S., Grange, S., Albu-Schäffer, A., Conti, F., Hirzinger, G.: The sigma.7 haptic interface for mirosurge: a new bi-manual surgical console. In: IEEE/RSJ International Conference on Intelligent Robots and Systems, IROS 2011, pp. 3023–3030 (2011). doi:10.1109/IROS.2011.6094433
13. Yoon, J., Ryu, J.: Design, fabrication, and evaluation of a new haptic device using a parallel mechanism. IEEE/ASME Trans. Mechatron. **6**(3), 221–233 (2001). doi:10.1109/3516.951360

A Family of Non-overconstrained 3-DoF Reconfigurable Parallel Manipulators

Matteo-Claudio Palpacelli, Luca Carbonari, Giacomo Palmieri and Massimo Callegari

Abstract An extensive conceptual production of three degrees of freedom translating and rotating parallel kinematics machines has been already provided in the scientific literature. In this paper an attempt to gather the two mentioned mobilities in a single reconfigurable machine is made by means of a spherical lockable joint for a family of fully parallel kinematics machines. An overconstrained configuration is exploited for each manipulator of the family in order to realize the transition between the non-overconstrained kinematics associated to the translational and rotational mobilities. Kinematic synthesis is carried out by means of Lie group algebra.

1 Introduction

Parallel kinematics machines (PKMs) find today a widespread use in automated processes where high speeds and accelerations, or otherwise high thrusts, are required, even though conventional serial robots are preferred in most cases because of their larger workspace. Several attempts were made by researchers to deal with the complexity of analytic models of full mobility PKMs [13, 28]. Alternatively, a low degree of mobility is often proposed, having recognized that a large part of industrial operations can be realized by means of subgroups of displacements, like pure translations, pure rotations or Schönflies motions.

One of the most successful kinematic machines adopted in industry is the Delta robot [6], a parallel manipulator usually devoted to pick and place operations because

M.-C. Palpacelli (✉) · L. Carbonari · G. Palmieri · M. Callegari
Department of Industrial Engineering and Mathematical Sciences,
Polytechnic University of Marche, v. Brecce Bianche, 60131 Ancona, Italy
e-mail: m.palpacelli@univpm.it

L. Carbonari
e-mail: l.carbonari@univpm.it

G. Palmieri
e-mail: g.palmieri@univpm.it

M. Callegari
e-mail: m.callegari@univpm.it

J. Lenarčič and J.-P. Merlet (eds.), *Advances in Robot Kinematics 2016*,
Springer Proceedings in Advanced Robotics 4, DOI 10.1007/978-3-319-56802-7_20

of its high dynamics performance. It belongs to the family of translating parallel manipulators (TPMs), in particular of fully parallel manipulators of pure translation, where the number of actuated legs that connect the moving platform to the fixed base is equal to the platform degrees of freedom. An elegant demonstration of the translational mobility of the Delta robot is provided by Hervé et al. [12] by means of Lie Groups. The paper also shows how to synthesize overconstrained TPMs by choosing a suitable sequence of joints for each leg of the robot. A complete list of combinations of one degree-of-freedom (1-DoF) kinematic pairs is presented in both the mentioned paper and a more recent work of the same author [21]. Later works extend the study to non-overconstrained TPMs [17, 22], orthogonal TPMs [26] and fully-isotropic TPMs [9]. Several examples of TPMs are currently available in the literature [2, 16, 27].

Similarly, rotational PKMs received great attention. Group theory and Screw theory have been extensively used to synthesize fully parallel manipulators of pure rotation [7, 15, 18], sometimes called spherical parallel manipulators (SPMs). The Agile eye proposed by Gosselin [10] is the major example of SPM, where all bodies, and in particular the moving platform, move on spherical surfaces. Many other examples can be found in the scientific literature [1, 4, 11]. SPMs usually need specific geometric conditions to be fulfilled by the kinematics of their legs, which are mainly realized with sequences of revolute joints. In more detail, the revolute axes must intersect at a common point, which results to be the centre of rotation. However, the synthesis of SPMs presents a complexity greater than TPMs, because rotations do not commute and a sequence of revolute joints used in the mechanics of a leg cannot be changed in their order. A different behaviour is provided instead by prismatic pairs, which on the contrary can be reversed in the order.

The existence of 3-DoF PKMs of pure translation and rotation that share the same leg sequence of joints while having different arrangements of their axes suggested further study, aimed at the design of a reconfigurable machine that could modify its mobility with small local changes in its mechanical structure. Algebraic geometry was used by the authors to analyse the working modes of a manipulator with 3-CPU kinematics [3], a PKM with three identical legs that consist of two links connected by a prismatic pair. Each leg is connected to the fixed base with a cylindrical joint, whose linear displacement is actuated, and to the mobile platform with a universal joint. This latter was replaced by a lockable spherical joint (S_r) realized with a series of three revolute axes [24], which can be activated or locked according to user needs. The investigation resulted in a reconfigurable 3-CPU manipulator [25], whose mechanical design is shown in Fig. 1. Two different U-joint configurations can be alternately enabled, obtaining respectively a translational and a rotational 3-CPU. In the following it is shown how the S_r-joint can be exploited to generate a whole family of reconfigurable PKMs.

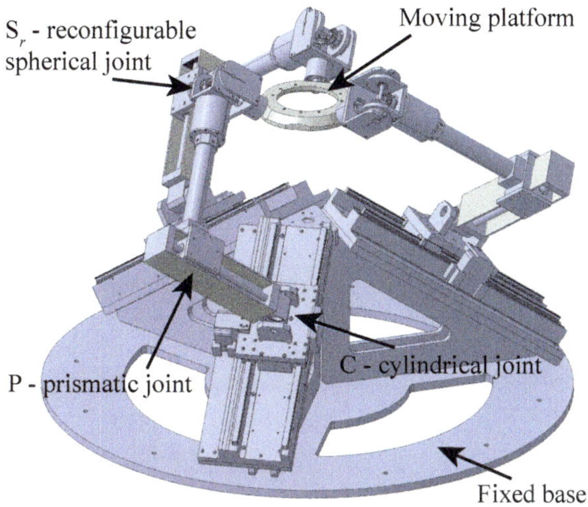

Fig. 1 Mechanical design of the 3-CPU reconfigurable manipulator

2 Lockable Spherical Joint

In this section a description of the reconfigurable spherical joint is presented. As already mentioned the S_r-joint is thought of as a sequence of three revolute joints, similarly to robotic roll-pitch-roll wrists. The idea is to endow the joint with a switch device, which will allow to lock alternately one axis between two of the three available, resulting in two different U-joint configurations.

The quadrants I and III shown in Fig. 2 are associated to the different U-joint configurations, whereas II and IV refer respectively to a revolute (R) and a spherical (S) joint configuration. The joint is based on a bevel gear coupling. A sliding cursor c (red) is driven in four different positions by an actuator, which confers to the joint the four different operating modes.

The relevant rigid bodies that make up the S_r-joint are indicated in the sketch of Fig. 2. Member a (green) is a hollow cylinder, which the cursor c can slide within. A second member b (blue) consists of a hollow cylinder with a C-shaped flange rigidly connected at its top. Member b can rotate about its axis with respect to a because of roller bearings. Angle ψ refers to such rotation. A square ring e (orange), solid with a bevel gear, can rotate with respect to body b by means of two roller bearings. Angle θ refers to such rotation. A body d (grey), with a second gear that completes the bevel gear connection with e, can rotate with respect to b about its axis. A further rotation φ is allowed between the last body f (black) and e about an axis always orthogonal to the axis of θ. Body f can be considered fixed to the frame, while body a is the output member of the chain. Cursor c moves within a and b. It has a cylindrical shape with an external gear and an internal one at its left extreme, as shown in Fig. 2. The former is used to engage the teeth of cursor c with those of internal gears realized within the cylinders a and b, whereas the latter is used to connect c with an external gear realized at the right extreme of d. It is now possible

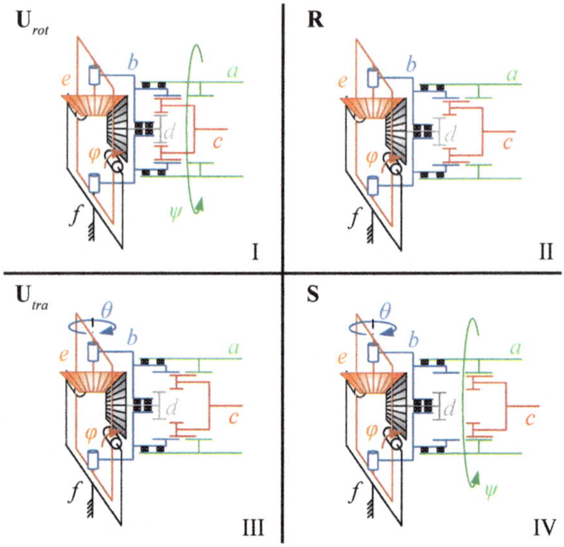

Fig. 2 Sketch of the lockable spherical joint

to analyse the four configurations: in I cursor *c* mechanically engages with *b* and *d*, so that they globally behave like a rigid body. It follows that the rotation θ between *b* and *e* is prevented, whereas the other two are free to rotate. This configuration is called U_{rot}. Its meaning will be clear in the following section. When the cursor *c* is moved to the right there exists a phase, II in the figure, where all elements *a*, *b*, *c*, *d* are rigidly joined. In this case also the rotation ψ is prevented, giving rise to a revolute joint, given by the only rotation φ.

A further motion of cursor *c* to the right provides a disengagement between *d* and *c*, while *a*, *b* and *c* still remain a whole body. Therefore, configuration III is obtained, also called U_{tra}: the joint gains the rotation θ, but rotation ψ is still prevented. In this case, while θ changes, body *d* rotates about its axis because of the bevel gear connection. The joint globally behaves like a universal joint, different with respect to the other mentioned before. Finally, cursor *c* can be disengaged from *b* with a further displacement, giving rise to a spherical joint: all rotations φ, θ and ψ are possible. This last configuration is not exploited for the family of manipulators presented in the following section. Summarizing:

- configurations I and III confer to the mechanism the mobility of a universal joint with two different arrangements of the last axis of rotation;
- configuration II represents a revolute joint;
- configuration IV is an additional mode, which can be exploited as a spherical joint.

Some other examples of lockable joints were proposed and analysed in the literature [5, 8]. However, even if they show higher versatility, they do not generally deal with some issues related to their integration in a physical prototype: during the transition between different operation modes a reconfigurable manipulator generally gains degrees of freedom, becoming temporarily underconstrained. Therefore an external manual intervention is needed to hold the robot in its pose. As shown in the following section, configuration II of the S_r-joint makes the manipulator overconstrained, preventing the robot from moving during the transition when the axes of motors are braked.

3 Family of Reconfigurable 3-DoF PKMs

The joint described in the previous section, with small changes in its functional design and in the sequence of revolute axes, was already used to realize a reconfigurable PKM of kinematics 3-CPU. The universal joint that is used to connect the three legs with the mobile platform is actually a S_r-joint. It was showed that, when the joint is arranged in one of the U-joint configurations, the robot behaves like a TPM, whereas the other U-joint configuration makes the robot a SPM [25]. A downgrade of all the S_r-joints to R-joints, allowed only when the robot is in its home configuration with the three rotation axes (ψ, θ, φ) mutually orthogonal, gives rise to an overconstrained 3-CPR manipulator. The transient phase needed to pass from one mobility to the other can be managed without any external intervention on the machine, which stays in its home pose when the actuated axes are braked.

Starting from the concept exposed above, new architectures with analogous features can be looked for. The study is narrowed to 3-DoF fully parallel manipulators, where the three legs are connected to the mobile platform with the S_r-joint presented in Sect. 2. Each leg in a non-overconstrained PKM provides a well-defined 5-DoF kinematic bond, which can be obtained by different mechanical generators. Prismatic (P), revolute (R), cylindrical (C) and universal (U) joints, together with the S_r-joint, are chosen as kinematic pairs of the leg.

With reference to Fig. 2 and for all the architectures that will be investigated, body a is rigidly joined to the mobile platform of the manipulator, whereas body f to the second link of each leg. The mechanical generators must be chosen so that, when the lockable joints are all in configuration III, the U_{tra}-joint makes the tripod a TPM, when in configuration I, the U_{rot}-joint makes the tripod a SPM.

The algebraic structure of Lie groups and subgroups can be conveniently used to synthesize TPMs and SPMs. After a brief analysis of the kinematic bond of the 3-CPU architecture, which already demonstrated a property of reconfigurability [25], further kinematics with similar properties can be found out. A sketch of the CPU kinematics developed in a plane π is shown in Fig. 3, where Fig. 3a refers to the U_{tra} configuration and Fig. 3b to the U_{rot}. The mobile platform is indicated as MP and the fixed platform as FP. The unit vector $\mathbf{u}$ represents the axis of the C-joint, which connects the first link of the leg with FP. The unit vector $\mathbf{w}$, orthogonal to $\mathbf{u}$, indicates the direction of the P-joint between the two links of the leg. The axes of $\mathbf{u}$ and $\mathbf{w}$ intersect at point J. They both generate the plane π with normal the unit vector $\mathbf{v}$. The second link of the leg belongs to π. It is connected to MP by means of the S_r-joint of Sect. 2: in Fig. 3a it is configured in the U_{tra} mode, namely with the first axis (rotation φ of Fig. 2) aligned with $\mathbf{v}$ and the last revolute axis (rotation

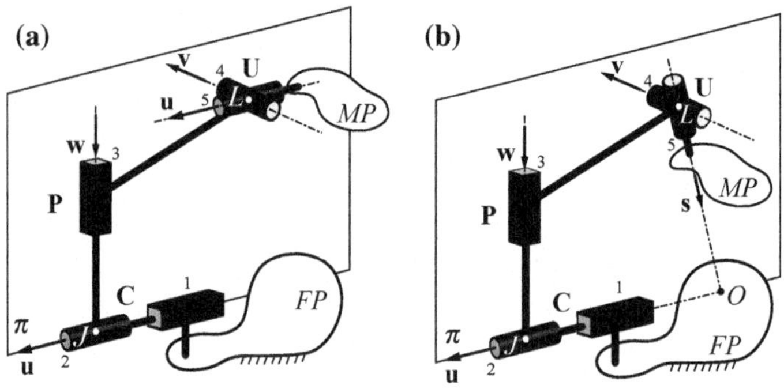

Fig. 3 Leg kinematics CPU: **a** Configuration U_{tra}, **b** configuration U_{rot}

θ) directed as $\mathbf{u}$, whereas in Fig. 3b it is configured in the U_{rot} mode, with the same direction of the first axis, but with the last axis (rotation ψ) identified by the unit vector $\mathbf{s}$ that is directed toward the point O. The latter is given by the intersection with the axis of the C-joint. Finally point L is the center of the U-joint.

Said $\{C(J, \mathbf{u})\}$, $\{T(\mathbf{w})\}$ and $\{S(L)\}$ respectively the Lie subgroups generated by C, P and S joints, the mechanical generator of a 6-DoF kinematic bond is given by their product $\{C(J, \mathbf{u})\} \cdot \{T(\mathbf{w})\} \cdot \{S(L)\}$. It is easy to prove that such product generates the improper Lie subgroup $\{D\}$. The subgroup $\{C(J, \mathbf{u})\}$ can be thought of as the product of a prismatic pair and a revolute pair, which gives $\{T(\mathbf{u})\} \cdot \{R(J, \mathbf{u})\}$, whereas the spherical joint can be generated by the product $\{R(L, \mathbf{v})\} \cdot \{R(L, \mathbf{u})\} \cdot \{R(L, \mathbf{s})\}$ of three revolute pairs with linear independent axes intersecting at the common point L, because of the product closure in the subgroup $\{S(L)\}$. It should be mentioned that such assumption is valid only for small finite displacements around the identity, which corresponds to the home configuration of the manipulator. It follows that the generator of the 6-DoF bond is given by $\{T(\mathbf{u})\}\{R(J, \mathbf{u})\}\{T(\mathbf{w})\}\{R(L, \mathbf{v})\}\{R(L, \mathbf{u})\}\{R(L, \mathbf{s})\}$, where the dot representing the product is neglected for the sake of conciseness. An equivalent representation of the mechanical bond is $\{R(J, \mathbf{u})\}\{G(\mathbf{v})\}\{R(L, \mathbf{u})\}\{R(L, \mathbf{s})\}$, where a commutation between $\{T(\mathbf{u})\}$ and $\{R(J, \mathbf{u})\}$ is considered and the statement $\{G(\mathbf{v})\} = \{T(\mathbf{u})\}\{T(\mathbf{w})\}\{R(L, \mathbf{v})\}$ is assumed when the boundaries of the neighborhood are neglected.

It follows that when the S_r-joint is configured as the U_{tra} the mechanical bond loses the term $\{R(L, \mathbf{s})\}$, whereas in the U_{rot} configuration it loses $\{R(L, \mathbf{u})\}$. Summarizing, the 5-DoF mechanical bond is given by:

$$\{R(J, \mathbf{u})\}\{G(\mathbf{v})\}\{R(L, \mathbf{u})\} \quad (\text{U}_{tra} \text{ mode}) \tag{1}$$

$$\{R(J, \mathbf{u})\}\{G(\mathbf{v})\}\{R(L, \mathbf{s})\} \quad (\text{U}_{rot} \text{ mode}) \tag{2}$$

The mobility of the reconfigurable 3-CPU manipulator for both the configurations can be evaluated according to the procedures shown in [14, 20]. Some assumptions

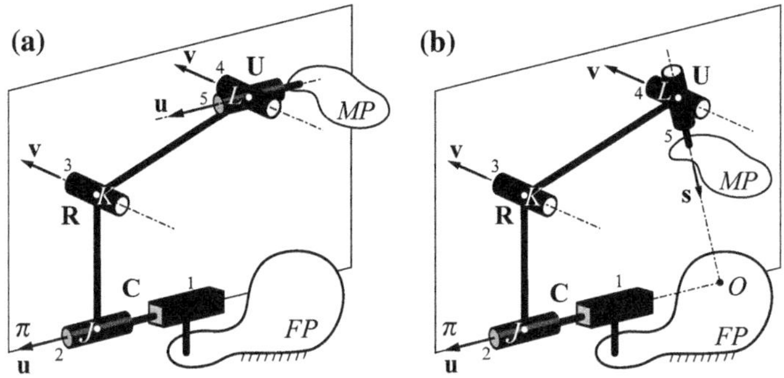

Fig. 4 Leg kinematics CRU: **a** Configuration U_{tra}, **b** configuration U_{rot}

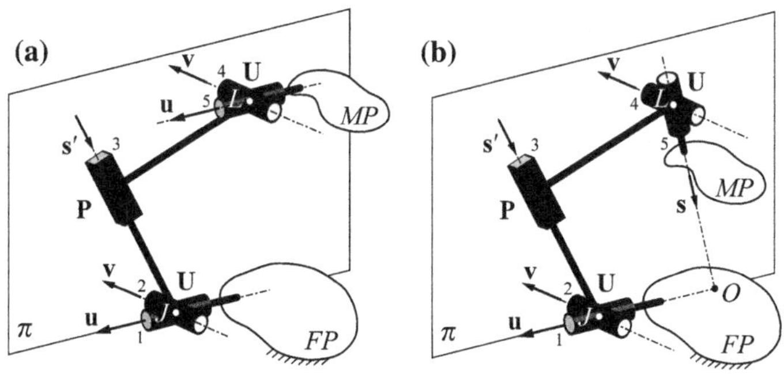

Fig. 5 Leg kinematics UPU: **a** Configuration U_{tra}, **b** configuration U_{rot}

are made: the three legs of the tripod belong to three orthogonal planes, the direction of the C-joint becomes the direction of **v** and **w** for the second and the third leg respectively, and the axis of the P-joint becomes the direction of **u** and **v**. The robot mobility results from the intersection of the mechanical bonds of the legs.

The results obtained for the 3-CPU manipulator allow to find out new architectures when different generators of the planar motion set $\{G(\mathbf{v})\}$ used in (1) and (2) are taken into account. The reconfigurable fully parallel manipulators of kinematics 3-CRU, 3-UPU, 3-URU, 3-RPaPaU are obtained, without claiming to be exhaustive. The sketch of their leg kinematics is shown in Figs. 4, 5, 6 and 7. The mentioned kinematics are obtained as follows:

- 3-CRU of Fig. 4 - $\{G(\mathbf{v})\} = \{T(\mathbf{u})\}\{R(K, \mathbf{v})\}\{R(L, \mathbf{v})\}$
- 3-UPU of Fig. 5 - $\{G(\mathbf{v})\} = \{R(J, \mathbf{v})\}\{T(\mathbf{s})\}\{R(L, \mathbf{v})\}$
- 3-URU of Fig. 6 - $\{G(\mathbf{v})\} = \{R(J, \mathbf{v})\}\{R(K, \mathbf{v})\}\{R(L, \mathbf{v})\}$
- 3-RPaPaU of Fig. 7 - $\{G(\mathbf{v})\} = \{T(Pl_{\mathbf{v}})\}\{R(L, \mathbf{v})\}$

where $J \neq K \neq L$. A small index from 1 to 5 is indicated in the figures according to the sequence of joints from the fixed base to the mobile platform. In order to have a non-overconstrained manipulator, the parallelogram Pa of the last kinematic

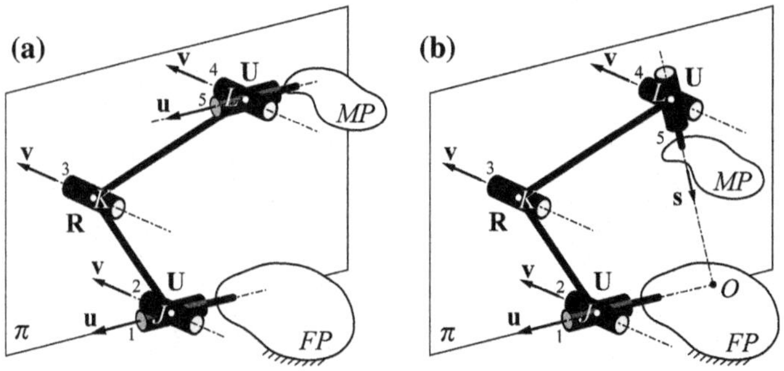

Fig. 6 Leg kinematics URU: **a** Configuration U_{tra}, **b** configuration U_{rot}

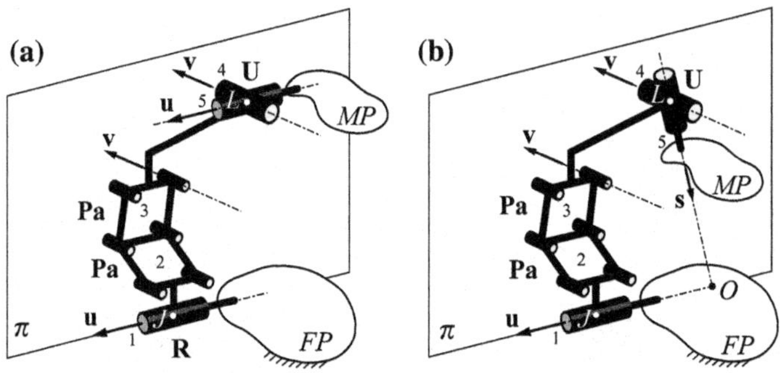

Fig. 7 Leg kinematics RPaPaU: **a** Configuration U_{tra}, **b** configuration U_{rot}

architecture should be designed without redundancies, namely two revolute joints should be replaced respectively with a spherical and a cylindrical joint.

The authors already proved that the reconfigurable 3-CPU and 3-CRU manipulators allow large finite motions around the home configuration, which turns to be an isotropic configuration both for the translational and rotational mobilities [23]. The other architectures proposed in this work should be analysed in terms of singularity maps and workspaces, in order to identify their kinematic performance.

Finally, it must be pointed out that the kinematic architectures that result from this study are not new if they are considered as one-operation-mode machines, as already mentioned in Sect. 1. On the contrary they are conceived to share a configuration, called home configuration, in which the lockable joint of Sect. 2 allows a smooth automatic transition between translation and rotation modes. Other studies in the literature have a more theoretical content, showing how it is possible to synthesize metamorfic manipulators with lockable joints, but without giving full information about how to solve the change of configuration automatically in a physical device. An example is the work proposed by Kong and Jin [19], who obtained an architecture very similar to the reconfigurable 3-URU here presented, even if by following a different approach.

4 Conclusions

A family of 3-DoF reconfigurable fully parallel manipulators is presented. A lockable spherical joint is used for all the kinematic architectures to change the configuration of a universal joint, which is used to connect the legs of the manipulator to its mobile platform. The joint reconfiguration results in a change of mobility of the mobile platform, which can have motions of pure translation and rotation. The novelty of the paper is mainly in the smooth transition between the mentioned working modes, which is allowed only at a specific pose of the manipulator, called home configuration. An overconstrained kinematics for each manipulator of the family is exploited so that the transition can be driven automatically. All the kinematic architectures proposed in this work share the same reconfigurability, however further study is needed to investigate their behaviour for large finite motions and to find out their singularity maps.

References

1. Callegari, M., Marzetti, P.: Inverse dynamics model of a parallel orienting device. In: 8th International IFAC Symposium on Robot Control, SYROCO, vol. 8, PART 1. Bologna, Italy (2006)
2. Callegari, M., Palpacelli, M.C., Principi, M.: Dynamics modelling and control of the 3-RCC translational platform. Mechatronics **16**(10), 589–605 (2006)
3. Carbonari, L., Callegari, M., Palmieri, G., Palpacelli, M.C.: A new class of reconfigurable parallel kinematic machines. Mech. Mach. Theory **79**, 173–183 (2014)
4. Chaker, A., Mlika, A., Laribi, M., Romdhane, L., Zeghloul, S.: Synthesis of spherical parallel manipulator for dexterous medical task. Front. Mech. Eng. **7**(2), 150–162 (2012)
5. Chen, W., Zhang, J., Quan, J., Lv, T.: A novel spherical joint designed for metamorphic mechanism. 2008 IEEE Conference on Robotics. Automation and Mechatronics, pp. 976–981. Chengdu, China (2008)
6. Clavel, R.: DELTA, a fast robot with parallel geometry. In: 18th International Symposium on Industrial Robot, pp. 91–100. Lausanne, Switzerland (1988)
7. Fang, Y., Tsai, L.W.: Structure synthesis of a class of 3-DOF rotational parallel manipulators. IEEE Trans. Robot. Autom. **20**(1), 117–121 (2004)
8. Gan, D., Dai, J., Dias, J., Seneviratne, L.: Reconfigurability and unified kinematics modeling of a 3rTPS metamorphic parallel mechanism with perpendicular constraint screws. Robot. Comput. Integr. Manufact. **29**(4), 121–128 (2013)
9. Gogu, G.: Structural synthesis of fully-isotropic translational parallel robots via theory of linear transformations. Eur. J. Mech. A/Solids **23**(6), 1021–1039 (2004)
10. Gosselin, C.M., Wang, J.: Singularity loci of a special class of spherical three-degree-of-freedom parallel mechanisms with revolute actuators. Int. J. Robot. Res. **21**(7), 649–659 (2002)
11. Gregorio, R.D.: The 3-RRS wrist: a new, simple and non-overconstrained spherical parallel manipulator. J. Mech. Des. **126**(5), 850–855 (2004)
12. Hervé, J., Sparacino, F.: Structural synthesis of ‘parallel’ robots generating spatial translation. In: Fifth International Conference on Advanced Robotics, ‘Robots in Unstructured Environments’, 91 ICAR, vol. 1, pp. 808–813 (1991)
13. Huang, X., Liao, Q., Wei, S.: Closed-form forward kinematics for a symmetrical 6–6 stewart platform using algebraic elimination. Mech. Mach. Theory **45**(2), 327–334 (2010)

14. Huynh, P., Hervé, J.: Equivalent kinematic chains of three degree-of-freedom tripod mechanisms with planar-spherical bonds. J. Mech. Des. **127**(1), 95–102 (2005)
15. Karouia, M., Hervé, J.: Enumération de mécanismes parallèles sphériques isostatiques. In: 16th Congrès Français de Mécanique, vol. 1, pp. 1–6. Nice, France (2003)
16. Kong, X., Gosselin, C.: Kinematics and singularity analysis of a novel type of 3-CRR 3-DOF translational parallel manipulator. Int. J. Robot. Res. **21**(9), 791–798 (2002)
17. Kong, X., Gosselin, C.: Type synthesis of 3-DOF translational parallel manipulators based on Screw theory. J. Mech. Des. **126**(1), 83–92 (2004)
18. Kong, X., Gosselin, C.: Type synthesis of three-degree-of-freedom spherical parallel manipulators. Int. J. Robot. Res. **23**(3), 237–245 (2004)
19. Kong, X., Jin, Y.: Type synthesis of 3-dof multi-mode translational/spherical parallel mechanisms with lockable joints. Mech. Mach. Theory **96**(Part 2), 323–333 (2016)
20. Lee, C.C., Hervé, J.: Translational parallel manipulators with doubly planar limbs. Mech. Mach. Theory **41**(4), 433–455 (2006)
21. Lee, C.C., Hervé, J.: Type synthesis of primitive Schoenflies-motion generators. Mech. Mach. Theory **44**(10), 1980–1997 (2009)
22. Li, Q., Hervé, J.: Structural shakiness of nonoverconstrained translational parallel mechanisms with identical limbs. IEEE Trans. Robot. **25**(1), 25–36 (2009)
23. Palpacelli, M., Carbonari, L., Palmieri, G., Callegari, M.: Mobility analysis of non-overconstrained reconfigurable parallel manipulators with 3-CPU/3-CRU kinematics. In: Ding, X., Kong, X., Dai, J. (eds.) Advances in Reconfigurable Mechanisms and Robots II, Mechanisms and Machine Science, pp. 189 – 200. Kluwer Academic Publishers, Dordrecht (2016)
24. Palpacelli, M.C., Carbonari, L., Palmieri, G.: Details on the design of a lockable spherical joint for robotic applications. J. Intell. Robot. Syst. **81**(2), 169–179 (2016)
25. Palpacelli, M.C., Carbonari, L., Palmieri, G., Callegari, M.: Analysis and design of a reconfigurable 3-dof parallel manipulator for multimodal tasks. IEEE/ASME Trans. Mechatron. **20**(4), 1975–1985 (2015)
26. Yu, J., Dai, J., Bi, S., Zong, G.: Numeration and type synthesis of 3-DOF orthogonal translational parallel manipulators. Progr. Nat. Sci. **18**(5), 563–574 (2008)
27. Zeng, Q., Ehmann, K., Cao, J.: Tri-pyramid robot: design and kinematic analysis of a 3-DOF translational parallel manipulator. Robot. Comput. Integr. Manufact. **30**(6), 648–657 (2014)
28. Zhou, W., Chen, W., Liu, H., Li, X.: A new forward kinematic algorithm for a general stewart platform. Mech. Mach. Theory **87**, 177–190 (2015)

Dealing with Redundancy of a Multiple Mobile Coil Magnetic Manipulator: A 3RPR Magnetic Parallel Kinematics Manipulator

Baptiste Véron, Arnaud Hubert, Joel Abadie and Nicolas Andreff

Abstract This paper presents a magnetic manipulation system composed of three mobile electromagnets. This system is used to control the position and the orientation of a capsule embedding a small permanent magnet in the horizontal plane. The kinematico-magnetic redundancy of the system is dealt with by imposing the planar $3R\underline{P}R$ parallel kinematics constraints. The resulting controller is demonstrated *in silico*.

1 Introduction

The main existing systems dedicated to contactless manipulation of a magnetic object can be divided in several categories. In [14], we introduced a kinematic criterion which clusters most of the existing systems into two categories: those using *static electromagnets* [2, 7, 8, 11, 15] and those using *mobile permanent magnet(s)* [1, 3–5, 9, 10].

For the remaining systems, a third category emerges: systems using *mobile electromagnets*. This category has been very little studied so far. More, most of the systems that belong to this category have a limited number of degrees of freedom per electromagnet and use a classical architecture with coils in Helmholtz and Maxwell configuration [16, 17].

We propose here to study a system with 3 mobile electromagnets used to control motion of magnetic capsule in the plane. Unlike what is done on most of the systems found in the literature, both movements and supplied currents of the coils are controlled here, which results in a complex non-linear control problem. Specifically, the system is *kinematico-magnetically* redundant, because it possesses 6 inputs (3 currents +3 electromagnet orientations in the horizontal plane) for only 3 outputs (position and orientation of the magnetic capsule in the plane). One way to deal with

---

This work was partly supported by ANR Labex ACTION ("ANR-11-LABX-01-01").

B. Véron · A. Hubert · J. Abadie · N. Andreff (✉)
Institute FEMTO-ST, UBFC/CNRS, Besançon, France
e-mail: Nicolas.Andreff@femto-st.fr

J. Lenarčič and J.-P. Merlet (eds.), *Advances in Robot Kinematics 2016*,
Springer Proceedings in Advanced Robotics 4, DOI 10.1007/978-3-319-56802-7_21

this redundancy, presented here for the first time, is to impose a kinematic constraint and convert this system into a $3R\underline{P}R$ magnetic parallel manipulator, where mechanical prismatic actuators are replaced by magnetic contactless actuators. Potential interests for such an architecture are: (i) it can work in a cluttered environment without the arms sweeping the workspace and (ii) it reduces the ratio between the displaced mass and the manipulator masses.

The system studied is described in Sect. 2. Then, the system model and control law is explained in Sects. 2.1 and 3. Finally, results obtained in simulation are shown in Sect. 4 with emphasis on kinematic issues.

2 System Description

Our system is composed of a permanent magnet placed inside a capsule which is controlled in the horizontal plane (3 degrees of mobility). The control is performed by three electromagnets ($n = 3$) placed in an original architecture presented in [12]. As shown in Fig. 1, each electromagnet has one kinematic degree of freedom: a rotation around the vertical axis.

The system control diagram is presented in Fig. 2. It is a closed loop control composed with a *Perception* block for detecting the capsule current position and orientation. This data is provided to a *Trajectory* block where it is compared with the time-varying desired position to determine the desired accelerations for following this trajectory. The $\begin{bmatrix} m_c \\ \underline{\underline{I_c}} \end{bmatrix}$ block computes the efforts to be applied to the capsule thanks to Newton's law. Finally, the *Controller* block computes the system inputs (currents in the coils $\mathbf{I}$, motion of the coils $\boldsymbol{\Delta\beta}$).

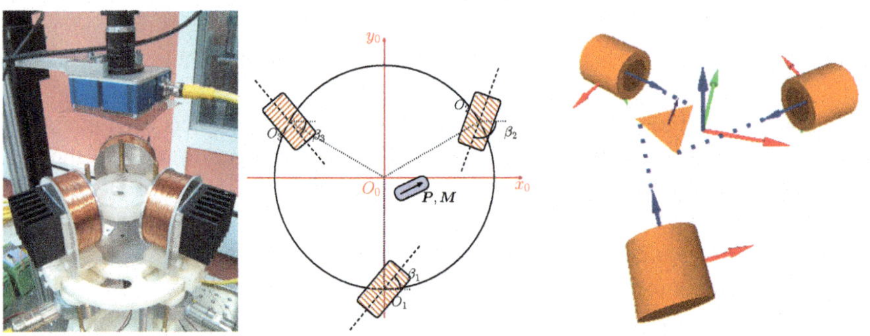

(a) Real system. (b) Schematic description (top-view). (c) A $3R\underline{P}R$ magnetic parallel manipulator.

Fig. 1 System description

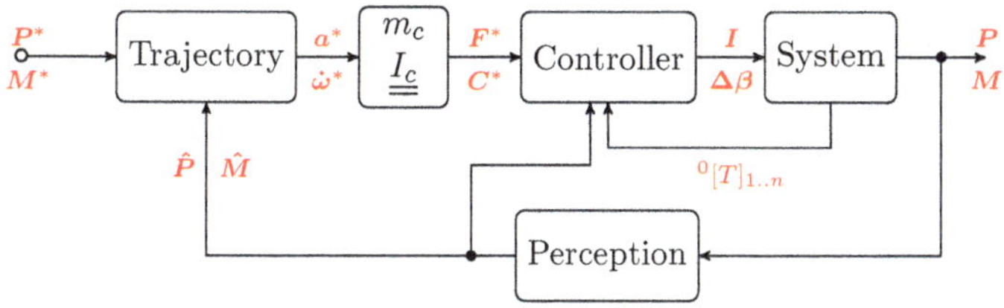

Fig. 2 Control diagram of the system

This control law relies on the direct electromagnetic model (which is established following the methodology explained in [8] while complementing the model with the coils mobility) and deals with the redundancy.

2.1 Direct Electromagnetic Model

Most of the literature assumes that the magnetic field ${}^{i}\boldsymbol{B}_i$ produced at the capsule position ${}^{i}\boldsymbol{P}$ by the ith electromagnet is proportional to the current I_i flowing through the electromagnet:

$$ {}^{i}\boldsymbol{B}_i = I_i \cdot {}^{i}\boldsymbol{b}_i({}^{i}\boldsymbol{P}) \tag{1}$$

with ${}^{i}\boldsymbol{b}_i({}^{i}\boldsymbol{P})$ the magnetic field per current unit created by electromagnet i.

A global reference frame $\mathcal{F}_0$ is defined at the system centre (see Fig. 1b). Each electromagnet orientation is defined by an angle β_i, thus the rotation matrix ${}^{0}\mathbf{R}_i = Rot(\beta_i, z)$ represents the transformation between the local reference frame $\mathcal{F}_i$ and $\mathcal{F}_0$. As a result, the magnetic field ${}^{0}\boldsymbol{B}_i$ produced by an electromagnet is computed in the global frame as:

$$ {}^{0}\boldsymbol{B}_i({}^{0}\boldsymbol{P}, \beta_i) = I_i \cdot {}^{0}\mathbf{R}_i \cdot {}^{i}\boldsymbol{b}_i({}^{0}\mathbf{R}_i^T\, {}^{0}\boldsymbol{P} + {}^{0}\boldsymbol{t}_i) \tag{2}$$

with ${}^{0}\boldsymbol{t}_i = \boldsymbol{O}_0\boldsymbol{O}_i$, the translation vector defining the origin of $\mathcal{F}_i$.

Unlike the model used in [8], Eq. (2) clearly shows the dependence of the magnetic field to the coil variable poses, thanks to ${}^{0}\mathbf{R}_i = Rot(\beta_i, z_0)$. The notation ${}^{0}\boldsymbol{B}_i({}^{0}\boldsymbol{P}, \beta_i)$ is simplified by ${}^{0}\boldsymbol{B}_i$ in the sequel.

The interaction between this field and a magnetic capsule creates efforts on this capsule given by [6]:

$$ {}^{0}\boldsymbol{F}_i = V \cdot \boldsymbol{\nabla}({}^{0}\boldsymbol{M} \cdot {}^{0}\boldsymbol{B}_i) \tag{3}$$

$$ {}^{0}\boldsymbol{C}_i = V \cdot {}^{0}\boldsymbol{M} \wedge {}^{0}\boldsymbol{B}_i \tag{4}$$

with $\boldsymbol{\nabla}$ the gradient operator, $\wedge$ the cross product, V the volume of the magnet inside the capsule, and ${}^{0}\boldsymbol{M}$ its magnetisation.

It is interesting to note that the permanent magnet moment is fixed inside the capsule, thus its magnetisation ${}^0\boldsymbol{M}$ is a good indication of the capsule orientation. Moreover, we assume that the magnetic fields produced by the system are not powerful enough to modify this magnetisation. More, the Jacobian matrix of the magnetic vector ${}^0\boldsymbol{B}_i$ is defined as:

$$ {}^0\mathbf{J}_{Bi} = \left[\frac{\partial\, {}^0\boldsymbol{B}_i}{\partial x} \; \frac{\partial\, {}^0\boldsymbol{B}_i}{\partial y} \; \frac{\partial\, {}^0\boldsymbol{B}_i}{\partial z} \right] \tag{5} $$

Thus, Eqs. (3) and (4) can be expressed in a matrix form as:

$$ {}^0\boldsymbol{F}_i = V \cdot {}^0\mathbf{J}_{Bi}^T \cdot {}^0\boldsymbol{M} \tag{6} $$

$$ {}^0\boldsymbol{C}_i = V \cdot [{}^0\boldsymbol{M}]_\wedge \cdot {}^0\boldsymbol{B}_i \tag{7} $$

with $[{}^0\boldsymbol{M}]_\wedge$ the skew-symmetric matrix associated with the vector cross-product. On our system, the electromagnets are considered far enough from each other so that the coupling between them can be neglected. Thus, air and water being linear mediums for magnetic fields, the superposition principle applies and the overall magnetic field ${}^0\boldsymbol{B}(\boldsymbol{\beta}, {}^0\boldsymbol{P}, \boldsymbol{I})$ produced by the system is the sum of the magnetic fields produced by each electromagnet:

$$ {}^0\boldsymbol{B}(\boldsymbol{\beta}, {}^0\boldsymbol{P}, \boldsymbol{I}) = \sum_{i=1}^{3} I_i \cdot {}^0\mathbf{R}_i \cdot {}^i\boldsymbol{b}_i({}^0\mathbf{R}_i^{-1}\, {}^0\boldsymbol{P} + {}^0\boldsymbol{t}_i) \tag{8} $$

with $\boldsymbol{\beta} = \begin{pmatrix}\beta_1 & \beta_2 & \beta_3\end{pmatrix}^T$ the vector representing the electromagnets configuration and $\boldsymbol{I} = \begin{pmatrix}I_1 & I_2 & I_3\end{pmatrix}^T$ the vector gathering the supplied currents.

Similarly, the gradient of the overall magnetic field is the sum of the gradients produced by each electromagnet:

$$ {}^0\mathbf{J}_B(\boldsymbol{\beta}, {}^0\boldsymbol{P}, \boldsymbol{I}) = \sum_{i=1}^{3} {}^0\mathbf{J}_{Bi} = \sum_{i=1}^{3} I_i \cdot {}^0\mathbf{J}_{bi} \tag{9} $$

with ${}^0\mathbf{J}_{bi}$, the Jacobian matrix of the magnetic field per current unit ${}^0\boldsymbol{b}_i$. The total efforts produced on the capsule are thus given by:

$$ {}^0\boldsymbol{F} = V \sum_{i=1}^{3} I_i \cdot {}^0\mathbf{J}_{bi}^T \cdot {}^0\boldsymbol{M} \tag{10} $$

$$ {}^0\boldsymbol{C} = V\,[M]_\wedge \sum_{i=1}^{3} I_i\, {}^0\mathbf{R}_i\, {}^i\boldsymbol{b}_i({}^0\mathbf{R}_i^{-1}\, {}^0\boldsymbol{P} + {}^0\boldsymbol{t}_i) \tag{11} $$

Introducing $\mathcal{B} = [{}^0\mathbf{R}_1 \cdot \boldsymbol{b}_1 \ {}^0\mathbf{R}_2 \cdot \boldsymbol{b}_2 \ {}^0\mathbf{R}_3 \cdot \boldsymbol{b}_3]$ and $\mathcal{J} = [{}^0\mathbf{J}_{b1}^T \cdot {}^0\boldsymbol{M} \, {}^0\mathbf{J}_{b2}^T \cdot {}^0\boldsymbol{M} \, {}^0\mathbf{J}_{b3}^T \cdot {}^0\boldsymbol{M}]$, leads to express (10) and (11) in matrix form as:

$$
\begin{aligned}
{}^0\boldsymbol{F} &= V \cdot \mathcal{J} \cdot \boldsymbol{I} &&\triangleq \mathcal{A}_{\mathcal{F}}(\boldsymbol{\beta}, {}^0\boldsymbol{P}, {}^0\boldsymbol{M}) \cdot \boldsymbol{I} \qquad (12)\\
{}^0\boldsymbol{C} &= V \cdot [M]_\wedge \cdot \mathcal{B} \cdot \boldsymbol{I} &&\triangleq \mathcal{A}_{\mathcal{C}}(\boldsymbol{\beta}, {}^0\boldsymbol{P}, {}^0\boldsymbol{M}) \cdot \boldsymbol{I} \qquad (13)
\end{aligned}
$$

Matrices $\mathcal{A}_{\mathcal{C}}$ and $\mathcal{A}_{\mathcal{F}}$, which depend on the capsule position $\boldsymbol{P}$ and magnetisation ${}^0\boldsymbol{M}$, are computed from the magnetic fields ${}^0\boldsymbol{B}_i$ created by each coil. As shown in (2), these magnetic fields depend on the orientation of each coil $\boldsymbol{\beta}$.

Finally, these equations can be gathered into the direct electromagnetic model (D_{EM}):

$$
\begin{pmatrix} {}^0\boldsymbol{F} \\ {}^0\boldsymbol{C} \end{pmatrix} = \begin{bmatrix} \mathcal{A}_{\mathcal{F}} \\ \mathcal{A}_{\mathcal{C}} \end{bmatrix} \cdot \boldsymbol{I} = \mathcal{A}(\boldsymbol{\beta}, {}^0\boldsymbol{P}, {}^0\boldsymbol{M}) \cdot \boldsymbol{I} \qquad (14)
$$

This equation enlightens that the magnetic efforts linearly depend on the currents applied in the electromagnets. Each current modifies the efforts applied on the capsule. This model also highlights the non-linear dependence of the matrix $\mathcal{A}(\boldsymbol{\beta}, {}^0\boldsymbol{P}, {}^0\boldsymbol{M})$ to the capsule position and orientation, but also to the orientation of each electromagnet, the total magnetic field depending on the coils configuration. To simplify notations, we write: $\mathcal{A}(\boldsymbol{\beta}, {}^0\boldsymbol{P}, {}^0\boldsymbol{M}) = \mathcal{A}$ in the sequel and we note that this matrix is of size 6×3.

3 Control

Because of the actuation redundancy of the system, several control laws are admissible to control the capsule. The simplest way would be to keep the electromagnets static and to focus on the currents to apply the efforts allowing the capsule to follow a defined trajectory, as in most of the literature. But here, our aim is to optimize the capsule manipulability and to avoid singularities such as those shown in [13].

To find how to move the electromagnets, several strategies are possible because of the system redundancy (3 degrees of mobility in the plane *versus* 3 electromagnetic degrees of freedom plus 3 kinematic degrees of freedom). In this paper, we present a funny way to handle this redundancy by applying a kinematic constraint. Instead of aiming at the capsule center as in [12], we chose to mimic a planar $3R\underline{P}R$ parallel kinematics mechanism. Thereby, the coil axes must always aim at a virtual corner of the $3R\underline{P}R$ platform (Fig. 1c), replacing mechanical prismatic actuators by magnetic contactless actuators.

Once the angular errors $\boldsymbol{\Delta\beta}_{|k}$ between the current coil axis orientation $\boldsymbol{\beta}_{|k-1}$ and the desired one computed from the 3RPR kinematic constraint are known, a proportional control (with gain λ_β) is used to bring the motors to their next configuration:

$$
\boldsymbol{\beta}_{|k} = \boldsymbol{\beta}_{|k-1} + \lambda_\beta \boldsymbol{\Delta\beta}_{|k} \qquad (15)
$$

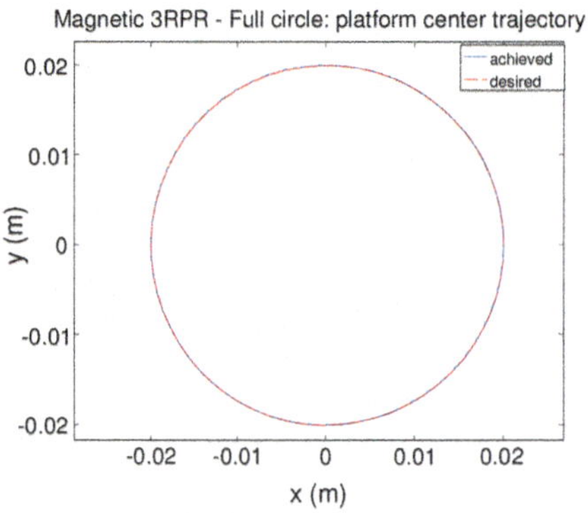

(a) Capsule trajectory.

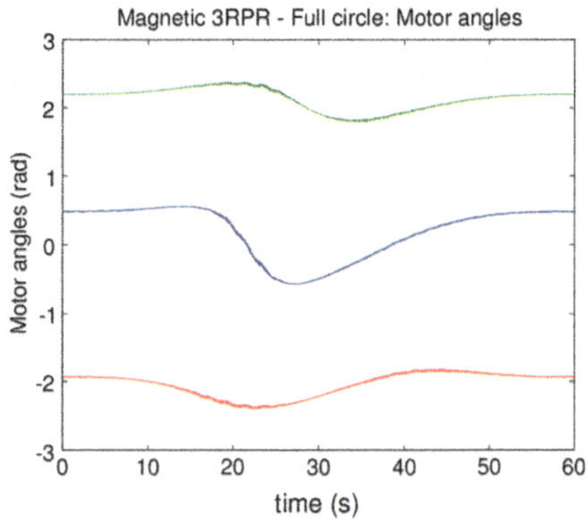

(b) Motor angles.

Fig. 3 Simulation result: **a** trajectory of the capsule and **b** evolution of the motor angles (*right*)

The D_{EM} is updated afterwards and the currents are computed by taking the pseudo-inverse of $\mathcal{A}$:

$$\boldsymbol{I} = \mathcal{A}(\boldsymbol{\beta}_{|k}, {}^0\boldsymbol{P}, {}^0\boldsymbol{M})^{\dagger} \cdot \begin{pmatrix} {}^0\boldsymbol{F}^* \\ {}^0\boldsymbol{C}^* \end{pmatrix} \tag{16}$$

Thus, the modification of the coils orientation allows first to have a better system configuration to realise the requested efforts, second to minimise the supplied current variations. This second point is important, especially if coils have a large number of turns, since it minimises the impact of the coil inductance on current control.

4 Results

This control law was implemented on our C++/OpenGL simulator. To make the simulation more realistic, noise on the capsule position detection (±0.5 mm, ±1°), the currents flowing in the coils (5%) and the coils orientation (±1°) was added.

In this simulation, the capsule follows a circle with its magnetisation tangent to the circle (Fig. 3a), while moving the coils according to the kinematic constraint (Fig. 3b). The trajectory is well performed, with a position error less than 0.3 mm (Fig. 4a) and an orientation error less than 0.3° (Fig. 4b).

5 Conclusions and Perspectives

A magnetic manipulation system with mobile electromagnets was presented in this article. A model of the system was established to compute the magnetic efforts. Unlike the models found in the literature, this one takes into account the mobility

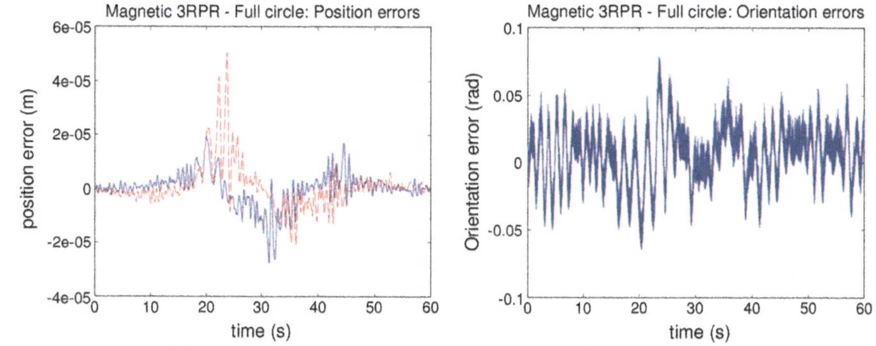

(a) Position errors. (b) Orientation error.

Fig. 4 Simulation result: **a** position errors and **b** orientation error along the trajectory

of each electromagnet composing the system. This yields a highly non-linear control problem, with several additional difficulties: redundancy, kinematico-magnetic couplings, among others.

This opens to the development of new control laws as the one presented here, where redundancy was handled by imposing virtual kinematic linkage. The coil motors thus behave as the first passive joints of a planar $3R\underline{P}R$ parallel kinematics manipulator, whereas the magnetic field created by each coils plays roughly the role of the prismatic actuators. Thereby, kinematic control and magnetic control are decoupled, and magnetic control reuses the literature results.

The effectiveness of this strategy was implemented and tested in simulation. Of course, singularities might occur in the system, but surely in different locations than the kinematic singularities. For instance, the system does not loose torsional rigidity in the kinematic singularity where all legs intersect, because in that case, the magnetic field always produces a torque. This opens up to new kinematico-magnetic analyses, and wider, to new possibilities in the design of manipulation systems. Also, such coupled systems deserve and support research in non-linear and redundant control.

Finally, an important hypothesis was made while developing the direct electromagnetic model: the electromagnets were considered far enough from each other so that coupling between them were neglected. In practice, this hypothesis might not always be true and opens research paths related to electromagnetics.

References

1. Carpi, F., et al.: Magnetically controllable gastrointestinal steering of video capsules. IEEE Trans. Biomed. Eng. **58**(2), 231–234 (2011)
2. Chiba, A., et al.: Magnetic actuator for a capsule endoscope navigation system. J. Magn. **12**(2), 89–92 (2007)

3. Ciuti, G., et al.: Robotic magnetic steering and locomotion of capsule endoscope for diagnostic and surgical endolumial procedures. Robotica **28**(2), 199–207 (2010)
4. Ciuti, G., et al.: Robotic versus manual control in magnetic steering of an endoscopic capsule. Endoscopy **42**(2), 148 (2010)
5. Clark, J., et al.: Magnetic guidance of cochlear implants: proof-of-concept and initial feasibility study. J. Med. Dev. **6**(3), 035,002 (2012)
6. Jackson, J.D.: Classical Electrodynamics, 3rd edn., Wiley, New Jersey (1999)
7. Kratochvil, B.E., et al.: Visual servoing and characterization of resonant magnetic actuators for decoupled locomotion of multiple untethered mobile microrobots. In: IEEE International Conference Robotics and Automation, pp. 1010–1015. IEEE, New York (2009)
8. Kummer, M., et al.: Octomag: An electromagnetic system for 5-dof wireless micromanipulation. IEEE Trans. Robot. **26**(6), 1006–1017 (2010)
9. Latcu, D.G., et al.: Robotic magnetic navigation for ablation of human arrhythmias: initial experience. Archiv. Cardiovasc. Dis. **102**(5), 419–425 (2009)
10. Mahoney, A.W., Abbott, J.J.: Managing magnetic force applied to a magnetic device by a rotating dipole field. Appl. Phys. Lett. **99**(13), 103–134 (2011)
11. Martel, S., et al.: Towards swarms of communication-enabled and intelligent sentotaxis-based bacterial microrobots capable of collective tacks in an aqueous medium. In: IEEE International Conference Robotics and Automation, pp. 2617–2622. IEEE, New York (2009)
12. Véron, B., Abadie, J., Hubert, A., Andreff, N.: Magnetic manipulation with several mobile coils towards gastrointestinal capsular endoscopy. In: New Trends in Mechanism and Machine Science, pp. 681–689. Springer, Berlin (2013)
13. Véron, B., Hubert, A., Abadie, J., Andreff, N.: Geometric analysis of the singularities of a magnetic manipulation system with several mobile coils. In: IEEE/RSJ International Conference Intelligent Robots and Systems (IROS), pp. 4996–5001. IEEE, New York (2013)
14. Véron, B., Hubert, A., Abadie, J., Andreff, N., Renaud, P.: Advocacy for multi mobile coil magnetic manipulation in active digestiveendoscopy. In: Workshop on Magnetically Actuated Multiscale Medical Robots, International Conference on Intelligent Robots and Systems (2012)
15. Xu, T.: Propulsion characteristics and visual servo control of scaled-up helical microswimmers. Ph.D. thesis, Université Pierre et Marie Curie-Paris VI (2014)
16. Yesin, K., Vollmers, K., Nelson, B.: Modeling and control of untethered biomicrorobots in a fluidic environment using electromagnetic fields. Int. J. Robot. Res. **25**(5–6), 527–536 (2006)
17. Yu, C., et al.: Novel electromagnetic actuation system for three-dimensional locomotion and drilling of intravascular microrobot. Sens. Actuators A Phys. **161**(1), 297–304 (2010)

A New Generic Approach for the Inverse Kinematics of Cable-Driven Parallel Robot with 6 Deformable Cables

Jean-Pierre Merlet

Abstract Cable-driven parallel robot (CDPR) have a kinematics that is usually complex as soon as there are possible deformation of the cable due to elasticity or cable mass. The classical approach to solve the kinematics in that case is to inject a cable model in the kinematics equations, that are then solved. According to the cable model this solving may be extremely complex and a change in the model requires to customize the solving algorithm. In this paper we consider the inverse kinematics problem of CDPR with 6 cables and exhibit a generic solving approach that will work for any cable model, provided that it satisfies a minimal assumption. We demonstrate it's use on a CDPR with catenary cables.

1 Introduction

Cable-driven parallel robot (CDPR) have the mechanical structure of the Gough platform with rigid legs except that the legs are cables whose length may be controlled. Numerous applications of CDPRs have been mentioned e.g. large scale maintenance studied in the European project Cablebot [7], rescue robot [6, 9] and transfer robot for elderly people [4] to name a few. We will assume that the output of the coiling system for cable i is a single point A_i, while the cable is connected at point B_i on the platform. A cable may be assumed to be mass-less and *non-deformable* i.e. the cable shape is the linear segment going from A to B and its length does not change whatever is the tension in the cable or may be *deformable* i.e. the previous assumptions on the cable shape and/or its lengths do not hold. For example Fig. 1 presents a robot with sagging cables. In this paper we will consider the inverse kinematics problem (IK) for CDPR having 6 cables. If we assume that the cables are non-deformable, then being given the pose of the platform the lengths of the cables are obtained directly as the 2-norm of the vector $\mathbf{A_iB_i}$ that is obtained directly from the platform pose. But as the cable may exert only a positive tension we have to consider the static equation obtained as

J.-P. Merlet (✉)
INRIA Sophia-Antipolis, Biot, France
e-mail: Jean-Pierre.Merlet@inria.fr

J. Lenarčič and J.-P. Merlet (eds.), *Advances in Robot Kinematics 2016*,
Springer Proceedings in Advanced Robotics 4, DOI 10.1007/978-3-319-56802-7_22

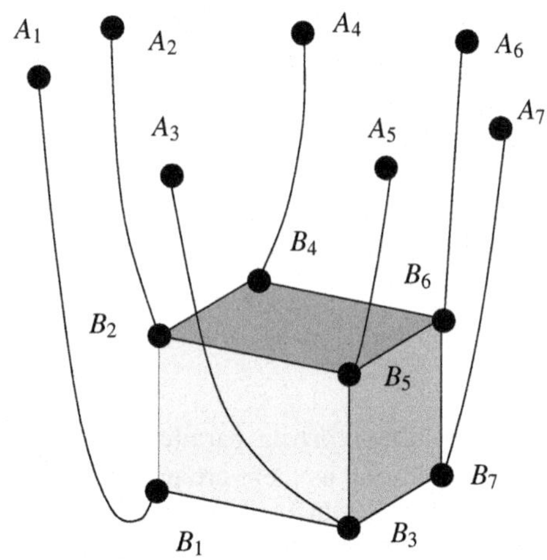

Fig. 1 Cable driven parallel robots with sagging cables

$$\mathscr{F} = \mathbf{J}^{-\mathbf{T}}\tau \quad (1)$$

where $\mathscr{F}$ is the external wrench applied on the platform, τ is the vector of the cable tensions and $\mathbf{J}^{-\mathbf{T}}$ is the transpose of the inverse kinematic jacobian of the robot, that is fully determined as soon as the pose of the platform is known. Equation (1) is a 6 dimensional linear system that may easily be solved to provide the cable tensions. If all these tensions are positive, then we have got a solution for the IK, otherwise the IK has no solution. If the cable are deformable there has been very few works addressing the IK solving: Riehl [8] and Hui [1] assume both Irvine sagging model [2] for the cable but their numerical solver provides only a single solution, if any. Using the same cable model we have exhibited a solving algorithm that allows one to calculate all the solutions [5] (and exhibit a case for which a CDPR has 3 solutions) but the solving algorithm is computer intensive. Simple linear elasticity model has been used to study the kinematics of a special configuration of CDPR [3]. But to the best of the author knowledge no upper bound on the number of solutions of the IK has even been provided and no other cable model has been studied. The purpose of this paper is to provide a generic solving approach that can be used whatever is the cable model and possibly allow to provide a (probably largely overestimated) bound on the number of solutions. An essential issue is the concept of cable model that is addressed in the next section.

2 Cable Model

We denote by L_0 the length of a cable before it is submitted to any deformation and by $\mathscr{P}$ a set of parameters that allows one to describe the physical properties of the cable with respect to deformation under tension. A cable model is a set of relations

$\mathbf{T}(A, B, L_0, \tau, \mathscr{P}) = 0$ that allows one to determine the cable force action τ at point B according to the values of A, B, L_0.

For using our IK solving approach the following assumptions on the cable model will be required:

1. the set $\mathbf{T}$ is constituted of continuous and differentiable functions,
2. the writing of $\mathbf{T}$ may involve new unknowns but the number of equations in $\mathbf{T}$ is such that for given $A, B, L_0, \mathscr{P}$ the system has only a discrete number of solutions
3. for each parameter in $\mathscr{P}$ there is a limit value such that the cable model will be asymptotically identical to the non-deformable cable model

As example of cable model we may mention the Irvine sagging cable model that is valid for elastic cable with mass. In this model we consider the vertical plane that includes the cable and assume that the cable is attached at point A with coordinates $(0, 0)$ while the other extremity is attached at point B with coordinates $(x_b \geq 0, z_b)$. The vertical and horizontal forces F_z, F_x are exerted on the cable at point B and the cable length at rest is L_0. With this notation the coordinates of B are related to the forces F_x, F_z [2] by the C^n functions:

$$x_b = F_x\left(\frac{L_0}{EA_0} + \frac{sinh^{-1}(F_z) - sinh^{-1}((F_z - \frac{\mu g L_0}{F_x})}{\mu g}\right) \tag{2}$$

$$z_b = \frac{\sqrt{F_x^2 + F_z^2} - \sqrt{F_x^2 + (F_z - \mu g L_0)^2}}{\mu g} + \frac{F_z L_0}{EA_0} - \frac{\mu g L_0^2}{2EA_0}$$

where E is the Young modulus of the cable material, μ its linear density, A_0 the surface of the cable cross-section and $F_x > 0$. For the IK problem the coordinates x_b, z_b are known, the L_0 have to be determined and two new unknowns are introduced, F_x, F_z, while this cable model provide two relations. Consequently this model satisfies assumption 1 and 2. Assume now that E goes to infinity and μ to 0. The limit values x_b^l, z_b^l of x_b, z_b are then

$$x_b^l = \frac{L_0 F_x}{\sqrt{F_x^2 + F_z^2}} \qquad z_b^l = \frac{L_0 F_z}{\sqrt{F_x^2 + F_z^2}} \tag{3}$$

which corresponds to a cable directed along the line A, B and exerting a force of amplitude $\sqrt{F_x^2 + F_z^2}$. Therefore this cable model also satisfies assumption 3.

3 The Continuity Model

Let us consider a CDPR with n cables and a cable model that involves p unknowns: consequently the IK has n (the cable lengths) plus np unknowns. As for the equations we have 6 equations coming from (1) and np equations coming from the cable model for a total of $6 + np$ equations. The system of IK equations is therefore square if $n = 6$ which is the case we are considering in this paper.

3.1 The Inverse Model

Assume that a cable model has been decided and a solving algorithm has allowed us to determine the IK solution(s) for a given pose and for given values of the cable model parameters $\mathscr{P}_d$. The cable model and IK equations being C^1 we know that for a "small" perturbation on $\mathscr{P}$ we will get IK solutions that are close to the initial one. Furthermore for a given IK solution **S** (i.e. a set of length for the 6 cables) with the initial $\mathscr{P}$ we also know that the Newton-Raphson algorithm with **S** as initial guess may converge toward the solution of the IK with the new values of $\mathscr{P}$, the convergence being ensured as soon as the perturbation is small enough provided that the system is not singular at **S**. The Kantorovitch theorem [10] allows one to determine the meaning of a small perturbation: provided that the jacobian of the new system has an inverse at **S** and that some conditions are satisfied for the norm of the equations at **S**, for the norm of the jacobian inverse and for the norm of the Hessian matrix of the system, then the theorem ensures that there is a single solution of the new system in a ball centered at **S** and guarantees that the Newton-Raphson scheme will converge toward this solution. Let $\mathscr{P}_s$ be the cable model parameters limit values and a linear iterative interpolation scheme defined by

$$\mathscr{P}_{k+1} = \mathscr{P}_k + \alpha(\mathscr{P}_s - \mathscr{P}_k)$$

initialized with $\mathscr{P}_0 = \mathscr{P}_d$. For the values $\mathscr{P}_d$ we assume that we know a set of n solutions $\mathscr{S}_0 = \{S_1, S_2, \ldots, S_n\}$. We will choose the positive α in such a way that the IK system obtained with the parameters $\mathscr{P}_{k+1}$ satisfies the conditions of the Kantorovitch theorem for the solutions obtained for the system whose parameter values are $\mathscr{P}_k$). If the conditions does not hold we divide α by 2. For example starting from $\mathscr{P}_0$ we set α to an arbitrary small value and test the condition of the Kantorovitch theorem for $\mathscr{P}_1$ and decrease α until they hold. At this stage we will use Newton to calculate the set $\mathscr{S}_1$ of the n solutions for the parameter set $\mathscr{P}_1$. We will stop this scheme when α is close to 1. To determine how α should be close to 1 to stop the process we look at the cable tensions and lengths for all solutions in $\mathscr{S}_k$ to determine what will be the non-deformable case to which will lead the parameters going to their limit values. Namely we compare all cable lengths L_0^i to the distance d_i between the A_i, B_i points and if $L_0^i > d_i$, then the i-th cable is slack otherwise it is under tension. This provide us a system of equations for each solution of $\mathscr{S}_k$ which should have as approximate solution the corresponding element in $\mathscr{S}_k$. If the Kantorovitch conditions hold for the system, then we will be able to determine the non-deformable configuration to which the deformable solution will lead, otherwise we increase α.

3.2 The Direct Model

As we have shown in the inverse model the manipulator ends up close to a configuration with non-deformable cables for which the platform is at the desired pose with some cables that are possibly slack. We may now revert the process to obtain the IK solutions for the deformable cables. We will consider all combinations of cables under tension in the set of 6 cables, assuming that they are non-deformable. We then solve their IK, retaining only the solutions for which the non slack cable have a positive tension.

There is clearly at most one solution when considering that all 6 cables are under tension. This can be checked by solving equation (1) that is a six-dimensional linear system in the 6 cable tensions. For the combinations with less than 6 cables under tension we will assume that the slack cable have 0 tension or a minimal one depending on the cable model. Hence if the CDPR has m cables under tension the system (1) is still a linear system, possible overconstrained, that may have a positive solution in terms of the tensions in the m cables.

After this processing we get feasible configurations for the non-deformable case. For each of them we have a set of valid lengths L_0 for the cables. $\mathscr{P}$ is set to a value $\mathscr{P}_i$ close to $\mathscr{P}_s$ and then we change the parameter using the iterative scheme:

$$\mathscr{P}_{k+1} = \mathscr{P}_k + \beta(\mathscr{P}_d - \mathscr{P}_k) \tag{4}$$

with $\mathscr{P}_0 = \mathscr{P}_i$. As in the inverse scheme we choose β small enough so that the Kantorovitch conditions are fulfilled for $\mathscr{P} = \mathscr{P}_{k+1}$. We then stop the process when $\mathscr{P}_{k+1} = \mathscr{P}_d$ and at this stage we have obtained the IK solution(s) for the CDPR with deformable cables.

Note that this scheme starts with a non-deformable cables state with possibly some slack cables. However during the iterations it may perfectly happen that an initially slack cable, which therefore does not support the platform, becomes supportive and vice-versa.

3.3 Maximum Number of Solutions

The inverse scheme shows that the IK solutions originates from an IK solution with non-deformable cables. Being given a distribution of slack and under tension cables there is always at most a single solution to the IK problem and hence the total number of solutions of the IK with deformable cables cannot exceed the total number of slack/under tension combinations. This number may be established as 1 (6 cables under tension) +6 (5 cables under tension) +15 (4 cables under tension) +20 (3 cables under tension) +15 (2 cables under tension) +6 (1 cable under tension) which amounts to a maximum of 63 solutions. However this number will be the real bound under some assumptions on singularities.

3.4 The Singularity Case and Workspace Calculation

Both the inverse and direct scheme assume that the jacobian of the system does not become singular. Such a case may occur if the cable model equations are singular or a singularity will occur if (1) become singular. If the unit vector $\mathbf{n_i}$ denotes the direction of action of cable i at point B_i ($\mathbf{n_i}$ may differ from the unit vector of the line going through A_i, B_i because of the cable deformation) and τ_i the tension in the cable at B_i then we define the interaction matrix $\mathbf{G}$ whose $i\text{-}th$ row $\mathbf{G_i}$ is defined by $\mathbf{G_i} = (\mathbf{n_i} \quad \mathbf{CB_i} \times \mathbf{n_i})^T$ so that the static of the CDPR may be written as

$$\mathscr{F} = \mathbf{G}\tau \tag{5}$$

A singularity will occur if the Plücker vectors $\mathbf{G_i}$ are dependent, a well known problem for the analysis of parallel robots. Such a singularity may be detected through an increase in the tension of some cables. Open issues regarding this aspect are:

- can we avoid a singularity by modifying the iterative scheme (4)?
- in the inverse scheme can we encounter a singularity that will prohibit us to converge toward a solution with non-deformable cables? If this is the case, then the approach may miss IK solutions. Should we consider complex values for the unknowns in order to avoid singularities?
- is possible to have IK solutions with deformable cables for which the inverse continuation problem does not lead to a non deformable cable IK configuration?

All these issues are quite complex and will be the subject of another paper(s). Our conjecture is that in general we will have only isolated singular points so that by using bifurcation theory at the singular point so that all IK solutions will originate from (possibly multiple) non-deformable configuration. If this conjecture is true it has an important practical consequence: *the reachable workspace of a CDPR with deformable cables, whatever the cable model, is identical to the reachable workspace of the same CDPR that has non-deformable cable.*

4 Example

We consider as example our large scale robot MARIONET-CRANE [6], probably the largest CDPR ever deployed, for which we will assume Irvine sagging cables. The IK problem has already been studied in [5] but we will correct some mistakes of this paper. We assume here that the external wrench applied on the platform is only the gravity. This robot is a suspended CDPR (i.e. there is no cable having a B point under the platform) with 6 cables, whose A_i, B_i coordinates are given in Table 1.

The cables characteristics are $E = 100^9\ \mathrm{N/m^2}$, $\mu = 0.079\ \mathrm{kg/m}$ and their diameter is 4 mm. For finding the IK solutions for the non deformable case we assume that the slack cables act along the vertical with a tension equal to $\mu g L_0/2$, where μ has

Table 1 Coordinates of the A_i and B_i points on the base and on the platform (in cm, by rows)

x	y	z	x	y	z	x	y	z	x	y	z
–325.9	–47.5	882.6	942.1	–348.2	1155.5	–10	–93	–3	10	–93	–3
953.8	379.7	1153.3	557.0	2041.4	870.4	27	50	–7	27	50	–7
–250.5	1681.0	864.9	–334.2	942.1	878.8	–27	50	–7	–27	50	–7

a very low value. With that assumption equation (1) is 6 dimensional linear system in the tensions of the cables that are supposed to be under tension and in the L_0 for the slack cables. We keep as potential IK solution the one for which the tensions and the L_0 are positive.

We are basically finding the same IK solution as in [5] except that for the pose $x = 400$, $y = 700$, $z = 200$ and a platform mass of 69 kg our new IK algorithm provides a solution although we have claimed that there was none. This can be explained as the IK solution leads to F_X values that are not included in the intervals we have provided in our previous algorithm. However we confirm that for $x = y = 0$, $z = 200$ there is no IK solution. It must also be noted that we have found cases with a singular configuration for the direct scheme but it appears when the F_x of cable(s) are close to 0 (corresponding to a singularity of the cable model). In that case we use a *simplified model*: we rewrite the equations with $F_x = 0$, $F_z = \mu g L_0/2$ (corresponding to a cable that acts vertically on the platform) and we remove the Irvine equations for the corresponding cable(s) so that this system is still square. Then we use the Newton scheme to solve the simplified system and go on decreasing E and increasing μ. After eachg successfull solving step of the simplified model we use Kantorovitch and Newton to get a solution of the full system, using the solution obtained for the simplified system and setting a small positive value for the F_x of the singular cables. If we succeed, then we switch to the full model.

The previous paper have shown an example with up to 3 solutions but our new algorithm has allowed to find examples with 5 solution, for example for $x = 96.733$, $y = 1138.33$, $z = 165$, $\mu = 0.004$ kg/m and a mass of 10 kg,the platform being horizontal. We get 5 IK solutions for non-deformable cables with the following cables that are not under tension: none, [3], [6], [3, 5], [4, 6] but 4 of them have an unreasonable cable lengths (for all of them one cable has a length over 105 m and may reach 4103 m). Note that if we increase μ to 0.079, then the branches none and [3] meet the same point for $\mu = 0.0042$ and apparently no solution can be found for larger values of μ so that we end up with only 3 solutions. This shows that apparently there may be a limit on the value of μ that may lead two branches to collapse and not generating an IN solution.

However a more rigourous singularity analysis has to be performed in order to guarantee that will not miss one of the IK solution. An extensive search on the x, y, z using a grid and a fixed orientation has provided at most 5 potential IK solutions for the non-deformable case. During this search the only singularity we have found were cable model singularity with one F_x going to 0. For a grid of over 150 000 points we get 1, 2, 3, 4, 5 potential solutions respectively in 76, 19, 5, 0.01 and 0.05% of the cases.

5 Conclusion

Finding all IK solutions of CDPR with deformable cables may be complex according to the used cable model. We propose in this paper a generic approach that allow to manage all cable models that satisfy minimal assumptions, while providing for the first time an upper bound for the maximal number of solutions, provided that our conjecture on singularity hold. Although it has be proven to be efficient even for a complex cable model, the issue of the crossing of singularity remains to be addressed. If the CDPR has more than 6 cables the IK equations have more unknowns than equations but we may still apply the method after having chosen specific tensions for the non-deformable case. At each step we may choose a close but different set of tensions that satisfy some optimality criterion and then solve the IK equations for the current values of the parameters. If the CDPR has less than 6 cables the procedure may also be used provided that we choose which dof of the platform has to be controlled.

References

1. Hui, L.: A giant sagging-cable-driven parallel robot of FAST telescope:its tension-feasible workspace of orientation and orientation planning. In: 14th IFToMM World Congress on the Theory of Machines and Mechanisms. Taipei (October, 27–30, 2015)
2. Irvine, H.M.: Cable Structures. MIT Press, Cambridge (1981)
3. Merlet, J.P.: The kinematics of the redundant N-1 wire driven parallel robot. In: IEEE International Conference on Robotics and Automation, pp. 2313–2318. Saint Paul (May, 14–18, 2012). http://www-sop.inria.fr/coprin/PDF/icra2012.pdf
4. Merlet, J.P.: MARIONET, a family of modular wire-driven parallel robots. In: ARK, pp. 53–62. Piran (June 28–July 1, 2010). http://www-sop.inria.fr/coprin/PDF/ark2010.pdf
5. Merlet, J.P.: On the inverse kinematics of cable-driven parallel robots with up to 6 sagging cables. In: IEEE International Conference on Intelligent Robots and Systems (IROS), pp. 4536–4361. Hamburg, Germany (September 28–October 2, 2015)
6. Merlet, J.P., Daney, D.: A portable, modular parallel wire crane for rescue operations. In: IEEE International Conference on Robotics and Automation, pp. 2834–2839. Anchorage (May, 3–8, 2010). http://www-sop.inria.fr/coprin/PDF/icra2010.pdf
7. Pott, A., et al.: IPAnema: a family of cable-driven parallel robots for industrial applications. In: 1st International Conference on Cable-driven Parallel Robots (CableCon), pp. 119–134. Stuttgart (September, 3–4, 2012)
8. Riehl, N., et al.: Effects of non-negligible cable mass on the static behavior of large workspace cable-driven parallel mechanisms. In: IEEE International Conference on Robotics and Automation, pp. 2193–2198. Kobe (May, 14–16, 2009)
9. Tadokoro, S., et al.: A portable parallel manipulator for search and rescue at large-scale urban earthquakes and an identification algorithm for the installation in unstructured environments. In: IEEE International Conference on Intelligent Robots and Systems (IROS), pp. 1222–1227. Kyongju (October, 17–21, 1999)
10. Tapia, R.: The Kantorovitch theorem for Newton's method. Am. Math. Mon. **78**(1.ea), 389–392 (1971)

Rolling Contact in Kinematics of Multifingered Robotic Hands

Lei Cui and Jian S. Dai

Abstract Rolling contact has been used in fine-manipulation by robot hands. It provides an opportunity to manipulate an object to a desired location even when the finger linkages do not have enough degrees of freedom. This paper was aimed to provide a systematic approach to fine-manipulation of multifingered robot hands with rolling contact. We established a Jacobian of a multifingered hand with point contact and formulated the forward and inverse characteristic equations in terms of joint velocities and contact trajectories. The results of the forward kinematics are useful for simulation and off-line programming, and the results of the inverse kinematics provide direct inputs to a kinematic controller.

1 Introduction

Humans use rolling contact to achieve fine manipulation when grasping small and smooth objects, for example rolling a coin between two fingers. The development of multifingered robotic hands has been aimed to imitate and replicate, ultimately to surpass, the functionality of the human hand. Thus, Rolling contact has been applied to in-hand fine manipulation for a grasped object to achieve a desired location, which may not be feasible if only the finger joints are used [1, 8].

A useful kinematic formulation of in-hand manipulation should consider computational efficiency and the ease with which it can be manipulated. Further, it should be flexible enough to admit a degree of coordinate independence, i.e. a given problem should not be confined to any specific choice of reference frames to carry out the kinematic analysis [2, 13].

Motivated in part by these considerations, the product of exponential formula [2, 9–11] has been used in analyzing kinematics of multifingered hands [4, 5, 9,

L. Cui (✉)
Department of Mechanical Engineering, Curtin University, Bentley, WA, Australia
e-mail: lei.cui@curtin.edu.au

J.S. Dai
Centre for Robotics Research, King's College London, London, UK
e-mail: jian.dai@kcl.ac.uk

J. Lenarčič and J.-P. Merlet (eds.), *Advances in Robot Kinematics 2016*,
Springer Proceedings in Advanced Robotics 4, DOI 10.1007/978-3-319-56802-7_23

12]. However, the kinematics of rolling contact is traditionally derived in terms of the coordinates of surface patches, thus these coordinates inevitably appear in the kinematic formulations. This to some extent compromises the initial motivation.

This work presents a systematic approach to the kinematics of in-hand manipulation with rolling contact, where the Darboux frame, which moves along a trajectory on each of a fingertip surface and an object surface, is applied to the forward and inverse kinematics of the moving object [3, 6, 7]. The combination of the product of exponential and the moving frame method yields an algebraic formulation, providing an alternative to a differential equation approach.

2 Forward Kinematics of In-Hand Manipulation with Rolling Contact

When the object surface and the fingertip surface roll on each other, the point of contact moves across both surfaces. The object undergoes a spin-rolling motion with respect to the fingertips. Each fingertip has to follow the object and maintain contact. In most cases, the motions of each finger are independent, so it suffices to study the motion of a single finger.

Let $(P - \mathbf{ijk})$ represent a frame fixed at the palm, $(T - \mathbf{i}'\mathbf{j}'\mathbf{k}')$ a tool frame fixed on the fingertip, S_1 to S_n the screws related to the n revolute joints, L and L' the contact trajectories respectively on the fingertip and on the object, the Darboux frame $(M - \mathbf{e}_1\mathbf{e}_2\mathbf{e}_3)$ at the contact point M, and $\mathbf{e}_1$ the tangent vector to the contact trajectories and $\mathbf{e}_3$ the common normal vector, as in Fig. 1.

The frame $(T - \mathbf{i}'\mathbf{j}'\mathbf{k}')$ on the fingertip with respect to the palm frame $(P - \mathbf{ijk})$ can be obtained by the product of exponential formula in terms of joint velocities [9] as

$$\mathbf{g}_{PT} = e^{S_1\eta_1} e^{S_2\eta_2} \dots e^{S_n\eta_n} \mathbf{g}_{PT}(0) \tag{1}$$

where η_1 to η_n are the joint angles of the finger joints and $\mathbf{g}_{PT}(0)$ is the initial location of the frame T.

The twist of the frame $(T - \mathbf{i}'\mathbf{j}'\mathbf{k}')$ with respect to the frame $(P - \mathbf{ijk})$ can be obtained as

$$\mathbf{S}_{PT} = \mathbf{J}(\eta)\dot{\eta} \tag{2}$$

where $\mathbf{J}(\eta) = \left[\mathbf{S}'_1\ \mathbf{S}'_2 \dots \mathbf{S}'_n\right]$, $\mathbf{S}'_i = Ad_{e^{S_1\eta_1}\dots e^{S_{i-1}\eta_{i-1}}} S_i$, $\dot{\eta} = \left[\dot{\eta}_1\ \dot{\eta}_2\ \dots\ \dot{\eta}_n\right]^T$.

The homogeneous transformation matrix of the frame $(M - \mathbf{e}_1\mathbf{e}_2\mathbf{e}_3)$ with respect to the frame $(T - \mathbf{i}'\mathbf{j}'\mathbf{k}')$ is

$$\mathbf{g}_{TM} = \begin{bmatrix} \mathbf{E} & \mathbf{M}^T \\ 0 & 1 \end{bmatrix} \tag{3}$$

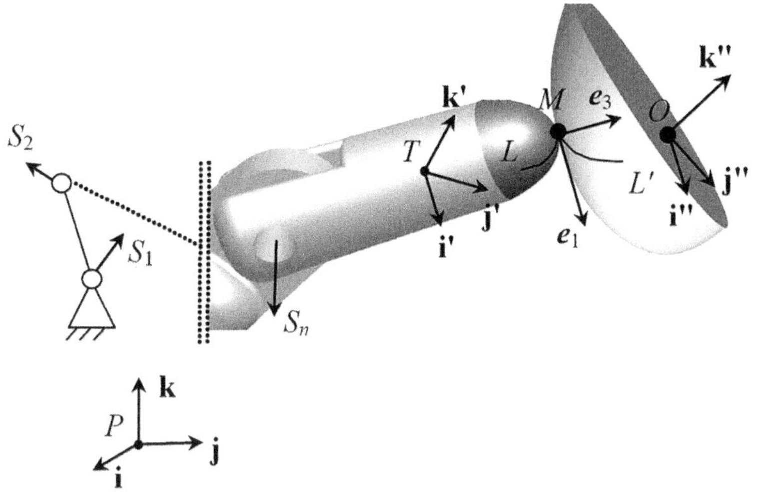

Fig. 1 Screws of the finger joints and the Darboux frame ($M - \mathbf{e}_1\mathbf{e}_2\mathbf{e}_3$) of the fingertip and the object surfaces

where $\mathbf{E} = \left[\mathbf{e}_1^T \ \mathbf{e}_2^T \ \mathbf{e}_3^T\right]$. The differentiation of the matrix $\mathbf{E}$ with respect to time t yields

$$\frac{d\mathbf{E}}{dt} = \sigma\mathbf{QE} \tag{4}$$

where the scalar σ is the rolling rate of the contact curve L, and

$$\mathbf{Q} = \begin{bmatrix} 0 & k_n & k_g \\ -k_n & 0 & \tau_g \\ -k_g & -\tau_g & 0 \end{bmatrix} \tag{5}$$

The scalars k_n, k_g, τ_g are the normal curvature, geodesic curvature, and geodesic torsion of the contact curve L on the fingertip, respectively. The differentiation of the vector of the contact point M with respect to time t yields

$$\frac{d\mathbf{M}}{dt} = \frac{d\mathbf{M}}{ds}\frac{ds}{dt} = \sigma\mathbf{e}_1 \tag{6}$$

Thus the differentiation of $\mathbf{g}_{TM}$ with respect to time t can be obtained as

$$\dot{\mathbf{g}}_{TM} = \sigma\begin{bmatrix} \mathbf{QE} & \mathbf{e}_1^T \\ \mathbf{0} & 0 \end{bmatrix} \tag{7}$$

The space velocity of the frame M with respect to T can be obtained as

$$\mathbf{V}_{TM} = \dot{\mathbf{g}}_{TM}\mathbf{g}_{TM}^{-1} = \sigma \begin{bmatrix} \mathbf{QE} & \mathbf{e}_1^T \\ \mathbf{0} & 0 \end{bmatrix} \begin{bmatrix} \mathbf{E}^T & -\mathbf{E}^T\mathbf{M} \\ \mathbf{0} & 1 \end{bmatrix} = \sigma \begin{bmatrix} \mathbf{Q} & \mathbf{e}_1^T - \mathbf{QM}^T \\ \mathbf{0} & \mathbf{0} \end{bmatrix} \tag{8}$$

It follows that the twist of the frame M with respect to the frame T is

$$\mathbf{S}_{TM} = \begin{bmatrix} \sigma\mathbf{e}_1 - \mathbf{M} \times \boldsymbol{\omega}_M & \boldsymbol{\omega}_M \end{bmatrix} \tag{9}$$

where

$$\boldsymbol{\omega}_M = \sigma(-\tau_g\mathbf{e}_1 + k_n\mathbf{e}_2 - k_g\mathbf{e}_3) \tag{10}$$

The angular velocity of the object with respect to the frame $(M - \mathbf{e}_1\mathbf{e}_2\mathbf{e}_3)$ [3] is

$$\boldsymbol{\omega}_O = \omega_1\mathbf{e}_1 + \omega_2\mathbf{e}_2 + \omega_3\mathbf{e}_3 \tag{11}$$

where $\omega_1 = -\sigma(\tau_g' - \tau_g)$, $\omega_2 = \sigma(k_n' - k_n)$, $\omega_3 = -\sigma(k_g' - k_g)$ and k_g', k_n', τ_g' are the geodesic curvature, normal curvature, and geodesic torsion of the curve L' on the object. The twist of the object frame $(O - \mathbf{i}''\mathbf{j}''\mathbf{k}'')$ with respect to the frame $(M - \mathbf{e}_1\mathbf{e}_2\mathbf{e}_3)$ is

$$\mathbf{S}_{MO} = \begin{bmatrix} \mathbf{O} \times \omega_M & \omega_1\mathbf{e}_1 + \omega_2\mathbf{e}_2 + \omega_3\mathbf{e}_3 \end{bmatrix} \tag{12}$$

where the vector $\mathbf{O}$ represents the vector $\mathbf{MO}$ with respect to the frame $(M - \mathbf{e}_1\mathbf{e}_2\mathbf{e}_3)$.

Hence the twist of the object with respect to the palm frame P can be obtained as

$$\mathbf{S}_{PO} = \mathbf{S}_{PT} + Ad_{\mathbf{g}_{PT}}\mathbf{S}_{TM} + Ad_{\mathbf{g}_{PM}}\mathbf{S}_{MO} \tag{13}$$

This gives the instantaneous motion of the object under the effects of rolling contact and the joints rates.

3 Inverse Kinematics of In-Hand Manipulation with Rolling Contact

The inverse kinematics is to obtain the joint rates and the contact trajectories given the twist of the object. Suppose the surfaces of the fingertip is parameterized as $\mathbf{r}(u, v)$ and the surface of the object as $\mathbf{r}'(\alpha, \beta)$. Let $\mathbf{e}_u$ represent the tangent vector to the u- curve and $\mathbf{e}_\alpha$ the unit tangent vector to the α-curve. The common normal vector is $\mathbf{e}_3$ and the angle between $\mathbf{e}_u$ and $\mathbf{e}_\alpha$ is θ, as in Fig. 2.

A general rigid object has 6 degrees-of-freedom (DOF). Hence the number of joints is at least 3 for a robotic finger to manipulate the object, since rolling contact provides extra 3 terms to the system. In the authors' previous work [7], these three

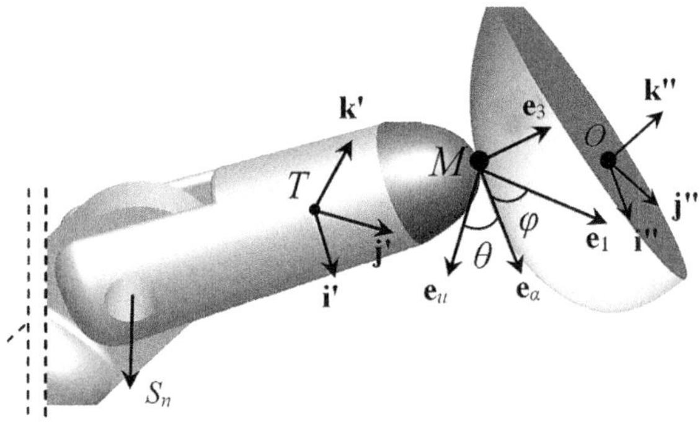

Fig. 2 The Darboux frames $(M - \mathbf{e}_u\mathbf{e}_v\mathbf{e}_3)$, $(M - \mathbf{e}_\alpha\mathbf{e}_\beta\mathbf{e}_3)$, and $(M - \mathbf{e}_1\mathbf{e}_2\mathbf{e}_3)$ at the contact point M

terms were defined as the rolling rate σ, the angle φ between the rolling direction $\mathbf{e}_1$ and $\mathbf{e}_\alpha$, and the complementary spin-speed ω_3'.

Hence the problem of inverse kinematics is formulated to obtain the joint rates $\dot{\eta}_1$ to $\dot{\eta}_n$, and the aforementioned three terms given the twist of the object $S_{obj} = [\mathbf{v}_O, \boldsymbol{\omega}_O]$.

The vectors $\mathbf{e}_1$, $\mathbf{e}_2$, and $\mathbf{e}_3$ can be obtained in the frame $M - \mathbf{e}_\alpha\mathbf{e}_\beta\mathbf{e}_3$ as

$$\begin{aligned} \mathbf{e}_1 &= \cos(\varphi + \theta)\mathbf{e}_\alpha + \sin(\varphi + \theta)\mathbf{e}_\beta \\ \mathbf{e}_2 &= -\sin(\varphi + \theta)\mathbf{e}_\alpha + \cos(\varphi + \theta)\mathbf{e}_\beta \end{aligned} \tag{14}$$

The normal curvature k_n, geodesic curvature k_g, and the geodesic torsion τ_g in the direction of $\mathbf{e}_1$ in terms of the curvatures of the fingertip surface can be obtained accordingly.

It follows from Eqs. 9, 14 and 15 that the twist $\mathbf{S}_{TM}$ is a 6×1 vector function in terms of φ and ω_3'

$$\mathbf{S}_{TM} = \begin{bmatrix} M_2\omega_{M_3}(\sigma, \varphi, \omega_3') - M_3\omega_{M_2}(\sigma, \varphi) \\ M_3\omega_{M_1}(\sigma, \varphi) - M_1\omega_{M_3}(\sigma, \varphi, \omega_3') \\ M_1\omega_{M_2}(\sigma, \varphi) - M_2\omega_{M_1}(\sigma, \varphi) \\ \omega_{M_1}(\sigma, \varphi) \\ \omega_{M_2}(\sigma, \varphi) \\ \omega_{M_3}(\sigma, \varphi, \omega_3') \end{bmatrix}^T \tag{15}$$

where M_1 to M_3 are the coordinates of the point M and ω_{M_1} to ω_{M_3} are the components of the angular velocity ω_M.

Similarly, it follows from Eq. 12 that the twist S_{MO} is a vector function in terms of φ and ω_3'

$$\mathbf{S}_{TM} = \begin{bmatrix} O_2\omega_{O_3} - O_3\omega_{O_2} \\ O_3\omega_{O_1} - O_1\omega_{O_3} \\ O_1\omega_{O_2} - O_2\omega_{O_1} \\ \omega_{O_1}(\sigma, \varphi) \\ \omega_{O_2}(\sigma, \varphi) \\ \omega_{O_3}(\sigma, \varphi, \omega_3') \end{bmatrix}^T \tag{16}$$

where O_1 to O_3 are the coordinates of the point O and ω_{O_1} to ω_{O_3} are the components of the angular velocity ω_O with respect to the frame $(M - \mathbf{e}_1\mathbf{e}_2\mathbf{e}_3)$. Hence, it follows from Eq. 13 the inverse kinematics is now formulated as a system of 6 nonlinear equations:

$$\dot{\eta}\mathbf{J}(\eta) + [f_1(\sigma, \varphi, \omega_3') \ldots f_6(\sigma, \varphi, \omega_3')] = \mathbf{S}_{obj} \tag{17}$$

where $f_i(\sigma, \varphi, \omega_3'$ are the elements yielded by the rolling contact. In these 6 scalar nonlinear equations, the number of unknowns is $n + 3$, namely $\dot{\eta}_1$ to $\dot{\eta}_n$, σ, φ, and ω_3'. Hence it takes at least 3 joints on the finger to have full manipulability, with the other 3 inputs from rolling contact.

These nonlinear equations will have to be solved by using numerical methods in most cases. Suppose there are n joints on the finger with $3 \leq n < 6$. Taking the reciprocity product on the both sides of Eq. 17 by the reciprocal matrix $\mathbf{J}^R$ of the Jacobian $\mathbf{J}$ removes the joint velocity $\dot{\eta}_1$ to $\dot{\eta}_n$, yielding $6 - n$ scalar equations that only contains the three variables from the rolling contact, namely σ_M, φ and ω_3'. If $n = 3$, three scalar nonlinear equations will be generated, which have a finite number of solutions. If $n \geq 4$, an infinite number of solutions exist, giving more options as to rolling motion.

4 An Example Problem of Forward Kinematics

A 3R robotic finger was used to illustrate the proposed approach. Suppose the finger is equipped with a spherical fingertip of radius R that maintains rolling contact with a unit ball. Establish a palm frame $(P - \mathbf{ijk})$ and a tool frame $(T - \mathbf{i}'\mathbf{j}'\mathbf{k}')$ in such a way that the point T is located at the center of the half-sphere of the fingertip. The radius of the half-sphere is assumed to be R, and the half-sphere is parameterized as $\mathbf{r} = R[\cos u \cos v, -\cos u \sin v, \sin u]$. The joint screws are represented by S_1 to S_3 and the contact point is M, as in Fig. 3.

Suppose the joint angles of S_1 to S_3 are η_1 to η_3, respectively. It follows from Eq. 2 that the twist of the frame T with respect to the frame P can be obtained straightforwardly. The unit ball can be parameterized as $\mathbf{r}' = (\cos\alpha \cos\beta, -\cos\alpha \sin\beta, \sin\alpha)$. Suppose the coordinates of the contact point M are (u_0, v_0) and (α_0, β_0) on the spherical fingertip and on the unit ball, respectively. Further, suppose the trajectory L on the fingertip is along a meridian and the trajectory L' on the object is also along a meridian. It follows that

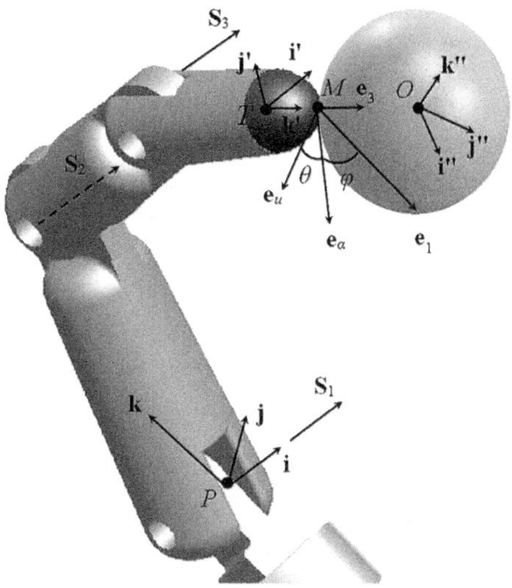

Fig. 3 A ball manipulated by a 3R finger with a spherical fingertip

$$\begin{aligned} \mathbf{e}_1 &= \left[-\sin u_0 \cos v_0,\ \sin u_0 \sin v_0\ \cos u_0\right] \\ \mathbf{e}_2 &= \left[-\sin v_0,\ \cos v_0\ 0\right] \\ \mathbf{e}_3 &= \left[-\cos u_0 \cos v_0,\ \cos u_0 \sin v_0\ -\sin u_0\right] \end{aligned} \tag{18}$$

The normal curvatures, geodesic curvatures, and the geodesic torsion of the curves L and L' are

$$\begin{aligned} k_n &= -\frac{1}{R}, k_g = 0, \tau_g = 0 \\ k_n' &= 1, k_g' = \tau_g' = 0 \end{aligned} \tag{19}$$

It follows from Eq. 9 that the twist S_{TM} with respect to the frame T is

$$\mathbf{S}_{TM} = \sigma_M \left[\sin u_0 \cos v_0,\ -\sin u_0 \sin v_0,\ -\cos u_0\ -\tfrac{\sin v_0}{R},\ -\tfrac{\cos v_0}{R},\ 0\right] \tag{20}$$

It follows from Eq. 12 that the twist S_{MO} with respect to the frame T is

$$\mathbf{S}_{MO} = \sigma_M \left[0,\ 0,\ 0\ (\tfrac{1}{R} + 1) \sin v_0,\ (\tfrac{1}{R} + 1) \cos v_0,\ 0\right] \tag{21}$$

The twist of the unit ball can be obtained from Eq. 13. This completes the forward kinematics of in-hand fine-manipulation.

5 Conclusions

We proposed a systematic approach to the forward and inverse kinematics of multi-fingered hands based on the product of exponential formula. This approach yielded a representation in forward kinematics that amounts to obtaining the twist of the grasped object given the joint rates and the contact trajectories. The inverse kinematics was formulated as a system of nonlinear algebraic equations, where the variables consists of the joint rates and contact parameters. The proposed approach provides an alternative to the traditional approach that requires numerical integration of a system of second-order differential equations.

References

1. Armour, R., Vincent, J.: Rolling in nature and robotics: a review. J. Bionic Eng. **3**(4), 195–208 (2006). doi:10.1016/S1672-6529(07)60003-1
2. Brockett, R.W.: Robotic Manipulators and the Product of Exponentials Formula. Springer, Berlin (1984)
3. Cui, L., Dai, J.S.: A darboux-frame-based formulation of spin-rolling motion of rigid objects with point contact. IEEE Trans. Robot. **26**(2), 383–388 (2010). doi:10.1109/TRO.2010.2040201
4. Cui, L., Dai, J.S.: Posture, workspace, and manipulability of the metamorphic multifingered hand with an articulated palm. ASME J. Mech. Robot. **3**(2), 021001 (2011). doi:10.1115/1.4003414
5. Cui, L., Dai, J.S.: Reciprocity-based singular value decomposition for inverse kinematic analysis of the metamorphic multifingered hand. ASME J. Mech. Robot. **4**(3), 034502 (2012). doi:10.1115/1.4006187
6. Cui, L., Dai, J.S.: From slidingrolling loci to instantaneous kinematics. Adjoin. Approach. Mech. Mach. Theory **85**(0), 161–171 (2015). doi:10.1016/j.mechmachtheory.2014.11.015
7. Cui, L., Dai, J.S.: A polynomial formulation of inverse kinematics of rolling contact. J. Mech. Robot. **7**(4), 041003 (2015). doi:10.1115/1.4029498
8. Kiss, B., Lévine, J., Lantos, B.: On motion planning for robotic manipulation with permanent rolling contacts. Int. J. Robot. Res. **21**(5–6), 443–461 (2002). doi:10.1177/027836402321261959
9. Murray, R.M., Li, Z., Sastry, S.S.: A Mathematical Introduction to Robotic Manipulation. CRC Press, Boca Raton (1994)
10. Park, F.C., Brockett, R.W.: Kinematic dexterity of robotic mechanisms. Int. J. Robot. Res. **13**(1), 1–15 (1994). doi:10.1177/027836499401300101
11. Park, F.C., Bobrow, J.E., Ploen, S.R.: A lie group formulation of robot dynamics. Int. J. Robot. Res. **14**(6), 609–618 (1995). doi:10.1177/027836499501400606
12. Romdhane, L., Duffy, J.: Kinestatic analysis of multifingered hands. Int. J. Robot. Res. **9**(6), 3–18 (1990). doi:10.1177/027836499000900601
13. Selig, J.: Lie Groups and Lie Algebras in Robotics. Springer, Berlin (2004)

Synergies Evaluation of the SCHUNK S5FH for Grasping Control

Fanny Ficuciello, Alba Federico, Vincenzo Lippiello and Bruno Siciliano

Abstract In this work, a study on postural synergies has been conducted on an under-actuated anthropomorphic hand, the SCHUNK Five-Fingered Hand (S5FH). Human hand grasps are mapped on the robotic hand using fingertips measurements, obtained with an RGBD camera sensor, and inverse kinematics. Since the S5FH is under-actuated, an approximate solution can be obtained using the differential kinematics mapping between the motor space and the Cartesian space and a closed-loop inverse kinematics (CLIK), based on a high-rectangular hand Jacobian that takes into account the mechanical synergies of the hand. The so-computed motor synergies have been tested for hand control during grasping. The motor current measurements have been used to limit the grasping forces trough a motor position control in the synergies subspace.

1 Introduction

Nowadays, in robotics and prosthetic applications, postural synergies have been widely recognized to be a powerful tool to plan grasps and control artificial hands using few parameters compared to the degrees of freedom (DOFs) of the hand itself [1]. Several methods have been proposed to compute the synergies subspace. In [2–4] the basis space of synergies is represented by a matrix of constant eigengrasps (basis of eigenvectors), while in [5] synergies are mapped directly from human to robotic hands using non-constant eigengrasps.

F. Ficuciello (✉) · A. Federico · V. Lippiello · B. Siciliano
Dipartimento di Ingegneria Elettrica E Tecnologie Dell'Informazione,
Università Degli Studi di Napoli Federico II, via Claudio 21, 80125 Napoli, Italy
e-mail: fanny.ficuciello@unina.it

A. Federico
e-mail: alb.federico@studenti.unina.it

V. Lippiello
e-mail: vincenzo.lippiello@unina.it

B. Siciliano
e-mail: bruno.siciliano@unina.it

J. Lenarčič and J.-P. Merlet (eds.), *Advances in Robot Kinematics 2016*,
Springer Proceedings in Advanced Robotics 4, DOI 10.1007/978-3-319-56802-7_24

The authors' previous work refers to fully actuated anthropomorphic hands [6–9]. In these works, the experimental results highlight that, despite the differences in kinematics and mapping methods, the first three synergies have some basic features that are preserved if the hand kinematics is anthropomorphic and if the grasps data set is suitably chosen to cover a large variety of human grasping postures [10, 11].

In this work, the method developed in [8] has been adapted and tested to evaluate the first three synergies on an under-actuated five-fingered hand suitable for service robot applications in the household domain, the S5FH depicted in Fig. 1. The hand possesses 20° of mobility and it is designed with "mechanical synergies" that regulate the kinematic couplings between the finger joints while decreasing the number of motors from 20 to 9.

One of the main problems of under-actuation is that the inverse kinematics problem that maps fingertips Cartesian space onto joints motor space does not necessarily have a closed-form solution, but in some cases only an approximate solution that minimizes the norm of the error can be obtained. Moreover, the human hand grasps cannot be accurately mapped onto the robotic hand since some information is unavoidably lost due to mechanical couplings between the joints.

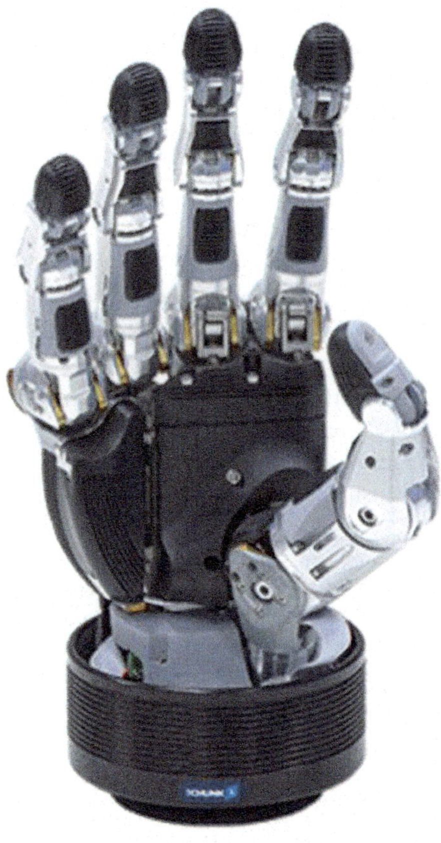

Fig. 1 The Schunk Five-Fingered Hand

A valuable possibility to obtain a solution to the inverse kinematics problem is to use the differential kinematics mapping between the motor space and the Cartesian space and a closed-loop inverse kinematics (CLIK) algorithm based on the high-rectangular hand Jacobian that takes into account the mechanical synergies of the hand.

In this work, a data set of grasps, measured on five human subjects and available from the authors' previous work, is used to evaluate the grasping capabilities of the robot hand in a synergy-based framework. For this purpose, a synergies Jacobian can be computed and suitably used in the CLIK algorithm to map the grasps from the human hand to the robotic hand. The details of the grasping data and mapping method can be found in [2, 8].

The results demonstrate that the computed synergies are suitable to control the hand in a three-dimensional subspace and the evaluated features of the first three synergies confirm the results obtained in [12], i.e. the grasping capabilities are very similar to those of the fully actuated anthropomorphic hands.

The paper is structured as follows: in Sect. 2 the hand kinematics and the mechanical synergies are described, while in Sect. 3 the method for synergies computation is briefly described and the computed synergies are analyzed. Section 4 reports the experimental results obtained using the synergy-based control for grasping actions and grasping forces regulations. Finally, Sect. 5 provides the conclusions and sketches future work.

2 The Schunk S5FH

The Schunk S5FH has an anthropomorphic structure very similar to the human hand for shape, size and overall for the cosmetic appearance. Indeed, the dimensions are of $1:1$ ratio with the human hand and the weight is of 1.3 kg. The control and power electronics are integrated in the wrist allowing an easy connection with market-standard industrial and lightweight robots. The current technology, however, does not allow arranging twenty or more motors within a mechanical structure with dimensions similar to those of the human hand while ensuring appropriate requirements of speed and strength. As a matter of fact, the S5FH has 20 joints and 9 DOFs led by servo motors. The reader can find the whole technical data, hardware and software specifications in [13, 14].

Hence, the number of motors is significantly lower than the number of joints and suitable motion couplings are obtained by means of mechanical synergies defined via mechanical transmissions.

Let $\boldsymbol{q}$ be the vector of the 20 joint angles describing the robotic hand configuration. The joint motions (Fig. 2) are coupled according to the mechanical synergies matrix, defined below. $\boldsymbol{m} \in \mathbb{R}^m$, with $m = 9$, is the vector of motor variables and $\boldsymbol{q}_0$ is an offset representing the vector of joint values when the motor positions are zero. In vector $\boldsymbol{q}$ the finger joints are pointed in progressive order from the thumb to the little finger using the subscripts t, i, m, r, l. Not all the fingers have the same number of joints

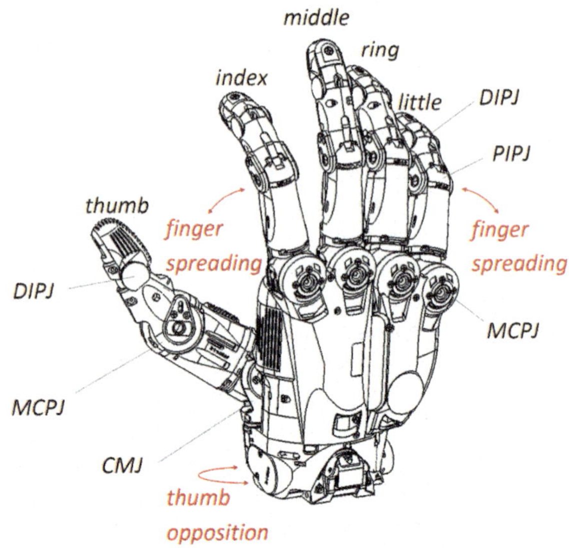

Fig. 2 The finger movements of the Schunk Five-Fingered Hand are illustrated

and motors. About the thumb, the opposition joint q_{t_o} is coupled with the q_{p_o} joint allocated into the palm that moves only the ring and the little finger with respect to the palm frame. The carpometacarpal flexion joint $q_{t_{cm}}$, metacarpophalangeal flexion joint $q_{t_{mcp}}$, distal interphalangeal flexion joint $q_{t_{dip}}$ are clearly indicated in Fig. 2 as well as the metacarpophalangeal flexion joint (DIP), proximal interphalangeal joint (PIP) and distal interphalangeal flexion joint (MCP) for the other fingers. Finally, the index, ring and little fingers have also coupled spread motion $(q_{i_s}, q_{r_s}, q_{l_s})$.

$$\underbrace{\begin{bmatrix} q_{t_o} \\ q_{t_{cm}} \\ q_{t_{mcp}} \\ q_{t_{dip}} \\ q_{i_s} \\ q_{i_{mcp}} \\ q_{i_{pip}} \\ q_{i_{dip}} \\ q_{m_{mcp}} \\ q_{m_{pip}} \\ q_{m_{dip}} \\ q_{p_o} \\ q_{r_s} \\ q_{r_{mcp}} \\ q_{r_{pip}} \\ q_{r_{dip}} \\ q_{l_s} \\ q_{l_{mcp}} \\ q_{l_{pip}} \\ q_{l_{dip}} \end{bmatrix}}_{\boldsymbol{q}} = \underbrace{\begin{pmatrix} 0.5 & 0 & 0 & 0 & 0 & 0 & 0 & 0 & 0 \\ 0 & 0.29 & 0 & 0 & 0 & 0 & 0 & 0 & 0 \\ 0 & 0.29 & 0 & 0 & 0 & 0 & 0 & 0 & 0 \\ 0 & 0.42 & 0 & 0 & 0 & 0 & 0 & 0 & 0 \\ 0 & 0 & 0 & 0 & 0 & 0 & 0 & 0 & 0.25 \\ 0 & 0 & 1 & 0 & 0 & 0 & 0 & 0 & 0 \\ 0 & 0 & 0 & 0.49 & 0 & 0 & 0 & 0 & 0 \\ 0 & 0 & 0 & 0.51 & 0 & 0 & 0 & 0 & 0 \\ 0 & 0 & 0 & 0 & 1 & 0 & 0 & 0 & 0 \\ 0 & 0 & 0 & 0 & 0 & 0.49 & 0 & 0 & 0 \\ 0 & 0 & 0 & 0 & 0 & 0.51 & 0 & 0 & 0 \\ 0.5 & 0 & 0 & 0 & 0 & 0 & 0 & 0 & 0 \\ 0 & 0 & 0 & 0 & 0 & 0 & 0 & 0 & 0.25 \\ 0 & 0 & 0 & 0 & 0 & 0 & 0.26 & 0 & 0 \\ 0 & 0 & 0 & 0 & 0 & 0 & 0.36 & 0 & 0 \\ 0 & 0 & 0 & 0 & 0 & 0 & 0.38 & 0 & 0 \\ 0 & 0 & 0 & 0 & 0 & 0 & 0 & 0 & 0.5 \\ 0 & 0 & 0 & 0 & 0 & 0 & 0 & 0.26 & 0 \\ 0 & 0 & 0 & 0 & 0 & 0 & 0 & 0.36 & 0 \\ 0 & 0 & 0 & 0 & 0 & 0 & 0 & 0.38 & 0 \end{pmatrix}}_{S_m} \underbrace{\begin{bmatrix} m_0 \\ m_1 \\ m_2 \\ m_3 \\ m_4 \\ m_5 \\ m_6 \\ m_7 \\ m_8 \end{bmatrix}}_{\boldsymbol{m}} + q_0, \tag{1}$$

3 Postural Synergies Computation

A data set of 36 grasping configurations, measured on five human hands, have been considered as in [8] and the same mapping method has been applied to the S5FH hand. The differential kinematics mapping between the mechanical synergies subspace and the Cartesian space, used in the CLIK algorithm, is represented by the following equation

$$\dot{x} = \boldsymbol{J}_{h_m} \dot{\boldsymbol{m}}, \tag{2}$$

where $\boldsymbol{J}_{h_m}$ is the mechanical synergies Jacobian and is computed as

$$\boldsymbol{J}_{h_m} = \boldsymbol{J}_h \boldsymbol{S}_m. \tag{3}$$

In (2), $\dot{\boldsymbol{x}}$ is the derivative of the position vector of the five fingertips $\boldsymbol{x} \in \mathbb{R}^{15}$, $\boldsymbol{J}_h$ is the (15×20) S5FH hand Jacobian, $\boldsymbol{S}_m$ is the (20×9) matrix of the mechanical synergies, and finally $\boldsymbol{m} \in \mathbb{R}^9$ is the vector of the motor angles. The CLIK algorithm for inverse kinematics resolution can be based on the transpose of the Jacobian $\boldsymbol{J}_h^T$, or on the pseudoinverse of the Jacobian $\boldsymbol{J}_h^\dagger$. When the CLIK algorithm is used for mapping gasps from the human to the robot hand, the desired fingertips position in the Cartesian space, $\boldsymbol{x}_d$, is constant and the required feedforward term of the velocities is null. In this work, the synergies subspace of the hand, constituted by the first three eigengrasps, has been computed with both solutions in the CLIK algorithm, namely $\boldsymbol{J}_h^T$ and $\boldsymbol{J}_h^\dagger$. Actually, these solutions do not lead to significant differences in the results. On the other hand, the use of the transpose Jacobian may be easier and more convenient for real-time implementation. Moreover, even in the presence of a variable Cartesian desired position, the latter solution does not require the addition of a feedforward term. For these reasons, in this paper the results obtained with $\boldsymbol{J}_h^T$ are considered (Fig. 3).

For the sake of brevity, the synergies subspace, resulting from computation, is schematically represented in Fig. 6. Here, the configurations that the hand assumes according to the patterns of the first, second and third synergies when their coefficients vary from minimum to maximum values are represented on the three principal axes of the eigengrasps. The minimum and maximum values of the synergy coefficients are in agreement with the positive direction of the arrows. The obtained results are very similar to those in [6] and the differences are due to the kinematic limitation introduced by the under-actuation. It is worth observing that the first synergy opens and closes the hand acting mainly on the flexion joints by moving them in the same direction. The second synergy generates opposite motions for the metacarpophalangeal flexion and proximal interfalangeal flexion joints. Obviously, this is true for the only two fingers that have no couplings on these joints (index and middle fingers). On the other hand, the third synergy influences mainly the thumb motion both for flexion and opposition.

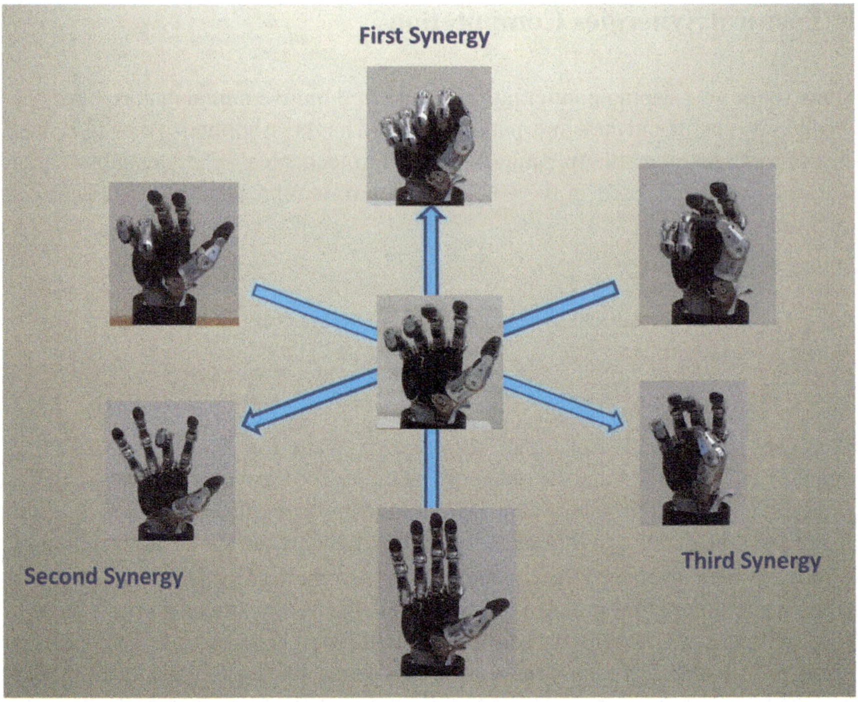

Fig. 3 The first three eigengrasps

4 Grasping Control in the Synergies Subspace

Once the (9×3) $\boldsymbol{S}_s$ synergy matrix has been computed, in order to test the efficiency of the mapping method, different grasps have been reproduced in the three-dimensional synergies subspace. Reproduced power grasps of spherical and cylindrical objects are represented in Fig. 4a, b. Actually, since mechanical synergies affect the mapping from the human hand, the projection of a grasp from the data set in the synergies subspace is not so effective as for the full-actuated anthropomorphic hands [6, 8]. Thus, the reproduction is not successful for all the grasps. This means that a control strategy is required to adjust the reference grasp in order to let the hand adapt to the object while moving in the synergies subspace. The kinematic control of the hand in the synergies subspace is again a CLIK algorithm, but in this case it is based on the synergies Jacobian given by $\dot{\boldsymbol{x}} = \boldsymbol{J}_{h_{m_s}} \dot{\boldsymbol{\sigma}}$, where $\boldsymbol{J}_{h_{m_s}} = \boldsymbol{J}_h \boldsymbol{S}_m \boldsymbol{S}_s$ and $\dot{\boldsymbol{\sigma}}$ are the synergy coefficients. The differential mapping between synergies coefficients and joint velocities is given by the following equation

$$\dot{\boldsymbol{q}} = \boldsymbol{S}_m \dot{\boldsymbol{m}} = \boldsymbol{S}_m \boldsymbol{S}_s \dot{\boldsymbol{\sigma}} .$$

(a) Spherical object (b) Cylindrical object

Fig. 4 Some examples of reproduced grasps

A simple strategy to modify the reference grasp can be adopted. The fingertips desired positions are modified in the control algorithm in order to reduce their distance with respect to the centroid of a virtual object computed as the centroid of the fingertips involved in the desired grasp. Moreover, in order to limit the grasping forces, the desired target of the CLIK algorithm is modified on the basis of the measured motor current and of a defined threshold that is related to the texture of the object. The experiments demonstrate that the synergies subspace is suitable for hand control in

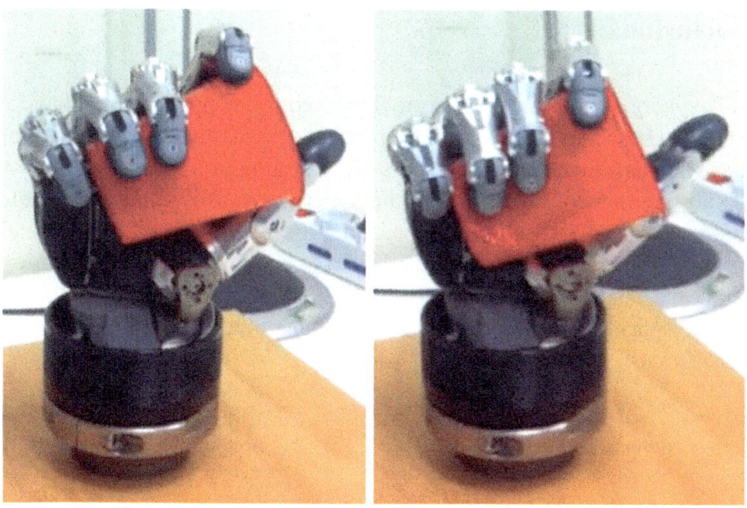

Fig. 5 A cylindrical object is grasped

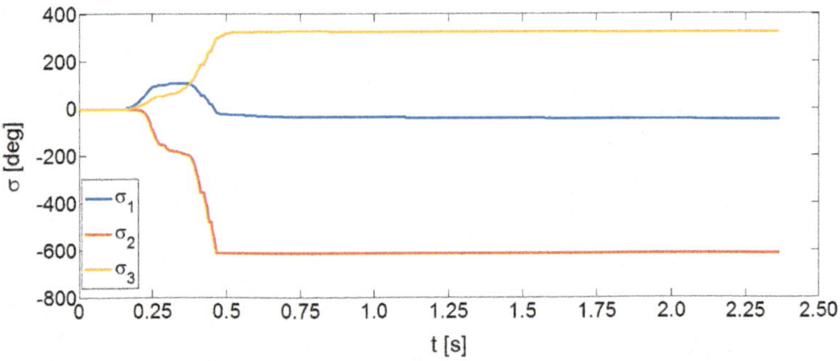

Fig. 6 Time history of synergy coefficients during grasping a cylindrical object with contact forces regulation

grasping a wide variety of objects, i.e. the algorithm is stable and effectively regulates the grasping forces by modifying the motor positions in the synergies subspace. Thus, to improve the grasping capabilities as future work, strategies based on quality indexes to close the hand toward the object in the synergies subspace will be tested. In Fig. 5 two different grasps of the same object are controlled in the synergies subspace using different current thresholds.

In Fig. 6, the synergies coefficient are reported for one of the cylindrical grasps described above. It is possible to observe that the control modifies the synergies coefficients until reaching a steady-state value which depends on the allowed motors current limits.

5 Conclusions

The S5FH synergies subspace has been computed by mapping human hand grasps using a method based on fingertips measurements. The features of the first three synergies have been evaluated and the results are very similar to those of fully actuated anthropomorphic hands. Furthermore, the synergies subspace has been tested for hand control using a CLIK algorithm based on the synergies Jacobian. The experiments have demonstrated that the method used to compute synergies provides good results since the hand can be successfully and stably controlled in a three-dimensional synergies subspace for grasping purpose while guaranteeing suitable regulation of the contact forces.

Acknowledgements This research has been partially funded by the EC Seventh Framework Programme (FP7) within RoDyMan project 320992. The authors are solely responsible for its content. It does not represent the opinion of the European Community and the Community is not responsible for any use that might be made of the information contained therein.

References

1. Wimboeck, T., Jan, B., Hirzinger, G.: Synergy-level impedance control for a multifingered hand. In: Proceedings IEEE/RSJ International Conference on Intelligent Robots and Systems, pp. 973–979. San Francisco (2011)
2. Ficuciello, F., Palli, G., Melchiorri, C., Siciliano, B.: Experimental evaluation of postural synergies during reach to grasp with the UB Hand IV. In: Proceedings IEEE/RSJ International Conference on Intelligent Robots and Systems, pp. 1775–1780, San Francisco (2011)
3. Geng, T., Lee, M., Hulse, M.: Transferring human grasping synergies to a robot. Mechatronics **21**(1), 272–284 (2011)
4. Sun, S., Rosales, C., Suarez, R.: Study of coordinated motions of the human hand for robotic applications. In: Proceedings of the Anchorage, Alaska, IEEE International Conference on Robotics and Automation, pp. 776–781 (2010)
5. Gioioso, G., Salvietti, G., Malvezzi, M., Prattichizzo, P.: Mapping synergies from human to robotic hands with dissimilar kinematics: An object based approach. In: IEEE International Conference on Robotics and Automation, Workshop on Manipulation Under Uncertainty, Shangai (2011)
6. Ficuciello, F., Palli, G., Melchiorri, C., Siciliano, B.: Postural synergies of the ub hand iv for human-like grasping. Robot. Autonom. Syst. **62**, 357–362 (2014)
7. Palli, G., Melchiorri, C., Vassura, G., Scarcia, U., Moriello, L., Berselli, G., Cavallo, A., Maria, G.D., Natale, C., Pirozzi, S., May, C., Ficuciello, F., Siciliano, B.: The DEXMART hand: mechatronic design and experimental evaluation of synergy-based control for human-like grasping. Int. J. Robot. Res. **33**, 799–824 (2014)
8. Ficuciello, F., Palli, G., Melchiorri, C., Siciliano, B.: A model-based strategy for mapping human grasps to robotic hands using synergies. In: Proceedings of the IEEE/ASME International Conference on Advanced Intelligent Mechatronics, Wollongong, Australia, 2013, pp. 1737–1742 (2013)
9. Ficuciello, F., Palli, G., Melchiorri, C., Siciliano, B.: Planning and control during reach to grasp using the three predominant UB Hand IV postural synergies. In: Proceeding IEEE International Conference on Robotics and Automation, Saint Paul, 2012, pp. 2255–2260 (2012)
10. Feix, T., Pawlik, R., Schmiedmayer, H., Romero, J., Kragic, D.: The generation of a comprehensive grasp taxonomy. In: Robotics, Science and Systems, Workshop on Understanding the Human Hand for Advancing Robotic Manipulation, Washington (2009)
11. Cutkosky, M.: On grasp choice, grasp models, and the design of hands for manufacturing tasks. IEEE Trans. Robot. Autom. **5**(3), 269–279 (1989)
12. Ruehl, S., Parliz, C., Heppner, G., Hermann, A., Roennau, A., Dillman, R.: Experimental evaluation of the schunk 5-finger gripping hand for grasping tasks. In: Proceedings of the Bali, Indonesia, IEEE International Conference on Robotics and Biomimetics pp. 2465–2470 (2014)
13. Schunk Hand Webpage. http://mobile.schunk-microsite.com/en/produkte/produkte/servoelektrische-5-finger-greifhand-svh.html
14. schunk_svh_driver. http://wiki.ros.org/schunk_svh_driver

In-Hand Manipulative Synthesis Using Velocity Subspaces

Neda Hassanzadeh, Shramana Ghosh and Nina Robson

Abstract The kinematic design of robots for tasks involving positions and its derivatives has been explored in the past in order to shape the trajectory of the robot at a given set of points. This approach has been successful for the synthesis of linkages for grasping applications; however defining a single velocity in the vicinity of a specified location might not be enough for a desired manipulation task. In the synthesis of multi-fingered robotic hands, it is interesting to ask whether a hand can be designed for a certain in-hand manipulation task that ensures contact and at the same time relative motion of the fingertips on the object surface in multiple desired directions. In this article we define a method for designing robotic hands that guide an object through a kinematic task with velocity specifications in the vicinity of key task positions. Given the mobility for a hand topology, the necessary velocities are derived at each task position to fully define a subspace of allowable directions for object manipulation. As an example, a multi-fingered robotic hand for an in-hand manipulation of a body with a known geometry has been designed. The proposed synthesis technique is the first step in creating tangent spaces in the vicinity of positions to successfully guide a body for desired manipulation tasks.

1 Introduction

The design of end-effector robotic tools is directed towards grasping actions or towards higher dexterity, usually anthropomorphic in structure (see for instance [1]). In this research, we focus on creating multi-fingered hand designs specially tailored to specific groups of manipulation tasks.

N. Hassanzadeh (✉)
Department of Mechanical Engineering, Idaho State University, Pocatello, ID, USA
e-mail: hassneda@isu.edu

S. Ghosh
Department of Mechanical Engineering, University of California, Irvine, CA, USA
e-mail: shramang@uci.edu

N. Robson
Department of Mechanical Engineering, Cal State Fullerton, Fullerton, CA, USA
e-mail: nrobson@fullerton.edu

J. Lenarčič and J.-P. Merlet (eds.), *Advances in Robot Kinematics 2016*,
Springer Proceedings in Advanced Robotics 4, DOI 10.1007/978-3-319-56802-7_25

We define a multi-fingered robotic hand as a series of common joints branching at least once in several other serial chains, ending in a finite set of end-effector links (the fingertips). The methodology developed in [2] for the design of multi-fingered hands for kinematic tasks, both for finite and infinitesimal motion, offers a systematic process to design innovative end-effectors for a simultaneous free-motion task of all the fingertips.

When the hand grasps an object, the constraints on the relative motion need to be taken into account. In this application, we focus on in-hand roll-slide, when the fingertip contact is kept throughout the motion [3]. In such motions, the geometry of robotic fingers cause the contact points to travel on the surface of the object's body as the configuration of the fingers change during the manipulation. The rolling motion of the fingertip on the body requires that the velocities of the contact points on the fingers and the body must remain the same. Including roll-slide constraints in the problem involves the application of differential geometry. The analysis takes as inputs the linear and angular relative velocities of the contact points on the finger and the body and outputs the parameterized contact point velocities on the surfaces of the finger and the body. Figure 1 shows an example of the progression of the finger and body contact coordinate frames for the rolling of a finger on the surface of a body.

Given a hand topology and the mobility for a generally-grasped object, we propose to use kinematic synthesis techniques in order to define a subspace of potential velocities at a given position, compatible with desired object manipulation directions. This is based on defining as many infinitesimal motion tasks at each position as general degrees of freedom of the hybrid fingers-body topology is. That will fully specify the subspace of velocities, yielding control over the allowable motion at discreet points of the workspace. This new technique is applied to a three-fingered hand with a wrist. Experimental motion capture data is used for defining the positions. As a next step, the fingers-body problem geometry is used to define the velocities in the desired directions in the vicinity of contact locations, which define the velocity subspaces.

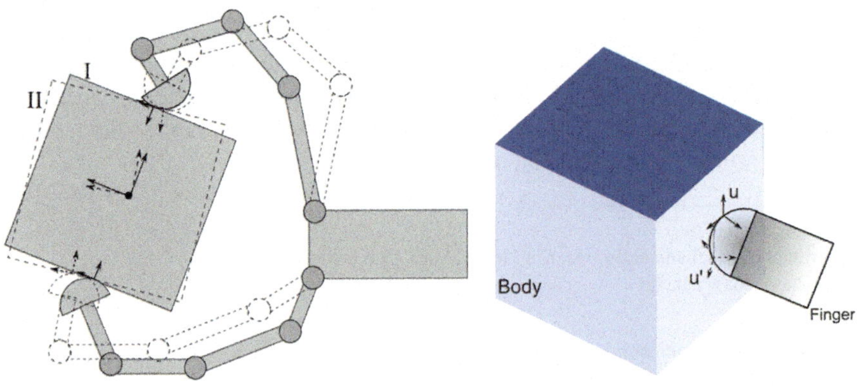

Fig. 1 *Left* An illustration of the planar rolling problem: the body is moved from configuration I to configuration II. *Right* The contact frame for the finger is u. After rolling occurs the new contact frame for the finger is u'. The same holds true for the body frame

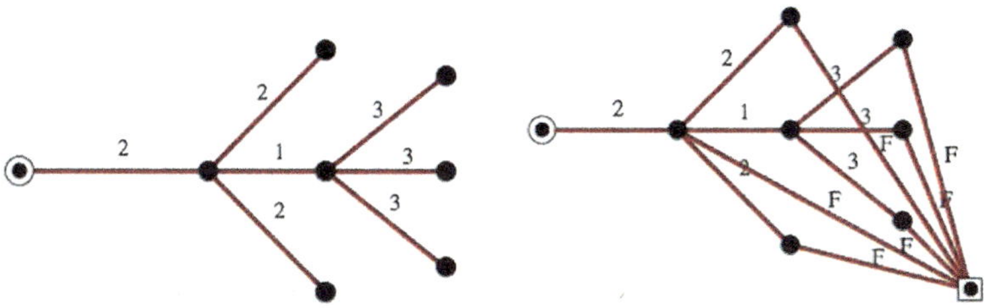

Fig. 2 A five-fingered tree topology with $j = \{0, 1, 1, 1, 3, 3, 3\}$ (*left*), and holding an object with F degree-of-freedom contacts (*right*)

2 Topology of the Hand-Object System

A tree topology for a kinematic chain has a set of common joints spanning several chains, possibly in several stages, and ending in multiple end-effectors.

Open hands not holding an object in the fingers can be represented as rooted tree graphs [2, 4], where all links are ternary or above. Figure 2 presents the compacted graph for a hand with five branches and two palms. The root vertex is indicated with a double circle. When the multi-fingered hand is holding an object, a set of constraints are created at the contact point between fingertip and object. Standard finger denominations, such as pointy finger or soft fingers [5], impose constraints on the motion that can be modeled as a joint or set of joints. During the grasp and manipulation of the grasped object, the topology is represented with a hybrid graph, as shown in Fig. 2, where the object is indicated as a square vertex.

A first approximation to the ability to manipulate when the hand is grasping an object can be calculated using standard mobility and connectivity formulas, see for instance [5]. Object and hand geometry are taken into account by calculating the subspaces of wrenches or twists created by the contacts on the object, see for instance [6].

3 Kinematic Synthesis for Subspaces of Velocities

3.1 Velocity Subspaces

The design methodology follows [2] and uses a kinematic task as input to create multi-fingered hand designs. The kinematic task consists of a set of simultaneous displacements for each fingertip, and velocities defined at those positions.

Given a hand topology with b end-effectors and n_c joints in branch c, solvable [7] for m_p positions and m_v velocities, the design equations seek to minimize the difference between the task and the motion of the robot. For displacements, this is created as the product of exponentials defining each serial chain from root to end-effector, while the velocities are created as the linear combination of the joint twists times the joint rate at a given position,

$$\hat{P}^i_{1k} = \prod_{j \in \{B_i\}} e^{\frac{\Delta\hat{\theta}^k_j}{2} \mathsf{S}_j},$$
$$\dot{\mathsf{P}}^i_k = \sum_{j \in \{B_i\}} \dot{\theta}^k_j \mathsf{S}^k_j, \quad i = 1, \ldots, b; \quad k = 2, \ldots, m, \tag{1}$$

where the number of end-effectors, or branches as root-to-fingertip chains, is indicated by b, and $\{B_i\}$ is the set of ordered indices of the joints belonging to branch i. Notice that some of the joints will be common to several branches. The joint axes at the kth position are S^k_j (at the reference configuration they are denoted as S_j), and $\hat{P}^i_{1k}$ are the relative task positions for each branch. For most topologies, this method yields many potential designs.

Equation 1 can be used to specify the twist for each fingertip at a given position. When several fingers are in contact with an object, the twist is compatible with the contact and the displacement of the object.

The maximum number of twists that can be specified for a serial chain at a given configuration, corresponds to the dimension of the twist subspace, which coincides with its mobility M for generally-oriented axes. Specifying M twists fully defines the subspace of potential velocities of the end-effector at a particular position. The obtained equations are linear and the coordinates of the axes can be written as a function of the joint rates,

$$\begin{Bmatrix} \mathsf{S}_1 \\ \vdots \\ \mathsf{S}_{b1} \end{Bmatrix} = \begin{bmatrix} \dot{\theta}^1_1 & \ldots & \dot{\theta}^1_{b1} \\ \vdots & & \vdots \\ \dot{\theta}^M_1 & \ldots & \dot{\theta}^M_{b1} \end{bmatrix}^{-1} \begin{Bmatrix} W_1 \\ \vdots \\ W_M \end{Bmatrix}, \tag{2}$$

for each branch i with number of joints b_i.

Equation (2) allows the definition of the tangent space at a given position, constraining the motion to create a controlled subspace of potential velocities for the in-hand manipulation and path-planning. In order to do so, the contact conditions, the type of the manipulative motion, as well as the geometry of the object-fingertip system must be considered.

3.2 Deriving Velocity Specifications from Task Geometry

To derive the position and velocity of the body at an instant $t = 0$ imposed by contact with the fingers, we follow Robson and Tolety [8]. Let the movement of a rigid body be defined by the parameterized set of 8-dimensional dual quaternions

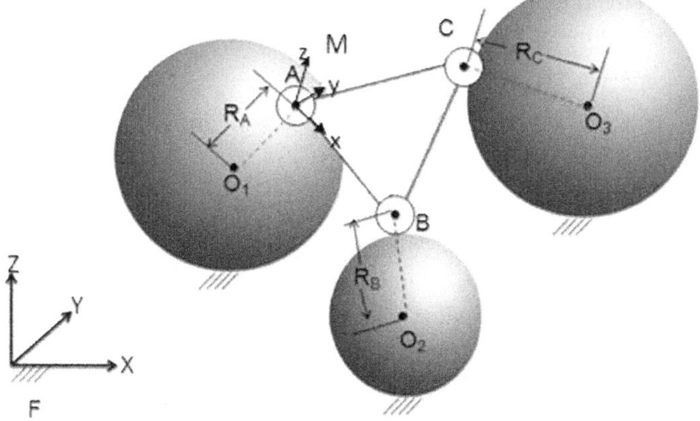

Fig. 3 A triangular shaped object is in contact with three spheres with radii of curvature R_A, R_B, R_C in the vicinity of the specified position. The moving frame M is located such that its origin coincides with $\mathbf{A}(\mathbf{t})$ and its x-axis is directed along the line $\mathbf{B} - \mathbf{A}$

$\hat{Q}(t) = \hat{q}(t) + \varepsilon\hat{q}^0(t)$, defined in terms of the screw axis of the displacement, $\mathsf{S}(t) = \mathbf{s}(t) + \varepsilon\mathbf{s}^0(t)$, and the rotation about and slide along the axis, $\phi(t)$ and $d(t)$.

A point $\mathbf{p}$ fixed with respect to the moving body traces a trajectory $\mathbf{P}(\mathbf{t})$ in a fixed coordinate frame F, given by:

$$\hat{P}(t) = \hat{Q}(t)\hat{p}\hat{Q}^*(t) = \left(\cos\frac{\hat{\theta}}{2} + \sin\frac{\hat{\theta}}{2}\mathsf{S}\right)(1 + \varepsilon\mathbf{p})\left(\cos\frac{\hat{\theta}}{2} - \sin\frac{\hat{\theta}}{2}\mathsf{S}\right), \tag{3}$$

where $\cos\frac{\hat{\theta}}{2} = \cos\frac{\theta}{2} - \varepsilon\frac{d}{2}\sin\frac{\theta}{2}$, $\sin\frac{\hat{\theta}}{2} = \sin\frac{\theta}{2} + \varepsilon\frac{d}{2}\cos\frac{\theta}{2}$, and the point has coordinates $\mathbf{p} = p_x i + p_y j + p_z k$

Consider a moving frame attached to a body to be grasped, shown in Fig. 3. Assume that while rolling/sliding along the object surface in the vicinity of a specified contact location(s), the fingertips are moving on spherical paths with radii R_i. When the body is in contact with n stationary frictionless fingers in an equilibrium grasp, there are two possible free motions: either roll-slide on the fingers, or escape from the fingers [9]. The mobility and locked-joints mobility of the hand-object system [5] is specified. The goal is to determine the movement $\hat{Q}(t)$ so that the moving body has n points with trajectories $\mathbf{n}(\mathbf{t})$ consistent with contact with n fingers.

3.3 Position Specification

Assuming that the contact of our moving body $\{M\}$ with the three fixed objects constrains the point trajectories $\mathbf{A}(\mathbf{t})$, $\mathbf{B}(\mathbf{t})$ and $\mathbf{C}(\mathbf{t})$ to follow spherical trajectories in the vicinity of the reference position at $t = 0$, the movement of $\{M\}$ in the vicinity of $t = 0$ can be expressed by the Taylor series expansion:

$$\hat{Q}(t) = \hat{Q}_0 + \hat{Q}_1 t + \frac{1}{2}\hat{Q}_2 t^2 + \cdots, \quad \text{where} \quad \hat{Q}_i = \frac{d^i \hat{Q}}{dt^i}. \tag{4}$$

In this equation, the subscript denotes the derivative: $\hat{Q}_1 = \dot{\hat{Q}}$, and so on. The dual quaternion $\hat{Q}_0 = \cos\frac{\hat{\theta}_0}{2} + \varepsilon \sin\frac{\hat{\theta}_0}{2}\mathbf{S}_0$ is calculated using the position of the moving frame $\{M\}$ with the coordinates of the contact points $\mathbf{A_0} = \mathbf{A(0)}$, $\mathbf{B_0} = \mathbf{B(0)}$ and $\mathbf{C_0} = \mathbf{C(0)}$. See [8, 10] for a similar approach.

3.4 Velocity Specification

Given a point $\mathbf{a}$ in the object expressed in the moving frame, its expression with respect to the fixed frame is given by $\hat{A} = \hat{Q}\hat{a}\hat{Q}^*$. The velocity of this point can be obtained in general as

$$\dot{\hat{A}} = (\dot{\hat{Q}}\hat{Q}^*)\hat{A}(\dot{\hat{Q}}\hat{Q}^*)^* = \hat{\Omega}\hat{A}\hat{\Omega}^*. \tag{5}$$

In order to satisfy force constraints at the prescribed positions, we determine directions of the velocity vectors $\dot{\mathbf{A}}$, $\dot{\mathbf{B}}$ and $\dot{\mathbf{C}}$ that are perpendicular to the contact forces $\mathbf{F}_A$, $\mathbf{F}_B$ and $\mathbf{F}_C$ passing along the direction of the radii of the three spheres with centers $\mathbf{O_1}$, $\mathbf{O_2}$, $\mathbf{O_3}$ (see Fig. 3). The equations defining these constraints are:

$$\dot{\mathbf{A}} = \Omega\hat{A}\hat{\Omega}^* = \hat{\Omega}_{O_1A}(\hat{A} - \hat{O}_1)\hat{\Omega}^*_{O_1A}, \quad \dot{\mathbf{B}} = \Omega\hat{B}\hat{\Omega}^* = \hat{\Omega}_{O_2B}(\hat{B} - \hat{O}_2)\hat{\Omega}^*_{O_2B},$$
$$\dot{\mathbf{C}} = \Omega\hat{C}\hat{\Omega}^* = \hat{\Omega}_{O_3C}(\hat{C} - \hat{O}_3)\hat{\Omega}^*_{O_3C}, \tag{6}$$

where $\hat{\Omega}_{O_1A}(\hat{A} - \hat{O}_1)\hat{\Omega}^*_{O_1A}$ yields the expression for the velocity of a point rotating about a fixed point, $\omega_{\mathbf{O_1A}} \times (\mathbf{A} - \mathbf{O_1})$, and similarly for the rest of the points.

The same formulation is used to derive a second velocity at $\mathbf{A(t)}$ in Fig. 3, by specifying another finger motion in a different desired direction. That results in the definition of three more spheres, which describe that motion in the vicinity of the contact location, with radii of curvature R_D, R_E and R_F and centers $\mathbf{O_4}$, $\mathbf{O_5}$ and $\mathbf{O_6}$, respectively. The process was repeated once again to derive a third velocity at $\mathbf{A(t)}$. The three derived velocities at $\mathbf{A(t)}$ define the twist subspace for the desired roll-slide motions of the fingertip along the body in the vicinity of the specified contact position.

4 Exact Synthesis of a Three-Fingered Non-anthropomorphic Mechanical Hand for In-Hand Manipulation Task

Rolling motion in three different directions of a small cube/dice was chosen as a desired manipulation task. A subject was asked to roll-slide the dice, using the thumb, index and middle fingers and keep the contact with the dice throughout the motion. The consecutive motion in the three directions was captured using a Vicon™ motion capture system, as shown in Figs. 4 and 5.

Two positions for each of the fingertips for the in-hand manipulation task were defined and shown in Table 1. The positions were chosen at the start and end of the motion trajectories, obtained from the motion capture system.

Using the task geometry from Sect. 3, three velocities, compatible with the three in-hand manipulation directions, were defined at the first task position of each fin-

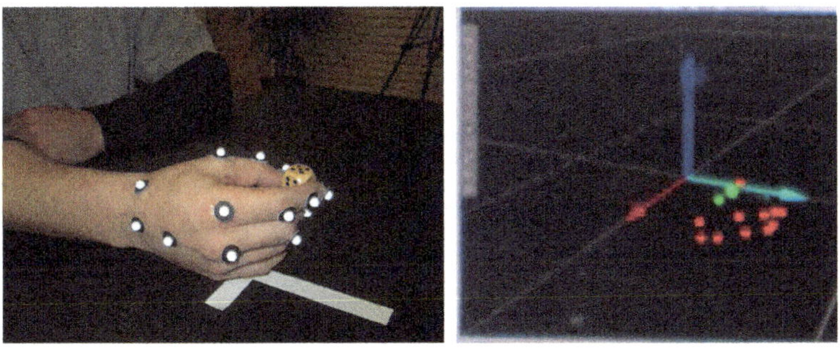

Fig. 4 *Left* Experimental setup of a human subject performing a rolling task. *Right* View from Vicon motion capture system of human hand performing rolling task

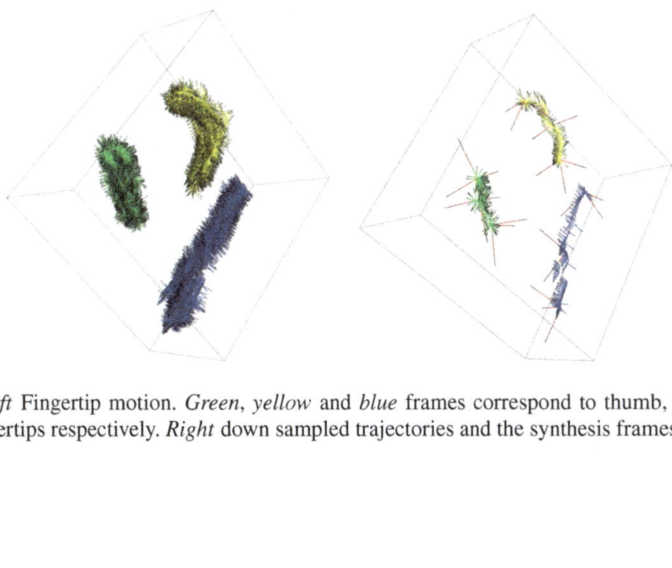

Fig. 5 *Left* Fingertip motion. *Green*, *yellow* and *blue* frames correspond to thumb, middle and index fingertips respectively. *Right* down sampled trajectories and the synthesis frames

Table 1 Finite positions selected for each fingertip

Fingertip	Position
Finger 1	{{0.522, 0.723, −0.358, 0.274804}, {−35.066, 116.668, 59.566, −162.493}}
	{{0.233, 0.785, 0.190, 0.542}, {76.771, 50.987, 119.006, −148.429}}
Finger 2	{{0.061, 0.729, −0.679, 0.057}, {−126.520, 120.326, 110.801, −84.442}}
	{{0.059, 0.761, −0.641, 0.077}, {−119.123, 117.141, 117.279, −90.694}}
Finger 3	{{−0.273, 0.810, −0.470, 0.221}, {−59.617, 97.518, 175.020, −59.467}}
	{{0.132, −0.249, −0.904, 0.322}, {−72.584, 194.181, −48.121, 44.531}}

Table 2 Twists for the index finger at start position

Twist	Values
1	{−0.064, −0.365, 0.535, 147.809, 439.141, 188.850}
2	{0, 0, 0, 2.539, −2.416, 1.440}
3	{−0.001, −0.003, −0.007, 0, 0, 0}

Table 3 Twists for the middle finger at start position

Twist	Values
1	{−0.547, 0.0216, 0.002, 322.429, 349.003, −71.279}
2	{0, 0, 0, 2.652, −2.074, 0.714}
3	{−0.001, −0.006, −0.008, 0, 0, 0}

Table 4 Twists for the thumb at start position

Twist	Values
1	{−0.106, 0.003, 0.153, 271.907, 346.412, 32.534}
2	{−0.001, −0.003, −0.007, 0, 0, 0}
3	{0, 0, 0, 0.930, −2.445, 1.684}

gertip. The velocities for the index and middle fingers, as well as the thumb in the start task position are presented in the Tables 2, 3 and 4.

4.1 *Topology Selection and Synthesis Results*

To select a suitable topology for this task, all the three-fingered hands with in-palm mobility and overall mobility being equal to 2 or 3 were investigated. When bounding the overall number of joints and adding solvability conditions for the hands, the topology presented in Table 5 was selected as candidate. This topology is solvable

Table 5 Candidate topology for synthesis

Parent-pointer array	Joint array	Tree graph
{0, 0, 1, 1}	{2, 3, 2, 2}	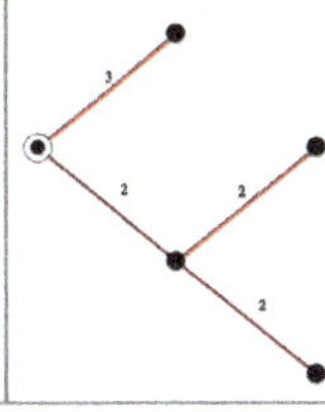

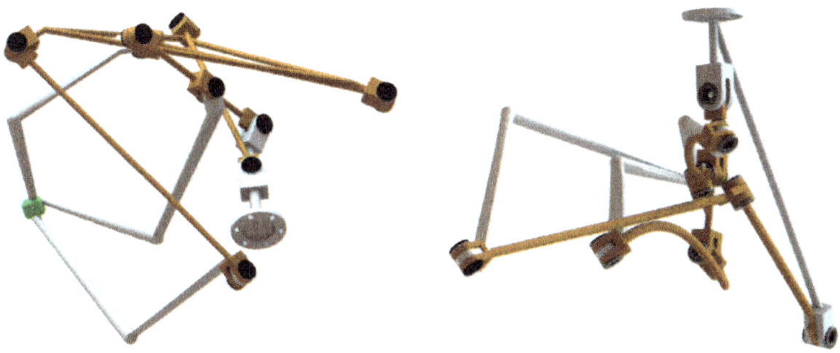

Fig. 6 Hand design for twist subspace, *left*; hand design for approximate trajectory synthesis, *right*

for a total of $m = 5$ exact positions, and the general mobility at the object is $M = 3$ for fingertip contacts with 2 degrees of freedom.

For the synthesis results, both approximate synthesis for ten positions and ten velocities of the experimental trajectory and exact synthesis using two positions and the three obtained velocities in the first position for each finger were used. The system of position and velocity design equations was solved using ArtTreeKS [2]. Figure 6, shows the resulting hand designs.

The analysis and further optimization of the obtained hand designs is subject of our future research.

5 Conclusions

This paper introduces a new technique for designing robotic hands with control of the allowable grasping and in-hand manipulation at a set of finite positions. Dimensional kinematic synthesis with positions and velocity subspaces is presented. The technique ensures not only the grasp positions of the fingertips, but also the in-hand manipulation in desired directions at key contact locations. The method is tested with a three-fingered hand design and the resulting twists subspaces are calculated and compared. Future work will be devoted to optimization of the obtained hand designs.

Acknowledgements This work is supported by the National Science Foundation (first and fourth authors: under Grant No. 1208385; second and third authors: Grant No. 1404011 and 2013–2908). The content is solely the authors' responsibility.

References

1. Mason, M., Srinivasa, S., Vazquez, A., Rodriguez, A.: Generality and simple hands. Int. J. Robot. Res. **31**(5), 688–703 (2010)
2. Simo-Serra, E., Perez-Gracia, A.: Kinematic synthesis using tree topologies. Mech. Mach. Theory **72 C**, 94–113 (2014)
3. Han, L., Trinkle, J.: Dextrous manipulation by rolling and finger gaiting. In: IEEE International Conference of Robotics and Automation (ICRA) (1998)
4. Tsai, L.W.: Mechanism Design: Enumeration of Kinematic Structures According to Function. CRC Press, Boca Raton (2001)
5. Salisbury, J., Roth, B.: Kinematic and force analysis of articulated mechanical hands. J. Mech. Design **105**(1), 35–41 (1983)
6. Murray, R.M., Li, Z., Sastry, S.S.: A Mathematical Introduction to Robotic Manipulation. CRC Press Inc., Boca Raton (1994)
7. Makhal, A., Perez-Gracia, A.: Solvable Multi-fingered Hands for Exact Kinematic Synthesis. Advances in Robot Kinematics. Ljubljiana, Slovenia (2014)
8. Robson, N., Tolety, A.: Geometric design of spherical serial chains with curvature constraints in the environment. In: ASME International Design Engineering Technical Conferences (2011)
9. Rimon, E., Burdick, J.W.: A configuration space analysis of bodies in contact. 1st order mobility. Mech. Mach. Theory **30**(6), 897–912 (1995)
10. Robson, N.P., McCarthy, J.M.: Kinematic Synthesis With Contact Direction and Curvature Constraints on the Workpiece, vol. 8, no. 48094, pp. 581–588. ASME, Las Vegas, NV (2007)

Synthesis of Linkages to Trace Plane Curves

Yang Liu and J. Michael McCarthy

Abstract Kempe's universality theorem introduced in 1876 has recently been proven to ensure that given any algebraic curve a mechanism exists that traces the curve. In this paper, we present two methods to simplify Kempe's linkages. One method uses gear trains, differentials and belt drives to replace his multiplicator, additor and translator linkages. A second method uses the Scotch yoke mechanism and a summing belt drive to generate a mechanical Fourier series that traces the curve. Examples are provided that demonstrate the two approaches.

1 Introduction

This paper considers the design of a mechanical device that guides a point along a specified curve. The goal is to find a middle ground between the synthesis of linkages using Kempe's construction and the synthesis of linkages using a set of points that approximate the desired curve, called path generation or path synthesis.

Kempe's construction [7] uses a set of standard linkages termed the Reversor, Additor, Multiplicator, and Translator that he combines to constrain the two joint angles of an RR planar chain so its end-point traces a specified algebraic curve. Artobolevskii [2] presents a synthesis theory that yields simpler linkages that trace curves up to degree four.

The approach presented in this paper increases the set of standard linkages used for the synthesis of curve-tracing mechanisms to include gear trains to add and multiply, and pulley and belt drives to translate values. In addition, the desired curve is approximated by its Fourier series representation. The result is physically realizable mechanical devices that trace complex plane curves.

Y. Liu (✉) · J.M. McCarthy
Robotics and Automation Laboratory, University of California, Irvine, CA 92697, USA
e-mail: liuy14@uci.edu

J.M. McCarthy
e-mail: jmmccart@uci.edu

J. Lenarčič and J.-P. Merlet (eds.), *Advances in Robot Kinematics 2016*,
Springer Proceedings in Advanced Robotics 4, DOI 10.1007/978-3-319-56802-7_26

2 Literature Review

Interest in the mechanical generation of plane curves is traced by Nolle [13, 14] and Koetsier [11, 12] to Watt's 1784 patent that describes his approximate straight-line linkage, and the associated parallel motion linkage, which he used in his design of a double acting steam engine; also see Hartenberg and Denavit [4]. In his 1877 book, Kempe [8] summarized the design theory for linkages that generate a straight line, and about the same time presented a construction that yields a linkage to trace a given algebraic curve, see [7].

Kempe's construction introduces a correspondence between linkages and algebraic curves, which has been formalized by Jordan and Steiner [5] and Kapovich and Millson [6], and termed *Kempe's Universality Theorem*. Saxena [17] provides a step-by-step description of this construction to obtain a linkage consisting of 48 links and 70 joints that traces a quadratic curve (Fig. 1). Gao et al. [3] showed that the number of bars in Kempe's construction is of order $O(n^4)$, where n is the degree of algebraic curve. Abbott [1] tightened this bound to $O(n^2)$. However, using the dynamic geometry software Cinderella, Kobel [10] provides a number of examples that illustrate the complexity of the resulting linkages.

Artobolevskii [2] states that the direct application of Kempe's construction "would lead ... to such complicated mechanisms that in practice they would be impossible to achieve." He then proceeds to provide a wide range of practical designs for linkages that generate algebraic curves through degree four and higher. For example, the eight-bar conograph linkage shown in Fig. 2 can be adjusted to trace any quadratic curve.

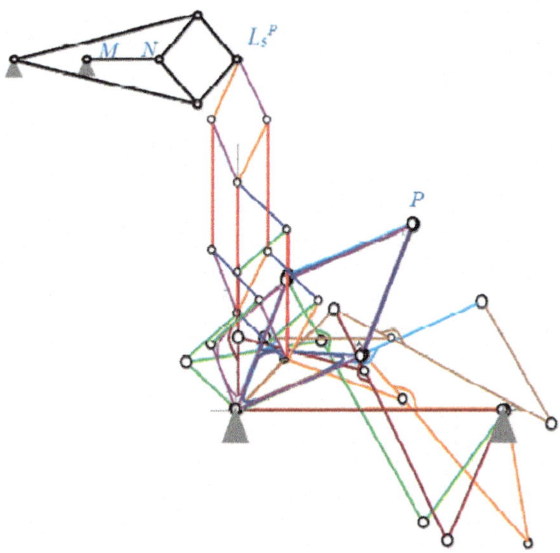

Fig. 1 The linkage that traces the quadratic curve presented in Saxena [17]

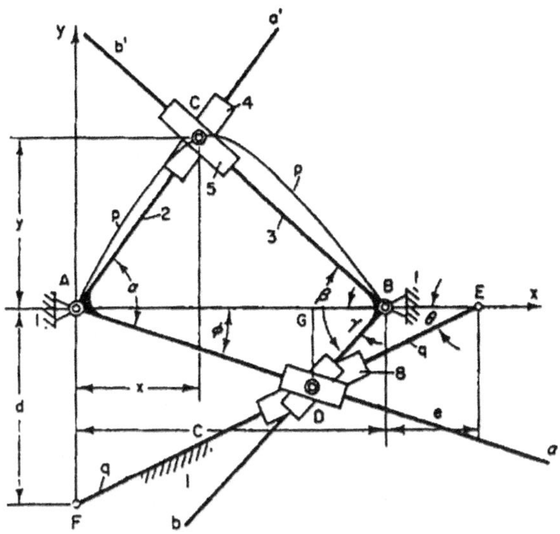

Fig. 2 Artobolevskii [2] shows that this conograph linkage can be adjusted to trace any quadratic plane curve

Roth and Freudenstein [16] introduced a different approach to linkage design for curve tracing. They used the loop equations of a four-bar linkage and nine accuracy points along a desired curve to obtain a system of equations that defined the dimensions of a linkage that guides a coupler point through the given points. Wampler et al. [18] obtained a complete solution for these synthesis equations using polynomial homotopy and showed that there are as many as 4326 distinct four-bar linkages that pass a curve though nine accuracy points. Kim et al. [9] obtained a similar set of synthesis equations for a six-bar linkage and showed that solutions can guide a coupler point though 15 accuracy points. Recent research by Plecnik [15] shows that the equations for 15-point six-bar path generation has a Bezout over 10^{46}, and, while individual solutions can be obtained, it is beyond our ability to compute a complete set of solutions for a given set of 15 accuracy points.

3 Kempe's Linkage with Gears and Pulleys

In this section, we introduce Kempe's method to design planar linkages to trace an algebraic curve. Then we modify his approach to simplify the resulting design.

Let $f(x, y) = 0$ be an algebraic curve. Kempe introduced planar serial chain formed from two revolute joints with link lengths L_1 and L_2 to trace this curve. Thus, the goal is to coordinate the angles θ and ϕ for this RR chain, so that x and y are given by,

$$\mathbf{P} = \begin{Bmatrix} x(\theta, \phi) \\ y(\theta, \phi) \end{Bmatrix} = \begin{Bmatrix} L_1 \cos\theta + L_2 \cos\phi \\ L_1 \sin\theta + L_2 \sin\phi \end{Bmatrix}, \tag{1}$$

such that

$$f(x(\theta,\phi), y(\theta,\phi)) = 0. \tag{2}$$

Kempe shows that this equation can always be reduced to the form,

$$f(\theta,\phi) = \Sigma_i^n A_i \cos(r_i\phi + s_i\theta + \alpha) - C = 0, \tag{3}$$

where $\alpha = 0$ or $\pi/2$, where the A_i and C are constants.

Rather than follow Kempe and introduce his multiplicator, additor, and translator linkages, we use gears, differentials and pulleys to perform these operations. For each r_i and s_i we perform the multiplication using a set of meshing gears, which means for n terms there are at most $g = 2n$ gear pairs. The addition of the terms $r_i\phi + s_i\theta$ are each performed by a gear differential, thus for n terms, we have at most $d = n$ differentials. Finally, we assemble Kempe's serial chain consisting of bars of lengths A_i. We constrain this serial chain to move along the line $x = C$ by a prismatic joint.

In order to obtain the constraint on θ and ϕ to trace the curve f, we connect the gears, differentials and joints of Kempe's serial chain using belts and pulleys. Each pair of gears requires one belt, differential requires two belts, and the n joints of the serial chain requires $n(n+1)/2$ belts. Finally, three belts are required to drive the RR chain. Thus, the number of belts can be estimated to be,

$$b = g + 2d + n(n+1)/2 + 3. \tag{4}$$

In order to demonstrate this procedure, we obtain the mechanism that traces the cubic curve,

$$f(x, y) = x^3 - y - 1 = 0. \tag{5}$$

Let $L_1 = L_2 = 1$ be the length of the RR chain that is to trace the curve, and substitute the resulting $x(\theta,\phi)$ and $y(\theta,\phi)$ into to $f(x, y)$ to obtain,

$$f(\theta,\phi) = \cos^3\theta + \cos^3\phi + 3\cos^2\theta\cos\phi + 3\cos^2\phi\cos\theta - \sin\theta - \sin\phi - 1 = 0. \tag{6}$$

The powers of cosine are reduced to first degree using the identities,

$$\cos^2\theta = \frac{1+\cos(2\theta)}{2}, \quad \cos^3\theta = \frac{3\cos\theta + \cos(3\theta)}{4}. \tag{7}$$

Similarly, the trigonometric sum and difference identities can be used to obtain

$$\begin{aligned} f(\theta,\phi) = {} & \frac{9}{4}\cos\theta + \frac{9}{4}\cos\phi + \frac{1}{4}\cos 3\theta + \frac{1}{4}\cos 3\phi + \frac{3}{4}\cos(2\theta - \phi) + \frac{3}{4}\cos(2\theta + \phi) \\ & + \frac{3}{4}\cos(2\phi - \theta) + \frac{3}{4}\cos(2\phi + \theta) + \cos(\frac{\pi}{2} + \theta) + \cos(\frac{\pi}{2} + \phi) = 1, \end{aligned} \tag{8}$$

which has $n = 10$ terms.

Table 1 Serial chain configuration

Link number	Link length	Phase offset (Degree)	Angular velocity
A_1	2.25	−60	θ
A_2	2.25	60	ϕ
A_3	0.25	−180	3θ
A_4	0.25	180	3ϕ
A_5	0.75	−180	$2\theta - \phi$
A_6	0.75	−60	$2\theta + \phi$
A_7	0.75	180	$2\phi - \theta$
A_8	0.75	60	$2\phi + \theta$
A_9	1	30	θ
A_{10}	1	150	ϕ

Examining (8) we see that Kempe's serial chain that constrains θ and ϕ has 10 links, which are listed in Table 1. This equation requires six gear pairs and four differentials. The number of belts are computed to be 72.

The initial configuration of the links in Kempe's serial chain can be determined by setting the initial position of $\mathbf{P} = (1, 0)$, so we have

$$\begin{Bmatrix} 1 \\ 0 \end{Bmatrix} = \begin{Bmatrix} \cos\theta + \cos\phi \\ \sin\theta + \sin\phi \end{Bmatrix}. \tag{9}$$

Solve this equation to obtain

$$\begin{aligned} &\theta = -60^\circ, \quad \phi = 60^\circ, \\ &\theta = 60^\circ, \quad \phi = -60^\circ. \end{aligned} \tag{10}$$

Both solutions work, so we pick the first solution. This defines each of the angles of the links in the Kempe's serial chain, see Table 1. This mechanical system traces the algebraic curve when the end of Kempe's serial chain is constrained to move along the line $x = 1$ by a prismatic joint, Fig. 3.

In order to compare our linkage to Kempe's construction, we count the components of elementary versions of Kempe's additor, multiplicator and translator linkages. The additor has six bars and is required for each addition including the constants. A multiplication by k requires a multiplicator with at least $m(k) = 2(k-2) + 6$ bars. We model the translator as a parallelogram linkage that requires three bars for each belt used in our design, which means t = 3b. Therefore, in order to estimate Kempe's linkage, we note that (8), requires $a = 6$ additors, $m(2) = 4$ multiplicators with $k = 2$, and $m(3) = 2$ with $k = 3$, thus

$$p = 6a + m(2)6 + m(3)8 + 3b = 36 + 24 + 16 + 216 = 292. \tag{11}$$

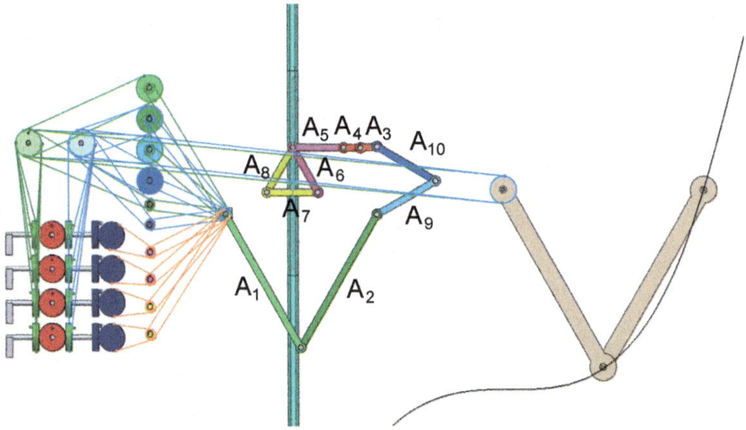

Fig. 3 Gear pairs, differentials and belt drives are used to provide the multiplications, additions and transmission necessary to constrain the RR chain to trace this curve

Thus, we can estimate Kempe's construction to require at least 292 parts for this example.

If we count the individual parts for our method, we have two gears per multiplication and four gears per addition, and two pulleys for each belt. Thus, the part count is

$$p = 2g + 4d + 2b + b = 12 + 15 + 144 + 72 = 244. \quad (12)$$

This comparison shows that the primary difference arises from the complexity of the multiplicator linkage. Our method simplifies this further by using the sizes of pulleys to perform the multiplication. This also shows the dominant role that the translator linkages play in the part count of Kempe's designs. It is our expectation that effective use of gears, differentials, belts and pulleys can simplify the application of Kempe's results to a wide range of algebraic curves.

4 Fourier Series Method

In this section, we provide another approach to the design of a mechanism to trace a plane curve. We assume the curve can be parameterized, then we compute its Fourier decomposition for each component function. We use an array of Scotch yoke mechanisms to generate individual terms and use a belt to add the terms of the Fourier series.

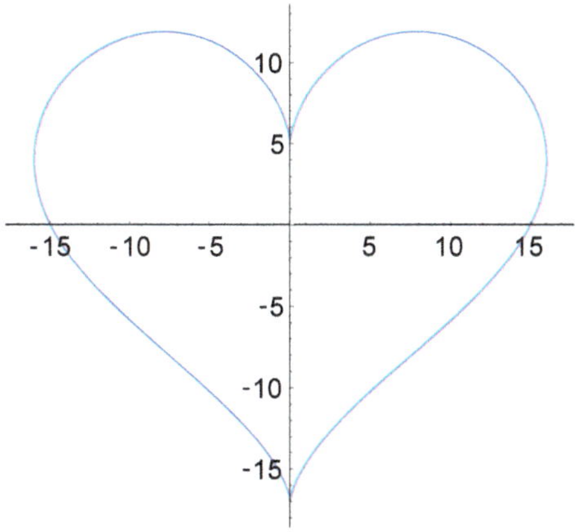

Fig. 4 The parameterized heart curve has a finite Fourier expansion for each coordinate function

In order to demonstrate this procedure, we obtain a mechanism that traces the heart curve (Fig. 4),

$$\begin{Bmatrix} x \\ y \end{Bmatrix} = \begin{Bmatrix} 16\sin^3 t \\ 13\cos t - 5\cos 2t - 2\cos 3t - \cos 4t \end{Bmatrix}. \tag{13}$$

Now reduce powers of sine to first degree using the identity,

$$\sin^3\theta = \frac{3\sin\theta - \sin 3\theta}{4} \tag{14}$$

The result is the equation,

$$\begin{Bmatrix} x \\ y \end{Bmatrix} = \begin{Bmatrix} 12\sin t - 4\sin 3t \\ 13\cos t - 5\cos 2t - 2\cos 3t - \cos 4t \end{Bmatrix}. \tag{15}$$

Reduce minus sign using shift angle properties of trigonometric function, and convert sine terms into cosine terms to obtain

$$\begin{Bmatrix} x \\ y \end{Bmatrix} = \begin{Bmatrix} 12\cos(\frac{-\pi}{2} + t) + 4\cos(\frac{-3\pi}{2} + 3t) \\ 13\cos t + 5\cos(\pi + 2t) + 2\cos(\pi + 3t) + \cos(\pi + 4t) \end{Bmatrix}. \tag{16}$$

The mechanical system that traces this heart curve is obtained by using two Scotch yoke mechanisms for the two terms of the x coordinate, and four of these mechanisms for the y coordinate. These terms are summed using belts. See Fig. 5.

This mechanism provides another approach to the design of mechanisms to trace a plane curve. We have also used this Fourier series approach to design a curve that fits an image generated by an array of points.

Fig. 5 Each Scotch yoke mechanism computes a term in the Fourier expansion of the x and y coordinate functions for the heart curve. There are two terms for x component and four terms for the y component

5 Conclusion

In this paper, we present two ways to assemble a mechanical system to trace plane curves. The first method uses Kempe's universality theorem that guarantees a mechanism exists for any algebraic plane curve. We use gear pairs, differentials and belt drives to simplify the resulting device. The second method uses a mechanical Fourier series constructed from Scotch yoke mechanisms to generate a parametrized plane curve. Kempe's formulation provides an exact representation of algebraic curves, while the Fourier formulation is exact for certain parameterized curves. These are early results in the formulation of a design methodology for mechanical systems that trace arbitrary plane curves.

References

1. Abbott, T.G.: Generalizations of Kempe's Universality Theorem. Massachusetts Institute of Technology (2008)
2. Artobolevskii, I.I.: Mechanisms for the Generation of Plane Curves. Pergamon Press, New York (1964)
3. Gao, X.S., Zhu, C.C., Chou, S.C., Ge, J.X.: Automated generation of Kempe linkages for algebraic curves and surfaces. Mech. Mach. Theory **36**, 1019–1033 (2001)
4. Hartenberg, R.S., Denavit, J.: Kinematic Synthesis of Linkages. McGraw-Hill, New York (1964)
5. Jordan, D., Steiner, M.: Configuration spaces of mechanical linkages. Discret. Comput. Geom. **22**, 297–315 (1999)

6. Kapovich, M., Millson, J.J.: Universality theorems for configuration spaces of planar linkages. Topology **41**(6), 1051–1107 (2002)
7. Kempe, A.B.: On a general method of describing plane curves of the nth degree by linkwork. Proc. Lond. Math. Soc. **VII**(102), 213–216 (1876)
8. Kempe, A.B.: How to Draw a Straight Line. Macmillan and Co, London (1877)
9. Kim, H.S., Hamid, S., Soni, A.H.: Synthesis of six-link mechanisms for point path generation. J. Mech. **6**, 447–461 (1971)
10. Kobel, A.: Automated generation of Kempe linkages for algebraic curves in a dynamic geometry system. Saarland University, Saarbrucken, Germany, Faculty of Natural Sciences and Technology, Department of Computer Science (2008)
11. Koetsier, T.: A contribution to the history of kinematics-i, Watt's straight-line linkages and the early French contributions to the theory of the planar 4-bar coupler curve. Mech. Mach. Theory **18**(1), 37–42 (1983)
12. Koetsier, T.: A contribution to the history of kinematics-ii, the work of English mathematicians on linkages during the period 1869–1878. Mech. Mach. Theory **18**(1), 43–48 (1983)
13. Nolle, H.: Linkage coupler curve synthesis: a historical review-i. Developments up to 1875. Mech. Mach. Theory **9**, 147–168 (1974)
14. Nolle, H.: Linkage coupler curve synthesis: a historical review-ii. Developments after 1875. Mech. Mach. Theory **9**, 325–348 (1974)
15. Plecnik, M.M.: Dimensional synthesis of six-bar linkages for function, motion and path synthesis. Ph.D. thesis, University of California, Irvine (2015)
16. Roth, B., Freudenstein, F.: Synthesis of path-generating mechanisms by numerical methods. J. Eng. Ind. **85**, 298–304 (August 1963)
17. Saxena, A.: Kempe's linkages and the universality theorem. Resonance **16**, 220–237 (March 2011)
18. Wampler, C.W., Morgan, A.P., Sommese, A.J.: Complete solution of the nine-point path synthesis problem for four-bar linkages. J. Mech. Des. **114**, 153–159 (1992)

Subject-Specific Model of Knee Natural Motion: A Non-invasive Approach

Michele Conconi, Nicola Sancisi and Vincenzo Parenti-Castelli

Abstract The capability to model human joint motion is a fundamental step towards the definition of effective treatments and medical devices, with an increasing request to adapt the devised models to the specificity of each subject. We present a new approach for the definition of subject-specific models of the knee natural motion. The approach is the result of a combination of two different techniques and exploits the advantages of both. It relays upon non invasive measurements based on which a kinematic model of the natural motion is built, suitable to be extended to the definition of static and dynamic models. Comparison of the model outcomes with in vitro measurements performed on one specimen shows promising results supporting the proposed approach.

1 Introduction

The natural motion of the knee is the motion of the joint in unloaded conditions. It is the joint starting condition before loads are applied, thus contributing in the determination of the tibio-femoral relative position in loaded conditions. For this reason, the knowledge of the natural motion is useful for all applications which aim at replicating or restoring the natural behaviour of the knee, such as lower-limb modelling, surgical planning and prosthesis design.

The modelling of the joint natural motion can be based on mean data taken from the literature, thus providing a representation of an average joint [3, 17, 24]. However, there is an increasing request of subject-specific models that would allow personalization of treatments and prosthesis geometry to the patient needs. In these cases, the subject-specific motion would be required.

M. Conconi (✉) · N. Sancisi · V. Parenti-Castelli
DIN - Department of Industrial Engineering, University of Bologna, Bologna, Italy
e-mail: michele.conconi@unibo.it

N. Sancisi
e-mail: nicola.sancisi@unibo.it

V. Parenti-Castelli
e-mail: vincenzo.parenti@unibo.it

J. Lenarčič and J.-P. Merlet (eds.), *Advances in Robot Kinematics 2016*,
Springer Proceedings in Advanced Robotics 4, DOI 10.1007/978-3-319-56802-7_27

An accurate estimation of the joint motion is difficult to obtain in vivo [16]: non-invasive techniques could be inaccurate (skin-markers) or too complicated (fluoroscopy) for standard practice, while more invasive techniques (bone-pins) are not acceptable in most cases. Thus, new solutions are needed to predict the joint motion with a good accuracy, based on non-invasive measurements.

In this study a new approach is presented which exploits two techniques with complementary advantages for the modelling of the knee natural motion. The first technique (T1), was originally developed and validated for the ankle joint [4] and is here tested on the knee. T1 predicts the joint motion by optimizing the articular load distribution, assuming this condition as representative of the joint behaviour in physiological working conditions. T1 only requires a 3D representation of the articular surfaces that can be obtained from standard in vivo images of the articulation. It is however not suitable for the characterization of the joint behaviour under generic working conditions.

The second technique (T2) models the knee as a one-degree-of-freedom (1-Dof) spatial mechanism, featuring the two articular contacts and the three isometric fibres of the anterior cruciate (ACL), posterior cruciate (PCL) and medial collateral (MCL) ligaments [17, 18]. T2 was very accurate to replicate the natural motion of specimens over the full flexion arc and can be easily extended to define more complex static and dynamic models that can take into account different loading conditions [22, 23], but a reference motion is needed to adjust the model parameters.

In this study we want to exploit the advantages of both techniques by combining them into a new approach (T1+T2) which allows the definition of subject-specific models of the knee (as T2 does) from non invasive observations of its natural motion (via T1).

The aim of this work is twofold: first, to evaluate the application of T1 to the knee articulation and, second, to test the applicability of T1+T2 on the same joint. To this purpose, a leg specimen is analyzed and the knee joint motion is obtained by T1 starting from magnetic resonance imaging (MRI) data. The motion resulting from T1, together with additional information about the anatomy of the joint specimen also taken from MRI, is used as an input for the definition of T2. Finally, the results of both T1 and T1+T2 are validated against in vitro experimental measurements of the joint natural motion.

2 Materials and Methods

2.1 T1 Technique

Biologic tissues are able to modify their structure in response to the mechanical environment to which they are exposed [2, 6, 12, 19]. Experimental evidence from the literature suggests that the aim of this process is the mechanical optimization of the tissues (functional adaptation). In particular, this process produces articular surfaces that, in physiological working conditions, optimize the contact load distribution or, equivalently, maximize the joint congruence [8, 13].

It is thus possible to identify the adapted motion as the envelope of the maximum congruence configurations (i.e., positions and orientations of all bones constituting the joint). In [5] a measure of joint congruence was proposed, based on the Winkler elastic foundation contact model [14]. This measure makes it possible to estimate the peak-pressure to resultant-force ratio from the geometry of the articulating surfaces at a given configuration, i.e., from a purely geometrical perspective. As a consequence, the adapted motion can be obtained starting solely from the knowledge of the shape of the articular surfaces.

As discussed in [4], the adapted motion should also keep the isometry of the joint main ligaments. This condition is verified during the natural motion, which for this reason can be taken as a good approximation of the adapted one. In the same study, T1 was used to determine the adapted motion of ten human ankles, providing good agreement with experimental measurements of the natural motion of the same specimens. Based on these results, T1 is here applied to determine the knee natural motion.

2.2 T2 Technique

Many studies showed that the natural motion of the tibia with respect to the femur is represented by a complex 1-Dof spatial path, i.e. the relative position and orientation of the tibia and femur is a function of a single motion parameter, for instance the flexion angle [17, 24]. Moreover, some fibres of the ACL, PCL and MCL proved to be almost isometric during this motion. From a mechanical point of view, this means that the natural motion can be reproduced by an appropriate 1-Dof mechanism. Three-dimensional parallel mechanisms were thus defined based on this concept. One of them [17, 18] featured three rigid links representing the ACL, PCL and MCL, while the contacts between tibial and femoral condyles were replaced by the contacts between two pairs of spheres, or, equivalently, by two rigid links connecting the sphere centres at each pair. The result was a 1-Dof 5-5 spatial parallel mechanism, which features two rigid bodies (the femur and tibia) interconnected by 5 binary links.

In previous studies, the initial geometry of the mechanism, namely the attaching points and lengths of the five rigid links was determined from knee specimens. This initial geometry was then optimized in order to best-fit the experimental natural motion of the corresponding specimens [22]. This approach has been extensively validated with very good agreement between model outcomes and corresponding experimental natural motion [17]. The same approach is applied here, but the motion obtained by T1 is used as a reference for the model definition instead of the subject experimental motion.

2.3 Data Acquisition and Processing

A single fresh-frozen lower-limb specimen from a donor (female, 63 years old, weight 68 kg, height 158 cm) was analyzed. The study was approved by the donor organization, which provided written consent. A surgeon declared the leg free from anatomical defects and removed the forefoot and the soft tissues external to the joint, leaving the knee joint capsule and ligaments intact.

A stereophotogrammetric system (Vicon Motion Systems Ltd) was used to measure the tibia and femur relative motion by means of two trackers directly fixed to the bones, thus introducing no soft tissue artefacts (Fig. 1a). The specimen was mounted on a test rig for in vitro analysis of the knee joint behaviour [7] which also allows measurement of the femur-tibia relative motion when no external forces are applied. In this condition, the joint is guided only by the knee passive structures, namely ligaments and contacts, and thus the natural motion can be registered. This experimental natural motion was used only for validation purposes, but it was not used for model definition.

A MRI of the knee was acquired using an isotropic three-dimensional fast spin-echo pulse sequence T2-weighted (3D-FSE-CUBE-T2) within a 1.5 T scanner. Articular surfaces and ligament insertions were then manually segmented using the free open-source software Medical Imaging Interaction Toolkit (MITK), obtaining 3D models of the femur and tibia including bone, cartilage and ligaments (Fig. 1b). In the same way, anatomical reperi were determined on the femur and tibia models, and were used to build anatomical reference systems [22] on both bones (Fig. 2).

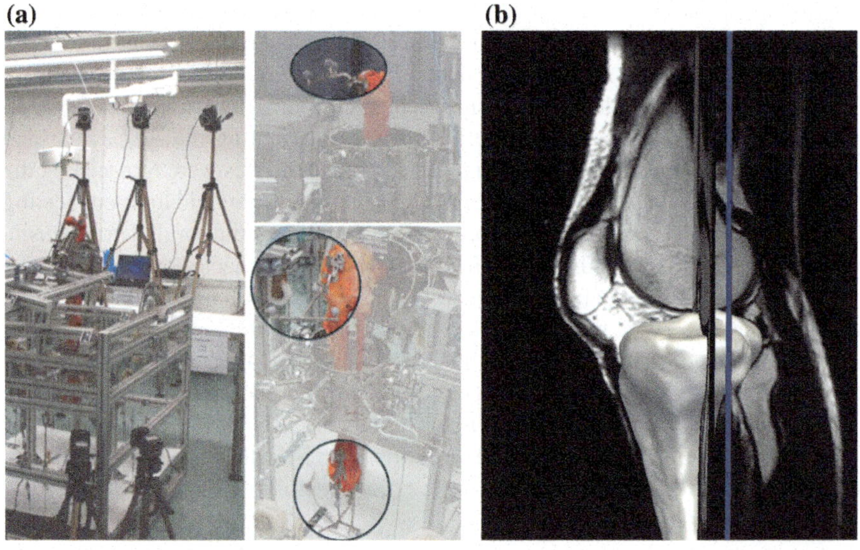

Fig. 1 **a** Stereophotogrammetric system for the measure of the bone relative motion. **b** Reconstruction of knee anatomy from MRI

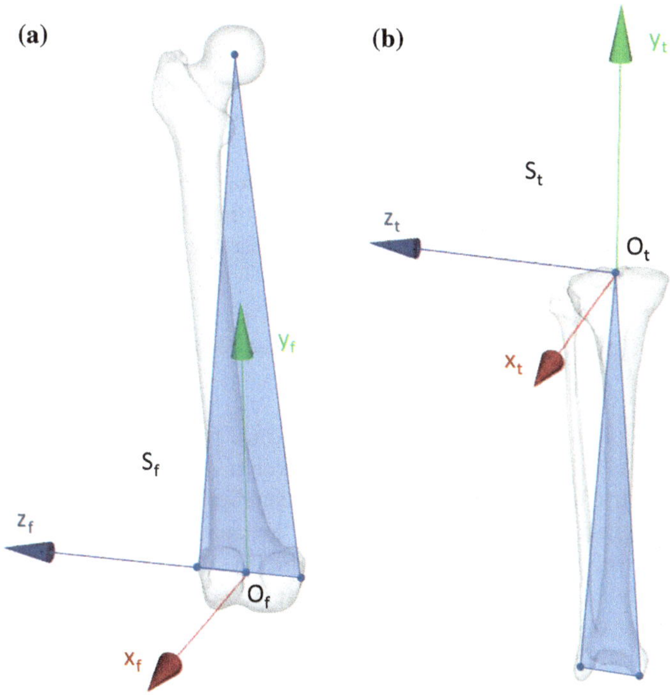

Fig. 2 Anatomical reference systems for the femur **a** and the tibia **b**. The tibia anatomical frame has origin in the tibia centre, i.e., the deepest point in the sulcus between the medial and lateral tibial intercondylar tubercles; x-axis orthogonal to the plane defined by the two malleoli and the tibia centre, anteriorly directed; y-axis directed from the midpoint between the malleoli to the tibia centre; z-axis as a consequence, according to the right hand rule. The femur anatomical frame has origin in the midpoint between the lateral and medial epicondyles; x-axis orthogonal to the plane defined by the two epicondyles and the hip joint centre, anteriorly directed; y-axis directed from the origin to the hip joint centre; z-axis as a consequence, according to the right hand rule

The relative motion of these reference systems was then expressed by means of a standard convention [10], both for the computed and experimental motions.

The anatomical 3D models of the femur and tibia, comprehensive of both bone and articular cartilage, were used within T1 for the evaluation of the knee joint congruence. Flexion angle was imposed and the other five motion components were obtained by maximizing the congruence; the procedure was repeated over the full flexion arc [4].

T2 definition was then performed based on the T1 motion and on the 3D femur and tibia models. The articular surfaces at the femur condyles and tibia plateaus used for congruence evaluation in T1 were approximated by best-fitting spheres in T2, and were then substituted by equivalent rigid links connecting the sphere centres. The most isometric fascicles of the ACL, PCL, MCL (i.e., the anteromedial, posteromedial, anterior fascicle respectively) were identified within the segmented ligament insertion areas [11]. The ligament isometric fibres were obtained as the pair of points (one on the femur, the other on the tibia insertion areas of isometric fascicles) that showed the minimum change in distance during the motion obtained by T1. The so-determined isometric fibres were then substituted by three rigid links. Finally, these preliminary mechanism parameters were adjusted by an optimization procedure [22] to best-fit the T1 motion. The final mechanism parameters were constrained to remain inside the experimental insertion areas and to have a maximum distance of 2 mm with respect to the preliminary parameters.

3 Results

The relative motion of the tibia and femur identified by T1 was consistent with the joint constraints. All ligaments indeed showed very small length changes during T1 motion: isometric fibre length excursions were smaller than 4% of the relevant fibre maximum length for the ACL, PCL, MCL and the lateral collateral ligament (LCL), in agreement with what reported in the literature [1]. The subsequent application of T2 made the ACL, PCL, MCL perfectly isometric, while the LCL showed a length change smaller than 2%. Figure 3 shows the add-abduction (AA), the in-external rotation (IE) and the antero-posterior (AP), proximo-distal (PD) and medio-lateral (ML) translation of the tibia, plotted versus the knee flexion angle as experimentally measured and computed by T1 and T1+T2.

In Table 1, the mean absolute errors (MAE) between T1 and experimental natural motions, between T1+T2 and T1 motions and between T1+T2 and experimental natural motions are presented.

4 Discussion

The tibio-femoral motion predicted by the combination of the two techniques T1 and T2 well replicates the experimental data. There are however some differences in the IE rotation and AP translation, for which the MAEs between the model and the natural motion are about 12 and 4.5 mm, respectively. Despite these quantitative differences, computed and experimental curves show a very similar trend, in particular for the IE rotations which differ essentially by a constant offset. The typical screw-home motion of the knee is therefore correctly predicted by the model, but at each flexion angle the configuration of the tibia results less internally rotated than in the natural motion.

It is worth mentioning that, despite ligaments and contacts do guide the knee natural motion on a 1-Dof spatial path, the IE rotation is less constrained than the other motion components. As a result, the knee shows the smallest stiffness about the IE axis [9, 15], which is thus the most sensitive among the knee motion components both for experimental measure and for numerical models. For what concern the AP

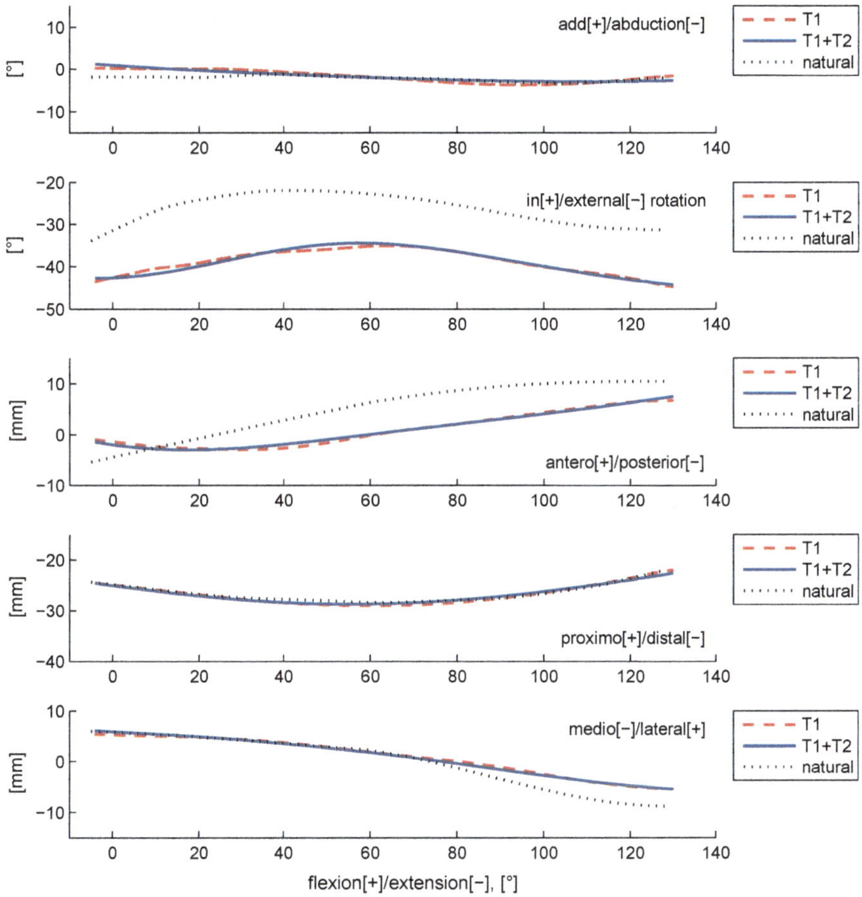

Fig. 3 Tibio-femoral relative motion as resulting from T1 (*red dashed*), T1+T2 (*blue continue*) and experimental natural motion (*black dotted*)

Table 1 MAE for each motion component between T1 and experimental motion, between T1+T2 and T1 motion and between T1+T2 and experimental motion

	AA [°]	IE [°]	AP [mm]	PD [mm]	ML [mm]
T1 versus exp.	0.90	12.24	4.55	0.32	1.22
T1+T2 versus T1	0.54	0.64	0.34	0.28	0.26
T1+T2 versus exp.	0.79	12.43	4.55	0.33	1.19

translations, variations in the IE rotation of the tibia are associated with AP displacements of the same bone. In fact, the tibiofemoral motion is close to a spherical one [21], whose centre does not coincide with the centre of the tibial anatomical reference system. As a result, an IE rotation of the tibia is associated with a translation of the origin of its reference system, mainly along the AP direction. It is thus reasonable that differences in the IE rotation are associated with differences in the AP translation.

Despite the above mentioned differences, both the motion computed by T1 and the experimental natural one respects the ligament isometry, producing length changes smaller than 4 and 5% respectively for the ACL, PCL, MCL and LCL. This result supports the analogy between adapted and natural motion, as validated for the ankle [4] and here hypothesized for the knee. It should also be stressed that only the ligament isometry during the T1 motion made it possible the subsequent application of T2. In fact, in general it is not possible to define a 5-5 mechanism that both follows a generic prescribed path and respects the joint anatomical constraints at the same time. A wider validation of T1 is therefore necessary in order to fully understand the relation between the natural and adapted motion of the knee joint, in terms of both the differences and analogies in terms of motion and ligament isometric behaviour shown in this study.

A similar combination of T1 and T2 was investigated in [20]. In that case however, only CT images of the knee were available, thus providing poor accuracy in the reconstruction of soft tissues that introduced some noise in the motion computed by T1. These limitations were overcome in this study by means of MRI of the articular surfaces. Moreover, the use of MRI makes the proposed approach less invasive with respect to CT images, not exposing the patient to ionizing radiation, and therefore more suitable for the in vivo clinical application.

5 Conclusion

The aim of this study was to test a new approach for the generation of subject-specific model of the natural motion of the knee joint based on non invasive measurements. This approach relays on two techniques defined as T1 and T2 that contribute to determine the final model. The advantages of both techniques are exploited: T1 provides an evaluation of the knee natural motion by non invasive measurements of the articular surfaces; then, based on this motion, T2 provides a mechanism which complies with the constraints imposed by the ligaments and articular contacts, and that can be easily extended to define more complex static and dynamic models.

The motion resulting from T1 fulfils the ligament isometry typical of the knee natural motion, thus making it possible the subsequent application of T2. The results of the combination of T1 and T2 are in good agreement with experimental data, although some differences were found.

Future work is therefore in progress on other specimens in order to further validate the proposed approach and to investigate whether the observed differences are common to all the knee joints, and in case to give a solid explanation of them.

References

1. Belvedere, C., Ensini, A., Feliciangeli, A., Cenni, F., D'Angeli, V., Giannini, S., Leardini, A.: Geometrical changes of knee ligaments and patellar tendon during passive flexion. J. Biomech. **45**(11), 1886–1892 (2012)
2. Benjamin, M., Ralphs, J.R.: Fibrocartilage in tendons and ligaments-an adaptation to compressive load. J. Anat. **193**(Pt 4), 481–494 (1998)
3. Blankevoort, L., Huiskes, R., De Lange, A.: The envelope of passive knee joint motion. J. Biomech. **21**(9), 705–720 (1988)
4. Conconi, M., Leardini, A., Parenti-Castelli, V.: Joint kinematics from functional adaptation: A validation on the tibio-talar articulation. J. Biomech. **48**(12), 2960–2967 (2015)
5. Conconi, M., Parenti-Castelli, V.: A sound and efficient measure of joint congruence. Proc. Inst. Mech. Eng. Part H J. Eng. Med. **228**(9), 935–941 (2014)
6. Eckstein, F., Hudelmaier, M., Putz, R.: The effects of exercise on human articular cartilage. J. Anat. **208**, 491–512 (2006)
7. Forlani, M., Sancisi, N., Conconi, M., Parenti-Castelli, V.: A new test rig for static and dynamic evaluation of knee motion based on a cable-driven parallel manipulator loading system. Meccanica (2015). doi:10.1007/s11012-015-0124-1
8. Frost, H.M.: An approach to estimating bone and joint loads and muscle strength in living subjects and skeletal remains. Am. J. Hum. Biol. **11**, 437–455 (1999)
9. Grood, E.S., Stowers, S.F., Noyes, F.R.: Limits of movement in the human knee. Effect of sectioning the posterior cruciate ligament and posterolateral structures. J. Bone Joint Surg. Am. **70**(1), 88–97 (1988)
10. Grood, E.S., Suntay, W.J.: A joint coordinate systenf for the clinical description of three-dimensional motions: application to the knee. J. Biomech. Eng. **135**, 136–144 (1983)
11. Harner, C.D., Baek, G.H., Vogrin, T.M., Carlin, G.J., Kashiwaguchi, S., Woo, S.L.: Quantitative analysis of human cruciate ligament insertions. Arthrosc. J. Arthrosc. Relat. Surg. **15**(7), 741–749 (1999)
12. Hayashi, K.: Biomechanical studies of the remodeling of knee joint tendons and ligaments. J.Biomech. **29**, 707–716 (1996)
13. Heegaard, J.H., Beaupre, G.S., Carter, D.R.: Mechanically modulated cartilage growth may regulate joint surface morphogenesis. J. Orthop. Res. **17**, 509–517 (1999)
14. Johnson, K.: Contact Mechanics. Cambridge University Press, Cambridge (1985)
15. Markolf, K.L., Mensch, J.S., Amstutz, H.C.: Stiffness and laxity of the knee-the contributions of the supporting structures. A quantitative in vitro study. J. Bone Joint Surg. Am. **58**(5), 583–594 (1976)
16. Masum, M.A., Pickering, M.R., Lambert, A.J., Scarvell, J.M., Smith, P.N.: A review: techniques for kinematic analysis of knee joints. In: Australian Biomedical Engineering Conference (ABEC) (2014)
17. Ottoboni, A., Parenti-Castelli, V., Sancisi, N., Belvedere, C., Leardini, A.: Articular surface approximation in equivalent spatial parallel mechanism models of the human knee joint. Proc. Inst. Mech. Eng. Part H J. Eng. Med. **224**(9), 1121–1132 (2010)
18. Parenti-Castelli, V., Di Gregorio, R.: Parallel mechanisms applied to the human knee passive motion simulation, pp. 333–344. Kluwer Academic Publishers, Pirano-Portoroz, Slovenia (2000)
19. Robling, A.G., Castillo, A.B., Turner, C.H.: Biomechanical and molecular regulation of bone remodeling. Ann. Rev. Biomed. Eng. **8**, 455–498 (2006)
20. Sancisi, N., Conconi, M., Parenti-Castelli, V.: Prediction of the subject-specific knee passive motion from non-invasive measurements. In: The 25th Congress of the International Society of Biomechanics - XXV ISB. Glasgow, UK (2015)
21. Sancisi, N., Parenti-Castelli, V.: A 1-dof parallel spherical wrist for the modelling of the knee passive motion. Mech. Mach. Theory **45**(4), 658–665 (2010)

22. Sancisi, N., Parenti-Castelli, V.: A novel 3d parallel mechanism for the passive motion simulation of the patella-femur-tibia complex. Meccanica **46**(1), 207–220 (2011)
23. Sancisi, N., Parenti-Castelli, V.: On the role of passive structures in the knee loaded motion, pp. 1–8. Springer, Berlin (2012)
24. Wilson, D., Feikes, J., Zavatsky, A., O'Connor, J.: The components of passive knee movement are coupled to flexion angle. J. Biomech. **33**(4), 465–473 (2000)

An Approach for Bone Pose Estimation via Three External Ellipsoid Pressure Points

Nikolas Bufe, Ansgar Heinemann, Peter Köhler and Andrés Kecskeméthy

Abstract The accurate reproduction of bone motion during normal gait using non-invasive external sensors is still an open issue: Using skin markers may lead to large artifacts due to skin sliding, while using newer technologies such as fluoroscopy allows only for short exposure in small regions due to radiation limits, and bone pins used in the past are today prohibited due to the risk of inflammations and pain. This paper presents a simple method for noninvasive bone motion estimation based on palpating prominent bone landmarks via tracked pressure foil planes, where three such landmarks suffice for bone pose estimation. Its mathematical formulation corresponds to determining the pose of a rigid body carrying three ellipsoids when the "pressure points", i.e. the perpendicular feet of the extremal distance points of the ellipsoids on the three pressure foil planes are given. In a previous paper, we showed that the planar case is akin to the 3PPR manipulator, but yielding instead of two solutions up to 64 complex and (up to now found) 48 real solutions. In this paper we treat the 3D case, which is solved numerically, and validate the concept by experimental measurements. It is shown that the method is numerically stable, yielding an accuracy of 0.8° for flexion/extension and 1.2° for abduction/adduction motion of the lower leg.

N. Bufe (✉) · A. Heinemann · P. Köhler · A. Kecskeméthy
University of Duisburg-Essen, Duisburg, Germany
e-mail: nikolas.bufe@uni-due.de

A. Heinemann
e-mail: ansgar.heinemann@uni-due.de

P. Köhler
e-mail: peter.koehler@uni-due.de

A. Kecskeméthy
e-mail: andres.kecskemethy@uni-due.de

J. Lenarčič and J.-P. Merlet (eds.), *Advances in Robot Kinematics 2016*,
Springer Proceedings in Advanced Robotics 4, DOI 10.1007/978-3-319-56802-7_28

1 Introduction

Bone motion tracking from external measurements is an important task in biomechanics, as it is indispensable for model validation and patient-specific objective diagnoses. Currently, most bone tracking systems use markers attached to the skin [9]. However, the skin moves with respect to the underlying bones during motion causing an error known as "soft tissue artifact" (STA). STAs can be substantial (up to 3 cm), as shown in comparisons with bone-pin measurements [8, 12, 17] and fluoroscopy [3, 7, 18]. Some authors have tried to minimize this error by using point clusters [7], but there still remains a systematic error when using skin markers for bone kinematics reconstruction [15]. Thus the problem of artifact-free noninvasive bone motion estimation during gait has remained unsolved.

This paper presents a novel approach for noninvasive bone motion estimation based on tracked external palpation of pressure points of prominent bone landmarks via pressure foils. The concept was first proposed in [13], and a numerical analysis of its planar counterpart was discussed in [5]. Later, the 2D case was formulated in [6] via Gröbner bases, and it was found that the general 2D case with three rigidly connected ellipses for which their "pressure points", i.e. the perpendicular feet of the ellipse extremal distance points (minimal or maximal) on the pressure lines are given, yields 64 complex solutions, for which up to now a case was found with 48 different real solutions [6]. This proves that (1) the geometric problem is solvable, but (2) that the solution is non-trivial, yielding more solutions than the general Steward-Gough platform (which maximally has 40 solutions). However, only one solution, namely the one bringing the bone landmarks as close as possible to the pressure points, is of interest. The present paper extends this solution to 3D.

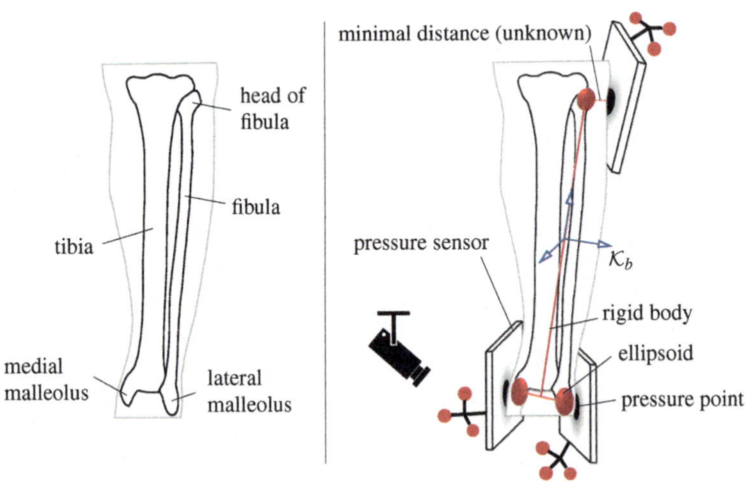

Fig. 1 Pose detection of a rigid-body with three pressure points on tracked pressure foil planes

Figure 1 shows the basic concept of the approach using the human shank as an example. Three pressure points, which are more or less rigid with respect to the tibia, can be easily palpated at the bone landmarks: head of fibula as well as the lateral and medial malleolus. The contact-relevant regions of the bone landmarks can be locally approximated by ellipsoids which are rigidly attached to a body-fixed frame $\mathcal{K}_b$. However, while the pressure points are known, the actual distances between the pressure points and the bone landmarks (representing soft tissue) are unknown. The geometric problem is to find the $\mathcal{K}_b$ poses fulfilling these conditions.

2 Formulation of the Constitutive Equations

Assume that the rigid body is palpable as an array of three rigid ellipsoids $\underline{\mathcal{E}}_i$ $i = 1, \ldots, 3$ which are rigidly attached to a body-fixed frame $\mathcal{K}_b$ (Fig. 2). The rigid-body pose is described by the rotation matrix $\mathbf{R}$ transforming body-fixed coordinates to coordinates in the inertial frame $\mathcal{K}_0$, and the position $\underline{r}_b$ from the origin of $\mathcal{K}_0$ to the origin of $\mathcal{K}_b$ in coordinates of $\mathcal{K}_b$. Each ellipsoid has a local coordinate frame $\mathcal{K}_{\underline{\mathcal{E}}_i}$ rigidly attached to it, whose position relative to the body-fixed frame $\mathcal{K}_b$ is given akin to the previous description by the vector $\Delta\underline{\bar{r}}_i$ and the rotation matrix $\mathbf{R}_i$. The surface of each ellipsoid $\underline{\mathcal{E}}_i$ with semi-principal axes of length a_i, b_i and c_i can be parameterized with respect to its local frame $\mathcal{K}_{\underline{\mathcal{E}}_i}$ as

$$\underline{\mathcal{E}}_i(u_i, v_i) = \begin{bmatrix} a_i \cos(u_i)\cos(v_i) \\ b_i \cos(u_i)\sin(v_i) \\ c_i \sin(u_i) \end{bmatrix} \tag{1}$$

where $-\pi/2 \le u_i \le \pi/2$ and $-\pi \le v_i \le \pi$.

Likely, let the pose of each pressure foil plane $\mathcal{P}_i$ be defined by a plane-fixed frame $\mathcal{K}_{\mathcal{P}_i}$ with in-plane orthonormal vectors $\underline{u}_{xi}$ and $\underline{u}_{yi}$. For an arbitrary point $\underline{r}_{ti}$ on the surface of $\underline{\mathcal{E}}_i$ it holds

$$\underline{r}_{ti} = \underline{r}_b + \mathbf{R}\,\Delta\underline{\bar{r}}_i + \mathbf{R}\mathbf{R}_i\,\underline{\mathcal{E}}_i(u_i, v_i)\,. \tag{2}$$

Let $\underline{r}_{ti}$ be the extremal ellipsoid point, i.e. with minimal or maximal distance to the pressure foil plane. Then this point must fulfill two sets of constraints:

(1) [Gradient orthogonality] For the gradient of $\underline{\mathcal{E}}_i$ at point $\underline{r}_{ti}$,

$$\underline{n}_{\underline{\mathcal{E}}_i} = \mathbf{R}\mathbf{R}_i \left(\frac{\partial \underline{\mathcal{E}}_i}{\partial u_i}(u_i, v_i) \times \frac{\partial \underline{\mathcal{E}}_i}{\partial v_i}(u_i, v_i) \right)\,. \tag{3}$$

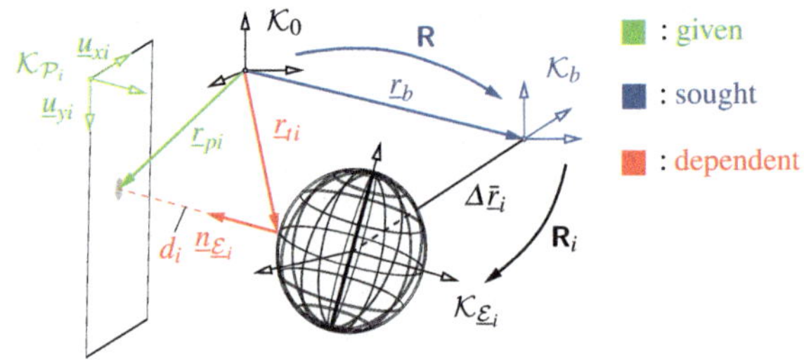

Fig. 2 Definition of the pressure point for a generic *ellipsoid*

it must hold

$$g_{i,x}: \quad \underline{n}_{\mathcal{E}_i}^{\mathrm{T}} \, \underline{u}_{xi} = 0 \tag{4}$$

$$g_{i,y}: \quad \underline{n}_{\mathcal{E}_i}^{\mathrm{T}} \, \underline{u}_{yi} = 0\,. \tag{5}$$

(2) [Distance orthogonality] Also, the distance vector from the measured pressure point $\underline{r}_{pi}$ on plane $\mathcal{P}_i$ to point $\underline{r}_{ti}$ must be normal to the plane $\mathcal{P}_i$:

$$d_{i,x}: \quad (\underline{r}_{ti} - \underline{r}_{pi})^{\mathrm{T}} \, \underline{u}_{xi} = 0 \tag{6}$$

$$d_{i,y}: \quad (\underline{r}_{ti} - \underline{r}_{pi})^{\mathrm{T}} \, \underline{u}_{yi} = 0\,. \tag{7}$$

Collecting all equations for the three ellipsoids gives a system of 12 scalar constraint equations for the 12 unknowns

$$\underline{q} = [\, \underline{r}_b \;\; \underline{\Phi} \;\; u_1 \;\; v_1 \;\; u_2 \;\; v_2 \;\; u_3 \;\; v_3 \,]^{\mathrm{T}}\,, \tag{8}$$

where $\underline{\Phi}$ are the rotation parameters defining the orientation of the rigid body frame $\mathcal{K}_b$, taken as roll-pitch-yaw angles in the present case.

As mentioned in the introduction, the general system of Eqs. (4)–(7) will yield a plurality of solutions, so that a direct numerical root solving of these equations is not be feasible due to the poor conditioning and the closeness of solutions with respect to each other. Thus, instead, the problem is solved as an unconstrained optimization problem with a cost function composed of the weighted sum of the squared constraint values and the squared distances $d_i^2 = \|\underline{r}_{ti} - \underline{r}_{pi}\|^2$

$$F(\underline{q}) = c_1 \sum_{i=1}^{3} \{\, g_{i,x}^2 + g_{i,y}^2 \,\} + c_2 \sum_{i=1}^{3} \{\, d_{i,x}^2 + d_{i,y}^2 \,\} + c_3 \sum_{i=1}^{3} {d_i}^2; \tag{9}$$

where the weights c_i are defined such as to balance between preciseness (first two terms) and feasibility (d_i), and also to homogenize units among terms.

3 Error Analysis

The procedure described above consists of four basic blocks (Fig. 3), each contributing to the overall error of the bone pose estimation algorithm. Here we discuss the individual error sources and transmissions. For marker tracking (Block 1), we used an A.R.T. camera system [2], yielding a resolution error below 0.4 mm for the camera volume used. For the pressure foil (Block 2), we used a 27.9 × 27.9 mm *Tekscan* sensor 5027 with $44 \times 44 = 1936$ cells, yielding a resolution of ±0.6 mm [1]. Each cell reports a binary pressure p^B from 0 to 255 which is scaled to a physical pressure range, in our case 0–34.5 N/cm^2. The pressure center (Block 3) is computed at each time step by five basic operations done using computer vision software (`OpenCV`, [4]): (1) smoothing the data using `cv::GaussianBlur` with a Gaussian kernel size $K = 9 \times 9$ pixels and a standard deviation of $\sigma = 2$, (2) determining the maximum binary pressure value $p^B_{\max}$ over all cells (function `cv::minMaxLoc`), (3) zeroing all cells with pressure below $(p^B_{\max} - \Delta p^B)$, Δp^B being the "thickness" of the highest pressure plateau (similar to a mesa in the mountains), in the present case chosen as 30; (4) determining the contours of all remaining areas (using `cv::findContours`) (see Fig. 4a), and (5) computing the center of pressure of the largest contour inner area.

The resolution error of block 3 can only be determined together with block 2. To this end, a pressure sensor foil was placed on a flat surface whose pose was tracked by the A.R.T. camera system w.r.t. the inertial frame $\mathcal{K}_0$. A thin foam layer was placed on top of the pressure foil in order to obtain a larger contact area, emulating skin-on-bone effects. Finally, a metal sphere was dragged along the foam using a *KUKA KR6/2* robot, so that the vector $_s\underline{r}_p$ remained constant during measurement (Fig. 4). The pose of the body-fixed frame $\mathcal{K}_S$ of the sphere was tracked with markers. The position vector $_s\underline{r}_p$ was determined by an initial calibration. The resulting coordinates of the center of pressure are displayed in Fig. 5. One can see that the total error is in the order of the pressure foil resolution of ±0.2mm, showing that the pressure contour algorithm contributes only marginally to the pressure point error.

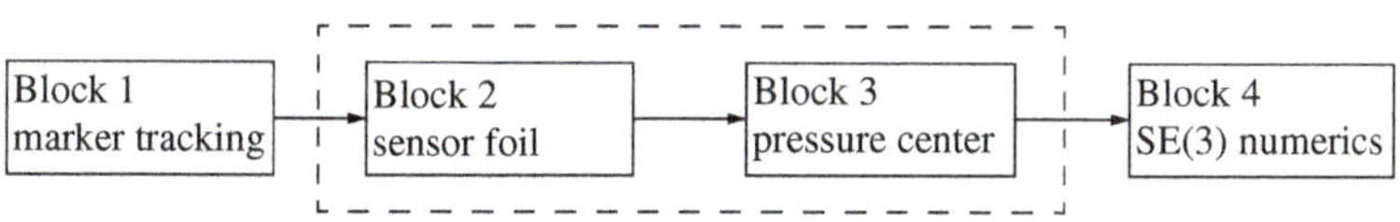

Fig. 3 Basic blocks for bone pose determination from pressure foil measurements

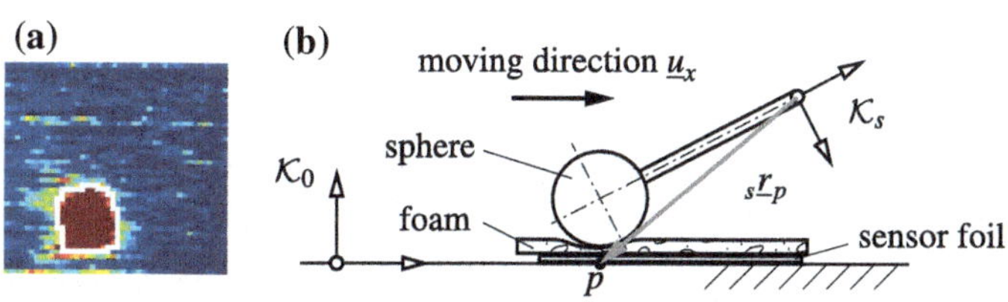

Fig. 4 **a** Peak pressure plateau with contour and **b** device for testing the pressure point

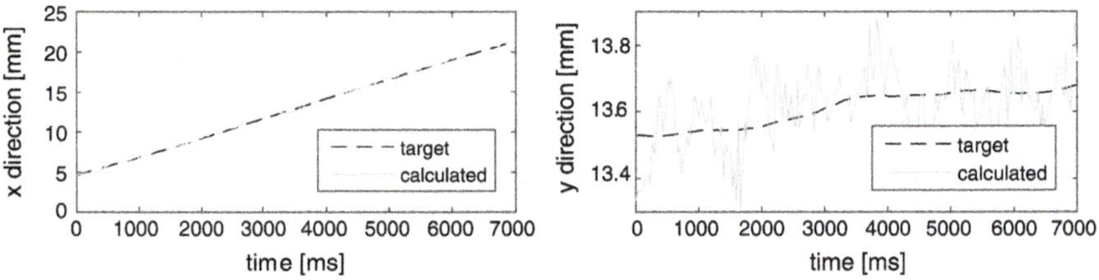

Fig. 5 Comparison between prescribed and computed pressure points

For error analysis of the fourth block, a simulation of bone pose estimation using the equations of Sect. 2 was carried out with a noisy input signal. For the solution of the equations the subroutine `LMDIF` of the numerical library `MINPACK` [11] was used. Starting from the known solution for noise-free data as initial guess, (a) the poses of each sensor were overlaid with Gaussian white noise with a signal-to-noise ratio of 32 and 48 dB for position $\underline{r}_{\mathcal{P}_i}$ and orientation $\mathcal{K}_{\mathcal{P}_i}$ respectively, which correspond to a deviation of approximately 0.1 mm in position and $\Delta\alpha = 1°$ in orientation, and (b) the relative position ${}_{\mathcal{P}_i}\underline{r}_{pi}$ of the pressure points on the sensors were overlaid with Gaussian white noise with a signal-to-noise ratio of 12 dB which corresponds approximately to the *Tekscan* foil cell spacing of 0.6 mm. For each of these variations, the new pose was computed by re-applying the optimization run. The deviations of the solutions were regarded as first-order variations Δr for translation and $\Delta \mathbf{R}$ for rotations. Both were then coupled using the concept of the characteristic length κ [16], yielding the equivalent translational change

$$\Delta s = \|\Delta \underline{r} + \kappa \cdot \Delta \underline{\varphi}\| \quad \text{with} \quad \kappa = \sqrt{2\ell_{\max}^2(1 - \cos(\Delta\alpha))}\,,$$

where $\ell_{\max}$ is the distance between the rigid body center $\mathcal{K}_b$ to the furthest ellipsoid center, giving in the present case $\kappa = 3.41$. By applying the noisy data as the input and computing the pose as the output, the relative error magnification for 7000 data points was computed. The result is shown in Fig. 6. It can be seen that for both cases the order of the output remains in the same order of magnitude as the input noise, and that there is no substantial amplification of errors in this block.

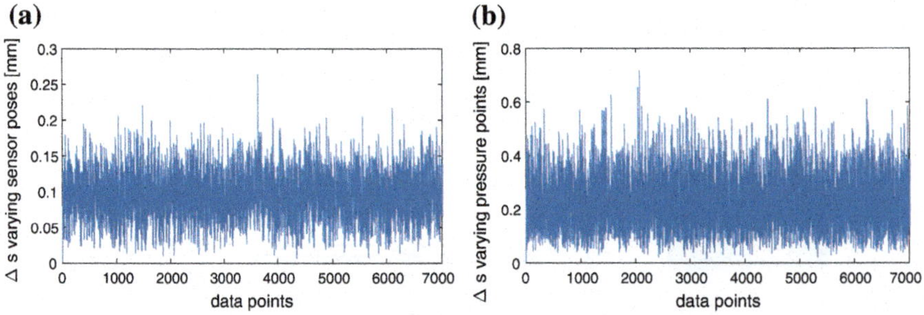

Fig. 6 Bone pose variations after Gaussian noise on **a** sensor pose and **b** pressure points

4 Experimental Verification

The approach has been tested in a self-experiment for the motion of the lower leg. The study was approved by the ethical committee at the authors' university. The test subject was a male of age 27 and was sitting on a chair and lifting and swaying the lower leg in flexion/extension (+105° – +135°) and adduction/abduction (−15° – +5°), respectively. The three ellipsoids where fitted manually to a CT scan of a typical tibia, and the positions of the ellipsoids in the tibia where determined by external measurements on the test persons with rulers. Three *Tekscan* pressure foil sensors 5027 were slightly ($\approx$8 N/cm^2) pressed on the skin at the landmarks head of fibula, lateral malleolus and medial malleolus (Figs. 1 and 7a, b). At the same time, a rail with reflective markers was tightly fixed on the anterior crest of the tibia by straps, so that near-zero relative motion between the rail and the tibia can be assumed (Fig. 7c) [19], and the rail be taken as representative for bone reference motion. Figure 8 shows the difference between pressure foil bone pose estimator and reference rail pose. For flexion/extension and adduction/abduction, the differences between the determined angle and the reference angle are below ±0.4° and ±0.6°, respectively. As the rail is not perfectly rigidly attached to the bone, even these discrepancies could emanate from the relative rail/bone motion, as the undulating discrepancy suggests. Thus the proposed approach seems to yield accurate results for bone motion tracking eliminating skin artifacts. This is remarkable, as knee STAs are known to be especially significant in the initial stage of a sit-to-stand cycle [14].

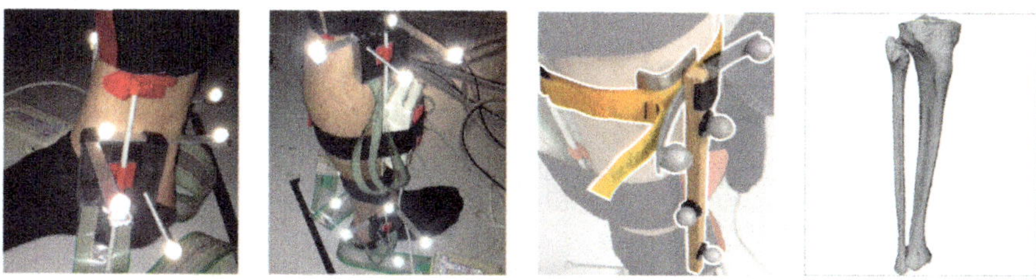

Fig. 7 Sensor placement on shank, rail on shin for reference and bone geometry from CT scan

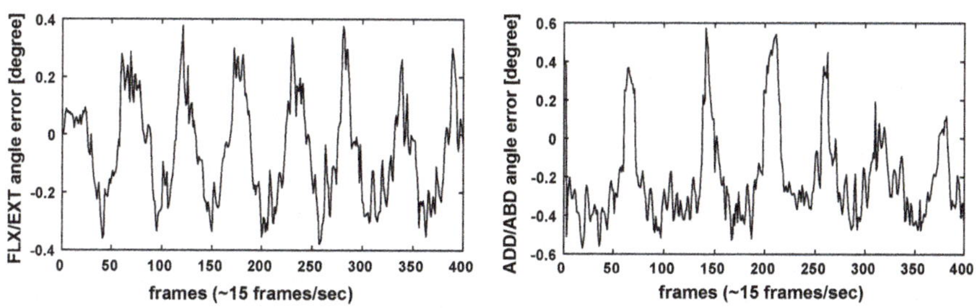

Fig. 8 Flexion-extension (*left*) and adduction-abduction (*right*) angle error for shank motion

5 Conclusions and Outlook

The paper presents a novel method for bone motion estimation using tracked pressure foils, which consists in determining the pose of a rigid body to which three rigid ellipsoids are attached when the "pressure points", i.e. the perpendicular feet of the minimal distance points of the ellipsoids on the three pressure foil planes, are given. The geometrical problem has a variety of solutions [6], from which the most feasible one is determined by optimization. The feasibility of the approach is verified by experiments, leading to an error of approximately $\pm 0.4°$ on the tibia for flexion/extension and $\pm 0.6°$ for abduction/adduction motion of the shank. For more precise estimators, sensors with higher resolution and smaller contact areas could be used. Moreover, the shape of the ellipsoids could be determined more precisely by slightly rotating the pressure foil sensor on the landmarks and fitting suitable ellipsoids to these palpating motions. By simplifying all ellipsoids to spheres and assuming that all three planes are perpendicular, the pose analysis problem becomes similar to the 3PPPS parallel manipulator presented in [10], which is shown to have exactly eight solutions. However, in the general three ellipsoid problem, the number of solutions is expected to be much larger, as already in the special planar case 64 complex solutions were found [6]. Also in this case, however, optimization may lead to a unique feasible solution. These topics could be handled in further research.

Acknowledgements The authors would like to thank Dr. med. A. Lazik-Palm from the University Hospital of Essen who provided us with valuable computer tomography images of the shank.

References

1. Tekscan pressure mapping sensor 5027. https://www.tekscan.com/products-solutions/pressure-mapping-sensors/5027. Accessed 10 Jan 2016
2. The webplatform of ART. http://www.ar-tracking.com/home/. Accessed: 2016-01-10
3. Quantification of soft tissue artefact in motion analysis by combining 3d fluoroscopy and stereophotogrammetry: a study on two subjects. Clin. Biomech. **20**(3), 320–329 (2005)
4. Bradski, G.: Dr. Dobb's Journal of Software Tools (2000)
5. Bufe, N.: Ein Ansatz zur 3D-Bewegungsbestimmung starrer Körper mit Hilfe von 2D-Druckpunktinformationen auf räumlich getrackten Druckmessfolien. Master's thesis, University of Duisburg-Essen, Germany (2013)
6. Bufe, N., Kecskeméthy, A.: Position analysis of a planar rigid-body tracked by three ellipse pressure points along straight lines. In: Proceedings of the 14th World Congress in Mechanism and Machine Science, Taipei, Taiwan (2015)
7. Cappello, A., Stagni, R., Fantozzi, S., Leardini, A.: Soft tissue artifact compensation in knee kinematics by double anatomical landmark calibration: performance of a novel method during selected motor tasks. IEEE Trans. Biomed. Eng. **52**(6), 992–998 (2005)
8. Cappozzo, A., Catani, F., Leardini, A., Benedetti, M., Croce, U.D.: Position and orientation in space of bones during movement: experimental artefacts. Clin. Biomech. **11**(2), 90–100 (1996)
9. Cappozzo, A., Croce, U.D., Leardini, A., Chiari, L.: Human movement analysis using stereophotogrammetry: part 1: theoretical background. Gait Posture **21**(2), 186–196 (2005)

10. Chen, C., Heyne, W., Jackson, D.: A new 6-dof 3-legged parallel mechanism for force-feedback interface. In: 2010 IEEE/ASME International Conference on Mechatronics and Embedded Systems and Applications (MESA), pp. 539–544 (2010)
11. Devernay, F.: C/C++ Minpack. http://devernay.free.fr/hacks/cminpack/ (2007)
12. Holden, J.P., Orsini, J.A., Siegel, K.L., Kepple, T.M., Gerber, L.H., Stanhope, S.J.: Surface movement errors in shank kinematics and knee kinetics during gait. Gait Posture **5**(3), 217–227 (1997)
13. Kecskeméthy, A.: Non-invasive sensor for the examination of the human or animal locomotor system. http://www.directorypatent.com/EP/2186477-a1.html
14. Kuo, M.Y., Tsai, T.Y., Lin, C.C., Lu, T.W., Hsu, H.C., Shen, W.C.: Influence of soft tissue artifacts on the calculated kinematics and kinetics of total knee replacements during sit-to-stand. Gait Posture **33**(3), 379–384 (2011)
15. Peters, A., Galna, B., Sangeux, M., Morris, M., Baker, R.: Quantification of soft tissue artifact in lower limb human motion analysis: a systematic review. Gait Posture **31**(1), 1–8 (2010)
16. Ranjbaran, F., Angeles, J., Kecskeméthy, A.: On the kinematic conditioning of robotic manipulators. In: IEEE International Conference on Robotics and Automation, vol. 4, pp. 3167–3172, Minneapolis (1996)
17. Reinschmidt, C., van den Bogert, A., Nigg, B., Lundberg, A., Murphy, N.: Effect of skin movement on the analysis of skeletal knee joint motion during running. J. Biomech. **30**(7), 729–732 (1997)
18. Sati, M., de Guise, J., Larouche, S., Drouin, G.: Quantitative assessment of skin-bone movement at the knee. Knee **3**(3), 121–138 (1996)
19. Tändl, M., Stark, T., Erol, N., Löer, F., Kecskeméthy, A.: An object-oriented approach for simulation of human gait motion based on motion tracking. Intern. J. Appl. Math. Comput. Sci. (AMCS) **19**(3), 469–483 (2009)

Robot Dynamics Constraint for Inverse Kinematics

Enrico Mingo Hoffman, Alessio Rocchi, Nikos G. Tsagarakis and Darwin G. Caldwell

Abstract Inverse Kinematics is a fundamental tool in Cartesian/Operational Space control. Recent approaches make use of Quadratic Programming Optimization to obtain desired joint velocities or accelerations from Cartesian references. QP based IK also permits to specify constraints to affect the solution. Constraints are fundamental and necessary when working with real robotic hardware since they prevent possible damages: joint limits, self collision avoidance and joint velocity limits are examples of such constraints. In this work we present a constraint to take into account joint torque limits based on the robot dynamics and force/torque sensor measurements. Despite the robot dynamics can be naturally expressed at acceleration level, our main goal is to specify this constraint in a *resolved motion rate control* IK. For this reason we formulate it also at the velocity level to be used in any IK QP based scheme. Hence, this formulation allows to generate dynamically feasible motions of the robot even in simple IK velocity based schemes. We apply this constraint to our humanoid robot COMAN while performing a Cartesian task which requires high torques in some joints. The constraint is developed inside the OpenSoT library.

1 Introduction

Inverse Kinematics (IK) is a fundamental step in robots control since it maps high level Cartesian commands into joint space commands. This step is in general highly non-linear, for this reason linearization through the robot Jacobian has been proposed and it is commonly used (named *Differential* IK):

E. Mingo Hoffman (✉) · A. Rocchi · N.G. Tsagarakis · D.G. Caldwell
Department of Advanced Robotics, Istituto Italiano di Tecnologia, Genova, Italy
e-mail: enrico.mingo@iit.it

A. Rocchi
e-mail: alessio.rocchi@iit.it

N.G. Tsagarakis
e-mail: nikos.tsagarakis@iit.it

D.G. Caldwell
e-mail: darwin.caldwell@iit.it

J. Lenarčič and J.-P. Merlet (eds.), *Advances in Robot Kinematics 2016*,
Springer Proceedings in Advanced Robotics 4, DOI 10.1007/978-3-319-56802-7_29

$$\dot{\mathbf{x}} = \mathbf{J}(\mathbf{q})\dot{\mathbf{q}} \tag{1}$$

where $\dot{\mathbf{q}}$ are joint space velocities, the Jacobian $\mathbf{J}$ (we will skip the dependency on the actual configuration $\mathbf{q}$ from now on) is expressed from a certain *base link* to a certain *distal link*, and operational space velocities $\dot{\mathbf{x}}$ of the *distal link* are expressed in the *base link* reference frame. A general and commonly used solution, for the redundant case of (1), is based on the Jacobian *pseudo-inverse* $\mathbf{J}^{\dagger}$:

$$\dot{\mathbf{q}}_{\mathrm{d}} = \mathbf{J}^{\dagger}\dot{\mathbf{x}}_{\mathrm{d}} + \left(\mathbf{I} - \mathbf{J}^{\dagger}\mathbf{J}\right)\dot{\mathbf{q}}_{0} \tag{2}$$

where $\dot{\mathbf{q}}_0$ is an arbitrary joint space velocity.

Recent approaches make use of Quadratic Programming (QP) Optimization that makes also possible to specify linear constraints for the IK to affect the solution:

$$\begin{aligned}
\dot{\mathbf{q}}_d = \underset{\dot{\mathbf{q}}}{\operatorname{argmin}} \quad & \left\|\mathbf{J}_n\dot{\mathbf{q}} - \dot{\mathbf{x}}_{\mathrm{n,d}}\right\| + \lambda \left\|\dot{\mathbf{q}}\right\| \\
s.t. \quad & \mathbf{A}_1\dot{\mathbf{q}} = \mathbf{A}_1\dot{\mathbf{q}}_1 \\
& \vdots \\
& \mathbf{A}_{n-1}\dot{\mathbf{q}} = \mathbf{A}_{n-1}\dot{\mathbf{q}}_{n-1} \\
& \mathbf{A}_{c,1}\dot{\mathbf{q}} \leq \mathbf{b}_{c,1} \\
& \vdots \\
& \mathbf{A}_{c,n}\dot{\mathbf{q}} \leq \mathbf{b}_{c,n}
\end{aligned} \tag{3}$$

where $\mathbf{A}$ matrices and $\mathbf{b}$ vectors are constraints. In (3), priorities are taken into account considering the previous solutions $\dot{\mathbf{q}}_i$, $i < n$ and constraints of the type $\mathbf{A}_i\dot{\mathbf{q}} = \mathbf{A}_i\dot{\mathbf{q}}_i$, $\forall i < n$, so that the optimality of all higher priority tasks is not changed by the current solution [5]. The second term in the cost function of (3) permits to handle kinematics singularities in order to avoid high joint velocities [6]. A similar structure can be used to solve the IK problem at the acceleration level [9, 11].

Many tasks and constraints have been presented in literature for the framework depicted in (3), examples are: joint limits, joint velocity limits, self collision avoidance [4], Cartesian velocity limits, minimum joint acceleration [2], *Capture Point* [8] and *Momentum Rate* control [3] for humanoid robot balancing.

In this work, a fundamental constraint for the IK step is presented: the robot dynamics. The computed velocities/accelerations in (3) may generate unfeasible motions, in terms of high joint torques, causing the damage of the robot. For this reason it is important to constrain the generated joint torques during the task execution. Despite this constraint is commonly used in *resolved acceleration control* IK schemes, it is not taken into account in simpler *resolved rate control* ones. The main goal of this work is to present the torque limit constraint expressed both at the acceleration and velocity level so that it can be applied to any IK scheme. Furthermore we use this constraint in a stack implementing a high level task in the simulation of a complex humanoid robot. This work follows the basic idea, presented in [2], to have a velocity control scheme that shares (approximately) the characteristics of an acceleration based scheme.

2 OpenSoT

OpenSoT is a framework developed at the *Istituto Italiano di Tecnologia* and aimed to control robots in Operational space [10]. OpenSoT implements the idea of decoupling atomic tasks/constraints descriptions and solvers to execute multiple tasks and achieve complex motion behaviors.

It employs a solver, based on the formulation in (3), implementing a cascade of QP problems, and a set of tasks and constraints in velocity space in order to solve a generic hierarchical inverse kinematics problem on a floating or fixed base robot. The IK solver consists of a state machine that hides all the complexity of the underneath QP solver based on a state-of-art library in QP resolution using the *active set* approach: *qpOASES* [1]. This yields the following features that make the implementation of OpenSoT unique and attractive:

- Demonstrates high modularity through the separation of task descriptions, control schemes and solvers maximizing customization, flexibility and expandability.
- Provides user friendly interfaces for defining tasks, constraints and solvers to promote integration and cooperation in the emerging field of whole-body hierarchical control schemes.
- Demonstrates computation efficiency to allow for real time performance implementations.
- Allows ease of use and application with arbitrary robots through the Universal and Semantic Robotic Description Formats (URDF and SRDF).

The architecture of OpenSoT encourages collaboration and helps integration and code maintenance.[1] With all this in mind, we developed a *library* of tasks and constraints and the robot dynamics constraint is part of the latter.

3 Robot Dynamics Constraint

One of the fundamental problem in IK is that some assigned Cartesian reference trajectories might be dynamically unfeasible by the robot. This means that the robot might get damaged since the required joint torques for a certain motion could be too high. Various techniques have been presented in the past to avoid this problem,

[1]The OpenSoT library is open-source and downloadable at https://github.com/robotology-playground/OpenSoT.

one of the most famous is the *Dynamic Filter* [12]. This technique basically uses an Inverse Dynamics step to filter the generated joint accelerations from the IK solution.

In this work we formulate the *Dynamic Filter* as a constraint. The dynamics of the robot can be written as:

$$\mathbf{M}(\mathbf{q})\ddot{\mathbf{q}} + \mathbf{C}(\mathbf{q}, \dot{\mathbf{q}})\dot{\mathbf{q}} + \mathbf{G}(\mathbf{q}) = \boldsymbol{\tau} - \mathbf{J}_c^T \mathbf{f}_c \tag{4}$$

where $\mathbf{M}(\mathbf{q})$ is the joint space inertia matrix, $\mathbf{C}(\mathbf{q}, \dot{\mathbf{q}})$ takes into account centrifugal and Coriolis terms, $\mathbf{G}(\mathbf{q})$ are the gravity torques, $\boldsymbol{\tau}$ are the joint torques and $\mathbf{J}_c^T \mathbf{f}_c$ are the torques due to contacts (that we measure from the force/torque sensors).

Considering an acceleration level control and taking into account that each joint can provide $\left[\underline{\boldsymbol{\tau}}, \ \overline{\boldsymbol{\tau}}\right]$, it is possible to write the constraint as:

$$\mathbf{D}(\mathbf{q}, \dot{\mathbf{q}}) + \underline{\boldsymbol{\tau}} \leq \mathbf{M}(\mathbf{q})\ddot{\mathbf{q}} \leq \mathbf{D}(\mathbf{q}, \dot{\mathbf{q}}) + \overline{\boldsymbol{\tau}} \tag{5}$$

with $\mathbf{D}(\mathbf{q}, \dot{\mathbf{q}}, \mathbf{f}_c) = -\left(\mathbf{C}(\mathbf{q}, \dot{\mathbf{q}})\dot{\mathbf{q}} + \mathbf{G}(\mathbf{q})\right) - \mathbf{J}_c^T \mathbf{f}_c$.

Despite the constraint is naturally described at the acceleration level, in this work we are considering velocity level control, so it is possible to approximate the joint acceleration $\ddot{\mathbf{q}}$ as:

$$\ddot{\mathbf{q}} \simeq \frac{\dot{\mathbf{q}}^* - \dot{\mathbf{q}}}{\Delta T} \tag{6}$$

then the constraint can be rewritten at the velocity level as:

$$\Delta T \left(\mathbf{D}(\mathbf{q}, \dot{\mathbf{q}}) + \underline{\boldsymbol{\tau}}\right) + \mathbf{M}(\mathbf{q})\dot{\mathbf{q}} \leq \mathbf{M}(\mathbf{q})\dot{\mathbf{q}}^* \leq \Delta T \left(\mathbf{D}(\mathbf{q}, \dot{\mathbf{q}}) + \overline{\boldsymbol{\tau}}\right) + \mathbf{M}(\mathbf{q})\dot{\mathbf{q}} \tag{7}$$

where $\dot{\mathbf{q}}^*$ are the new joint velocities references.

A similar idea was presented also in [7] but contact forces were not taken in consideration, while they are fundamental when working with floating base robots. Practically speaking, it is useful to have a scaling factor $\sigma \in (0, 1]$ in front of the constraint, which allows to smoothen the solution as the robot approaches its dynamic limits:

$$\sigma \left(\Delta T \left(\mathbf{D}(\mathbf{q}, \dot{\mathbf{q}}) + \underline{\boldsymbol{\tau}}\right) + \mathbf{M}(\mathbf{q})\dot{\mathbf{q}}\right) \leq \mathbf{M}(\mathbf{q})\dot{\mathbf{q}}^* \leq \sigma \left(\Delta T \left(\mathbf{D}(\mathbf{q}, \dot{\mathbf{q}}) + \overline{\boldsymbol{\tau}}\right) + \mathbf{M}(\mathbf{q})\dot{\mathbf{q}}\right) \tag{8}$$

4 Experiments

In this section we will show the application of the robot dynamics constraint into a complex IK problem to perform a Cartesian task with the simulated model of our humanoid robot COMAN (in Fig. 1). The task consists of moving both arms downwards generating a whole body squat motion. To show the effect of the dynamics constraint we highly reduce the available torques at the legs joints of 60%: from 50

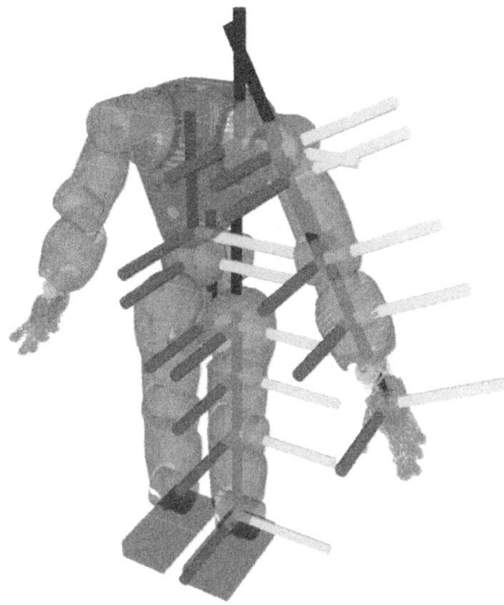

Fig. 1 COMAN robot kinematics and reference frames

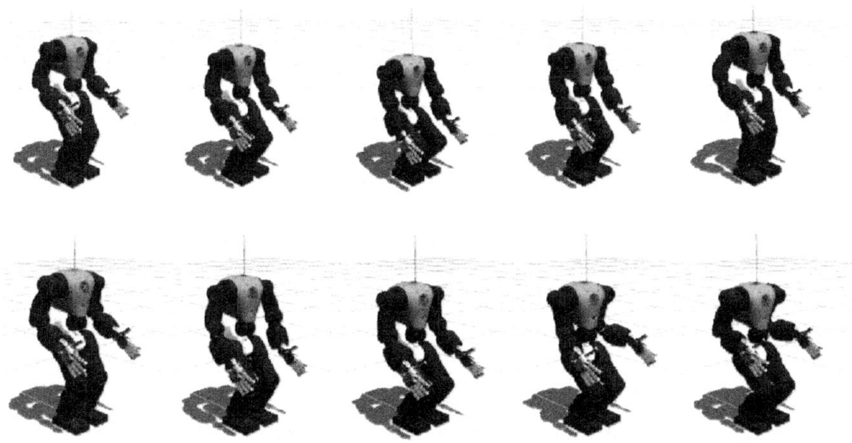

Fig. 2 COMAN performing the squat motion. The *upper* sequence results in an unfeasible motion considering the imposed torque limits while the second results in a feasible one

to 20 Nm. We will show, in particular, that the joint torque at the knee is bounded in the limits. Apart from the robot dynamics constraint, we consider joint limits, joint velocity limits (up to 0.6 $\left[\frac{\text{rad}}{\text{sec}}\right]$).

For the robot dynamics constraint we are using $\sigma = 0.85$ and we are filtering the sensed (simulated) wrenches at the force/torque sensors using a simple filter:

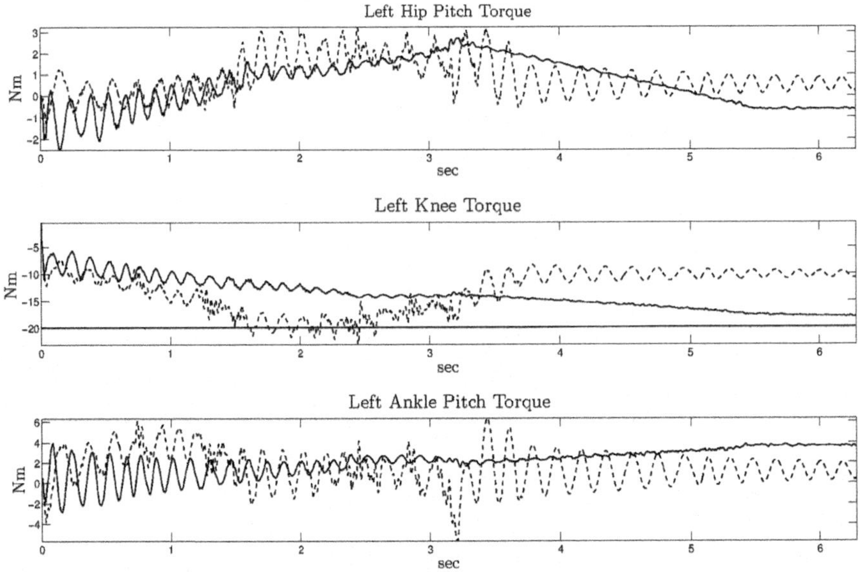

(a) Left leg

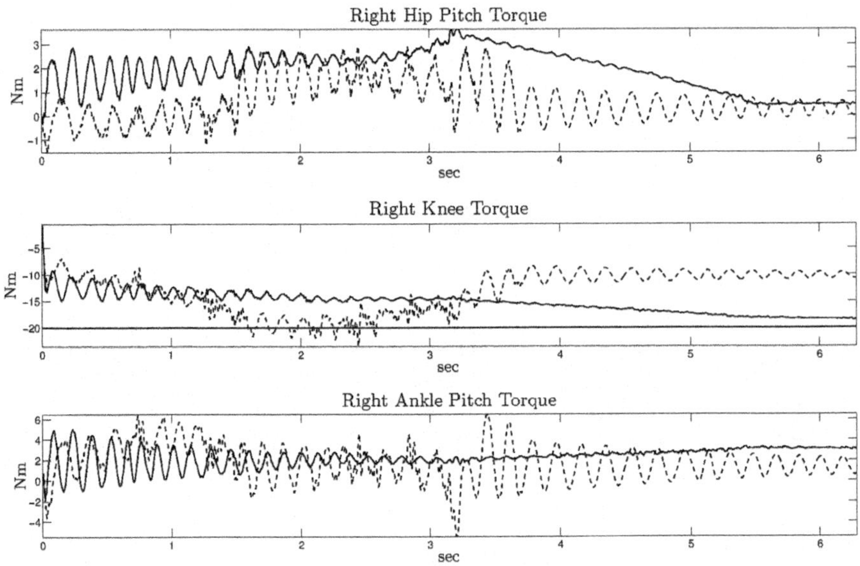

(b) Right leg

Fig. 3 Measured torques on the joints of the pitch joints in the legs while performing the task without (*dashed lines*) and with (*continuous lines*) the robot dynamics constraint. The *constant line* shows the limit on the torque of the knee joint

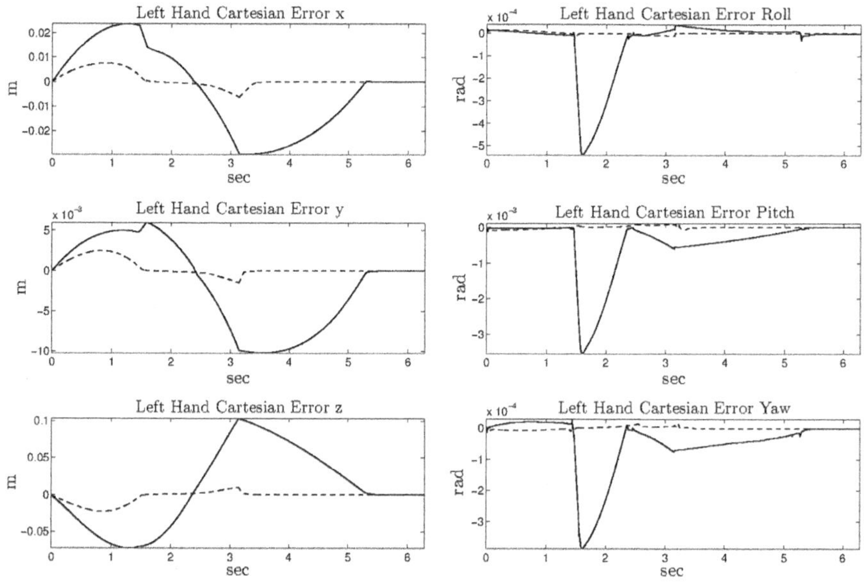

Fig. 4 Cartesian error on the *left* hand while performing the task without (*dashed lines*) and with (*continuous lines*) the robot dynamics constraint

$$\mathbf{w}_\mathrm{t} += (\mathbf{w}_\mathrm{t} - \mathbf{w}_\mathrm{t\text{-}1})\,0.9 \tag{9}$$

The Cartesian task consists of a linear trajectory for the left and right hands, from the initial pose, to 0.18 m down and then back again. Desired joint trajectories are sent to the robot open-loop integrating the results obtained from the IK:

$$\mathbf{q}_\mathrm{d} = \mathbf{q} + \dot{\mathbf{q}}\Delta T \tag{10}$$

Measured joint velocities and force/torques at the ankles are used as feedback.

In Fig. 2 it can be observed the final motion performed by the robot when the robot dynamics constraint is not active (upper sequence) and when it is active (lower sequence).

Cartesian errors are shown in Fig. 4. Despite the higher Cartesian errors when using the robot dynamics constraint, the robot exceeds the imposed torque limits, at the joint knees, trying to keep the Cartesian error small when the robot dynamics constraint is not used. Figure 3a and b shows in particular that the torques at left and right knees, respectively, remains in the imposed limits when using the robot dynamics constraint, while exceeds when not using it.

5 Conclusions

In this paper we have formulated the joints torque limit constraint at the velocity level, for fixed/floating base robots, to filter dynamically unfeasible motions. We presented the theoretical formulation and we showed results in simulation using our humanoid robot COMAN considering a whole body task involving also other constraints such as joint limits and joint velocity limits. We show that the robot dynamics constraint can make the task dynamically feasible and it is able to keep the torque at the knee joint on the given boundary limits. We think this is fundamental when working on real hardware as well as joint limits and joint velocity limits. This constraint is fundamental when Cartesian trajectories references are *aggressive*. Future works will consider the application of such constraint in more complicated tasks, investigate the effect of the tuning of the σ parameter as well as the interaction with other constraints and the test on the real robot.

Acknowledgements The research leading to these results has received funding from the European Union Seventh Framework Programme [FP7-ICT-2013-10] under grant agreements n.611832 WALKMAN.

References

1. Ferreau, H.J., Kirches, C., Potschka, A., Bock, H.G., Diehl, M.: qpoases: A parametric active-set algorithm for quadratic programming. Math. Program. Comput. 1–37 (2013)
2. Flacco, F., De Luca, A.: Discrete-time redundancy resolution at the velocity level with acceleration/torque optimization properties. Robot. Auton. Syst. **70**, 191–201 (2015)
3. Herzog, A., Righetti, L., Grimminger, F., Pastor, P., Schaal, S.: Momentum-based balance control for torque-controlled humanoids. Comput. Res. Repos. **1**, 1–7 (2013)
4. Kanehiro, F., Lamiraux, F., Kanoun, O., Yoshida, E., Laumond, J.P.: A local collision avoidance method for non-strictly convex polyhedra. In: Proceedings of Robotics: Science and Systems IV (2008)
5. Mansard, N., Stasse, O., Evrard, P., Kheddar, A.: A versatile generalized inverted kinematics implementation for collaborative working humanoid robots: The stack of tasks. In: International Conference on Advanced Robotics, 2009, ICAR 2009, pp. 1–6. IEEE (2009)
6. Nakamura, Y.: Advanced Robotics: Redundancy and Optimization, 1st edn. Addison-Wesley Longman Publishing Co., Inc., Boston (1990)
7. Park, K.C., Chang, P.H., Kim, S.H.: The enhanced compact qp method for redundant manipulators using practical inequality constraints. In: Proceedings of the 1998 IEEE International Conference on Robotics and Automation, 1998, vol. 1, pp. 107–114. IEEE (1998)
8. Ramos, O., Mansard, N., Souères, P.: Whole-body motion integrating the capture point in the operational space inverse dynamics control. In: IEEE-RAS International Conference on Humanoid Robots (Humanoid 2014), Madrid, Spain (2014)
9. Ramos, O.E., Mansard, N., Stasse, O., Benazeth, C., Hak, S., Saab, L.: Dancing humanoid robots: Systematic use of OSID to compute dynamically consistent movements following a motion capture pattern. IEEE Robot. Autom. Mag. **22**(4), 16–26 (2015)
10. Rocchi, A., Hoffman, E.M., Caldwell, D.G., Tsagarakis, N.G.: Opensot: A whole-body control library for the compliant humanoid robot coman. In: 2015 IEEE International Conference on Robotics and Automation (ICRA), pp. 1093–1099. IEEE (2015)

11. Saab, L., Ramos, O.E., Keith, F., Mansard, N., Souères, P., Fourquet, J.Y.: Dynamic Whole-Body motion generation under rigid contacts and other unilateral constraints. IEEE Trans. Robot. **29**(2), 346–362 (2013)
12. Yamane, K.: Simulating and generating motions of human figures. STAR/Springer tracts in advanced robotics. Springer, Berlin (2004). http://opac.inria.fr/record=b1119062. Evolution of the author's PhD

Path Planning in Kinematic Image Space Without the Study Condition

Martin Pfurner, Hans-Peter Schröcker and Manfred Husty

Abstract This article proposes a new dual quaternion based approach for motion interpolation. The highlight is that dual quaternions act in the usual way on points, even if the Study condition is not fulfilled. This induces a fibration of kinematic image space into straight lines that describe the same rigid body displacement. This allows to use standard interpolation schemes for (piecewise) rational curves in a linear space rather than on the curved Study quadric.

1 Introduction

Motion design is generally considered to be a more demanding task than curve design. One reason is that usual point models of SE(3), the group of rigid body displacements, are curved. Hence the motion – a curve in a kinematic image space – must not only satisfy approximation, interpolation, or fairness conditions, it is also constrained to lie on a curved manifold. One example is [4] where variational motion design is done in a Euclidean space of dimension twelve but a back-projection step onto the six-dimensional motion group is required. Other examples come from interpolation in the dual quaternion model of SE(3) [2, 3, 12] where the quadratic Study condition is always taken into account. *It is the highlight of this paper that the Study condition can actually be ignored for the purpose of motion design, thus allowing direct motion interpolation in a projective space.* In this way, it is possible to use standard (piecewise) rational interpolation schemes to generate motions of arbitrary smoothness. In this article we introduce the necessary theoretical background, discuss properties of the proposed approach and present a simple example.

M. Pfurner (✉) · H.-P. Schröcker · M. Husty
Unit for Geometry and CAD, University of Innsbruck, Innsbruck, Austria
e-mail: martin.pfurner@uibk.ac.at

H.-P. Schröcker
e-mail: hans-peter.schroecker@uibk.ac.at

M. Husty
e-mail: manfred.husty@uibk.ac.at

J. Lenarčič and J.-P. Merlet (eds.), *Advances in Robot Kinematics 2016*,
Springer Proceedings in Advanced Robotics 4, DOI 10.1007/978-3-319-56802-7_30

2 Kinematic Mapping

A rigid body displacement is often described by a homogeneous 4×4 matrix $\mathbf{M}$, acting on a point $\mathbf{x}$ in the moving frame according to $\mathbf{x}' = \mathbf{M}\mathbf{x}$. Here, $\mathbf{x}'$ is the image point in the base frame, the lower right 3×3 sub matrix of $\mathbf{M}$ is a proper orthogonal matrix corresponding to the orientation of the moving frame with respect to the base frame, and the first column of $\mathbf{M}$ contains the vector connecting the origins of the two frames representing the translational part of the transformation. Using Study's kinematic mapping κ (see [5, 13]), the displacement given by $\mathbf{M}$ is mapped to a point $\mathbf{d} = [x_0, x_1, x_2, x_3, y_0, y_1, y_2, y_3]^T$ in seven dimensional projective space P^7. Its coordinates are called the displacement's *Study parameters.* They fulfill the quadratic *Study condition*

$$x_0 y_0 + x_1 y_1 + x_2 y_2 + x_3 y_3 = 0, \tag{1}$$

its zero set is the *Study quadric* $S_6^2 \subset P^7$.

Conversely, a point on S_6^2 minus the *exceptional three space* $E\colon x_0 = x_1 = x_2 = x_3 = 0$ yields the matrix

$$\mathbf{M} := \kappa^{-1}(\mathbf{d}) = \frac{1}{\Delta}\begin{bmatrix} 1 & 0 & 0 & 0 \\ t_1 & x_0^2 + x_1^2 - x_3^2 - x_2^2 & -2x_0x_3 + 2x_2x_1 & 2x_3x_1 + 2x_0x_2 \\ t_2 & 2x_2x_1 + 2x_0x_3 & x_0^2 + x_2^2 - x_1^2 - x_3^2 & -2x_0x_1 + 2x_3x_2 \\ t_3 & -2x_0x_2 + 2x_3x_1 & 2x_3x_2 + 2x_0x_1 & x_0^2 + x_3^2 - x_2^2 - x_1^2 \end{bmatrix} \tag{2}$$

where $\Delta = x_0^2 + x_1^2 + x_2^2 + x_3^2$ and

$$\begin{aligned} t_1 &= 2x_0y_1 - 2y_0x_1 - 2y_2x_3 + 2y_3x_2, \\ t_2 &= 2x_0y_2 - 2y_0x_2 - 2y_3x_1 + 2y_1x_3, \\ t_3 &= 2x_0y_3 - 2y_0x_3 - 2y_1x_2 + 2y_2x_1. \end{aligned} \tag{3}$$

However, the Study condition (1) is not necessary for $\mathbf{M}$ to describe a rigid body displacement. Thus, the range of κ^{-1} may be extended to $P^7 \setminus E$. By doing this we lose injectivity but get rid of the non-linear Study condition. Properties of this extended inverse kinematic map are investigated in the following section.

3 Fibers of the Extended Inverse Kinematic Map

We consider the map $\kappa^{-1}\colon P^7 \setminus E \to \mathrm{SE}(3)$, defined in (2). Since it is not injective, we should look at its fibers, the sets of points in P^7 with the same image in SE(3). Given $\mathbf{a} = [a_0, \dots, a_7] \in P^7 \setminus E$ we have to solve the matrix equation

$$\mathbf{M}(\mathbf{a}) = \mathbf{M}(\mathbf{b}) \tag{4}$$

for $\mathbf{b} = [b_0, \ldots, b_7]$. Because the restriction of Study's kinematic map κ to the spherical motion group SO(3) is a bijection [6, Sect. 2] we immediately get $[a_0, a_1, a_2, a_3] = [b_0, b_1, b_2, b_3]$. Now, b_4, b_5, b_6, b_7 are obtained by solving the *linear system* arising from (3) and (4). It can be shown that this system of three linear equations in four unknowns has solutions if and only if $\mathbf{a} \notin E$. In this case the solution of the corresponding homogeneous system is $(b_4, b_5, b_6, b_7) = \lambda(a_0, a_1, a_2, a_3)$ with $\lambda \in \mathbb{R}$ so that the inhomogeneous system has the solution set

$$\{\mathbf{a} + \lambda(0, 0, 0, 0, a_0, a_1, a_2, a_3) \mid \lambda \in \mathbb{R}\}. \tag{5}$$

In projective sense, this is the parametric equation of a straight line characterized by

Theorem 1 *The fiber of point* $\mathbf{a} = [a_0, \ldots, a_7] \in P^7 \setminus E$ *with respect to the extended inverse kinematic map* κ^{-1} *is a straight line through* $\mathbf{a}$ *that intersects the exceptional generator* E *in* $[0, 0, 0, 0, a_0, \ldots, a_3]$.

4 Path Planning Using the Extended Inverse Kinematic Map

For a typical path planning task several poses of the end effector frame are given, possibly with higher order derivatives (velocity, acceleration, ...). The problem is to find a valid path such that the manipulator is able to guide its end effector frame through these poses. We propose to treat this as a (piecewise) rational curve interpolation problem in P^7 and map the solution to SE(3) via κ^{-1}. This is possible with or without prescribing parameter values. The fact that interpolation takes place in a projective space rather than an affine space gives degrees of freedom that may be used for additional optimization. Motion interpolation of this type is applicable for path planning in Cartesian space of parallel or serial manipulators with six degrees of freedom. If the inverse kinematics has closed form solutions, we automatically get a motion parametrization in joint space. This is in particular true for Stewart–Gough manipulators.

Because κ^{-1} is quadratic, the degree of any point path is at most twice the degree of the interpolant in P^7. Thus, we can achieve a geometric continuity of order n for the motion with point paths of degree at most $2(n + 1)$. At possible intersection points of interpolant and exceptional generator E, the map κ^{-1} becomes singular and a degree reduction of the paths occurs.

In this paper we consider an example of the simplest case, a linear interpolant ($n = 0$), in more detail. Its paths are at most of degree two. According to [8] this type of rational motion is either a translation in constant direction, a rotation about a fixed axis, a quadratic translation or a Darboux motion [1, Sect. 9.3]. It can be shown that quadratic translations do not occur. The generic case is that of Darboux motions which turn out to be of the special *vertical* type [11]. The projection on the spherical motion component yields a rotation with fixed direction. It corresponds to linear interpolation in (non-dual) quaternion space which has already be considered, for example in [9].

5 Path Planning for SG-Platforms

This section presents a numerical example for a path planning application. We take a 3-3 Stewart–Gough manipulator with given start and end pose of the platform. The anchor points in the base frame are

$$\mathbf{b}_1 = [1, 0, 0, 0], \quad \mathbf{b}_2 = \tfrac{1}{3}[3, 200\sqrt{3}, 600, 0], \quad \mathbf{b}_3 = \tfrac{1}{3}[3, -200\sqrt{3}, 600, 0]; \tag{6}$$

the anchor points in the moving frame are

$$\overline{\mathbf{p}}_1 = [1, 0, 100, 0], \quad \overline{\mathbf{p}}_2 = [1, -50\sqrt{3}, -50, 0], \quad \overline{\mathbf{p}}_3 = [1, 50\sqrt{3}, -50, 0]. \tag{7}$$

The legs of the parallel platform connect the point $\mathbf{b}_i$ with the points $\mathbf{p}_j$ and $\mathbf{p}_k$, for any three distinct indices $i, j, k \in \{1, 2, 3\}$. Start and end pose are given as

$$\begin{aligned} \mathbf{p_S} &= [-1.969, -0.759, 1.149, -0.554, -39.282, 204, 98.154, 63.692], \\ \mathbf{p_E} &= [-0.861, -1.169, 1.578, -0.362, -66.147, 220.447, 106.460, -90.362] \end{aligned} \tag{8}$$

and the linear motion in P^7 connecting them can be parameterized by $\mathbf{m} = (1 - \lambda)\mathbf{p_S} + \lambda\mathbf{p_E}$ with $\lambda \in [0, 1]$. The mapping $\kappa^{-1}(\mathbf{m}) = \mathbf{M}$ yields

$$\mathbf{M} = \frac{1}{25\gamma}\begin{pmatrix} 25\gamma & \mathbf{0} \\ \mathbf{t} & \mathbf{A_0} + \lambda\mathbf{A_1} + \lambda^2\mathbf{A_2} \end{pmatrix} \tag{9}$$

where $\gamma = 110821\lambda^2 - 203420\lambda + 416500$,

$$\mathbf{t} = \begin{pmatrix} 8293675\lambda^2 + 908782\lambda + 41650000 \\ 5372659\lambda^2 - 10882886\lambda + 29155000 \\ 24120195\lambda^2 - 29264718\lambda + 62475000 \end{pmatrix},$$

$$\mathbf{A_0} = \begin{pmatrix} 193694.118 & -268865.882 & -252321.177 \\ 29976.471 & 295556.471 & -291924.706 \\ 367500 & 117600 & 156800 \end{pmatrix},$$

$$\mathbf{A_1} = \begin{pmatrix} -309436.235 & 26810.165 & 69673.553 \\ -245088.941 & -259464.141 & 2367.812 \\ -47460 & -7056 & -423808 \end{pmatrix},$$

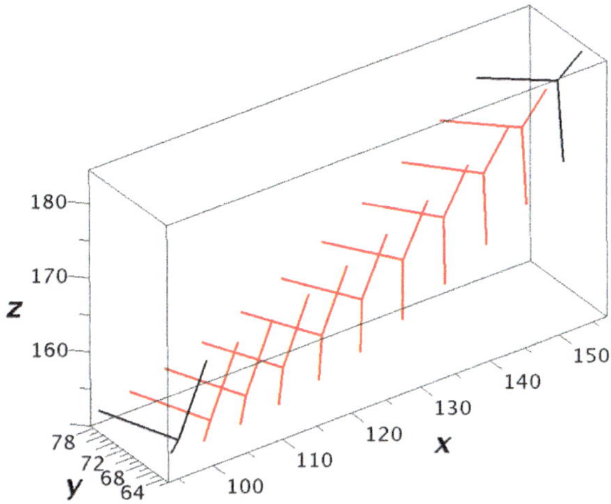

Fig. 1 Motion of the platform coordinate frame

$$\mathbf{A_2} = \begin{pmatrix} 80490.918 & -53290.586 & 54429.952 \\ 5074.071 & 82761.743 & 73525.790 \\ -76005 & -50910.720 & 62551.040 \end{pmatrix}.$$

In Fig. 1, 11 discrete poses of the platform's motion between $\mathbf{p_S}$ and $\mathbf{p_E}$ with respect to the base frame are shown.

The paths of the anchor points of the platform are parametrized by $\mathbf{p_i}(\lambda) = \mathbf{M}\overline{\mathbf{p}}_\mathbf{i}$, $i = 1 \ldots 3$, for example

$$\mathbf{p_1} = \frac{1}{5\gamma} \begin{pmatrix} 1 \\ 7(2471602\lambda^2 + 1648393\lambda + 358540) \\ 21935078\lambda^2 + 20334377\lambda + 9980810 \\ 3(15483078\lambda^2 + 11336647\lambda + 4206650) \end{pmatrix}. \tag{10}$$

It is easy to compute the leg length as distance between corresponding anchor points. The squared distance of $\mathbf{b}_1$ and $\mathbf{p}_2$ is

$$l_1^2 = \frac{1}{\sqrt{\gamma}}(230970295100000\lambda^4 + 233944277500000\lambda^3 + 122850542200000\lambda^2 + 24493035330000\lambda + 2567030405000). \tag{11}$$

Plots of the joint functions of l_1 and l_2 are shown in Fig. 2 and their first time derivative, the joint velocity functions, in Fig. 3.

Remark 1 The presented construction offers degrees of freedom for possible optimization:

- The vectors in (8) are unique only up to scalar multiplication, that is, $\mathbf{m} = (1 - \lambda)\sigma\mathbf{p}_S + \lambda\varepsilon\mathbf{p}_E$ with $[\sigma, \varepsilon] \in P^1$. Here, we chose $[\sigma, \varepsilon]$ such that the resulting

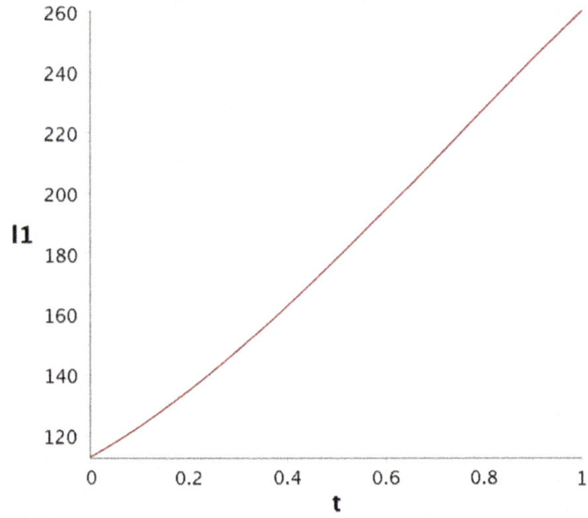

Fig. 2 Explicit joint functions of leg l_1

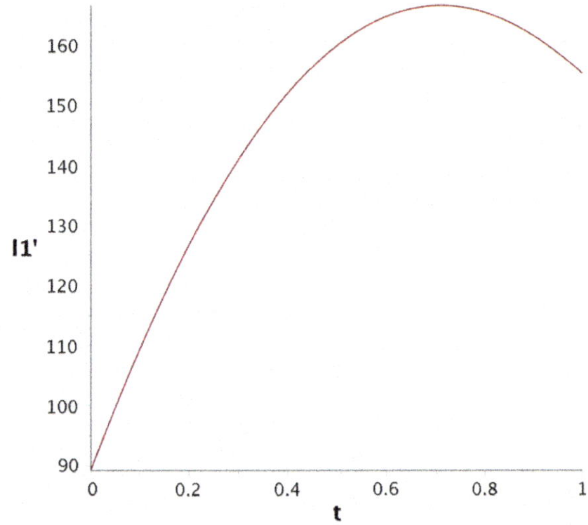

Fig. 3 First time derivative of the explicit joint function of leg l_1

parameterization minimizes the maximal joint velocity of the first leg. Other optimization is obviously possible and might be more relevant in practical applications.

- We may also replace $\mathbf{p}_S$ or $\mathbf{p}_E$ by any point of the shape (5) on their respective fibers. In general, this gives a different interpolating Darboux motion and yet further variables for optimization. We did not use them in our example.

Because the paths of the platform anchor points are given in parametric form, the Jacobian matrix can also be computed depending on the motion parameter. Its determinant reads

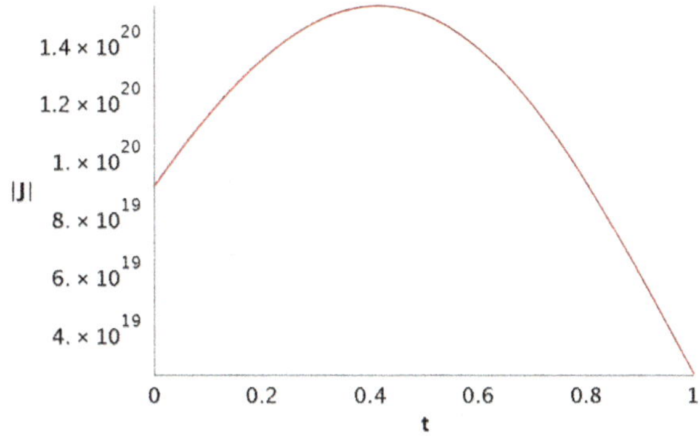

Fig. 4 Determinant of the Jacobian of the manipulator along the path

$$
\begin{aligned}
|J| = &-\frac{1}{\gamma^4}82944000000\sqrt{3}(28467084749878776795358693440\lambda^8 \\
&+ 743978108553268042728662046 4\lambda^7 + 3009884293952398044013325872\lambda^6 \\
&- 6978549515800829192770757528\lambda^5 - 10657661639071658458227857926\lambda^4 \\
&- 7021871831411348585713842061\lambda^3 - 2502177832360314067874514040\lambda^2 \\
&- 464246599952835104361411000\lambda - 25197137764017298716000000).
\end{aligned} \tag{12}
$$

From the plot of the Jacobian function in Fig. 4 we see that the designed path is free of singularities.

6 Discussion and Conclusion

We have demonstrated that the Study condition is not necessary for motion interpolation in the dual quaternion model of SE(3). It is possible to use existing algorithms for direct interpolation of motion data in projective space P^7. As an example, we created a path connecting two poses and related it to the joint parameters of a Stewart–Gough platform.

We can only speculate, why the Study condition has so far been omnipresent in dual quaternion based motion interpolation approaches, for example in [2, 3, 12]. One reason may be analogy to the representation of SO(3) by points of the quaternion unit sphere [6, Sect. 2]: The Study condition is what remains of the unit norm condition after projectivization. Another reason could be mathematical aesthetics. The points of Study quadric minus E are in bijection to SE(3) while our extended inverse kinematic map is no longer injective.

We have not dealt in detail with interpolation algorithms in P^7 so that it is probably too early for a fair comparison with existing work. What we can already claim is the possibility to generate arbitrarily smooth interpolants to given Hermite motion data by standard spline interpolation. By design, such approaches will yield paths of a slightly higher degree than those coming from geometric motion interpolation [7, 10, 12]. However, in contrast to these papers existence of interpolants can be guaranteed, their computation is straightforward and additional degrees of freedom may be used for optimization. This is certainly a worthy topic of further investigations.

In this paper we did not deal with avoidance of singularities or obstacles. It is possible to include this in the proposed method in particular if the obstacles or singularities are described by algebraic equations. In this context, (piecewise) rationality of paths is a great advantage because intersection with algebraic hypersurfaces immediately yields a univariate polynomial. In order to fully profit from our construction, it is necessary to extend the varieties of forbidden poses by the fibration described in Theorem 1. This gives rise to numerous interesting questions for future research.

Acknowledgements This work was supported by the Austrian Science Fund (FWF): I 1750-N26, Kinematic analysis of lower-mobility parallel manipulators using efficient algebraic tools.

References

1. Bottema, O., Roth, B.: Theoretical Kinematics. North-Holland Series in Applied Mathematics and Mechanics, vol. 24. North-Holland Publishing Company, Amsterdam (1979)
2. Gfrerrer, A.: On the construction of rational curves on hyperquadrics. Grazer Math. Berichte **340**, 1–69 (2000)
3. Gfrerrer, A.: Study's kinematic mapping. In: Lenarcic, J., Stanisic, M. (eds.) Recent Advances in Robot Kinematics, pp. 7–16. Kluwer Academic Publishers, Dordrecht (2000)
4. Hofer, M., Pottmann, H., Ravani, B.: From curve design algorithms to the design of rigid body motions. Visual Comput. **20**(5), 279–297 (2004)
5. Husty, M., Pfurner, M., Schröcker, H.P.: Algebraic methods in mechanism analysis and synthesis. Robotica **25**(6), 661–675 (2007)
6. Husty, M., Schröcker, H.P.: Algebraic geometry and kinematics. In: Emiris, I.Z., Sottile, F., Theobald, T. (eds.) Nonlinear Computational Geometry. The IMA Volumes in Mathematics and its Applications, vol. 151, pp. 85–107. Springer, Berlin (2010). Chap. Algebraic Geometry and Kinematics
7. Jaklic, G., Jüttler, B., Krajnc, M., Vitrih, V., Žager, E.: Hermite interpolation by rational g^k motions of low degree. J. Comput. Appl. Math. **240**, 20–30 (2013)
8. Jüttler, B.: Über zwangläufige rationale Bewegungsvorgänge. Österreich. Akad. Wiss. Math.-Natur. Kl. S.-B. II **202**(1–10), 117–232 (1993)
9. Kecskeméthy, A., Hiller, M.: Einflüsse der Parametrisierung auf die Interpolation von Drehbewegungen. Z. Angew. Math. Mech. **73**(4–5), T129–T131 (1993)
10. Krajnc, M., Pockaj, K., Vitrih, V.: Construction of low degree rational Lagrange motions. J. Comput. Appl. Math. **256**, 92–103 (2014)
11. Rad, T.D., Scharler, D.F., Schröcker, H.P.: The kinematic image of RR, PR, and RP dyads. Submitted for publication (2016)
12. Schröcker, H.P., Jüttler, B.: Motion interpolation with Bennett biarcs. In: Kecskeméthy, A., Müller, A. (eds.) Proceedings of Computational Kinematics (CK 2009), pp. 101–108. Springer, Berlin (2009)
13. Study, E.: Geometrie der Dynamen. B. G. Teubner, Leipzig (1903)

The 2D Orientation Interpolation Problem: A Symmetric Space Approach

Yuanqing Wu, Andreas Müller and Marco Carricato

Abstract In this paper, we propose a novel construction of Bézier curves of two-dimensional (2D) orientations using the geometry of real projective plane $\mathbb{R}P^2$. Unlike the commonly adopted unit 2-sphere model S^2, $\mathbb{R}P^2$ is naturally embedded in the 3D special orthogonal group SO(3). It is also a symmetric space that is equipped with a particular class of isometries called geodesic symmetry, which allows us to generate any geodesics using the exponential map of SO(3). We implement the generated geodesics to construct Bézier curves for direction interpolation.

1 Introduction

The 2D direction space (or orienting region [17]) refers to the set of all spatial directions attainable by a reference unit 3D vector, and may be naturally identified with a subset or the entirety of 2D unit sphere S^2. It should be distinguished from the 3D orientation space SO(3) of a rigid body [12]: two rotation matrices $\mathbf{R}, \mathbf{R}' \in SO(3)$ take the initial direction, say the north pole $\mathbf{z} = (0, 0, 1)^T \in S^2$, to the same direction $\pi(\mathbf{R}) = \mathbf{R}\mathbf{z} \in S^2$ so long as they differ by a spin $\mathbf{R}'' \in SO(3)$ about $\mathbf{z}$, i.e. $\mathbf{R} = \mathbf{R}'\mathbf{R}''$. The problem of motion design [19] on the direction space S^2 arises in applications where a decomposition of SO(3) into tilt (changing direction) and torsion (spin) [1] is required or preferred, such as motion planning for two to three degrees-of-freedom (3-DoF) robotic wrists, tool path generation for multi-axis machine tools, and trajectory generation for unmanned aerial vehicles (Fig. 1).

Solving the aforementioned motion design problem poses several difficulties. First, the motion design problem on S^2 is not equivalent to the interpolation problem

Y. Wu (✉) · M. Carricato
Department of Industrial Engineering (DIN), University of Bologna, Bologna, Italy
e-mail: yuanqing.wu@unibo.it

M. Carricato
e-mail: marco.carricato@unibo.it

A. Müller
Institut Für Robotik, Johannes Kepler Universität Linz, Linz, Austria
e-mail: a.mueller@jku.at

J. Lenarčič and J.-P. Merlet (eds.), *Advances in Robot Kinematics 2016*,
Springer Proceedings in Advanced Robotics 4, DOI 10.1007/978-3-319-56802-7_31

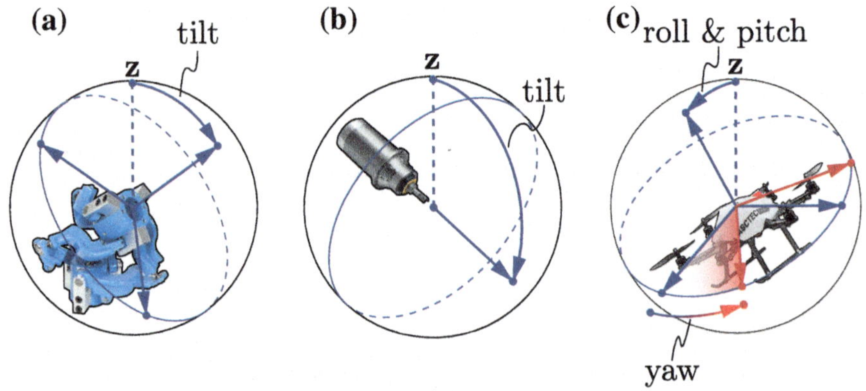

Fig. 1 The tilt motion of a **a** robotic wrist or **b** a five-axis machine spindle, and **c** the collective roll/pitch motion of an UAV are a few relevant applications of motion design on S^2 (photos from internet)

on S^2 [2–4]. In order to compute the inverse kinematics and/or equations of motion for a robot or machine under consideration, we need to designate a 2D submanifold M of SO(3) that (when acting on **z**) can reach any direction in S^2. Besides, although motion design on SO(3) [2, 6, 7, 13–16, 21] and interpolation on S^2 [2, 3] have been extensively studied in the past, existing methods may not be directly applied to general submanifolds of SO(3): the group structure of SO(3), its double covering by unit quaternions (S^3) and the availability of exponential coordinates are usually exploited in its motion design, which may no longer be available for submanifolds of SO(3).

Fortunately, there is a special 2D submanifold of SO(3), which is diffeomorphic to the real projective plane $\mathbb{RP}^2$, that retains properties similar to those of SO(3). We may solve the motion design problem on S^2 by reusing almost any SO(3) motion design method for $\mathbb{RP}^2$ with a slight modification. In particular, $\mathbb{RP}^2$ is complete (a geodesic segment can be infinitely extended), totally geodesic (sharing the same form of geodesics as SO(3)) and auto-parallel (sharing the same formulation for computing acceleration as SO(3)). In fact, this is a direct consequence of $\mathbb{RP}^2$ being a *symmetric subspace* of SO(3) [23]: $\mathbb{RP}^2$ is nothing but the exponential image of a 2D vector subspace of $\mathfrak{so}(3)$ (Lie algebra of SO(3)).

The paper is organized as follows. In Sect. 2, we give a brief review of SO(3) and $\mathbb{RP}^2$, and investigate explicit expressions for geodesics on $\mathbb{RP}^2$. In Sect. 3, we present a de Casteljau type cubic Bézier interpolation algorithm for $\mathbb{RP}^2$ similar to that for SO(3) [2, 15]. Finally in Sect. 4, we conclude our paper with a discussion of future works.

2 Model Space for Motion Design on S^2

We give a brief review of SO(3) and its unit quaternion representation, which will prepare the reader for a brief introduction of the real projective plane $\mathbb{R}P^2$ as the 2D model space for motion design on S^2.

The reader may refer to [12] for further analysis of SO(3) in robot kinematics and dynamics, and [23] for a brief introduction to its symmetric space characterization. SO(3) is a 3D compact Lie group with Lie algebra $\mathfrak{so}(3)$ given by the 3D vector space of all 3×3 real skew-symmetric matrices:

$$\mathfrak{so}(3) = \left\{ \widehat{\mathbf{w}} \in \mathbb{R}^{3\times3} \middle| \, \widehat{\mathbf{w}}^T = -\widehat{\mathbf{w}} \right\} \tag{1}$$

with $\widehat{\mathbf{w}}$ being induced by a 3D vector $\mathbf{w} = (\mathrm{w}_1, \mathrm{w}_2, \mathrm{w}_3)^T$ via $\widehat{\mathbf{w}}\mathbf{v} = \mathbf{w} \times \mathbf{v}$, $\forall \mathbf{v} \in \mathbb{R}^3$. The orthogonal matrix representing a rotation about unit axis $\mathbf{w} \in S^2$ with angle $\theta \in \mathbb{R}$ is given by the Rodriguez formula:

$$e^{\theta\widehat{\mathbf{w}}} = \mathrm{I} + \sin\theta\widehat{\mathbf{w}} + (1 - \cos\theta)\widehat{\mathbf{w}}^2 \qquad \|\mathbf{w}\| = 1 \tag{2}$$

where $e^{(\cdot)}$ denotes the exponential map $\exp : \mathfrak{so}(3) \to \mathrm{SO}(3)$. The same rotation is represented by two unit quaternions $\underline{\mathbf{q}}, -\underline{\mathbf{q}} \in S^3$:

$$\underline{\mathbf{q}} = e^{\frac{\theta}{2}\mathbf{w}} = \cos\frac{\theta}{2} + \sin\frac{\theta}{2}\mathbf{w} \qquad \|\mathbf{w}\| = 1 \tag{3}$$

where $\mathbf{w} = \mathrm{w}_1\mathbf{i} + \mathrm{w}_2\mathbf{j} + \mathrm{w}_3\mathbf{k}$ and $\mathbf{i}, \mathbf{j}, \mathbf{k}$ are quaternionic units that admit the following associative multiplication rules (see for example [20]):

$$\mathbf{ii} = \mathbf{jj} = \mathbf{kk} = -1 \qquad \mathbf{ij} = -\mathbf{ji} = \mathbf{k},\ \mathbf{jk} = -\mathbf{kj} = \mathbf{i},\ \mathbf{ki} = -\mathbf{ik} = \mathbf{j} \tag{4}$$

This two-to-one correspondence $\rho : \pm e^{\frac{\theta}{2}\mathbf{w}} \mapsto e^{\theta\widehat{\mathbf{w}}}$ between S^3 and SO(3) illustrates the fact that SO(3) is diffeomorphic to the 3D real projective space $\mathbb{R}P^3$, which arises from identifying antipodal points of S^3.

The geodesic between two orientations $\mathbf{R}, \mathbf{R}' \in \mathrm{SO}(3)$ is explicitly given by [15]:

$$\gamma(t, \mathbf{R}, \mathbf{R}') = \exp(t \log(\mathbf{R}'\mathbf{R}^T))\mathbf{R},\ t \in [0, 1] \tag{5}$$

If the initial velocity $\widehat{\omega} \in \mathfrak{so}(3)$ is specified instead of the end point $\mathbf{R}'$, we denote the geodesic by:

$$\gamma(t, \mathbf{R}, \widehat{\omega}) = \exp(t\widehat{\omega})\mathbf{R},\ t \in \mathbb{R} \tag{6}$$

Other than being a Lie group, SO(3) is also a *symmetric space* [5, Chap. IV. 6]. Roughly speaking, a symmetric space M can be reflected isometrically onto itself about each point $p \in \mathrm{M}$. These reflection maps (denoted S_p, $p \in \mathrm{M}$) are usually referred to as *geodesic symmetries*, since they reverse the geodesics passing through

$p \in \mathrm{M}$. SO(3) becomes a symmetric space with $S_{\mathbf{R}}$, $\mathbf{R} \in \mathrm{SO}(3)$ given by $S_{\mathbf{R}}(\mathbf{R}') = \mathbf{R}\mathbf{R}'^T\mathbf{R}$, $\forall \mathbf{R}' \in \mathrm{SO}(3)$. In this case, it is easy to see that:

$$S_{\mathbf{R}}(\gamma(t, \mathbf{R}, \widehat{\omega})) = \gamma(t, \mathbf{R}, -\widehat{\omega}) = \gamma(-t, \mathbf{R}, \widehat{\omega}) \tag{7}$$

The reason for us to introduce the concept of symmetric space is that the 2D submanifold we propose for motion design on S^2 is a symmetric subspace of SO(3).

A *symmetric subspace* N of a symmetric space M is a submanifold that is closed under geodesic symmetry [10]:

$$S_p(q) \in \mathrm{N} \qquad \forall p, q \in \mathrm{N} \tag{8}$$

SO(3) admits, up to conjugation, a unique symmetric subspace N given by the exponential image $\exp\widehat{\mathfrak{p}}$ of a 2D vector subspace $\widehat{\mathfrak{p}}$ of $\mathfrak{so}(3)$ defined by:

$$\widehat{\mathfrak{p}} = \mathrm{span}(\widehat{\mathbf{x}}, \widehat{\mathbf{y}}) \qquad \mathbf{x} = (1, 0, 0)^T, \mathbf{y} = (0, 1, 0)^T \tag{9}$$

Several useful geometric properties of $\mathrm{N} = \exp\widehat{\mathfrak{p}}$ along with those reported in [23] are summarized below without proof:

Properties of $\mathbb{RP}^2$

1. N *corresponds, under unit quaternion representation, to the unit 2-sphere* S^2:

$$S^2 = \{a_0 + a_1\mathbf{i} + a_2\mathbf{j} + 0\mathbf{k} \in S^3\} \tag{10}$$

   More precisely, N *is isometrically diffeomorphic to the real projective plane* $\mathbb{RP}^2$, *which arises from identifying antipodal points of* S^2. *Hereafter, we shall denote* N *simply by* $\mathbb{RP}^2$.
2. *The exponential map* $\exp : \widehat{\mathfrak{p}} \to \mathbb{RP}^2$ *is surjective, which allows us to take the square root of any* $\mathbf{R} \in \mathbb{RP}^2$:

$$\mathbf{R}^{1/2} = e^{\widehat{\mathbf{w}}/2} \qquad \mathbf{R} = e^{\widehat{\mathbf{w}}} \in \mathbb{RP}^2, \widehat{\mathbf{w}} \in \widehat{\mathfrak{p}} \tag{11}$$

3. *The spatial (right-trivialized) velocity space of* $\mathbb{RP}^2$ *at* $\mathbf{R} \in \mathbb{RP}^2$ *is given by* [23]:

$$Ad_{\mathbf{R}^{1/2}}\widehat{\mathfrak{p}} = span((\mathbf{R}^{1/2}\mathbf{x})^\wedge, (\mathbf{R}^{1/2}\mathbf{y})^\wedge) \tag{12}$$

   where the Adjoint transformation $Ad_{\mathbf{R}^{1/2}}$ *is defined by* $Ad_{\mathbf{R}^{1/2}}(\widehat{\mathbf{w}}) = \mathbf{R}^{1/2}\widehat{\mathbf{w}}\mathbf{R}^{-1/2} = (\mathbf{R}^{1/2}\mathbf{w})^\wedge, \forall\hat{\mathbf{w}} \in \mathfrak{so}(3)$. *This is referred to as the* half-angle property *in* [23].
4. $\mathbb{RP}^2$ *is a* totally geodesic submanifold *of* SO(3) [23]; *its geodesic, passing through* $\forall\mathbf{R} \in \mathbb{RP}^2$ *with velocity* $(\mathbf{R}^{1/2}\mathbf{w})^\wedge$, $\widehat{\mathbf{w}} \in \widehat{\mathfrak{p}}$, *is given by*:

$$\begin{aligned}\gamma(t, \mathbf{R}, \mathbf{R}^{1/2}\mathbf{w}) &= \exp(t(\mathbf{R}^{1/2}\mathbf{w})^\wedge)\mathbf{R} \\ &= \mathbf{R}^{1/2}e^{t\widehat{\mathbf{w}}}\mathbf{R}^{-1/2}\mathbf{R} = \mathbf{R}^{1/2}e^{t\widehat{\mathbf{w}}}\mathbf{R}^{1/2}\end{aligned} \qquad t \in [0, 1] \tag{13}$$

5. $\mathbb{RP}^2$ *is also an* auto-parallel submanifold *(see* [8, Vol. 2]*) of* SO(3): *computation of acceleration is exactly the same as that of* SO(3) (*see for example* [2, 14]).

The various properties of $\mathbb{RP}^2$ can be summarized into two useful propositions.

Proposition 1 *The geodesic between* $\mathbf{R}, \mathbf{R}' \in \mathbb{RP}^2$ *is given by:*

$$\begin{aligned}\gamma(t, \mathbf{R}, \mathbf{R}') &= \mathbf{R}^{1/2}\exp(t\log(\mathbf{R}^{-1/2}\mathbf{R}'\mathbf{R}^{-1/2}))\mathbf{R}^{1/2} \\ &= \exp(t\,\mathrm{Ad}_{\mathbf{R}^{1/2}}\log(\mathbf{R}^{-1/2}\mathbf{R}'\mathbf{R}^{-1/2}))\mathbf{R}\end{aligned} \qquad t \in [0, 1] \tag{14}$$

Proof This is a direct consequence of Eq. 13. □

We emphasize that the 2D direction space S^2 is not the same as that of the S^2 model for $\mathbb{RP}^2$ (unit quaternion in a hyperplane). In particular, the geodesics $\gamma(t)$ of $\mathbb{RP}^2$ do not map (by $\pi : \mathbb{RP}^2 \to S^2$) to great arcs on S^2. It can be seen from Fig. 2 that as the direction deviates further from $\mathbf{z}$, the projected geodesics $\pi(\gamma(t)) = \gamma(t)\mathbf{z}$ deviate further from great arcs.

Proposition 2 *Given a trajectory* $\mathbf{R}(t) = e^{\widehat{\mathbf{w}}(t)} \in \mathbb{RP}^2, t \in \mathbb{R}$ *where* $\widehat{\mathbf{w}}(t) \in \widehat{\mathfrak{p}}, \forall t \in \mathbb{R}$, *the right-trivialized velocity* $\widehat{\omega}(t)$ *and acceleration* $\widehat{\alpha}(t)$ *of* $\mathbf{R}(t)$ *are given by:*

$$\begin{aligned}\widehat{\omega}(t) &= \dot{\mathbf{R}}(t)\mathbf{R}(t)^T = \left(\mathrm{d}\exp_{\mathbf{R}(t)}\dot{\mathbf{w}}(t)\right)^\wedge = \left(\int_0^1 \exp(u\widehat{\mathbf{w}}(t))\dot{\mathbf{w}}(t)\mathrm{du}\right)^\wedge \\ \widehat{\alpha}(t) &= \left(\frac{D^2}{\mathrm{dt}^2}\mathbf{R}(t)\right)\mathbf{R}(t)^T = \dot{\widehat{\omega}}(t)\end{aligned} \tag{15}$$

where d exp *denotes the right-trivialized differential of the exponential map, and* D/dt *denotes the covariant differentiation along* $\mathbf{R}(t)$. *In particular, we have by the half-angle property:*

$$\omega(t) \in e^{\widehat{\mathbf{w}}(t)/2}\mathfrak{p} \qquad \alpha(t) \in e^{\widehat{\mathbf{w}}(t)/2}\mathfrak{p} \tag{16}$$

Proof The expression for acceleration is exactly the same as that for SO(3) [2], since $\mathbb{RP}^2$ is an auto-parallel submanifold of SO(3). □

In summary, $\mathbb{RP}^2$, except for not being a Lie group, retains all properties of SO(3) pertaining to motion design. It is also easy to see that $\mathbb{RP}^2$ (when acting on $\mathbf{z}$) may reach any direction on S^2. This motivates us to designate $\mathbb{RP}^2$ as the model space for motion design on S^2. Note that $\mathbb{RP}^2$ characterizes Bonev's zero torsional rotation [1, 22], though he did not pursue the symmetric space properties of $\mathbb{RP}^2$. Interpolation on the real projective plane $\mathbb{RP}^2$ is studied in [9] without resorting to its symmetric space characterization. This is later pointed out in [11]. In neither case $\mathbb{RP}^2$ is treated as a symmetric subspace of SO(3), nor is exponential coordinates utilized.

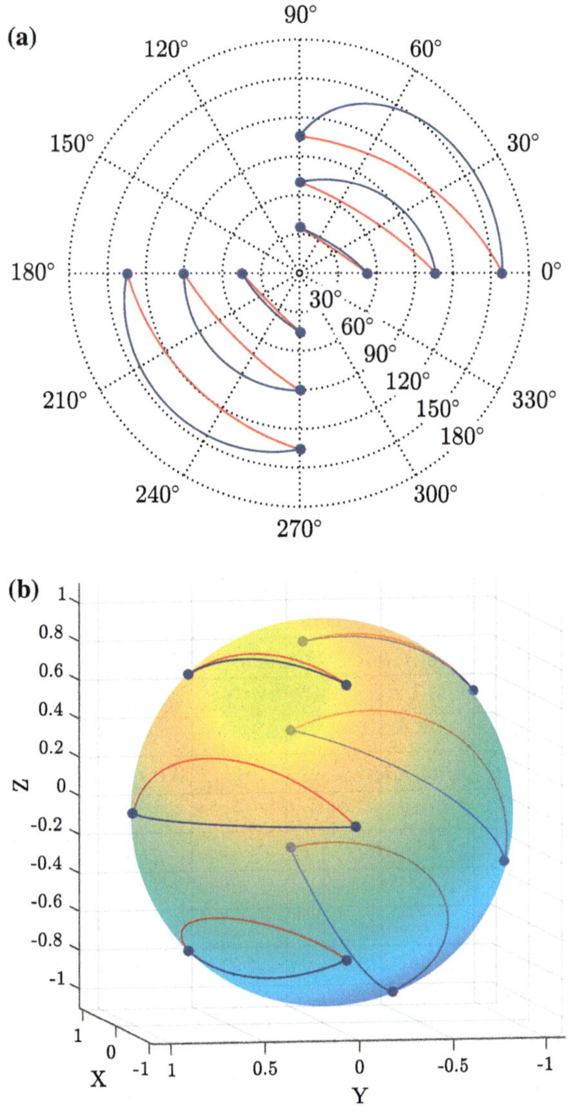

Fig. 2 Geodesics of $\mathbb{R}\mathrm{P}^2$ (*red*) in comparison with great arcs on S^2 (*blue*): **a** on $\widehat{\mathrm{p}}$, **b** on S^2

3 Motion Design on S^2 Using $\mathbb{R}\mathrm{P}^2$

Since $\mathbb{R}\mathrm{P}^2$ retains almost identical properties of SO(3), it is theoretically possible to adapt any motion design methods for SO(3) onto $\mathbb{R}\mathrm{P}^2$. It suffice to make our point by investigating a de Casteljau algorithm for cubic Bézier interpolation between two directions. The following treatment follows closely that in [2].

Given $n+1$ control points $\mathbf{R}_i^0 = e^{\widehat{\mathbf{w}}_i} \in \mathbb{RP}^2$, $\widehat{\mathbf{w}}_i \in \widehat{\mathfrak{p}}$, $0 \le i \le n$, and a parameter value $t \in [0, 1]$, the de Casteljau algorithm computes a point $\mathbf{R}(t) \in \mathbb{RP}^2$ on the nth order Bézier curve in a recursive manner:

$$\begin{aligned}\mathbf{R}_j^k(t) &= \gamma(t, \mathbf{R}_j^{k-1}(t), \mathbf{R}_{j+1}^{k-1}(t)) \\ &= \exp(t\widehat{\mathbf{w}}_j^{k-1}(t))\mathbf{R}_j^{k-1}(t) \qquad 1 \le k \le n, 0 \le j \le n-k \\ \mathbf{R}(t) &= \mathbf{R}_0^n(t)\end{aligned} \tag{17}$$

In reference to Proposition 1 and Eq. 12,

$$\mathbf{w}_j^{k-1}(t) = (\mathbf{R}_j^{k-1}(t))^{1/2}\widetilde{\mathbf{w}}_j^{k-1}(t) \qquad \widetilde{\mathbf{w}}_j^{k-1}(t) \in \mathfrak{p} \tag{18}$$

and

$$(\widetilde{\mathbf{w}}_j^{k-1}(t))^\wedge = \log((\mathbf{R}_j^{k-1}(t))^{-1/2}\mathbf{R}_{j+1}^{k-1}(t)(\mathbf{R}_j^{k-1}(t))^{-1/2}) \tag{19}$$

See Fig. 3a for the case $n = 3$.

We shall first investigate the two-point *boundary value problem* (BVP) where velocity for both initial and end point are specified. Given $\mathbf{R}_0^0 = e^{\widehat{\mathbf{w}}_0}$, $\mathbf{R}_3^0 = e^{\widehat{\mathbf{w}}_3} \in \mathbb{RP}^2$ with $\mathbf{w}_0, \mathbf{w}_3 \in \mathfrak{p}$, and initial and end velocities $\omega_0 = e^{\widehat{\mathbf{w}}_0/2}\widetilde{\omega}_0$, $\omega_3 = e^{\widehat{\mathbf{w}}_3/2}\widetilde{\omega}_3$ with $\widetilde{\omega}_0, \widetilde{\omega}_3 \in \mathfrak{p}$, we need to compute the two remaining control points $\mathbf{R}_1^0$, $\mathbf{R}_2^0$. In reference to [2, Sect. 3], we have:

$$\widetilde{\mathbf{w}}_0^1 = \frac{\widetilde{\omega}_0}{3} \qquad \mathbf{w}_2^1 = \frac{\omega_3}{3} \tag{20}$$

$$\mathbf{R}_1^0 = \exp(\widehat{\mathbf{w}}_0/2)\exp((\widetilde{\omega}_0^1)^\wedge)\exp(\widehat{\mathbf{w}}_0/2) \qquad \mathbf{R}_2^0 = \exp(-\widehat{\mathbf{w}}_2^1)\mathbf{R}_3^0 \tag{21}$$

Next, for a *initial-value problem* (IVP), the initial acceleration $\alpha_0 = e^{\widehat{\mathbf{w}}_0/2}\widetilde{\alpha}_0$, $\widetilde{\alpha}_0 \in \mathfrak{p}$ instead of end point velocity is given. The second missing control point is then given by:

$$\mathbf{w}_1^1 = \mathbf{w}_0^1 + \frac{1}{6}\int_0^1 \exp(u\widehat{\mathbf{w}}_0^1)\alpha_0 \mathrm{du} \tag{22}$$

$$\mathbf{R}_2^0 = \exp(\widehat{\mathbf{w}}_1^1)\mathbf{R}_1^0 \tag{23}$$

The IVP algorithm can be easily implemented in a multi-point interpolation in a way similar to that proposed in [16], with suitably chosen initial acceleration, such as for minimum distance (Fig. 3b) and for minimum acceleration (Fig. 3c). The details are omitted due to page limit.

We also remark, although a direct interpolation on S^2 is technically more convenient, it does not respect the geometry of $\mathbb{RP}^2$ which we have to face when controlling a robot to follow a particular motion trajectory that results in the desired point

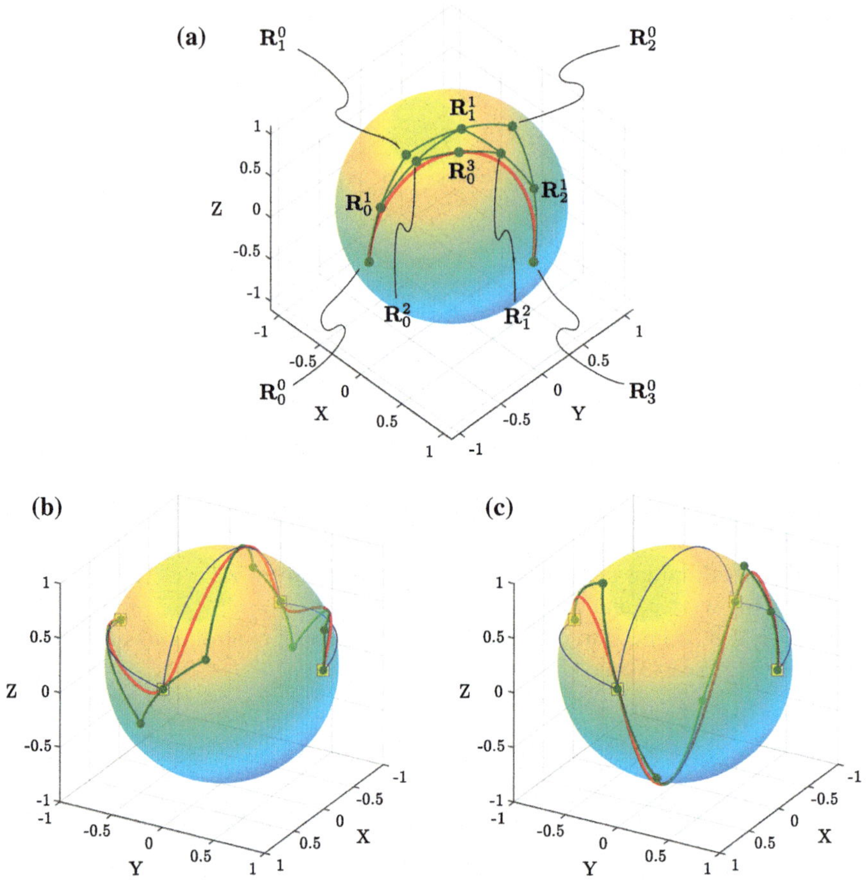

Fig. 3 **a** cubic Bézier curve on $\mathbb{R}\mathrm{P}^2$ (*red*) with control polygons (*green*); **b** and **c** minimum distance and acceleration cubic Bézier interpolation of four directions on S^2 (*red*) with control polygons (*green*) as compared to great arcs (*blue*)

trajectory on S^2. When the robot has a constant identity inertia tensor, for example, the aforesaid minimal distance and acceleration trajectory are also the minimum energy and minimum control effort trajectory, respectively.

4 Conclusion

In this paper, we have proposed a novel model space, the real projective plane $\mathbb{R}\mathrm{P}^2$ as a symmetric subspace of SO(3), for the motion design on S^2. We emphasize on the possibility of reusing existing interpolation methods for SO(3) on $\mathbb{R}\mathrm{P}^2$ due to the symmetric space and symmetric subspace structure of SO(3) and $\mathbb{R}\mathrm{P}^2$, respectively. We illustrate the use of this novel model space in direction interpolation using a Riemannian Bézier cubic IVP interpolant. Nevertheless, interpolants based on unit quaternions may be equally adapted to $\mathbb{R}P^2$. It may also be proved that the Riemannian cubic spline on $\mathbb{R}P^2$ admits exactly the same form as that on SO(3) [3, 14]. This opens up new applications, for example in optimal control problems on $\mathbb{R}P^2$ (see for example [18]) using the computationally convenient exponential coordinates. Our work may be carefully generalized to other symmetric subspaces of SE(3) [23], which we hope to report in the near future.

Acknowledgements The first and third author gratefully acknowledge the PRIN 2012 grant No. 20124SM Z88 and the MANET FP7-PEOPLE-ITN grant No. 607643. The second author acknowledges partial support of this work by the Austrian COMET-K2 program of the Linz Center of Mechatronics (LCM). This work is also in partial fulfillment to China National Natural Science Foundation Grant No.51375413, which supported Dr. Wu when he was working at HKUST and HKUST SRI.

References

1. Bonev, I., Zlatanov, D., Gosselin, C.: Advantages of the modified Euler angles in the design and control of PKMs. In: 2002 Parallel Kinematic Machines International Conference, pp. 171–188. Citeseer (2002)
2. Crouch, P., Kun, G., Leite, F.S.: The de Casteljau algorithm on Lie groups and spheres. J. Dyn. Control Syst. **5**(3), 397–429 (1999)
3. Crouch, P., Leite, F.S.: The dynamic interpolation problem: on Riemannian manifolds, Lie groups, and symmetric spaces. J. Dyn. Control Syst. **1**(2), 177–202 (1995)
4. Dietz, R., Hoschek, J., Jüttler, B.: An algebraic approach to curves and surfaces on the sphere and on other quadrics. Comput. Aided Geom. Des. **10**(3), 211–229 (1993)
5. Helgason, S.: Differential Geometry, Lie Groups, and Symmetric Spaces, vol. 80. Academic press, Dublin (1979)
6. Kang, I., Park, F.: Cubic spline algorithms for orientation interpolation. Int. J. Numer. Methods Eng. **46**(1), 45–64 (1999)
7. Kim, M.J., Kim, M.S., Shin, S.Y.: A general construction scheme for unit quaternion curves with simple high order derivatives. In: Proceedings of the 22nd Annual Conference on Computer Graphics and Interactive Techniques, pp. 369–376. ACM, New York (1995)
8. Kobayashi, S., Nomizu, K.: Foundations of Differential Geometry. Wiley, New York (1963)
9. Krakowski, K.A.: Envelopes of splines in the projective plane. IMA J. Math. Control Inf. **22**(2), 171–180 (2005)
10. Loos, O.: Symmetric Spaces. Benjamin (1969)
11. Meier, D.: Invariant higher-order variational problems: Reduction, geometry and applications. Ph.D. thesis, Imperial College London (2013)
12. Murray, R.M., Li, Z., Sastry, S.S., Sastry, S.S.: A Mathematical Introduction to Robotic Manipulation. CRC press, Boca Raton (1994)
13. Neubauer, M., Müller, A.: Smooth orientation path planning with quaternions using B-splines. In: 2015 IEEE/RSJ International Conference on Intelligent Robots and Systems (IROS), pp. 2087–2092 (2015)
14. Noakes, L., Heinzinger, G., Paden, B.: Cubic splines on curved spaces. IMA J. Math. Control Inf. **6**(4), 465–473 (1989)
15. Park, F., Ravani, B.: Bézier curves on Riemannian manifolds and Lie groups with kinematics applications. J. Mech. Des. **117**(1), 36–40 (1995)

16. Park, F.C., Ravani, B.: Smooth invariant interpolation of rotations. ACM Trans. Graph. (TOG) **16**(3), 277–295 (1997)
17. Paul, R.P., Stevenson, C.N.: Kinematics of robot wrists. Int. J. Robot. Res. **2**(1), 31–38 (1983)
18. Polpitiya, A.D., Dayawansa, W.P., Martin, C.F., Ghosh, B.K.: Geometry and control of human eye movements. IEEE Trans. Autom. Control **52**(2), 170–180 (2007)
19. Röschel, O.: Rational motion design-a survey. Comput. Aided Des. **30**(3), 169–178 (1998)
20. Selig, J.M.: Geometric Fundamentals of Robotics. Springer Science & Business Media, Berlin (2005)
21. Shoemake, K.: Animating rotation with quaternion curves. In: ACM SIGGRAPH Computer Graphics, vol. 19, pp. 245–254. ACM, New York (1985)
22. Wu, Y., Li, Z., Shi, J.: Geometric properties of zero-torsion parallel kinematics machines. In: 2010 IEEE/RSJ International Conference on Intelligent Robots and Systems (IROS), pp. 2307–2312. IEEE, New York (2010)
23. Wu, Y., Löwe, H., Carricato, M., Li, Z.: Inversion symmetry of the euclidean group: Theory and application to robot kinematics. IEEE Trans. Robot. **32**(2), 312–326 (2016)

Closure Polynomials for Strips of Tetrahedra

Federico Thomas and Josep M. Porta

Abstract A tetrahedral strip is a tetrahedron-tetrahedron truss where any tetrahedron has two neighbors except those in the extremes which have only one. Unless any of the tetrahedra degenerate, such a truss is rigid. In this case, if the distance between the strip endpoints is imposed, any rod length in the truss is constrained by all the others to attain discrete values. In this paper, it is shown how to characterize these values as the roots of a closure polynomial whose derivation requires surprisingly no other tools than elementary algebraic manipulations. As an application of this result, the forward kinematics of two parallel platforms with closure polynomials of degree 16 and 12 is straightforwardly solved.

1 Introduction

Let us consider the strip of tetrahedra in Fig. 1. Any such strip has two endpoints. In this case, P_a and P_b. If the distance between these two points is imposed, the length of any rod cannot be freely chosen. This paper is essentially devoted to obtain a closed-form solution for the length of any rod in a strip of tetrahedra, once the distance between its endpoints and the lengths of all other rods are known.

Although closure polynomials have been typically obtained on a case-by-case analysis, a common pattern can be identified for most cases. First, a set of loop equations involving both translation and orientation variables is derived. Then, translation variables are eliminated resulting in a system of trigonometric equations that is algebraized using the tangent half-angle substitution. Finally, elimination theory is used to obtain a univariate closure polynomial. Here we solve this problem departing from this standard approach. The proposed method can be summarized as follows. The distance between the strip endpoints is first derived by iterating a basic operation involving only two neighboring tetrahedra over the whole strip. This leads to a scalar

F. Thomas (✉) · J.M. Porta
Institut de Robòtica i Informàtica Industrial, CSIC-UPC, Barcelona, Spain
e-mail: fthomas@iri.upc.edu

J.M. Porta
e-mail: porta@iri.upc.edu

J. Lenarčič and J.-P. Merlet (eds.), *Advances in Robot Kinematics 2016*,
Springer Proceedings in Advanced Robotics 4, DOI 10.1007/978-3-319-56802-7_32

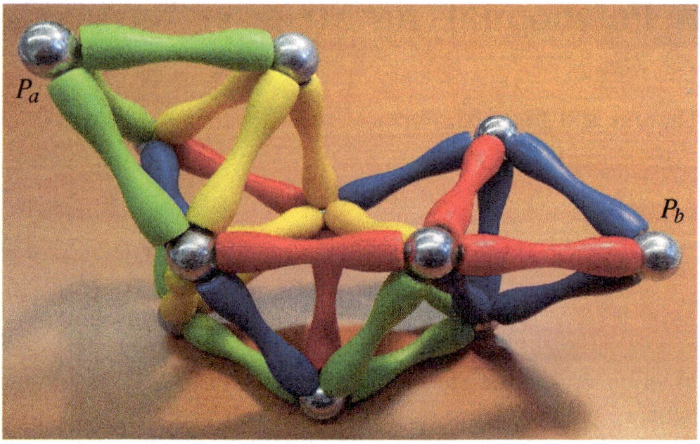

Fig. 1 A strip of eight tetrahedra whose endpoints are P_a and P_b. Observe that no triangular face is shared by more than two tetrahedra

equation containing radical terms. We will see how clearing these radicals is a trivial task, and how the resulting polynomial contains, in general, factor terms that correspond to singularities of the formulation that depend on the chosen variable length. Since these terms can be easily spotted beforehand, their elimination is just a matter of iterative polynomial division until a no null remainder is obtained. The result is the sought-after univariate closure polynomial obtained without variable eliminations or trigonometric substitutions.

Next, we detail this procedure and then we apply it to derive the minimal degree closure polynomial for two widely studied parallel platforms: the decoupled parallel platform, and a 4–4 platform with planar base and moving platform.

2 Obtaining the Closure Polynomials

Given a set of points, the valid distances between them can be characterized using the theory of Cayley–Menger determinants [1, 6, 8]. The Cayley–Menger bi-determinant of the two sets of points $P_{i_1}, \ldots, P_{i_n}$ and $P_{j_1}, \ldots, P_{j_n}$ is defined as

$$D(i_1, \ldots, i_n; j_1, \ldots, j_n) = 2\left(-\frac{1}{2}\right)^n \begin{vmatrix} 0 & 1 & \ldots & 1 \\ 1 & s_{i_1,j_1} & \ldots & s_{i_1,j_n} \\ 1 & \vdots & \ddots & \vdots \\ 1 & s_{i_n,j_1} & \ldots & s_{i_n,j_n} \end{vmatrix}, \qquad (1)$$

where $s_{i,j}$ stands for the squared distance between P_i and P_j.

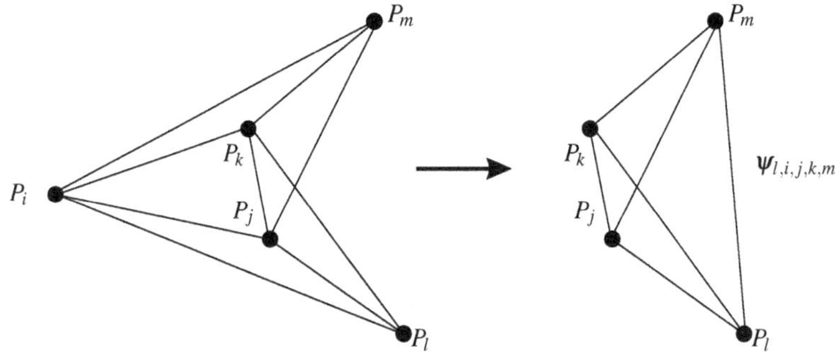

Fig. 2 Substitution rule

If the two sets of points are the same, then $D(i_1, \ldots, i_n) = D(i_1, \ldots, i_n; i_1, \ldots, i_n)$ is called the Cayley–Menger determinant of the involved set of points. The Cayley–Menger determinant $D(i_1, \ldots, i_n)$ is proportional to the squared volume of the simplex spanned by $P_{i_1}, \ldots, P_{i_n}$ in $\mathbb{R}^{n-1}$.

Now, let us suppose the two neighboring tetrahedra in Fig. 2-left belong to a strip of tetrahedra in $\mathbb{R}^3$. The squared distance between P_l and P_m can be expressed as (see [7] for details):

$$s_{l,m} = \frac{2}{D(i,j,k)} \left(D(i,j,k,l;i,j,k,m)\Big|_{s_{l,m}=0} \pm \sqrt{D(i,j,k,l)\,D(i,j,k,m)} \right). \tag{2}$$

where the $\pm$ sign accounts for the two possible solutions depending on the relative orientation between the two tetrahedra. To lighten the notation, (2) will be simply written as $s_{l,m} = \Psi_{l,i,j,k,m}$. If some of the distances involved in $\Psi_{l,i,j,k,m}$ are taken as variables, they will be made explicit in parenthesis. For example, if $s_{i,j}$ and $s_{i,k}$ are variables, we will write $s_{l,m} = \Psi_{l,i,j,k,m}(s_{i,j}, s_{i,k})$.

If one of the points in the set $\{P_i, P_j, P_k\}$ does not belong to any other tetrahedron in the strip, it can be removed from the strip provided that a rod connecting P_l and P_m is introduced with the double-valued length given by (2) (Fig. 2-right). This reduces the number of tetrahedra in the strip by one. Then, by repeating this operation until the strip contains only two tetrahedra, the distance between the tetrahedral strip endpoints is finally obtained as a 2^{n-1}–valued function, where n is the number of tetrahedra in the strip.

To obtain the closure condition as a polynomial in terms of a given rod length, the first step consists in taking the numerator of the rational form of the obtained function and then clearing radicals. As radicals will appear nested, they are cleared using an iterative process starting from the outer one. At each step of this process, the expressions involving a radical will have the general form

$$\alpha_0 + \alpha_1\sqrt{r} + \alpha_2(\sqrt{r})^2 + \alpha_3(\sqrt{r})^3 + \cdots = 0, \tag{3}$$

which can be rewritten as

$$(\alpha_0 + \alpha_2 r + \alpha_4 r^2 + \ldots) + \sqrt{r}(\alpha_1 + \alpha_3 r + \alpha_5 r^2 + \cdots) = 0. \tag{4}$$

This equation can be unfolded into two equations, one for each sign of $\sqrt{r}$. Since we are interested in the roots of both equations, we obtain their product, which can be written as

$$(\alpha_0 + \alpha_2 r + \alpha_4 r^2 + \ldots)^2 - r(\alpha_1 + \alpha_3 r + \alpha_5 r^2 + \cdots)^2 = 0. \tag{5}$$

While clearing radicals as explained above introduces no extraneous roots, one cannot expect for the obtained polynomial to be of minimal degree. This is due to the presence of singularities of the formulation. Indeed, let us suppose that the closure polynomial is expressed in terms of the squared rod length $s_{i,j}$. If a rod with variable length belongs to a shared face, this face degenerates for some values of $s_{i,j}$. When this happens, the three points defining the face get aligned and the tetrahedral strip can be decomposed into two parts so that one can freely rotate with respect the other about the axis defined by these three aligned points. As we will see, terms corresponding to these degenerate configurations will appear in the closure polynomial. They can be easily removed by iteratively dividing the closure polynomial by them until the remainder is not null.

3 Examples

Next, we apply the technique explained above to solve the forward kinematics of a decoupled platform and a 4–4 platform with planar base and platform (see Figs. 3 and 5, respectively). The decoupled platform owes its name to the fact that three legs permit the rotation of the platform about a point whose location is controlled

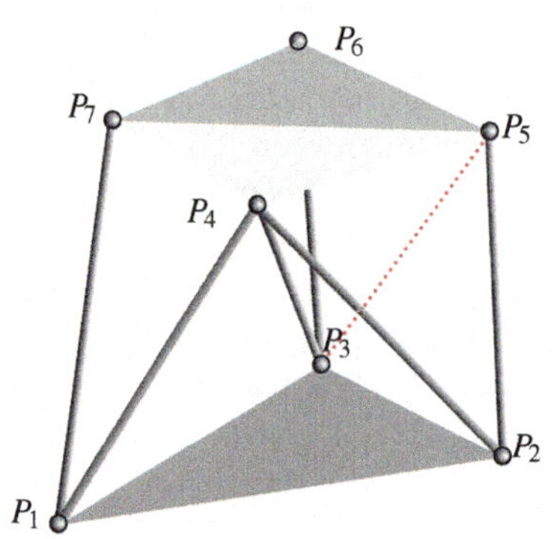

$$\mathbf{S} = \begin{pmatrix} 0 & 34 & 49 & 62 & ? & ? & 108 \\ 34 & 0 & 41 & 58 & 108 & ? & ? \\ 49 & 41 & 0 & 68 & ? & 126 & ? \\ 62 & 58 & 68 & 0 & 38 & 91 & 34 \\ ? & 108 & ? & 38 & 0 & 85 & 74 \\ ? & ? & 126 & 91 & 85 & 0 & 197 \\ 108 & ? & ? & 34 & 74 & 197 & 0 \end{pmatrix}$$

Fig. 3 Decoupled parallel manipulator, with non-planar moving platform, used as example by the other three. Since the forward kinematics for the translational part is trivial, the interest of this linkage lies in the spherical part for which a minimal closing polynomial of degree 8 on a squared variable was first derived in [2]. In [7], this derivation is simplified by using the closure polynomial of the so-called double banana. Despite the simpler derivation, variable eliminations were still necessary. For the chosen 4–4 platform with planar base and platform, a minimal 12th-degree closure polynomial was first derived in [3]. The derivation was far from trivial and applicable only to this particular platform. To properly compare our results with those reported in [3, 7], we use the same numerical examples.

First, let us consider the decoupled platform defined by the squared distance matrix $\mathbf{S}$ appearing in Fig. 3, where $s_{i,j} = \mathbf{S}(i, j)$. It can be topologically described as the

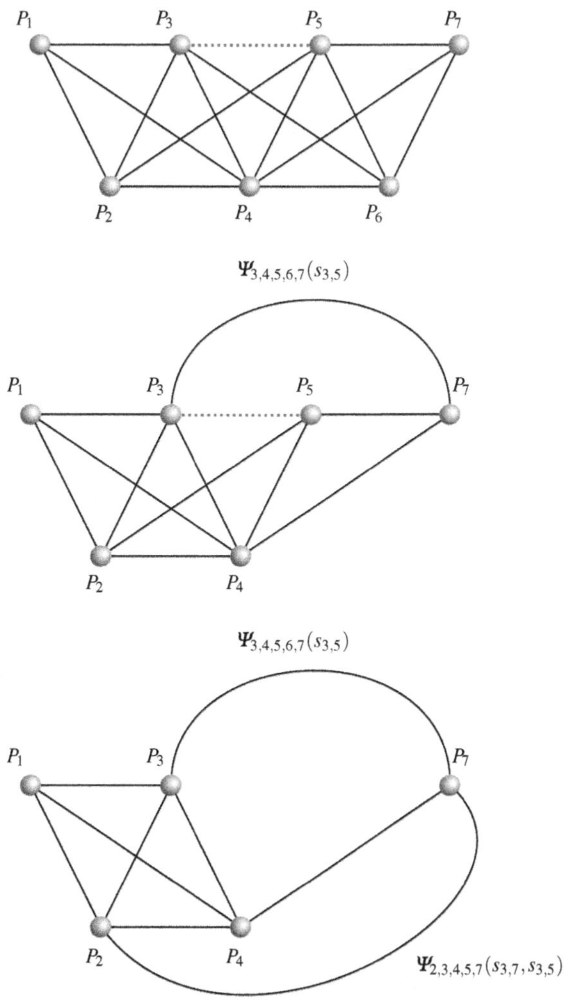

Fig. 4 The decoupled parallel platform in Fig. 3 can be topologically described as the strip of four tetrahedra in which the distance between P_3 and P_5 is variable and the distance between its endpoints, P_1 and P_7, is known. The application of the substitution rule presented in Sect. 2 to this strip (*top*) permits to sequentially eliminate P_6 (*center*) and P_5 (*bottom*)

strip of tetrahedra shown in Fig. 4-top. Applying the substitution presented in the previous section three times (see Fig. 4), we have

$$s_{3,7} = \Psi_{3,4,5,6,7}(s_{3,5}), \tag{6}$$

$$s_{2,7} = \Psi_{2,3,4,5,7}(s_{3,7}, s_{3,5}), \tag{7}$$

$$s_{1,7} = \Psi_{1,2,3,4,7}(s_{2,7}, s_{3,7}). \tag{8}$$

The numerator of the rational form resulting from substituting (6) in (7), and the result in (8), can be written as:

$$R_1 - 1346.0\, R_2 + 7899650\, R_3 + 24942632734\, s_{3,5} + 1402\, R_3\, {s_{3,5}}^2 - 323070338\, {s_{3,5}}^2 + 741658\, {s_{3,5}}^3 - 208500\, R_3\, s_{3,5} + 528767086008 = 0,$$

where

$$R_1 = \sqrt{-2027718\, {R_2}^2 + 4695768\, R_2\, R_3\, s_{3,5}^2 - 729124704\, R_2\, R_3\, s_{3,5} + \cdots},$$

$$R_2 = \sqrt{100464\, R_3^2\, s_{3,5}^2 - 19847712\, R_3^2\, s_{3,5} + 115799664\, R_3^2 - \cdots},$$

$$R_3 = \sqrt{-3481450\, s_{3,5}^2 + 806976100\, s_{3,5} - 27440188650}\,.$$

The full expressions for R_1 and R_2 are not included here due to space limitations.

Now, clearing the radicals as described in Sect. 2, we obtain a polynomial of 24th-degree. It is not of minimal degree because the rod connecting P_3 and P_5 belongs to the shared face defined by P_3, P_4, and P_5 which is singular when $D(3, 4, 5) = 0$, that is, when $s_{3,5}^2 - 214 s_{3,5} + 961 = 0$. By iteratively dividing the obtained polynomial by this singular factor until the remainder is not null, we get

$$s_{3,5}^{16} - 1.6652 \cdot 10^4\, s_{3,5}^{15} + 1.2722 \cdot 10^6\, s_{3,5}^{14} - 5.8952 \cdot 10^8\, s_{3,5}^{13} + 1.8487 \cdot 10^{11}\, s_{3,5}^{12} - 4.1525 \cdot 10^{13}\, s_{3,5}^{11} + 6.9146 \cdot 10^{15}\, s_{3,5}^{10} - 8.7384 \cdot 10^{17}\, s_{3,5}^{9} + 8.5338 \cdot 10^{19}\, s_{3,5}^{8} - 6.5533 \cdot 10^{21}\, s_{3,5}^{7} + 4.0715 \cdot 10^{23}\, s_{3,5}^{6} - 2.1848 \cdot 10^{25}\, s_{3,5}^{5} + 1.1165 \cdot 10^{27}\, s_{3,5}^{4} - 5.4256 \cdot 10^{28}\, s_{3,5}^{3} + 2.0923 \cdot 10^{30}\, s_{3,5}^{2} - 5.0066 \cdot 10^{31}\, s_{3,5} + 5.2479 \cdot 10^{32},$$

which coincides with the closure polynomial reported in [7], but obtained in a much simpler way.

As a second example, let us consider the 4–4 parallel platform appearing in Fig. 5. Its forward kinematics is known to have 24 solutions [3]. However, they can be split in two sets that are symmetric with respect to the base. Since the distance-based formulation is invariant to this symmetry, we will get a 12th-degree closure polynomial. The two sets of configurations are obtained in the coordinatization process using trilateration [4, 5, 8].

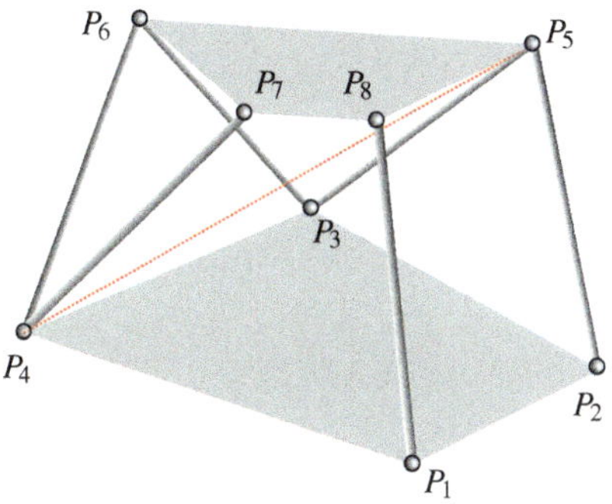

$$\begin{pmatrix} 0 & 10.198^2 & 19.799^2 & 18^2 & ? & ? & ? & 15.1657^2 \\ 10.198^2 & 0 & 16.4924^2 & s_{2,4} & 14.2917^2 & ? & ? & ? \\ 19.799^2 & 16.4924^2 & 0 & 14.5602^2 & 11.8770^2 & 10.8545^2 & ? & ? \\ 18^2 & s_{2,4} & 14.5602^2 & 0 & ? & 15.1719^2 & 15.7907^2 & ? \\ ? & 14.2917^2 & 11.8770^2 & ? & 0 & 6^2 & s_{5,7} & 4.4721^2 \\ ? & ? & 10.8545^2 & 15.1719^2 & 6^2 & 0 & 4.4721^2 & 5.6569^2 \\ ? & ? & ? & 15.7907^2 & s_{5,7} & 4.4721^2 & 0 & 2^2 \\ 15.1657^2 & ? & ? & ? & 4.4721^2 & 5.6569^2 & 2^2 & 0 \end{pmatrix}$$

Fig. 5 4–4 parallel manipulator used as example. Since the base and the moving platform are convex planar quadrilaterals, $s_{2,4}$ and $s_{5,7}$ are unambiguously determined by the other known distances

Applying the substitution presented in the previous section four times (see Fig. 6), we have that

$$s_{4,8} = \Psi_{4,5,6,7,8}(s_{4,5}), \qquad s_{2,8} = \Psi_{2,3,4,5,8}(s_{3,8}, s_{4,8}),$$
$$s_{3,8} = \Psi_{3,4,5,6,8}(s_{4,8}, s_{4,5}), \qquad s_{1,8} = \Psi_{1,2,3,4,8}(s_{2,8}, s_{3,8}, s_{4,8}).$$

After a proper sequence of forward substitutions in the above four equations, $s_{1,8}$ can be expressed only in terms of $s_{4,5}$. Since this parallel platform has planar base and platform, $\Psi_{4,5,6,7,8}$ and $\Psi_{1,2,3,4,8}$ are single-valued functions. Only $\Psi_{3,4,5,6,8}$ and $\Psi_{2,3,4,5,8}$ contribute with square roots to the obtained closure condition. Eliminating them as explained leads to a 52nd-degree polynomial in $s_{4,5}$. In this case, the rod connecting P_4 and P_5 belongs to two shared faces (the ones defined by $P_4P_5P_6$ and $P_3P_4P_5$), whose associated singular terms are $s_{4,5}^2 - 706.1251\, s_{4,5} + 5031.9580$, and ${s_{4,5}}^2 - 532.3731\, s_{4,5} + 37708.4160$. After iteratively dividing the obtained polynomial by these two factors until the remainder is not null, the following 12th-degree polynomial is obtained

$$s_{4,5}^{12} - 0.676 \cdot 10^3\, s_{4,5}^{11} - 3.873 \cdot 10^6\, s_{4,5}^{10} + 5.400 \cdot 10^9\, s_{4,5}^{9} - 9.858 \cdot 10^{12}\, s_{4,5}^{8}$$
$$- 2.327 \cdot 10^{15}\, s_{4,5}^{7} + 1.967 \cdot 10^{18}\, s_{4,5}^{6} - 7.316 \cdot 10^{20}\, s_{4,5}^{5} + 1.518 \cdot 10^{23}\, s_{4,5}^{4}$$
$$- 1.834 \cdot 10^{25}\, s_{4,5}^{3} + 1.257 \cdot 10^{27}\, s_{4,5}^{2} - 4.432 \cdot 10^{28}\, \mathrm{S}_{45} + 6.171 \cdot 10^{29}.$$

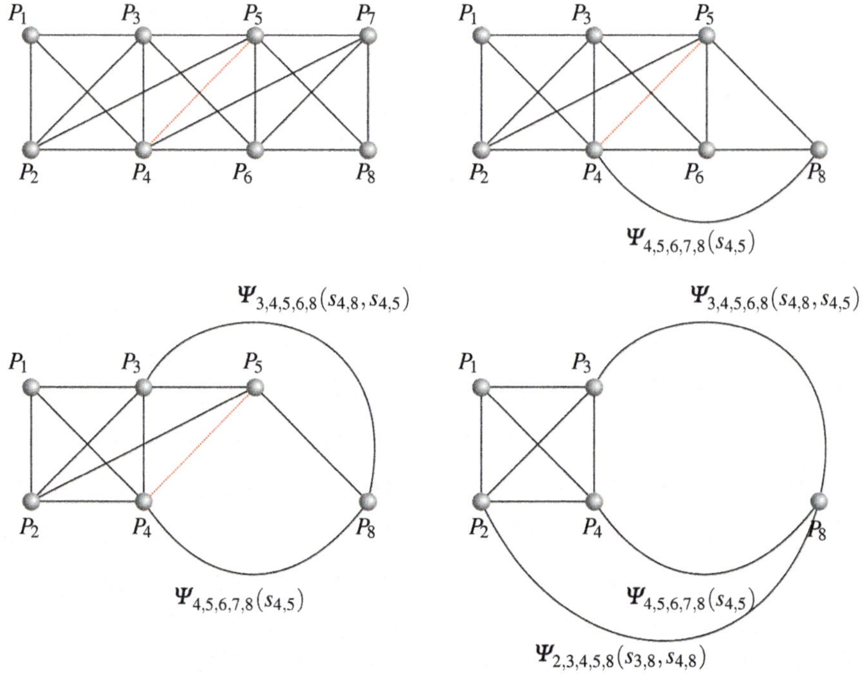

Fig. 6 The 4–4 parallel manipulator in Fig. 5 can be topologically described as the strip of five tetrahedra in which the distance between P_4 and P_5 is variable and the distance between its endpoints, P_1 and P_8, is known. The application of the substitution rule to this strip (*top left*) permits to sequentially eliminate P_7 (*top right*), P_6 (*bottom left*), and P_5 (*bottom right*)

This polynomial has six roots that lead to real configurations of the moving platform obtained by coordinatization via trilaterations [5, 8]. These roots and the corresponding configurations appear in Fig. 7. They coincide with the solutions reported in [3] obtained using an ad hoc intricate method.

4 Conclusions

It has been explained how to obtain closure polynomials for tetrahedral strips in terms of the involved rod lengths and the distance between the strip endpoints, and how this technique can be applied to solve some position analysis problems. However, this technique cannot incorporate orientation constraints between tetrahedra at different parts of the strip. As a consequence, if applied to a case in which such constraints are necessary, the obtained closure polynomial would not be of minimal-degree as some of its roots would violate these constraints. Despite this important limitation, it

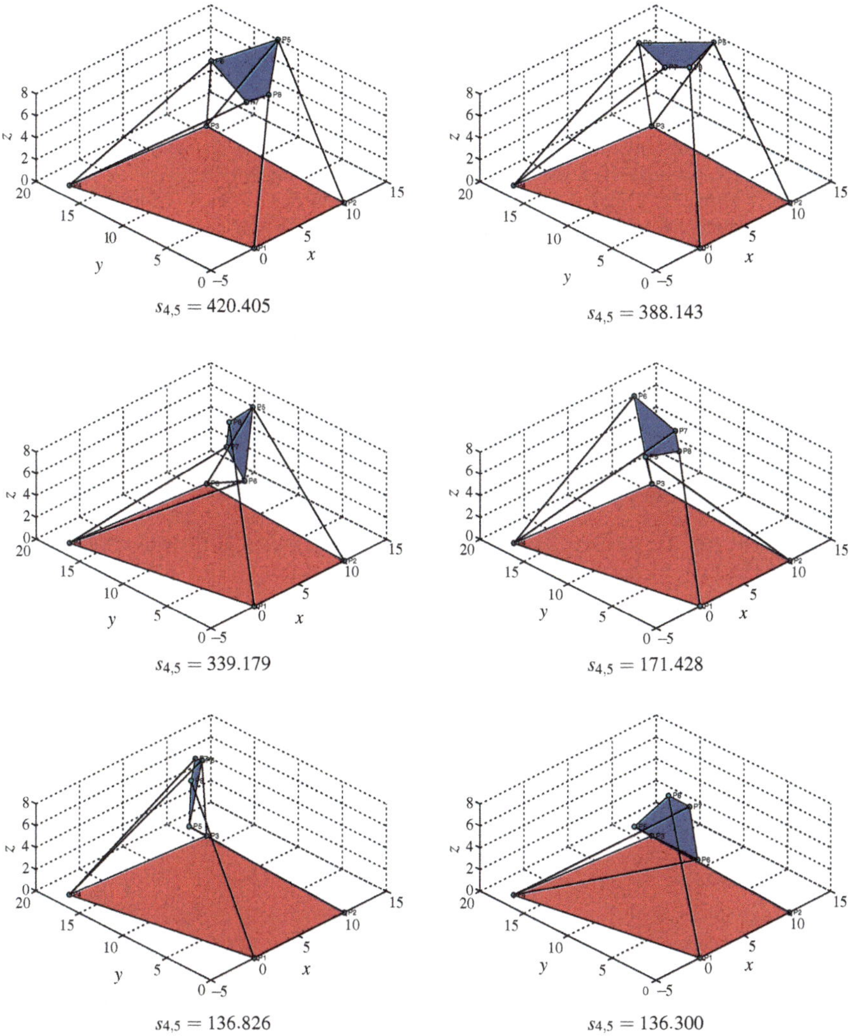

Fig. 7 Forward kinematics solutions of the 4–4 manipulator used as example. The mirror configurations with respect to the base are also solutions, but they are not represented

supersedes the method presented in [7] in scope and simplicity, thus providing a better starting point for a complete generalization to three dimensions of the techniques developed for the position analysis of planar linkages using Distance Geometry.

Acknowledgements This work has been partially supported by the Spanish Ministry of Economy and Competitiveness under project DPI2014-57220-C2-2-P.

References

1. Havel, T.: Some examples of the use of distances as coordinates for Euclidean geometry. J. Symb. Comput. **11**(5–6), 579–593 (1991)
2. Innocenti, C., Parenti-Castelli, V.: Echelon form solution of direct kinematics for the general fully-parallel spherical wrist. Mech. Mach. Theory **28**, 553–561 (1993)
3. Lin, W., Griffis, M., Duffy, J.: Forward displacement analyses of the 4–4 Stewart platforms. ASME J. Mech. Des. **114**(3), 444–450 (1992)
4. Porta, J.M., Ros, L., Thomas, F.: Inverse kinematics by distance matrix completion. In: International Workshop on Computational Kinematics (2005)
5. Porta, J.M., Ros, L., Thomas, F.: On the trilaterable six-degree-of-freedom parallel and serial manipulators. In: IEEE International Conference on Robotics and Automation, pp. 960–967 (2005)
6. Porta, J.M., Ros, L., Thomas, F., Torras, C.: A branch-and-prune algorithm for solving systems of distance constraints. In: IEEE International Conference on Robotics and Automation, pp. 342–348 (2003)
7. Rojas, N., Thomas, F.: The closure condition of the double banana and its application to robot position analysis. In: International Conference on Robotics and Automation, pp. 4641–4646 (2013)
8. Thomas, F., Ros, L.: Revisiting trilateration for robot localization. IEEE Trans. Robot. **21**(1), 93–101 (2005)

Robust Design of Parameter Identification

Aurélien Massein, David Daney and Yves Papegay

Abstract Quality of results computed during parameter identification problems relies on the selection of system's states while performing measurements. This choice usually does not take into account the uncertainty of states and of measures. For identifiability, classical methods focus only on the contribution of model errors on the uncertainty of parameters. We present an alternative approach that tackles this drawback: taking into account influence of all uncertainty sources in order to improve parameter identification robustness to uncertainties. A robotic application example that showcases the differences between approaches is developed as well.

1 Introduction

Design of experiments [2] is a way to improve results of a generic parameter identification problem. It is applied namely in robotics [8], for calibration [1], and in GPS area through dilution of precision [9].

In such identification problem, unknown parameters are related with states of the system, and with measured outputs through a model prone to uncertainties. Inaccuracy sources of the model are model discrepancy, measurement errors and inexactitude of system's states. Observation of different sets of measurements for different states of the system provides a way to compute unknown parameters through a regression analysis process. In this process choice of states is crucial to enhance parameters identification. It refers to identifiability of the model and identifiability criteria.

In the non-linear case [5], unknown parameters are classically computed by an iterative optimisation algorithm, starting from an initial estimation. During this iterative

A. Massein (✉) · D. Daney
Inria Bordeaux - Sud-Ouest, Talence, France
e-mail: aurelien.massein@inria.fr

D. Daney
e-mail: david.daney@inria.fr

Y. Papegay
Inria Sophia Antipolis Méditerranée, Valbonne, France
e-mail: yves.papegay@inria.fr

J. Lenarčič and J.-P. Merlet (eds.), *Advances in Robot Kinematics 2016*,
Springer Proceedings in Advanced Robotics 4, DOI 10.1007/978-3-319-56802-7_33

process, states and measures remain unchanged and influence of their uncertainties are neglected. At the end of the process, all uncertainties have the same order of magnitude. Despite this last remark, in the literature, choice of states is done with the same assumption that uncertainties on states and measures are negligible, by evaluating identifiability criteria to the so-called identification matrix. In this paper, we promote the idea to take into account the different sources of inaccuracy –namely all uncertainty sources– when selecting states for parameter identification. Hence, we are applying similarly identifiability criteria to what we call the *uncertainty* matrix, a matrix describing the contribution of all uncertainties to errors of the model.

In the next section, we are describing with more details the parameter identification problem, the regression analysis process, the iterative optimisation process, and identifiability criteria. We carefully define the identification matrix and the *uncertainty* matrix, and how they are used for selection of identification states of the system. The last section is devoted to a pedagogical application, that clearly shows the difference between the classical and the proposed approach of states selection.

This application concerns the localisation of a source by a mobile robot. In this application, measures are taken at regular time step, and the selection of states correspond to a trajectory determination, that can be easily visualized. Another interest of this example is the ability to perform the states selection incrementally, that allows some enhancements of the identification method.

2 Parameter Identification

2.1 Model Definition

We consider a system in step $k = 1 \dots K$, depending on unknown parameters x, system's states $u = \left[u_1 \ \dots \ u_k\right]^T$, and measured outputs $m = \left[m_1 \ \dots \ m_k\right]^T$, through the following model:

$$f_k\left(u_k, x, m_k\right) = 0 \tag{1}$$

As we consider that system's states and measured outputs are prone to uncertainties, we distinguish each variable x of its actual value x^*, and denote uncertainty on it by Δx. Once plugged into the model, the k-th observation of the system provides a set of $Dim(f)$ equations:

$$f_k\left(u_k^*, x^*, m_k^*\right) = f_k\left(u_k + \Delta u_k, x^*, m_k + \Delta m_k\right) = 0 \tag{2}$$

We need to acquire p system observations, such that the number of equations will be greater than the number of unknown parameters ($K \cdot Dim(f) > Dim(x)$). The resulting system is usually widely over-constrained.

2.2 Regression Analysis

Based on the set of these observations leading to residual errors,

$$f_k\left(u_k, x^*, m_k\right) = \varepsilon_k \qquad \text{with} \qquad k = 1 \dots K \tag{3}$$

a regression analysis estimator provides a solution $\hat{x}$ minimizing a given criterion. In the classical case of a Non-Linear Least Squares estimator, the criterion is the sum of the squares of residual errors:

$$\sum_{k=1}^{K} f_k\left(u_k, \hat{x}, m_k\right)^T \cdot f_k\left(u_k, \hat{x}, m_k\right) \tag{4}$$

Method of optimisation is based on a first-order linear approximation of the model:

$$\begin{gathered} f\left(u+\Delta u, x+\Delta x, m+\Delta m\right) \approx f\left(u, x, m\right) + J_u \cdot \Delta u + J_x \cdot \Delta x + J_m \cdot \Delta m \\ \text{or} \\ \Delta f \approx J_u \cdot \Delta u + J_x \cdot \Delta x + J_m \cdot \Delta m \end{gathered} \tag{5}$$

with $\Delta u = [\Delta u_1, \dots, \Delta u_p]^T$, $\Delta m = [\Delta m_1, \dots, \Delta m_p]^T$ and $\Delta f = [\Delta f_1, \dots, \Delta f_p]^T$.

Starting from an initial estimate x_0 of x, the Non-Linear Least Squares method performs several solving steps to reduce Δx. As $J_u \cdot \Delta u + J_m \cdot \Delta m$ is assumed to be negligible compared to $J_x \cdot \Delta x$, see [4], j-th step consists in solving the following linear system –with Δf and J_x computed from previous estimation x_j– in the unknown variables x_{j+1}:

$$\Delta f[x_j] = J_x[x_j] \cdot \Delta x_{j+1} \qquad \text{with} \qquad \Delta x_{j+1} = x_{j+1} - x_j \tag{6}$$

Iterative process ends when Δx is sufficiently small, and is the same order of magnitude of Δu and Δm - see [4] for the stop condition.

At the end of this iterative process, one classically considers that the quality of the final estimation $\hat{x}$ relies on the numerical quality of the Jacobian matrix J_x, denoted by *Identification Matrix*. Numerical quality is precisely defined in terms of identifiability criteria (see below). As identification matrix depends on system's states, poses of the system for measurements are selected by optimisation of these criteria.

2.3 Identifiability Criteria

Identifiability criteria have been widely studied in optimal design of experiments -being called optimality criteria- and in robotics calibration problems -being called observability criteria-. Sun and Hollerbach have synthesized in [10] those criteria and showed their similarities as some observability indexes have an optimality criterion counterpart.

Given a rectangular matrix M such that $Y = M \cdot X$, identifiability criteria quantifies how uncertainties on X propagate to uncertainties on Y through M. The objective is to minimize some observability -or optimality- indexes, denoted O, associated to M, by choosing the state variable $\hat{u}$ which parametrize the matrix M, such that:

$$\hat{u} = \arg\min_{u} O(M(u)) \tag{7}$$

Three criteria are popular, related to the singular values of the matrix M, obtained and sorted by a Singular Value Decomposition (and denoted by $\sigma_L \leq \cdots \leq \sigma_1$):

- D-Optimality: $O_1 = \prod_{l=1}^{L} \sigma_l{}^{1/(np)}$. This index corresponds to the determinant of $M^T.M$ here, $\sqrt{\det(M^T.M)} = \prod_{l=1}^{L} \sigma_l$. The sensitivity of Y with respect to X is decreased when O_1 is minimized.
- Inverse of the condition number: $O_2 = \sigma_L/\sigma_1$
- E-Optimality: $O_3 = \sigma_L$

Sun and Hollerbach argued that O_1 (D-optimality) is the best criterion for a parameter estimation of an unscaled model or a model without a convincing scaling approach has to be minimized [10]. Fedorov and Leonov stated that D-optimal (O_1) designs are most popular among theoretical and applied researchers in optimal experimental designs [2].

We can denote briefly that O_2 which related to condition number is for balancing parameter uncertainties importance, whereas O_3 is for reducing the worst parameter's uncertainty.

2.4 Uncertainty Matrix

At the end of the iterative process, a linear approximation of f in the neighborhood of the final estimation $\hat{x}$ of x is given by the p following equations:

$$f_k\left(u_k^*, x^*, m_k^*\right) \approx f\left(u_k, \hat{x}, m\right) + J_u \cdot \Delta u_k + J_x \cdot \Delta x + J_m \cdot \Delta m_k \tag{8}$$

that can be written synthetically:

$$\Delta f \approx J_u \cdot \Delta u + J_x \cdot \Delta x + J_m \cdot \Delta m \tag{9}$$

or by introducing what we denote by U, the *Uncertainty Matrix* such that:

$$\Delta x = -J_x^+ \cdot \begin{bmatrix} J_u \ J_m \ -I_p \end{bmatrix} \cdot \begin{bmatrix} \Delta u \\ \Delta m \\ \Delta f \end{bmatrix} = U \cdot \begin{bmatrix} \Delta u \\ \Delta m \\ \Delta f \end{bmatrix} \tag{10}$$

J_x^+ being the pseudo-inverse of J_x.

We then claim that it is interesting to apply identifiability criteria to this uncertainty matrix U –minimizing $O(U)$– when selecting states. To be efficient, at this point, we would benefit of normalizing the uncertainties as done in [6, 7] or alternatively [3].

3 Source Localisation Application

In this section, we illustrate the difference between using the identifiability matrix and using the uncertainty matrix in the previously described states selection process, on a pedagogical two-dimensional source localisation application.

3.1 Problem Statement

We aim to localize accurately and step-efficiently a fixed source with the help of a mobile robot.

The mobile robot we consider has an embedded sensor measuring its direction with respect to a source $x = \begin{bmatrix} x_S \ y_S \end{bmatrix}^T$, with a fixed sampling frequency and prone to bounded uncertainties. Our workspace is two dimensional and free of obstacle: wherever we are we can get a measure at each sampling step k and move anywhere. Speed of the robot is supposed constant and a constant distance r separates two consecutive measurement positions.

We describe the robot motion in polar coordinates. To move robot from position $u_k = \begin{bmatrix} x_k \ y_k \end{bmatrix}^T$ at step k onto the next position u_{k+1}, we need a leading direction α_{k+1} as written below:

$$u_{k+1} = r \begin{bmatrix} \cos \alpha_{k+1} \\ \sin \alpha_{k+1} \end{bmatrix} + u_k \tag{11}$$

At each measurement step, the goniometric sensor on the robot provides the azimuth m_k –with respect to a fixed reference frame– such that

$$\begin{bmatrix} \sin m_k \\ -\cos m_k \end{bmatrix} \cdot (x - u_k) = 0 \tag{12}$$

In the equations of the model, the position of the source x, the position u_k and the measured azimuth m_k play respectively the roles of unknown parameters x, states u and measures m. Expressed at the k-th step, the model f_k is precisely given by the previous equations. Please note that u_k depends on α_k.

Hence first-order linear approximation of the model expressed as:

$$\begin{bmatrix} \sin m_k \\ -\cos m_k \end{bmatrix} \cdot \Delta x + \begin{bmatrix} -\sin m_k \\ \cos m_k \end{bmatrix} \cdot \Delta u_k + \begin{bmatrix} \cos m_k \\ -\sin m_k \end{bmatrix} \cdot (x - u_k) \cdot \Delta m_k \tag{13}$$

The difference between robot and source position $x - u_k$ is equal to ρ_k, the distance between u_k and x, modulo $\cos(m_k^* - m_k)$ which approximation is considered equal to 1.

3.2 States Choices

The expressions of the identifiability matrix and of the uncertainty matrix are respectively (at the k-th step):

$$J_{x_k} = \begin{bmatrix} jx_1 \\ \dots \\ jx_k \end{bmatrix} \quad \text{with} \quad jx_k = \begin{bmatrix} \sin m_k, -\cos m_k \end{bmatrix} \tag{14}$$

$$U_k = J_{x_k}^+ \cdot N_k \tag{15}$$

$$\text{where} \quad N_k = \begin{bmatrix} n_1 & 0 & \dots & 0 \\ 0 & n_2 & 0 & \vdots \\ \vdots & 0 & \ddots & 0 \\ 0 & \dots & 0 & n_k \end{bmatrix} \quad \text{with} \quad n_k = \begin{bmatrix} \sin m_k & -\cos m_k & -\rho_k & 1 \end{bmatrix} \tag{16}$$

Then, the optimal next direction α_{k+1} for the mobile robot, is defined with the help of the identifiability criterion O_1 using either $J_{x_{k+1}}$ or U_{k+1}.

3.3 Experiment and Results

In our experiment, source x is at position $x = \begin{bmatrix} 0 \ 1 \end{bmatrix}^T$, and the initial mobile robot position is $u_1 = \begin{bmatrix} 0 \ 0 \end{bmatrix}^T$. The fixed motion step is $r = 0.01$, with a relative error lower than 1% and we have the same error on the direction α_k about 0.1°. Finally, the uncertainty of the measured angle Δm is uniformly distributed and bounded by $\Delta m = \pm 10°$.

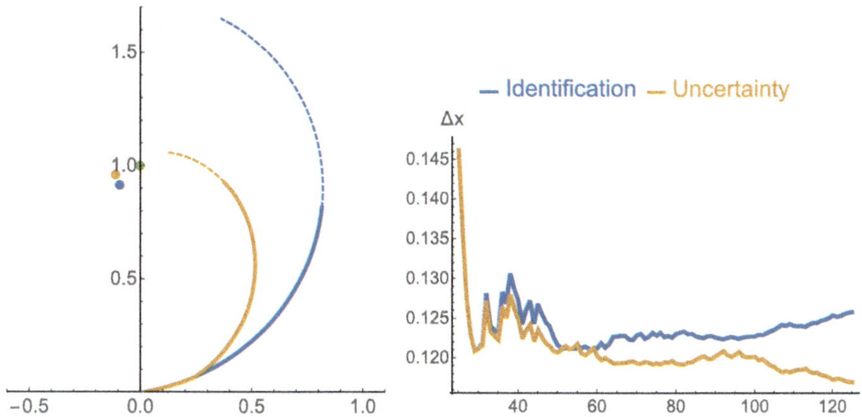

(a) Robot motions (b) Parameters Difference

Fig. 1 Source localisation application

In a first experiment, we select each step α_{k+1} according to the maximization of $O_1(J_{x_{k+1}})$. In the second, we select it to minimize $O_1(U_{k+1})$.

The obtained results are presented in Fig. 1a and b.

The first one displays the motion of the mobile robot into the euclidean workspace step-by-step: blue motion curve stands for a gradient determinant optimisation on *identification matrix* J_x and orange motion curve for one on *uncertainty matrix* U.

The second one presents the error $\|x^* - \hat{x}\|$ as a function of the number of steps in the two experiments with the same color code.

Note that the blue motion favours a circle approach centered on the source position whereas the orange motion favours a spiral approach on it: By neglecting uncertainties on measures (blue case), we intuitively want to change the angle m as much as possible between two measurement positions, see [11], for a better conditioning of J_{x_k}. On the orange trajectory, we also take into account the measurement errors, so we try to become closer to the source to minimize the influence of such errors.

4 Conclusion

We have introduced a new matrix to improve parameter estimation robustness to all uncertainties, in complement to classical identification optimisation. We applied our matrix optimisation in a source localisation application and demonstrated its possible application in a practical case. Results outperformed classical identification optimisation.

This application highlights that, to improve parameter estimation accuracy, we need to take into account more uncertainties than classically ones held in the *identification matrix*. Our proposed *uncertainty* matrix can be used in design of parameter identification problems, can improve significantly estimations accuracy and their robustness to uncertainties.

References

1. Daney, D., Papegay, Y., Madeline, B.: Choosing measurement poses for robot calibration with the local convergence method and tabu search. Int. J. Robot. Res. **24**(6), 501–518 (2005). doi:10.1177/0278364905053185. http://ijr.sagepub.com/content/24/6/501
2. Fedorov, V.V., Leonov, S.L.: Optimal Design for Nonlinear Response Models. CRC Press (2013)
3. Gayral, T.: Étalonnage d'un instrument d'observation spatial actif. phdthesis, Université Nice Sophia Antipolis (2013). https://tel.archives-ouvertes.fr/tel-00950898/document
4. Gayral, T., Daney, D.: A sufficient condition for parameter identifiability in robotic calibration. In: Thomas, E., Perez Gracia, A. (eds.) Computational Kinematics. Mechanisms and Machine Science, vol. 15, pp. 131–138. Springer, Netherlands (2014)
5. Hollerbach, J., Khalil, W., Gautier, M.: Model identification. In: Prof, B.S., Prof, O.K. (eds.) Springer Handbook of Robotics, pp. 321–344. Springer, Berlin (2008)
6. Hollerbach, J.M.: Advances in robot calibration. In: In 6th International Symposium on Robotics Research, pp. 319–326
7. Hollerbach, J.M., Wampler, C.W.: The calibration index and taxonomy for robot kinematic calibration methods. Int. J. Robot. Res. **15**(6), 573–591 (1996)
8. Kelly, A.: Precision dilution in triangulation based mobile robot position estimation. In: Intelligent Autonomous Systems, vol. 8, pp. 1046–1053 (2003). http://www.frc.ri.cmu.edu/~alonzo/pubs/papers/ias8.pdf
9. Langley, R.B.: Dilution of precision. GPS World **10**(5), 52–59 (1999)
10. Sun, Y., Hollerbach, J.: Observability index selection for robot calibration. In: IEEE International Conference on Robotics and Automation, 2008, ICRA 2008, pp. 831–836 (2008). doi:10.1109/ROBOT.2008.4543308
11. Takeda, Y., Shen, G., Funabashi, H.: A dbb-based kinematic calibration method for in-parallel actuated mechanisms using a fourier series. J. Mech. Des. **126**(5), 856–865 (2004)

Isotropic Design of the Spherical Wrist of a Cable-Driven Parallel Robot

Angelos Platis, Tahir Rasheed, Philippe Cardou and Stéphane Caro

Abstract Because of their mechanical properties, parallel mechanisms are most appropriate for large payload to weight ratio or high-speed tasks. Cable driven parallel robots (CDPRs) are designed to offer a large translation workspace, and can retain the other advantages of parallel mechanisms. One of the main drawbacks of CDPRs is their inability to reach wide ranges of end-effector orientations. In order to overcome this problem, we introduce a parallel spherical wrist (PSW) end-effector actuated by cable-driven omni-wheels. In this paper we mainly focus on the description of the proposed design and on the appropriate placement of the omni-wheels on the wrist to maximize the robot dexterity.

1 Introduction

Several applications could benefit from CDPRs endowed with large orientation workspaces, such as entertainment and manipulation and storage of large and heavy parts. This component of the workspace is relatively small in existing CDPR designs. To resolve this problem, a parallel spherical wrist (PSW) end-effector is introduced and connected in series with the translational 3-DOF CDPR to provide an unbounded singularity-free orientation workspace.

A. Platis · T. Rasheed
IRCCyN, École Centrale de Nantes, 1 rue de la Noë, 44321 Nantes, France
e-mail: Angelos.Platis@eleves.ec-nantes.fr

T. Rasheed
e-mail: Tahir.Rasheed@eleves.ec-nantes.fr

P. Cardou
Laboratoire de robotique, Département de génie mécanique, Université Laval,
Quebec City, QC, Canada
e-mail: pcardou@gmc.ulaval.ca

S. Caro (✉)
CNRS–IRCCyN, 1 rue de la Noë, 44321 Nantes, France
e-mail: stephane.caro@irccyn.ec-nantes.fr

J. Lenarčič and J.-P. Merlet (eds.), *Advances in Robot Kinematics 2016*,
Springer Proceedings in Advanced Robotics 4, DOI 10.1007/978-3-319-56802-7_34

This paper focuses on the kinematic design and analysis of a PSW actuated by the cables of a CDPR providing the robot independent translation and orientation workspaces. CDPRs are generally capable of providing a large 3-dofs translation workspace, normally needed four cables, which enable the user to control the point where all of them are concentrated [2, 5].

Robots that can provide large orientation workspace have been developed using spherical wrist in the past few years that allows the end-effector to rotate with unlimited rolling, in addition to a limited pitch and yaw movements [3, 10]. Eclipse II [9] is an interesting robot that can provide unbounded 3-dofs translational motions, however its orientation workspace is constrained by structural interference and rotation limits of the spherical joints.

Several robots have been developed in the past having decoupled translation and rotational motions. One interesting concept of such a robot is that of the Atlas Motion Platform [6] developed for simulation applications. Another robot with translation motions decoupled from orientation motions can be found in [11]. The decoupled kinematics are obtained using a triple spherical joint in conjunction with a 3-UPS parallel robot.

In order to design a CDPR with a large orientation workspace, we introduce a parallel spherical wrist (PSW) end-effector actuated by cable-driven omni-wheels. In this paper we mainly focus on the description of the proposed design and on the appropriate placement of the omni-wheels on the wrist to maximize the robot dexterity.

2 Manipulator Architecture

The end-effector is a sphere supported by actuated omni-wheels as shown in Fig. 1. The wrist contains three passive ball joints at the bottom and three active omni-wheels being driven through drums. Each cable makes several loops around each drum. Both ends are connected to two servo-actuated winches, which are fixed to the base. When two servo-actuated winches connected to the same cable turn in the same direction, the cable circulates and drives the drum and its associated omni-wheel. When both servo-actuated winches turn in opposite directions, the length of the cable loop changes, and the sphere centre moves. To increase the translation workspace of the CDPR, another cable is attached, which has no participation in the omni-wheels rotation. The overall design of the manipulator is shown in Fig. 2.

We have in total three frames. First, the CDPR base frame ($\mathscr{F}_0$), which is described by its center O_0 having coordinates x_0, y_0, z_0. Second, the PSW base frame ($\mathscr{F}_1$), which has its center O_1 at the geometric center of the sphere and has coordinates x_1, y_1, z_1. Third, the spherical end-effector frame ($\mathscr{F}_2$) is attached to the end-effector. Its centre O_2 coincides with that of the PSW base frame ($O_2 \equiv O_1$) and its coordinates are x_2, y_2, z_2.

Exit points A_i are the cable attachment points that link the cables to the base. All exit points are fixed and expressed in the CDPR reference frame $\mathscr{F}_0$. Anchor points

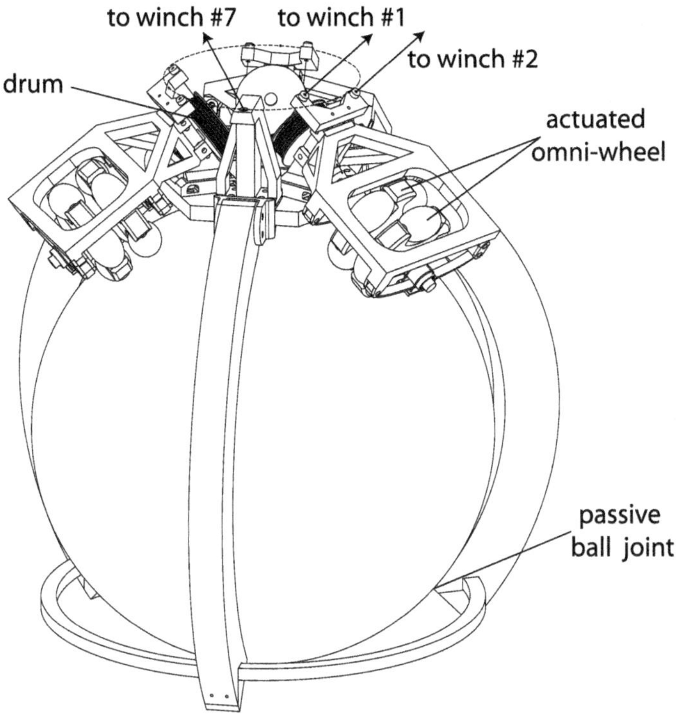

Fig. 1 Isotropic design of the parallel spherical wrist

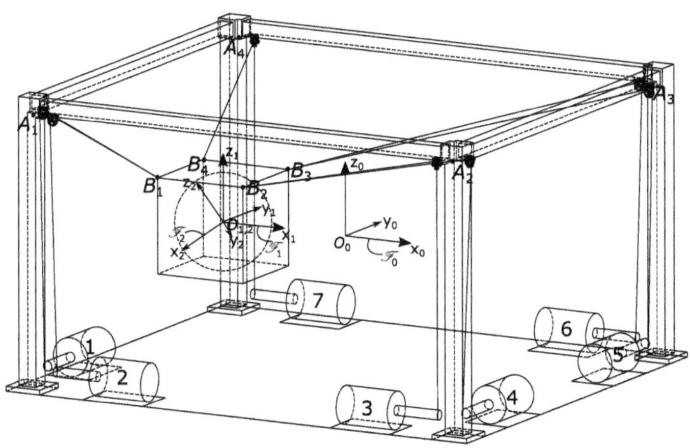

Fig. 2 Concept idea of the manipulator

B_i are the platform attachment points. These points are not fixed as they depend on the vector P, which is the vector that contains the pose of the moving platform expressed in the CDPR reference frame $\mathscr{F}_0$. The remaining part of the paper aims at finding the appropriate placement of the omni-wheels on the wrist to maximise the robot dexterity.

3 Kinematic Analysis of the Parallel Spherical Wrist

3.1 Parameterization

To simplify the parameterization of the parallel spherical wrist, some assumptions are made. First, all the omni-wheels are supposed to be normal the sphere. Second, the contact points of the omni-wheels with the sphere lie in the base of an inverted cone where its end is the geometrical center of the sphere parametrized by angle α. Third, the three contact points form an equilateral triangle as shown in [6, 7]. Fourth, the angle between the tangent to the sphere and the actuation force produced by the ith actuated omni-wheel is named β_i, $i = 1, 2, 3$, and $\beta_1 = \beta_2 = \beta_3 = \beta$. Figure 3 illustrates the sphere, one actuated omni-wheel and the main design variables of the parallel spherical wrist. Π_i is the plane tangent to the sphere and passing through the contact point G_i between the actuated omni-wheel and the sphere.

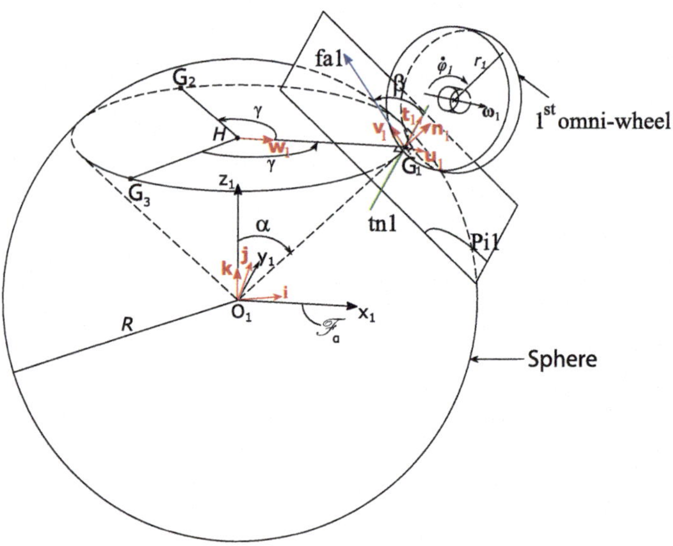

Fig. 3 Parameterization of the parallel spherical wrist

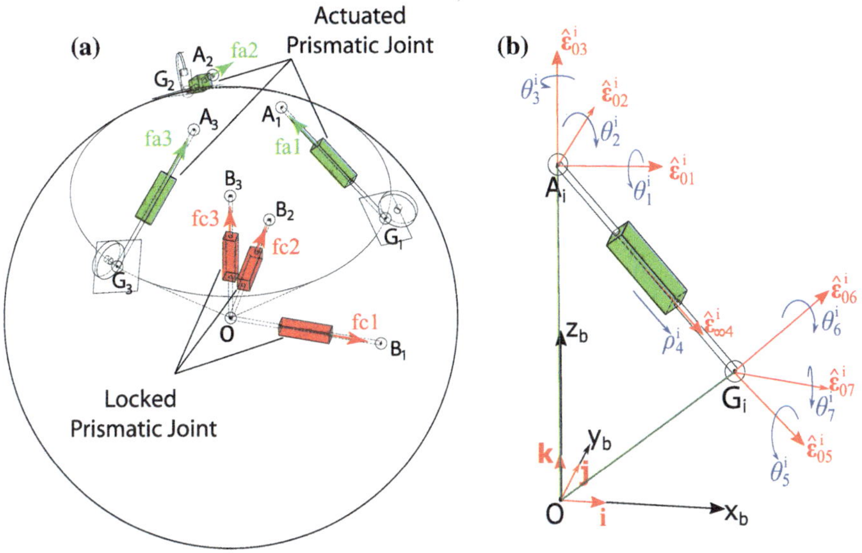

Fig. 4 **a** Actuation and constraint wrenches applied on the end-effector of the spherical wrist. **b** Virtual i^{th} leg with actuated prismatic joint

$\boldsymbol{\omega}_i$ denotes the angular velocity vector of the ith actuated omni-wheel. $\mathbf{s}_i$ is a unit vector along the tangent line $\mathcal{T}$ that is tangent to the base of the cone and coplanar to plane Π_i. $\mathbf{w}_i$ is a unit vector normal to $\mathbf{s}_i$. $\mathbf{f}_{ai}$ depicts the transmission force lying in plane Π_i due to the actuated omni-wheel. α is the angle defining the altitude of contact points G_i ($\alpha \in [0, \pi]$). β is the angle between the unit vectors $\mathbf{s}_i$ and $\mathbf{v}_i$ ($\beta \in [-\frac{\Pi}{2}, \frac{\Pi}{2}]$). As the contact points G_i are the corners of an equilateral triangle, the angle between the contact point G_1 and the contact points G_2 and G_3 is equal to γ. R is the radius of the sphere. r_i is radius of the ith actuated omni-wheel. $\dot{\varphi}_i$ is the angular velocity of the omni-wheel. $\mathbf{u}_i$, $\mathbf{v}_i$, $\mathbf{n}_i$ are unit vectors at point G_i and $\mathbf{i}, \mathbf{j}, \mathbf{k}$ are unit vectors along x_2, y_2, z_2 respectively.

In order to analyze the kinematic performance of the parallel spherical wrist, an equivalent parallel robot (Fig. 4) having six virtual legs is presented, each leg having a spherical, a prismatic and another spherical joints connected in series. Three legs have an actuated prismatic joint (green), whereas the other three legs have a locked prismatic joints (red). Here, the kinematics of the spherical wrist is analyzed with screw theory and an equivalent parallel robot represented in Fig. 4.

3.2 Kinematic Modeling

Figure 4a represents the three actuation forces $\mathbf{f}_{ai}$, $i = 1, 2, 3$ and the three constraint forces $\mathbf{f}_{ci}$, $i = 1, 2, 3$ exerted by the actuated omni-wheels on the sphere. The three constraint forces intersect at the geometric center of the sphere and prevent the latter from translating. The three actuation forces generated by the three actuated omni-wheels allow us to control the three-dof rotational motions of the sphere. Figure 4b depicts a virtual leg corresponding to the effect of the ith actuated omni-wheel on the sphere. The kinematic model of the PSW is obtained by using the theory of reciprocal screws [4, 8] as follows:

$$\mathbf{A}\,\mathbf{t} = \mathbf{B}\,\dot{\boldsymbol{\varphi}} \tag{1}$$

where $\mathbf{t}$ is the sphere twist, $\dot{\boldsymbol{\varphi}} = \left[\dot{\varphi}_1\ \dot{\varphi}_2\ \dot{\varphi}_3\right]^T$ is the actuated omni-wheel angular velocity vector. $\mathbf{A}$ and $\mathbf{B}$ are respectively the forward and inverse kinematic Jacobian matrices of the PSW and take the form:

$$\mathbf{A} = \begin{bmatrix} \mathbf{A}_{r\omega} & \mathbf{A}_{rp} \\ \mathbf{0}_{3\times3} & \mathbf{I}_3 \end{bmatrix} \tag{2}$$

$$\mathbf{B} = \begin{bmatrix} \mathbf{I}_3 \\ \mathbf{0}_{3\times3} \end{bmatrix} \tag{3}$$

$\mathbf{I}_3$ and $\mathbf{0}_{3\times3}$ are the 3×3 identity and zeros matrices, and the matrices $\mathbf{A}_{r\omega}$ and $\mathbf{A}_{rp}$ are:

$$\mathbf{A}_{r\omega} = \begin{bmatrix} R(\mathbf{n}_1 \times \mathbf{v}_1)^T \\ R(\mathbf{n}_2 \times \mathbf{v}_2)^T \\ R(\mathbf{n}_3 \times \mathbf{v}_3)^T \end{bmatrix} \ \ and\ \ \mathbf{A}_{rp} = \begin{bmatrix} \mathbf{v}_1^T \\ \mathbf{v}_2^T \\ \mathbf{v}_3^T \end{bmatrix} \tag{4}$$

As the contact points on the sphere form an equilateral triangle, $\gamma = 2\pi/3$. As a consequence, matrices $\mathbf{A}_{r\omega}$ and $\mathbf{A}_{rp}$ are expressed as functions of the design parameters α and β:

$$\mathbf{A}_{r\omega} = \frac{R}{2}\begin{bmatrix} -2C\alpha C\beta & -2S\beta & 2S\alpha C\beta \\ C\alpha C\beta + \sqrt{3}S\beta & S\beta - \sqrt{3}C\alpha C\beta & 2S\alpha C\beta \\ C\alpha C\beta - \sqrt{3}S\beta & S\beta + \sqrt{3}C\alpha C\beta & 2S\alpha C\beta \end{bmatrix} \tag{5}$$

$$\mathbf{A}_{rp} = \frac{1}{2}\begin{bmatrix} -2C\alpha S\beta & 2C\beta & 2S\alpha S\beta \\ C\alpha S\beta - \sqrt{3}C\beta & -(\sqrt{3}C\alpha S\beta + C\beta) & 2S\alpha S\beta \\ C\alpha S\beta + \sqrt{3}C\beta & \sqrt{3}C\alpha S\beta - C\beta & 2S\alpha S\beta \end{bmatrix} \tag{6}$$

where C and S denote the cosine and sine functions, respectively.

3.3 Singularity Analysis

As matrix $\mathbf{B}$ cannot be rank deficient, the parallel spherical wrist meets singularities if and only if (iff) matrix $\mathbf{A}$ is singular. From Eqs. (5) and (6), matrix $\mathbf{A}$ is singular iff:

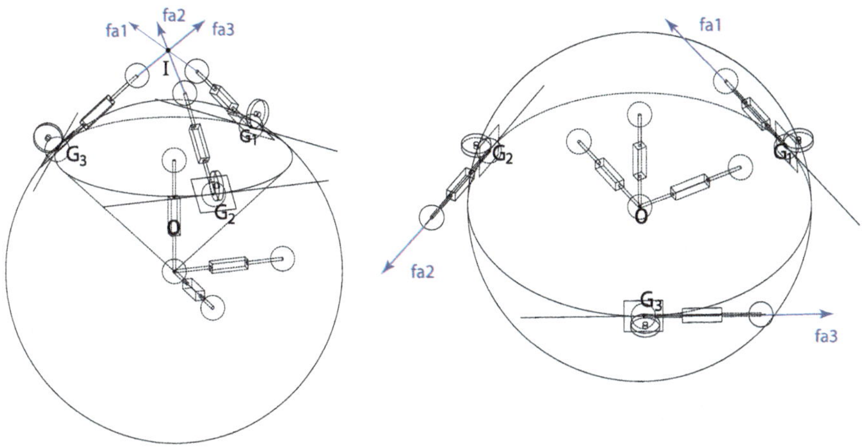

(a) $\beta = \pm\pi/2$ (b) $\alpha = \pi/2$ and $\beta = 0$

Fig. 5 Singular configurations of the parallel spherical wrist

$$\det(\mathbf{A}) = \frac{3\sqrt{3}}{2} R^3 S\alpha C\beta(1 - S^2\alpha C^2\beta) = 0 \tag{7}$$

namely, if $\alpha = 0$ or π; if $\beta = \pm\pi/2$; if $\alpha = \pi/2$ and $\beta = 0$ or $\pm\pi$.

Figure 5a, b represent two singular configurations of the parallel spherical wrist under study. The three actuation forces $\mathbf{f}_{a1}$, $\mathbf{f}_{a2}$ and $\mathbf{f}_{a3}$ intersect at point I in Fig. 5a. The PSW reaches a parallel singularity and gains an infinitesimal rotation (uncontrolled motion) about an axis passing through points O and I in such a configuration. The three actuation forces $\mathbf{f}_{a1}$, $\mathbf{f}_{a2}$ and $\mathbf{f}_{a3}$ are coplanar with plane (X_1OY_1) in Fig. 5b. The PSW reaches a parallel singularity and gains two-dof infinitesimal rotations (uncontrolled motions) about an axes that are coplanar with plane (X_1OY_1) in such a configuration.

3.4 Kinematically Isotropic Wheel Configurations

This section aims at finding a good placement of the actuated omni-wheels on the sphere with regard to the manipulator dexterity. The latter is evaluated by the condition number of reduced Jacobian matrix $\mathbf{J}_\omega = r\mathbf{A}_{r\omega}^{-1}$ which maps angular velocities of the omni-wheels $\dot{\boldsymbol{\varphi}}$ to the required angular velocity of the end-effector $\boldsymbol{\omega}$. From Eqs. (5) and (6), the condition number $\kappa_F(\alpha, \beta)$ of $\mathbf{J}_\omega$ based on the Frobenius norm [1] is expressed as follows:

$$\kappa_F(\alpha, \beta) = \frac{1}{3}\sqrt{\frac{3S^2\alpha C^2\beta + 1}{S^2\alpha C^2\beta(1 - S^2\alpha C^2\beta)}} \tag{8}$$

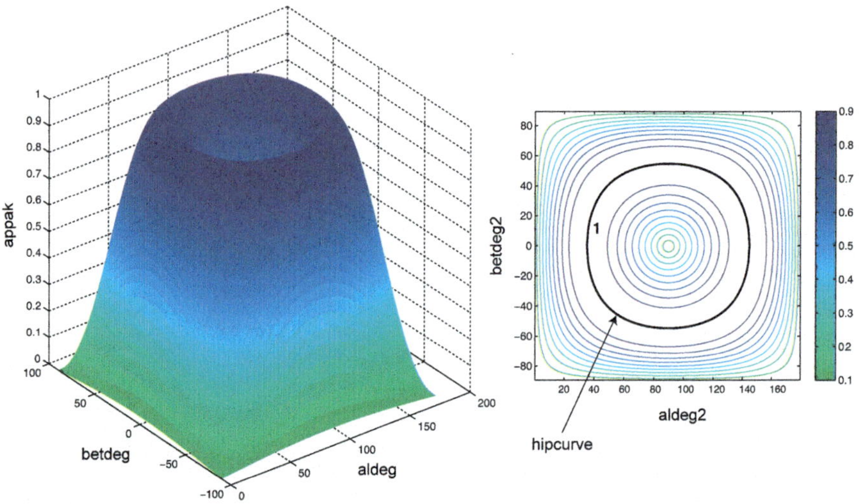

Fig. 6 Inverse condition number of the forward Jacobian matrix **A** based on the Frobenius norm as a function of design parameters α and β

Figure 6 depicts the inverse condition number of matrix **A** based on the Frobenius norm as a function of angles α and β. $\kappa_F(\alpha, \beta)$ is a minimum when its partial derivatives with respect to α and β vanish, namely,

$$\dot{\kappa}_\alpha(\alpha, \beta) = \frac{\partial \kappa}{\partial \alpha} = \frac{C\alpha(3S^2\alpha C^2\beta - 1)(S^2\alpha C^2\beta + 1)}{18S^3\alpha C^2\beta(S^2\alpha C^2\beta - 1)^2\kappa} = 0 \quad (9)$$

$$\dot{\kappa}_\beta(\alpha, \beta) = \frac{\partial \kappa}{\partial \beta} = -\frac{S\beta(3S^2\alpha C^2\beta - 1)(S^2\alpha C^2\beta + 1)}{18S^2\alpha C^3\beta(S^2\alpha C^2\beta - 1)^2\kappa} = 0 \quad (10)$$

and its Hessian matrix is semi-positive definite. As a result, $\kappa_F(\alpha, \beta)$ is a minimum and equal to 1 along the hippopede curve, which is shown in Fig. 6 and defined by the following equation:

$$3S^2\alpha C^2\beta - 1 = 0 \quad (11)$$

This hippopede curve amounts to the isotropic loci of the parallel spherical wrist.

Figure 7 illustrates some placements of the actuated omni-wheels on the sphere leading to kinematically isotropic wheel configurations in the parallel spherical wrist. It should be noted that the three singular values of matrix $\mathbf{A}_{r\omega}$ are equal to the ratio between the sphere radius R and the actuated omni-wheel radius r along the hippopede curve, namely, the velocity amplification factors of the PSW are the same and constant along the hippopede curve.

If the rotating sphere were to carry a camera, a laser or a jet of some sort, then the reachable orientations would be limited by interferences with the omni-wheels. Therefore, a designer would be interested in choosing a small value of alpha, so as to

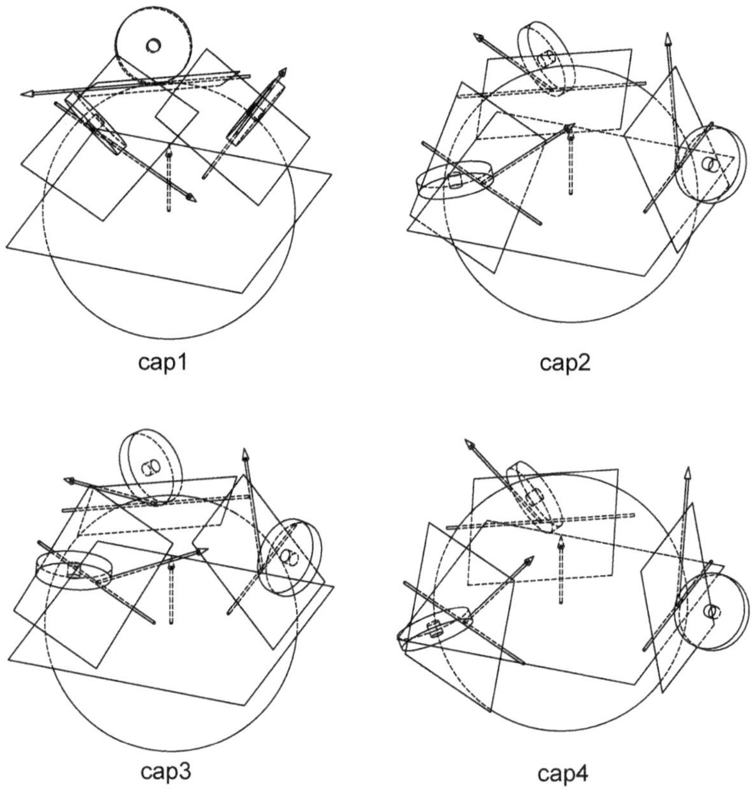

Fig. 7 Kinematically isotropic wheel configurations in the parallel spherical wrist

maximize the field of view of the PSW. As a result, the following values have been assigned to the design parameters α and β:

$$\alpha = 35.26^\circ \quad (12)$$

$$\beta = 0^\circ \quad (13)$$

in order to come up with a kinematically isotropic wheel configuration in the parallel spherical wrist and a large field of view. The actuated omni-wheels are mounted in pairs in order to ensure a good contact between them and the sphere. A CAD modeling of the final solution is represented in Fig. 1.

4 Conclusion

This paper presents the novel concept of mounting a parallel spherical wrist in series with a CDPR, while preserving a fully-parallel actuation scheme. As a result, the actuators always remain fixed to the base, thus avoiding the need to carry electric power to the end-effector and minimizing its size, weight and inertia. Another original contribution of this article is the determination of the kinematically isotropic wheel configurations in the parallel spherical wrist. These configurations allow the designer to obtain a very good primary image of the design choices. To our knowledge, these isotropic configurations were never reported before, although several researchers have studied and used omni-wheel-actuated spheres. Future work includes the development of a control scheme to drive the end-effector rotations while accounting for the displacements of its centre, and also making a small scale prototype of the robot.

References

1. Angeles, J.: Fundamentals of Robotic Mechanical Systems: Theory, Methods and Algorithms, 3rd edn. Springer, Berlin (2007)
2. Bahrami, A., Bahrami, M.N.: Optimal design of a spatial four cable driven parallel manipulator. In: 2011 IEEE International Conference Robotics and Biomimetics (ROBIO), pp. 2143–2149 (2011)
3. Bai, S., Hansen, M.R.: Modelling of a spherical robotic wrist with Euler parameters. In: 12th IFToMM World Congress (2007)
4. Ball, R.S.: A Treatise on the Theory of Screws, vol. 23(41). Cambridge University Press, Cambridge (1900)
5. Hadian, H., Fattah, A.: Kinematic isotropic configuration of spatial cable-driven parallel robots. Int. J. Intell. Mech. Robot. **1**(4), 61–86 (2011)
6. Hayes, M.J.D., Langlois, R.G.: Atlas motion platform: full-scale prototype. In: CSME International Congress (2012)
7. Hayes, M.J.D., Langlois, R.G., Weiss, A.: Atlas motion platform generalized kinematic model. Meccanica **46**, 17–25 (2011)
8. Hunt, K.H.: Kinematic Geometry of Mechanisms, vol. 23(41). Clarendon Press, Oxford (1978)
9. Kim, J., Hwang, J.C., Kim, J.S., Park, R.C.: Eclipse-ii: a new parallel mechanism enabling continuous 360-degree spinning plus three-axis translational motions. Robot. Autom. **4**, 3274–3279 (2001)
10. Wu, G., Caro, S., Bai, S., Kepler, J.: Dynamic modeling and design optimization of a 3-dof spherical parallel manipulator. Robot. Auton. Syst. **62**, 1377–1386 (2014)
11. Yime, E., Moreno, H., Saltarén, R.: A novel 6-dof parallel robot with decoupled translation and rotation. In: 13th World Congress in Mechanism and Machine Science (2011)

Dynamic Recovery of Cable-Suspended Parallel Robots After a Cable Failure

Alessandro Berti, Marc Gouttefarde and Marco Carricato

Abstract This paper studies how emergencies and failures can be managed in cable-driven parallel robots, in particular in the case of a redundant cable-suspended robot subjected to a cable breakdown. The objective is to present and test via numerical simulation the feasibility of an emergency strategy that allows the robot platform to be dynamically recovered to a safe position. Preliminary results, based on a simplified robot with a point-mass platform suspended by 4 cables, show that the proposed strategy may be an effective way to guide the platform from an unstable pose determined by the cable failure to a new static equilibrium pose.

1 Introduction

Cable-driven parallel robots (CDPRs) control the pose of the end-effector by cables. Consequently, they benefit from peculiar characteristics, such as a potentially large workspace, relatively lightweight support structures, ease of assembly, and superior modularity and reconfigurability. However, cables can only pull but not push on the mobile platform, which makes the study of CDPRs challenging. CDPRs are *fully constrained* if, once the actuators are locked, the mobile platform pose is completely determined. They are *underconstrained* [1, 2, 4] if the platform is movable when the cable lengths are assigned. In a cable-suspended robot (CSPR), all cables lie above the moving platform. A CSPR may be affected by several types of malfunctioning. This paper focuses on CSPR behavior in case of cable ruptures. The rupture of one or more cables generally causes the CSPR static workspace to change significantly. Indeed, after a cable failure, the platform pose may be located in a statically inadmissible part

A. Berti · M. Carricato (✉)
DIN, University of Bologna, Bologna, Italy
e-mail: marco.carricato@unibo.it

A. Berti
e-mail: alessandro.berti10@unibo.it

M. Gouttefarde
LIRMM - Université de Montpellier and CNRS, Montpellier, France
e-mail: marc.gouttefarde@lirmm.fr

J. Lenarčič and J.-P. Merlet (eds.), *Advances in Robot Kinematics 2016*,
Springer Proceedings in Advanced Robotics 4, DOI 10.1007/978-3-319-56802-7_35

of the workspace. In that case, even if the actuators are locked, the mobile platform will start moving towards a new statically admissible pose, which may be quite far from the pose where the cable breakdown took place.

Only few previous works deal with the topic investigated in this paper. Gosselin [5–7] presents an effective approach to control both planar and spatial CDPRs (with 3 dofs) outside their static workspace. Roberts [11] discusses a procedure for the design of "fault-tolerant cable-suspended systems", i.e. a method capable to assess if a robot in a certain pose may tolerate a cable loss. Notash [10] presents a method to recover the lost wrench after a cable failure, but only under the assumption that the robot remains overconstrained after the failure has occurred.

This paper proposes an approach that aims at safely recovering the mobile platform after a cable failure, even if the breakdown causes the current platform pose to lie outside the static workspace. To the best of our knowledge, this issue is investigated here for the first time. Results are obtained via numerical simulation. The paper is organized as follows. Sections 2, 3 and 4 describe, respectively, the CSPR model, the recovery strategy, and the control scheme. Section 5 presents simulation examples and discusses results. Section 6 concludes the paper.

2 CSPR Model

The simplified CSPR studied in this paper has 4 cables, whose lengths are varied by motorized winches. The cables exit from 4 coplanar points A_i, $i = 1, \dots, 4$, fixed on the robot support structure and connected to a point-mass mobile platform P (Fig. 1). The position vector of P in the fixed reference frame $Oxyz$ is $\mathbf{p}$, and the platform mass is denoted by m. O is located on the ground, with the z-axis pointing upward. Cables are assumed to be massless, and each one of them is considered to be a straight-line segment from A_i to P.

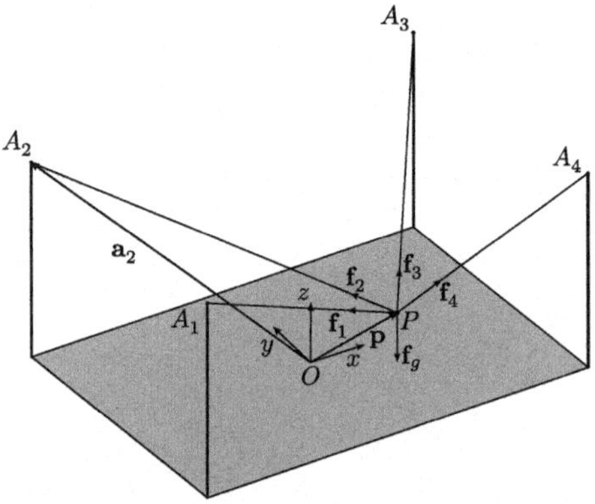

Fig. 1 Geometric model of the CSPR

The robot geometry and dimensions are inspired by the CoGiRo prototype [8, 9]. The position vectors $\mathbf{a}_i$ of points A_i in $Oxyz$ are $\mathbf{a}_1 = [-7.246, -5.174, 5.480]^\mathrm{T}$, $\mathbf{a}_2 = [-7.232, 5.305, 5.480]^\mathrm{T}$, $\mathbf{a}_3 = [7.253, 5.278, 5.480]^\mathrm{T}$ and $\mathbf{a}_4 = [7.232, -5.202, 5.480]^\mathrm{T}$. Cable lengths are defined as:

$$\rho_i = \|\mathbf{p} - \mathbf{a}_i\| \quad i = 1, \ldots, 4 \tag{1}$$

Differentiating Eq. (1) with respect to time yields the cable-length time derivatives, which can be expressed in matrix form as

$$\dot{\rho} = \mathbf{J}\dot{\mathbf{p}} \tag{2}$$

where

$$\mathbf{J} = \left[\ldots \frac{\mathbf{p} - \mathbf{a}_i}{\rho_i} \ldots\right]^\mathrm{T} \tag{3}$$

By letting the wrench matrix be $\mathbf{W} = -\mathbf{J}^\mathrm{T}$, the platform dynamics is expressed as

$$\mathbf{W}\mathbf{f} = m\left(\ddot{\mathbf{p}} - \mathbf{f}_g\right) \tag{4}$$

with $\mathbf{f}_g = [0, 0, -g]^\mathrm{T}$.

The cable-force array $\mathbf{f}$ is computed by a simplified elastic model based on Hooke's law, i.e.

$$\mathbf{f} = \mathbf{K}(\rho - \rho_0) \tag{5}$$

where ρ_0 is the array of the unstrained cable lengths and $\mathbf{K}$ is a diagonal matrix whose elements are the cable stiffnesses, namely

$$k_i = E_i S_i / \rho_i \tag{6}$$

with E_i and S_i being, respectively, the cable's Young modulus and cross-sectional area.

If $\mathbf{t}_m$ is the array of motor torques assigned by the controller, the winch motors' angular accelerations are

$$I_m \ddot{\mathbf{q}} = \mathbf{t}_m - r\mathbf{f} \tag{7}$$

where I_m is the motor inertia, $\ddot{\mathbf{q}}$ is the array of motor accelerations, and r is the winch drum radius. I_m and r are assumed to be equal for all winches. ρ_0 is computed as $\rho_0 = -r\mathbf{q}$, namely by the numeric integration of the left-hand side of Eq. (7). The integration of Eq. (4) provides the simulated motion of the robot.

3 Recovery Strategy

A cable failure may cause the CSPR mobile platform to lie outside its (new) static workspace. The main purpose of this work is to find a recovery strategy that can safely drive the platform to a statically admissible rest configuration. Gosselin [7] showed that a 2-dof planar CSPR can be dynamically controlled outside its static workspace. In [5, 6], these results were extended to a spatial 3-dof 3-cable CSPR with a point-mass platform. The recovery strategy proposed here is based on the results of [5, 6]. When cable k breaks, the new static workspace is defined as the prism Π whose base is the triangle formed by the projections on the xy plane of points $\mathbf{a}_i$, $i = 1, \ldots, 4$, $i \neq k$, and whose height is equal to the (common) height of the cable exit points, i.e. a_z. The platform position is statically feasible only if P lies inside Π. If P lies outside Π and the actuators are locked, P will uncontrollably move towards a new stable equilibrium within Π.

The strategy proposed in this paper is the following. Assuming that it is possible to determine when a cable failure occurs, the controller tries to recover the mobile platform towards the new static workspace by following a trajectory whose definition is based on the approach presented in [5, 6]. Then, as soon as the platform enters the new static workspace, the controller aims at stopping it. For simplicity, the whole recovery trajectory is designed such that the path followed by $\mathbf{p}$ lies in a horizontal plane. We define three main stages during the recovery phase (Fig. 2):

- (b) a cable breaks down;
- (r) the platform reaches its new static workspace (triangle $A_1A_2A_3$ in Fig. 2);
- (s) the platform reaches a complete stop.

Quantities relative to each stage are denoted by subscripts (b), (r), or (s), respectively. The first part of the trajectory, which immediately follows the cable breakdown, between stages (b) and (r), leads the platform to its new static workspace. To this end, the tensions in the remaining cables must be kept positive by the inertial load on the platform. Thus, according to [5, 6], this part of the trajectory may consist of two sinusoids with appropriate amplitudes and frequency along the x and y axes, thus forming the following ellipse:

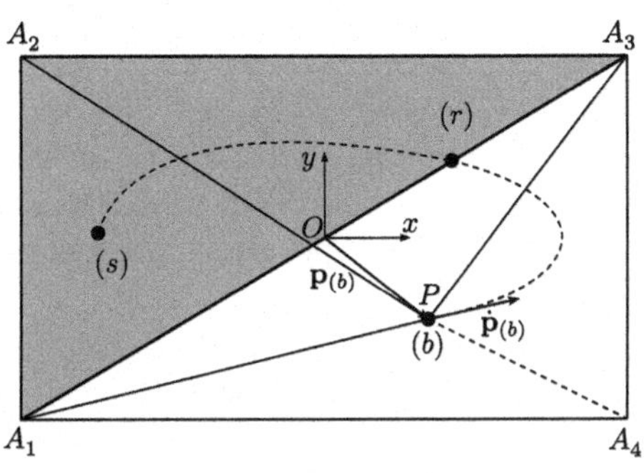

Fig. 2 The 3 main stages of the recovery phase

$$\Gamma_{(b)-(r)} = [x, y, z]^{\mathrm{T}} = [x_c + r_x \sin(\omega_n t + \alpha_x), y_c + r_y \sin(\omega_n t + \alpha_y), z_{(b)}]^{\mathrm{T}} \quad (8)$$

where x_c and y_c are the coordinates of the center C of the ellipse, r_x and r_y are the amplitudes of the sinusoids along x and y, respectively, α_x and α_y are their phases, and ω_n is the oscillation frequency. ω_n can be taken, according to [5, 6], as the natural frequency of the robot after the cable breakdown, namely as

$$\omega_n = \sqrt{g / \left(a_z - z_{(b)}\right)} \quad (9)$$

The center of the ellipse C must lie inside the new static workspace, i.e. in Π. The choice of C has a significant influence on the performance of the recovery operation. Currently, coordinates x_c and y_c are not automatically computed, but their values are tuned before the simulation. The remaining parameters, i.e. r_x, r_y, α_x and α_y, must be chosen so as to ensure that the platform position and velocity are continuous in (b). Thus, assuming that time t is zero in (b), the following equalities must hold

$$\begin{cases} x_c + r_x \sin(\alpha_x) = x_{(b)}, \; y_c + r_y \sin(\alpha_y) = y_{(b)} \\ \omega_n r_x \cos(\alpha_x) = \dot{x}_{(b)}, \quad \omega_n r_y \cos(\alpha_y) = \dot{y}_{(b)} \end{cases} \quad (10)$$

which allow r_x, r_y, α_x and α_y to be computed.

As soon as P begins moving along the elliptic path, the controller starts checking whether or not it lies in the new static workspace. Once the platform reaches stage (r), the controller begins following the second part of the recovery trajectory, i.e. the one leading to the complete stop of the platform in (s). This part is formed by two 7-degree polynomials along the x and the y axis:

$$\Gamma_{(r)-(s)} = [x, y, z]^{\mathrm{T}} = \left[\sum_{i=0}^{7} c_{ix} t^i, \sum_{i=0}^{7} c_{iy} t^i, z_{(b)} \right]^{\mathrm{T}} \quad (11)$$

The 16 unknown coefficients in $\Gamma_{(r)-(s)}$ are obtained by imposing continuity of position, velocity, acceleration and jerk in (r), and canceling out velocity, acceleration and jerk in (s), provided that the position $\mathbf{p}_{(s)}$ is assigned by the controller on the basis of the environment characteristics.

The duration of the second part of the recovery trajectory and the location of $\mathbf{p}_{(s)}$ must be chosen so as to maintain positive tensions in the remaining three cables, and to avoid collisions with the robot support structure. If necessary, the second part of the trajectory can be conveniently segmented, with additional control points added, and 7-degree polynomials used to lead P from one control point to the next one.

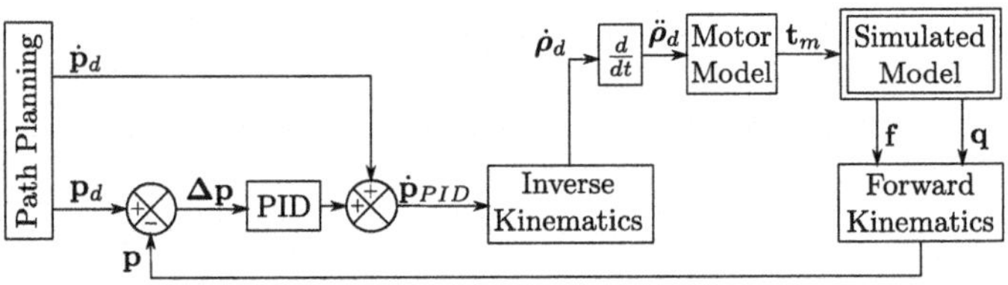

Fig. 3 The control and simulation scheme adopted during the recovery trajectory

4 Recovery Control Scheme

We assume that it is possible to estimate the cable tension values and, consequently, to detect when a cable breaks down. At the moment of a cable failure, the controller checks whether the actual platform pose and the remaining part of its trajectory are inside the new static workspace, in which case the robot performs an emergency stop along this trajectory [3]. If the actual pose is outside the new static workspace, or it is impossible to stop the mobile platform before it exits the new static workspace, the controller starts a recovery strategy. The new static workspace is known, since, for any 3-cable configuration, it can be computed offline. When a cable breakdown is detected, the controller computes the parameters of the elliptic trajectory in Eq. (8).

If cable tensions can be estimated (i.e. measured by sensors), we should know which cables are taut at every controller cycle. The input of the direct problem are the motor angular rotations measured at the ith simulation step. The solution of the direct problem returns the "real" platform pose $\mathbf{p}$. With the pose $\mathbf{p}$, the error with respect to the pose $\mathbf{p}_d$ coming from the trajectory planner at the simulation step $i+1$ is computed, and through a PID controller we obtain the corrected velocity $\dot{\mathbf{p}}_d$ that permits to follow the desired trajectory.

Inverse differential kinematics is then used to compute the motor velocities. The motor model yields the corresponding motor torque, which is the input to the simulated robot, thus yielding the new motor angular position that allow the control loop to restart (Fig. 3). According to the results of the numerical simulations conducted so far, it is not difficult to keep 3 cables in tension throughout the whole recovery trajectory.

5 Simulation Results

Test 1. The example presented here simulates a recovery trajectory after the failure of cable number 4. The platform mass is $m = 400$ kg and, at stage (b), the platform position and velocity are, respectively, $\mathbf{p}_{(b)} = [0.830, -2.781, 1.915]^{\mathrm{T}}$ and $\dot{\mathbf{p}}_{(b)} = [1.984, 0.490, 0.055]^{\mathrm{T}}$. The center of the ellipse $\Gamma_{(b)-(r)}$ is set as $C = [-1.0, 0.5, 1.915]^{\mathrm{T}}$ and the platform position at rest is $\mathbf{p}_{(s)} = [-5.0, -1.0, 1.915]^{\mathrm{T}}$. The duration of the entire recovery phase is $t_{(s)} = 4$ s. This test shows that the

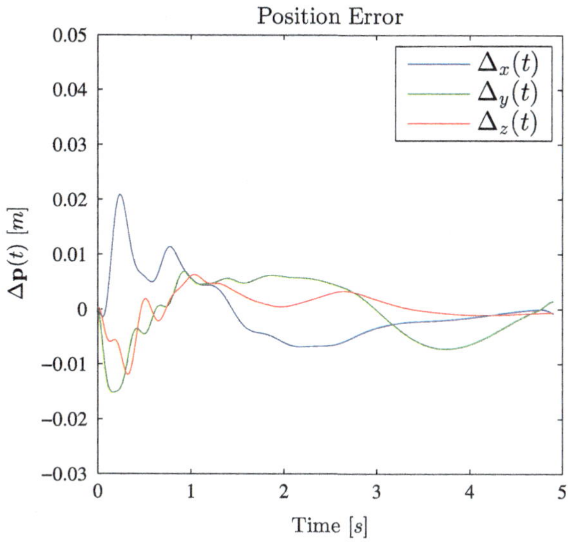

Fig. 4 Positioning error with respect to the recovery trajectory provided by the controller

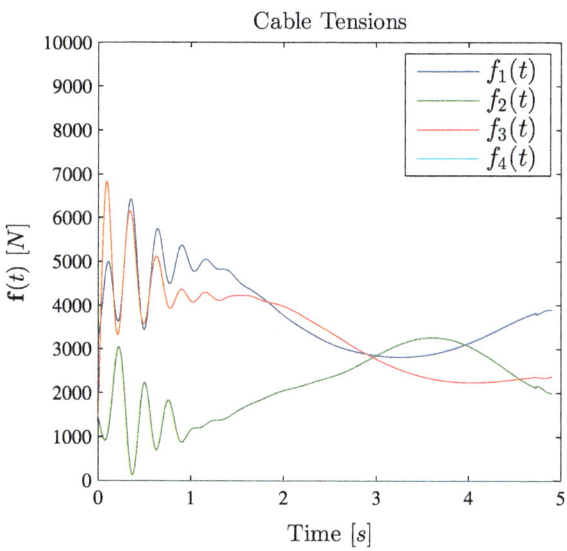

Fig. 5 Cable tensions during the recovery trajectory

controller can lead the platform along the recovery trajectory with a positioning error of a few centimeters (Fig. 4). It also shows that the tensions of the remaining active cables are positive throughout the recovery trajectory (Fig. 5). The cable cross-sectional area and elasticity are $S_i = 16.4\ \text{mm}^2$ and $E_i = 76$ GPa, respectively. The winch drum radius is $r = 67.5$ mm, and the motor inertia is $I_m = 0.062\ \text{kg m}^2$.

Test 2. The same approach described for a 4-cable robot may be adapted and applied to the 8-cable model of the CoGiRo prototype [8, 9]. Indeed, if the CoGiRo

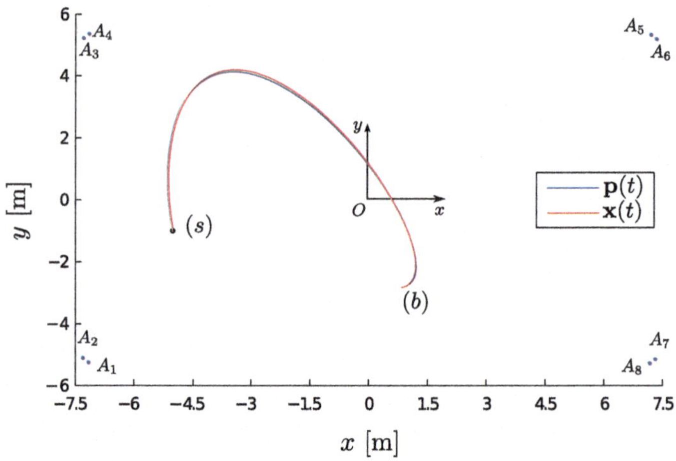

Fig. 6 Tests 1-2: Recovery trajectories for both models

exit points are approximated (taken 2 by 2) to lie at the vertices of a rectangle and connected to the platform center of mass P by 4 cables (with doubled cross section and Young modulus), the 4-cable robot simulated in Test 1 is obtained, and the recovery strategy proposed in this paper may be used to design the trajectory of P. This trajectory can then be used to compute, by the inverse kinematics, the cable lengths of the actual 8-cable robot, which are fed to the CoGiRo controller. Details are not reported due to space limitations. Figure 6 refers to a simulation that replicates the same conditions on stage (b) in Test 1 for the 8-cable CoGiRo model. For consistency, when cable 7 breaks, cable 8 is also deactivated, thus leading to a 6-cable robot lying outside its static workspace. The recovery strategy brings the robot to halt in the same position as in Test 1, with zero final orientation of the platform. Figure 6 shows that there is a very good match between the trajectory followed by the simplified and the non-simplified model.

6 Conclusions, Open Issues and Future Works

This paper focused on the issue of cable failure (breakdown) in cable-suspended parallel robots (CSPRs). A recovery strategy was introduced for a spatial 3-dof CSPR with 4 cables and a point-mass mobile platform. This strategy consists in planning a trajectory that, depending on which cable breaks down, may partly lie outside of the failed CSPR static workspace. Simulation results showed that the mobile platform may be recovered to a safe position, thereby preventing the mobile platform from falling down or making large uncontrolled oscillations.

A number of issues remain to be dealt with in order to make the proposed recovery strategy fully automated and reliable.

(a) The recovery strategy is notably based on the results presented in [5, 6]. The extension of [5, 6] to 6-cable 6-dof CSPRs is an open issue.

(b) In some circumstances, the recovery strategy may be counterproductive. For example, if a cable breaks when the platform is close to the ground, it may be safer to let it fall down. Criteria to determine whether or not the recovery strategy should intervene need to be investigated.

(c) Some parameters of the recovery trajectory, such as the center of the ellipse $\Gamma_{(b)-(r)}$, the location of the point where the platform has to be stopped, and the duration $t_{(s)}$ of the trajectory $\Gamma_{(r)-(s)}$, are currently provided by the user. The objective of part of our ongoing works is to make the controller capable of determining these parameters autonomously.

(d) The feasibility of the proposed recovery strategy in real-time must be assessed, in order to proceed with experiments on a CSPR prototype.

Acknowledgements This work was supported by Italian Ministry of Education, Universities and Research by the PRIN grant No. 20124SMZ88.

References

1. Abbasnejad, G., Carricato, M.: Direct geometrico-static problem of underconstrained cable-driven parallel robots with n cables. IEEE Trans. Rob. **31**(2), 468–478 (2015)
2. Berti, A., Merlet, J.P., Carricato, M.: Solving the direct geometrico-static problem of underconstrained cable-driven parallel robots by interval analysis. Int. J. Robot. Res. **35**(6), 723–739 (2016)
3. Berti, A., Merlet, J.P., Carricato, M.: Workspace analysis of redundant cable-suspended parallel robots. In: Pott, A., Bruckmann, T. (eds.) Cable-Driven Parallel Robots, pp. 41–53. Springer (2015)
4. Carricato, M., Merlet, J.P.: Stability analysis of underconstrained cable-driven parallel robots. IEEE Trans. Rob. **29**(1), 288–296 (2013)
5. Gosselin, C.: Global planning of dynamically feasible trajectories for three-DOF spatial cable-suspended parallel robots. In: Bruckmann, T., Pott, A., (eds.) Cable-Driven Parallel Robots, pp. 3–22. Springer (2013)
6. Gosselin, C., Foucault, S.: Experimental determination of the accuracy of a three-DOF cable-suspended parallel robot performing dynamic trajectories. In: Pott, A., Bruckmann, T. (eds.) Cable-Driven Parallel Robots, pp. 101–112. Springer (2015)
7. Gosselin, C., Ren, P., Foucault, S.: Dynamic trajectory planning of a two-DOF cable-suspended parallel robot. In: 2012 IEEE International Conference on Robotics and Automation, pp. 1476–1481 (2012)
8. Gouttefarde, M., Collard, J.F., Riehl, N., Baradat, C.: Geometry selection of a redundantly actuated cable-suspended parallel robot. IEEE Trans. Rob. **31**(2), 501–510 (2015)
9. Lamaury, J., Gouttefarde, M.: Control of a large redundantly actuated cable-suspended parallel robot. In: 2013 IEEE International Conference on Robotics and Automation, pp. 4659–4664 (2013)
10. Notash, L.: Failure recovery for wrench capability of wire-actuated parallel manipulators. Robotica **30**(6), 941–950 (2012)
11. Roberts, R.G., Graham, T., Lippitt, T.: On the inverse kinematics, statics, and fault tolerance of cable-suspended robots. J. Robot. Syst. **15**(10), 581–597 (1998)

Workspace and Interference Analysis of Cable-Driven Parallel Robots with an Unlimited Rotation Axis

Andreas Pott and Philipp Miermeister

Abstract A drawback of many cable-driven parallel robots is a relatively small orientation workspace. In this paper, two design variants for cable-driven parallel robots with nine and twelve cables are proposed that allow for large rotations. It is shown that the platform can perform a 360° rotation while maintaining positive tension in all cables and without collisions amongst the cables. Furthermore, workspace studies of the total orientation workspace are provided. Surprisingly, this family of cable robot is capable to perform an unlimited rotation within a translational workspace of reasonable size. Finally, the efficiency and computation time of force distribution algorithms is compared for cable robots having twelve cables.

1 Introduction

Cable-driven parallel robots, in the following simply called cable robots, mainly consist of a light weight platform, cables, and winches and therefore allow to design systems with an exceptional good power to mass ratio compared to conventional kinematics. The workspace of the robot mainly depends on the winch positions and platform anchor points. Cable robots have a relatively small rotational workspace compared to their translational workspace. Prototypes such as the CoGiRo [4] are able to rotate the platform by around 90° which is considered to be a large orientation workspace for a cable robot. Handling tasks such as conveyer belt pick-and-place applications demand for a larger rotational workspace which can yet only be achieved by additional actuated axes on the platform.

A. Pott (✉)
Institute for Control Engineering and Manufacturing Units, University of Stuttgart, Stuttgart, Germany
e-mail: andreas.pott@isw.uni-stuttgart.de

A. Pott · P. Miermeister
Fraunhofer IPA, Stuttgart, Germany

J. Lenarčič and J.-P. Merlet (eds.), *Advances in Robot Kinematics 2016*,
Springer Proceedings in Advanced Robotics 4, DOI 10.1007/978-3-319-56802-7_36

In this paper, an approach is presented where the desired rotation is actuated by cables. This keeps the mass of the system low and avoids the need for power supply. Beside that, additional cables increase the safety of the system and even may be used to increase the translational workspace.

2 Cable Robot with Unlimited Rotation

The basic idea of a cable-driven parallel robot with one unlimited rotation axis was recently presented [7]. In this paper, we detail the properties of such robots by proposing a configuration with more cables that has the unlimited rotation property of the platform within a reasonably large wrench-feasible workspace. Additionally, we show that the robot does not suffer from cable–cable collisions in this workspace. Considering the connection of many cables to the platform in a spatial robot, it seems to be clear from intuition that large rotation is impossible for a cable-driven robot. However, this intuition is wrong and one way to achieve this effect is to use a platform that has the form of a crank shaft. Fixing three cables to each end of the shaft is basically a generic 2R3T design. Then, one uses an eccentric connection point on the shaft to independently control the rotation of the shaft (see Fig. 1). In this simple example, three cables share a common anchor point at each end of the shaft as well as on the crank. From a kinematic point of view, this yields exactly the desired rotational mobility of the platform. As shown in [7], each additional degree-of-freedom can be actuated on the platform and requires at least one extra cable. However, for a cable robot, we might need to add more cables for a degree-of-freedom to achieve a preferred geometry of the platform.

In the following, two archetypic robot designs are presented where the geometry of the robot is characterized by the proximal anchor points $\mathbf{a}_i$ and the distal anchor points $\mathbf{b}_i$. The number of cables is denoted by m.

The *endless Z9* design is a minimal realization of the idea sketched above to achieve the rotation capabilities of the robot. The geometry of the robot consists of three triangular levels on different heights that respectively share a common distal

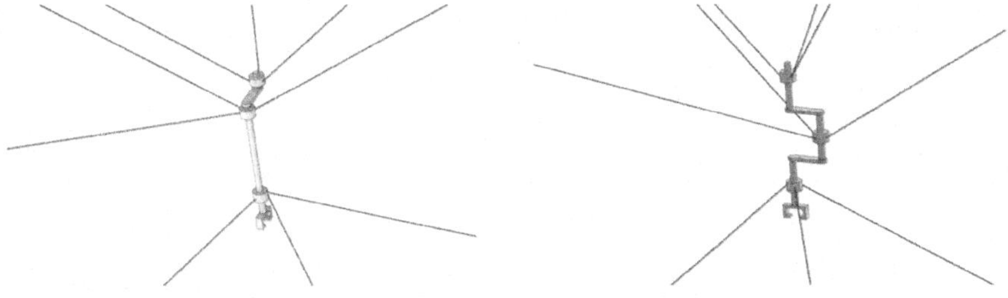

Fig. 1 Design of two cable robot with an unlimited yaw rotation around its z-axis in the form of a lever (*left*) and crank shaft (*right*) with nine cables

Table 1 Geometry data for the base $\mathbf{a}_i$ and platform $\mathbf{b}_i$ anchor points

Endless Z9	Base $\mathbf{a}_i$			Platform $\mathbf{b}_i$		
i	x	y	z	x	y	z
1	$-\frac{r_B}{2}$	$\frac{\sqrt{3}}{2}r_B$	H_B	r_P	0	H_P
2	r_B	0	H_B	r_P	0	H_P
3	$-\frac{r_B}{2}$	$-\frac{\sqrt{3}}{2}r_B$	H_B	r_P	0	H_P
4	$-\frac{r_B}{2}$	$\frac{\sqrt{3}}{2}r_B$	h_B	0	0	h_P
5	r_B	0	h_B	0	0	h_P
6	$-\frac{r_B}{2}$	$-\frac{\sqrt{3}}{2}r_B$	h_B	0	0	h_P
7	$-\frac{r_B}{2}$	$\frac{\sqrt{3}}{2}r_B$	0	0	0	0
8	r_B	0	0	0	0	0
9	$-\frac{r_B}{2}$	$-\frac{\sqrt{3}}{2}r_B$	0	0	0	0
Endless Z12	Base $\mathbf{a}_i$			Platform $\mathbf{b}_i$		
i	x	y	z	x	y	z
1	$-r_B$	r_B	H_B	r_P	0	H_P
2	r_B	r_B	H_B	r_P	0	H_P
3	r_B	$-r_B$	H_B	r_P	0	H_P
4	$-r_B$	$-r_B$	H_B	r_P	0	H_P
5	$-r_B$	r_B	h_B	0	0	h_P
6	r_B	r_B	h_B	0	0	h_P
7	r_B	$-r_B$	h_B	0	0	h_P
8	$-r_B$	$-r_B$	h_B	0	0	h_P
9	$-r_B$	r_B	0	0	0	0
10	r_B	r_B	0	0	0	0
11	r_B	$-r_B$	0	0	0	0
12	$-r_B$	$-r_B$	0	0	0	0

anchor point $\mathbf{b}_i$. The structure in layers shall avoid collisions amongst these groups of cables. Since all cables within a group share a common distal anchor point, the cables cannot interfere in any other point within the robot frame [11]. The endless Z9 is a 9-3 cable configuration. A parametric representation of this geometry is given in Table 1.

The *endless Z12* is a cable robot in a 12-3 cable configuration (see Fig. 2). In order to match the mostly desired shape of a rectangular robot frame, the proximal anchor points are moved to the surface of a box providing also a larger possible workspace for the robot. The design has again three layers on the platform and on the base, where for this robot each layer consists of four cables. Again, given a proper geometry for the platform, cables on different layers shall not collide. For the sample robot, we have chosen the parameters as follows: $r_B = 2$, $r_P = 0.3$, $H_B = 3$, $h_B = 2$, $H_P = 0.5$, and $h_P = 0.3$.

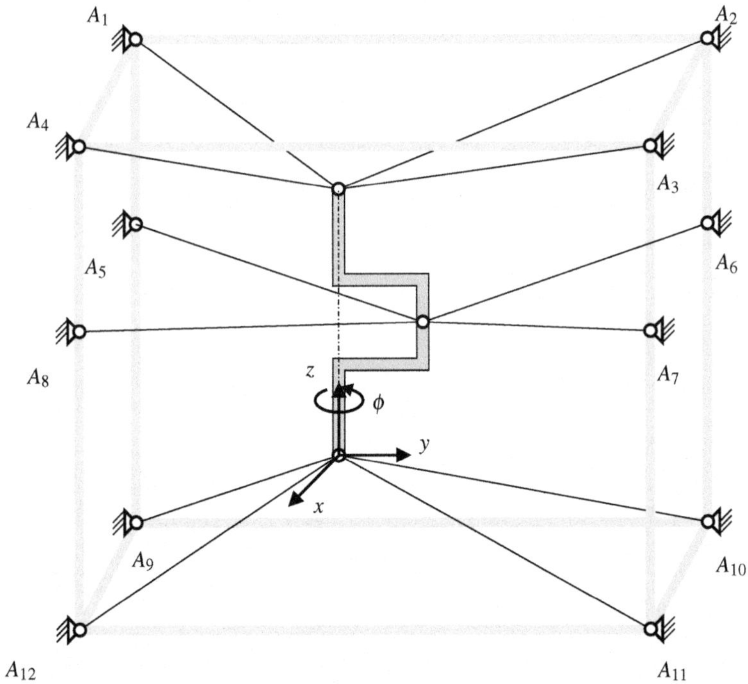

Fig. 2 The endless Z12 robot architecture with $m = 12$ cables

3 Kinematic Properties

In the following, we briefly present the modelling and tools used to analyze the cable robots. The standard kinematic modelling for cable-driven parallel robots is used for the analysis in this paper. The platform position $\mathbf{r}$ and rotation $\mathbf{R}$ can be controlled by changing the cable length $\mathbf{l}_i$ according to the inverse kinematics

$$\mathbf{l}_i = \mathbf{a}_i - \mathbf{r} - \mathbf{R}\mathbf{b}_i, \tag{1}$$

where vectors $\mathbf{a}_i$ relate to the cable's outlet points at the winch side and $\mathbf{b}_i$ are the distal anchor points on the mobile platform. Considering the platform as a free floating body, a stable platform position is characterized by the force and torque equilibrium

$$\mathbf{A}^{\mathrm{T}}\mathbf{f} + \mathbf{w} = \mathbf{0}, \tag{2}$$

where $\mathbf{f}$ and $\mathbf{w}$ denote the cable forces and external wrench, respectively, while $\mathbf{A}^{\mathrm{T}}$ relates to the well-known structure matrix

$$\mathbf{A}^{\mathrm{T}} = \begin{bmatrix} \mathbf{u}_1 & \cdots & \mathbf{u}_m \\ \mathbf{b}_1 \times \mathbf{u}_1 & \cdots & \mathbf{b}_m \times \mathbf{u}_m \end{bmatrix}. \tag{3}$$

The unit vectors $\mathbf{u}_i$ describe the direction of the cables. Cables can only resist pulling forces so that Eq. 2 must be fulfilled under the constraint of positive forces $f_i > 0$ for $i = 1 \ldots, m$. Under the consideration of the minimal and maximal allowed cable forces f_{min} and f_{max}, it is possible to determine the set of allowed platform poses for a given wrench $\mathbf{w}$ using Eq. (2), that is to find a positive solution for the cable forces $\mathbf{f}$ within the allowed range $f_{\mathrm{min}} \leq f_i \leq_{\mathrm{max}}$.

Cable robots with more cables than the six degree-of-freedom are over-constrained systems and therefore multiple valid force distributions for a single stable platform pose exist. This holds true especially for the endless Z robots at hand which exhibit a high degree-of-redundancy. Thus, for the determination of the cable force, different methods are considered that are capable of dealing with such highly redundant robots. For this study, we focus on the (advanced) closed-form method [10], the Dykstra method [5], and the wrench-set method [1]. However, only little can be found in the literature that analyze appropriateness and applicability of force distribution methods for robots with $m = 12$ cables. It is known that the computation time and also the convergence can degenerate if the degree-of-redundancy increases. Therefore, different approaches are used in this study to assess also the feasibility for this highly redundant case. Following the discussion in [10], we use amongst others a least squares approach

$$\mathbf{f} = \mathbf{f}_{\mathrm{m}} - \mathbf{A}^{+\mathrm{T}} \left(\mathbf{w} + \mathbf{A}^{\mathrm{T}}\mathbf{f}_{\mathrm{m}}\right) \tag{4}$$

where $\mathbf{A}^{+\mathrm{T}}$ is the Moore–Penrose pseudo inverse of $\mathbf{A}^{\mathrm{T}}$ and $\mathbf{f}_{\mathrm{m}}$ is the medium feasible force distribution $\mathbf{f}_{\mathrm{m}} = (\mathbf{f}_{\mathrm{min}} + \mathbf{f}_{\mathrm{max}})/2$.

The concept of wrench-feasibility poses is recalled above and is applied to the workspace here. The *wrench-feasible workspace* (WFW) was defined in [2, 3, 12] as follows: The wrench-feasible workspace is the set $\mathscr{W}$ of poses $(\mathbf{r}, \mathbf{R})$ of the mobile platform. For any wrench $\mathbf{w} \in \mathscr{Q}$ there exists a vector of cable tension $\mathbf{f} \in [f_{\mathrm{min}}, f_{\mathrm{max}}]^m$ such that Eq. (2) is fulfilled. The pose $(\mathbf{r}, \mathbf{R})$ is called *wrench-feasible* if it allows at least one solution $\mathbf{f} \in [f_{\mathrm{min}}, f_{\mathrm{max}}]^m$. To test if a pose belongs to the wrench-feasible workspace, the methods mentioned above to compute force distributions are employed. For studying the workspace of the endless Z robots, one is interested in the *total orientation workspace* [6], i.e. the set of all positions where every orientation $\mathbf{R} \in \mathscr{R}$ is wrench-feasible. In this study the orientation set $\mathscr{R} = \{\mathbf{R} \in \mathrm{SO}_3 \mid \mathbf{R} = \mathbf{R}_{\mathrm{Z}}(\phi) \, \forall \, \phi \in [0, 2\pi]\}$ contains a full rotation around the z-axis of the platform.

For the workspace assessment, the hull method is used [9] which allows for very accurate computation of the workspace border also taking into account sets of orientations $\mathscr{R}$. The hull methods uses a triangulation of a small sphere around the estimated center of the workspace and inflates this region using line search until the border of the workspace is found. However, similar results as presented here can be achieved by simple sampling the workspace with discrete positions.

For cable robots with many cables, the problem of colliding cables becomes an issue, especially if large rotation angles are considered. A very interesting technique to calculate the regions of cable interference within the constant orientation workspace was presented by Perreault [8]. Through purely geometric considerations, it is possible to determine the loci of cable–cable interference from the geometry of the frame $\mathbf{a}_i$ and the relative geometry of the mobile platform $\mathbf{b}_i$. The main concept of this approach is the simple fact that two cables can interfere only if the corresponding anchor points $\mathbf{a}_i, \mathbf{a}_j, \mathbf{b}_i, \mathbf{b}_j$ lie in a common plane. Since the anchor points on the frame are fixed in space, the plane can be constructed as follows: As a model of the possible interference region, one computes the normals of the connection lines between pairs of proximal and distal anchor points from

$$\mathbf{a}_{ij} = \frac{\mathbf{a}_j - \mathbf{a}_i}{||\mathbf{a}_j - \mathbf{a}_i||_2} \quad , \quad \mathbf{b}_{ij} = \frac{\mathbf{b}_j - \mathbf{b}_i}{||\mathbf{b}_j - \mathbf{b}_i||_2} \quad \text{for} \quad i, j = 1, \ldots, m \quad i \neq j \tag{5}$$

If $\mathbf{a}_{ij}$ and $\mathbf{b}_{ij}$ are not parallel, one can construct two triangles [8]

$$T_{ij}^{+} : \mathbf{x} = \mathbf{a}_j - \mathbf{b}_i + \lambda \mathbf{a}_{ij} + \nu \mathbf{b}_{ij} \tag{6}$$

$$T_{ij}^{-} : \mathbf{x} = \mathbf{a}_j - \mathbf{b}_i - \lambda \mathbf{a}_{ij} - \nu \mathbf{b}_{ij} \tag{7}$$

with $\lambda, \nu > 0$. Exploiting the normalized length of the vectors $\mathbf{a}_{ij}$ and $\mathbf{b}_{ij}$, one practically chooses a metric length for λ and ν in the range of the size of the robot to receive finitely large triangles with the critical interference region. Note, that common anchor points as used in the robot geometries above decrease the collision region from triangles to lines. The lines and triangles can be used for visual or automatic detection of cable–cable interference. For many robot designs, one can see from first glance, if the triangles are within the workspace of interest or outside. Further information on dealing with parallel vectors can be found in the paper [8].

Relaxing the considerations for the fixed orientation, the vector $\mathbf{b}_{ij}$ is transformed by the rotation matrix $\mathbf{R}_Z(\phi)$ in order to study a full rotation of the platform. This leads to the collision area given by the following parametric volume

$$\widehat{T}_{ij}^{+} : \mathbf{x} = \mathbf{a}_j - \mathbf{R}_Z(\phi)\mathbf{b}_i + \lambda \mathbf{a}_{ij} + \nu \mathbf{R}_Z(\phi)\mathbf{b}_{ij} \tag{8}$$

$$\widehat{T}_{ij}^{-} : \mathbf{x} = \mathbf{a}_i - \mathbf{R}_Z(\phi)\mathbf{b}_j - \lambda \mathbf{a}_{ij} - \nu \mathbf{R}_Z(\phi)\mathbf{b}_{ij}, \tag{9}$$

where for $\mathbf{R}_Z(\phi)$ is the elementary rotation matrix around the z axis.

4 Results

For the endless Z12 cable robots with the parameters given in Table 1, a test trajectory was computed where the platform simply performs a full rotation around its z-axis with angle ϕ at the position $\mathbf{r} = [0, 0, 1.5]^{\mathrm{T}}$ m. The Dykstra method was used to

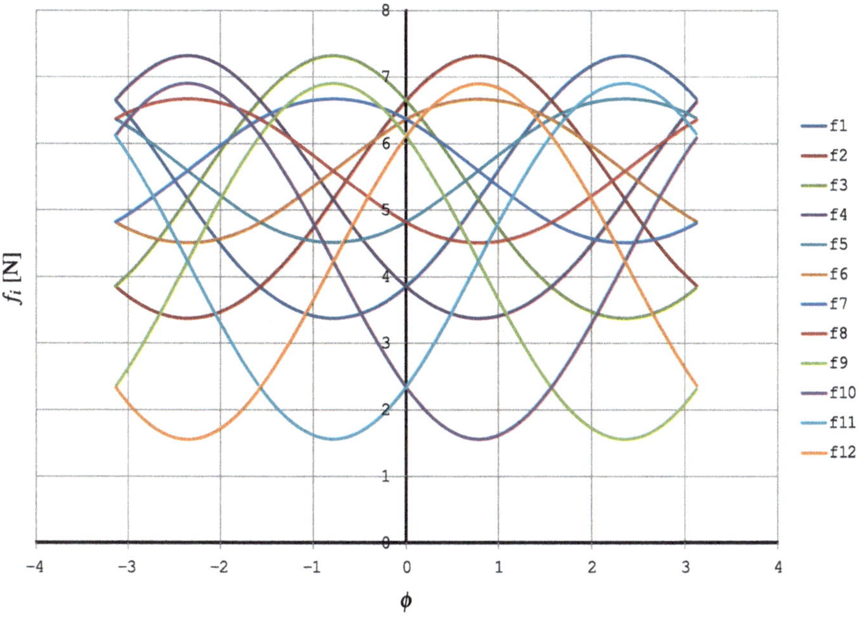

Fig. 3 Possible positive force distribution for a 2π rotation of the mobile platform at position $\mathbf{r} = [0, 0, 1.5]^{\mathrm{T}}$

compute force distribution for all angles in the given range $\phi \in [-\pi, \pi]$ which force limits $f_i \in [f_{\mathrm{min}}, f_{\mathrm{max}}] = [1, 10]$ N. The resulting forces are shown in Fig. 3 where one can easily see that all 12 cables can be kept under tension and no cable violated the given bounds for the cable forces. Thus, the orientation workspace of the sample pose indeed includes a full rotation of the platform.

Interestingly, the full rotation maneuver is possible at different positions. To quantitatively study this property of the robot, the wrench-feasible total orientation workspace of the robot was computed using the hull algorithm for the rotation set $\mathscr{R}$ given above. Using the Dykstra method and the same force limits as given above, a significantly large workspace was found (see Fig. 4).

The region of convergence and the computation time heavily differ amongst the considered methods for force distribution. As conjectured in [10], the closed-form method performs excellently in terms of computation time but is rather limited concerning the region where force distributions can be computed. The advanced closed-form method slows down the computation time by around a factor of four in average. From the analysis of the complexity of the algorithm at most, a slow-down of the degree-of-redundancy r is expected and the measured computation time is consistent with the expectations. Using the advanced closed-form method, the size of the computable workspace is largely extended by a factor of around eight. Finally, the Dykstra as reference method provides similar results in terms of workspace size compared to the advanced closed-form method. However, the alternative projections used

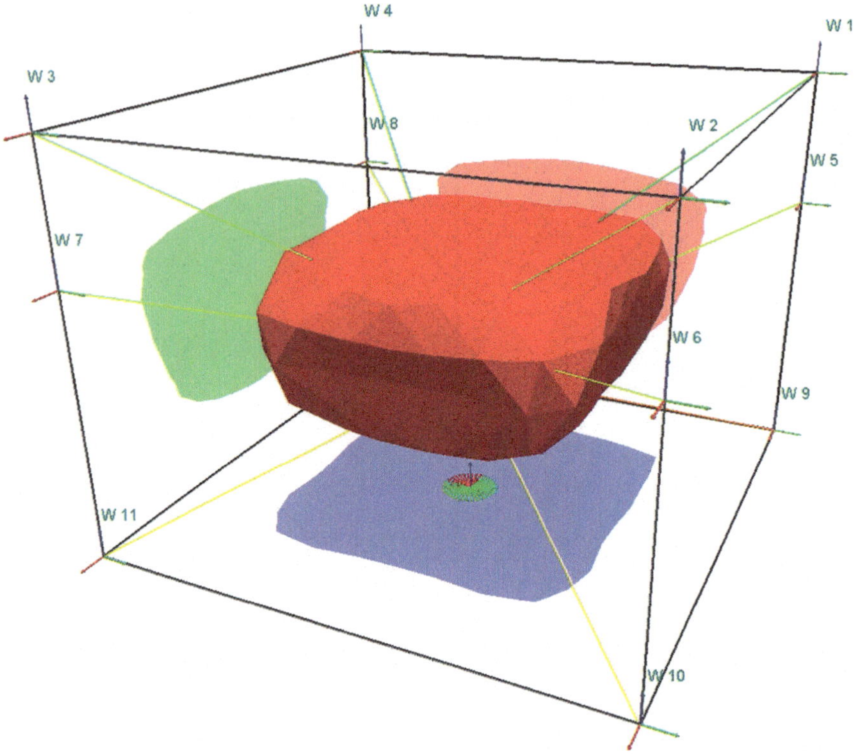

Fig. 4 Total orientation workspace of the cable robot endless Z12 computed with the Dykstra method

Table 2 Workspace volume and computation time of the endless Z12 computed with different force distribution algorithms

Method	Computation time (s)	Volume (m^3)	Max. cable force
Closed-form	1.292	0.889	$f_{max} = 10$
Closed-form	1.338	4.204	$f_{max} = 100$
Advanced closed-form	5.449	7.168	$f_{max} = 10$
Advanced closed-form	5.162	21.062	$f_{max} = 100$
Dykstra	66.747	7.250	$f_{max} = 10$
Dykstra	76.062	20.632	$f_{max} = 100$

in Dykstra's iterative scheme are rather inefficient for highly redundant cable robots and lead to computation times that are one order of magnitude larger. Furthermore, the consideration of a larger region of feasible cable forces adds additional efforts to the Dykstra scheme where the computation time for closed-form and advanced closed-form remains almost constant (Table 2).

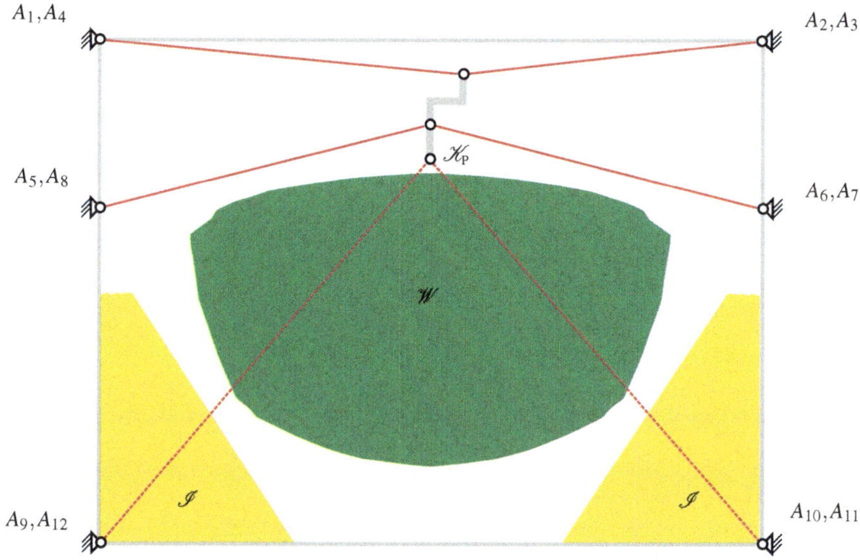

Fig. 5 Lateral view in the xz-plane of the region of cable–cable interference and total orientation workspace $\mathscr{W}$ of the endless Z12 robot and the region of cable–cable interference $\mathscr{I}$

Using the technique described above, the cable–cable interference was studied. For this analysis, the following geometric parameters were used for the endless Z12 design: $r_B = 2$, $r_P = 0.3$, $H_B = 3$, $h_B = 2$, $H_P = 0.5$, and $h_P = 0.2$. Note, that the z coordinates of the points $\mathbf{b}_5$–$\mathbf{b}_8$ are smaller in order to avoid collisions between the cable groups 1–4 and 5–8. It can be seen from Fig. 5 that the total orientation workspace and the region of interference is separated and thus cable–cable interference is avoided throughout the workspace.

5 Conclusion

In this paper, design archetypes of cable-driven parallel robots with a huge orientation workspace are proposed and studied. Surprisingly, such robots can execute a full rotation of their platform while maintaining tension in the cables and avoiding collisions amongst the cables. Also the size of the total orientation workspace is surprisingly large. However, the effect comes at the costs of employing a large number of actuators and additional challenges to cope with many cable. In the future, it is planned to experimentally study such designs. We conjecture that there exists an eight or even seven cable design that also has the unlimited rotation orientation workspace but until now no such configuration is known.

Acknowledgements The first authors would like to thank the German Research Foundation (DFG) for financial support of the project within the Cluster of Excellence in Simulation Technology (EXC 310/1) at the University of Stuttgart.

References

1. Bouchard, S., Moore, B., Gosselin, C.: On the ability of a cable-driven robot to generate a prescribed set of wrenches. J. Mech. Robot. **2**(1), 1–10 (2010)
2. Ebert-Uphoff, I., Voglewede, P.A.: On the connections between cable-driven parallel manipulators and grasping. In: IEEE International Conference on Robotics and Automation, pp. 4521–4526. New Orleans (2004)
3. Gouttefarde, M., Merlet, J.P., Daney, D.: Wrench-feasible workspace of parallel cable-driven mechanisms. In: IEEE International Conference on Robotics and Automation, pp. 1492–1497. Roma (2007)
4. Gouttefarde, M., Collard, J.F., Riehl, N., Baradat, C.: Geometry selection of a redundantly actuated cable-suspended parallel robot. IEEE Trans. Rob. **31**(2), 501–510 (2015)
5. Hassan, M., Khajepour, A.: Minimum-norm solution for the actuator forces in cable-based parallel manipulators based on convex optimization. In: IEEE International Conference on Robotics and Automation, pp. 1498–1503 (2007)
6. Merlet, J.P.: Parallel Robots, 2nd edn. Springer, Berlin (2006)
7. Miermeister, P., Pott, A.: Design of cable-driven parallel robots with multiple platforms and endless rotating axes. In: Proceedings of the Second Conference on Interdisciplinary Applications in Kinematics, pp. 21–29. Springer (2014)
8. Perreault, S., Cardou, P., Gosselin, C., Otis, M.J.D.: Geometric determination of the interference-free constant-orientation workspace of parallel cable-driven mechanisms. ASME J. Mech. Robot. **2**(3), 031016 (2010)
9. Pott, A.: Forward kinematics and workspace determination of a wire robot for industrial applications. In: Advances in Robot Kinematics (ARK), pp. 451–458. Springer (2008)
10. Pott, A.: An improved force distribution algorithm for over-constrained cable-driven parallel robots. In: Computational Kinematics, pp. 139–146. Springer (2013)
11. Verhoeven, R.: Analysis of the workspace of tendon-based stewart platforms. Ph.D. thesis, University of Duisburg-Essen, Duisburg, Germany (2004)
12. Verhoeven, R., Hiller, M.: Estimating the controllable workspace of tendon-based stewart platforms. In: Advances in Robot Kinematics (ARK), pp. 277–284. Springer (2000)

Elasto-Static Model for Point Mass Sagged Cable-Suspended Robots

Erika Ottaviano, Vincenzo Gattulli and Francesco Potenza

Abstract In this paper, a model is presented for the elasto-static problem of planar point mass robots suspended by m-cables. In particular, each cable configuration is described by an elastic catenary and static equations and compatibility conditions for the system are given, thus the $2m$ force reaction unknowns can be evaluated. The proposed formulation has been used to solve the direct problem and it is suitable for investigating the influence of elastic catenary on the end-effector exact positioning. The model allows evaluating the relation between end-effector position and the involvement of each cable in sustaining the payload.

1 Introduction

In recent years, cable-driven robots are attracting the attention of the scientific community as well as industry because their potentiality in applications related to manipulation for which large workspace, low inertia, high payload to robot mass ratio, transportability, economy in construction, and maintenance are required. Since cables are wounded around drums, the workspace may be very large [2] allowing the application to cable-driven scaffold systems for aircraft maintenance [14], cable-driven camera used for big entertainment or sport events [17], and contour crafting system [3]. A class of cable driven robots is the under-constrained or cable-suspended type, in which the moving platform acts in a crane configuration, [13]. Indeed, if all the attachment points on the fixed base are located above the workspace then cables do not clutter the portion of the robot workspace located below the platform. This

E. Ottaviano (✉)
DICeM University of Cassino, Via Di Biasio 43, 03043 Cassino (Fr), Italy
e-mail: ottaviano@unicas.it

V. Gattulli · F. Potenza
DICEAA - University of L'Aquila, Piazzale Pontieri, 1,
67040 Monteluco di Roio (AQ), Italy
e-mail: vincenzo.gattulli@univaq.it

F. Potenza
e-mail: francesco.potenza@univaq.it

J. Lenarčič and J.-P. Merlet (eds.), *Advances in Robot Kinematics 2016*,
Springer Proceedings in Advanced Robotics 4, DOI 10.1007/978-3-319-56802-7_37

occurrence reduces drastically the possible interference among cables, end-effector, and environment, but a crucial issue is that due to the geometric configuration of the system the pose capability is strictly related to the gravity and then to the solution of the static problem. Moreover, external disturbances on the end-effector determine complex dynamics involving cable vibrations.

Works on statics and dynamics of cable-suspended robots are reported in [4, 6, 14] evidencing the importance of suspended cable vibrations, which can be attenuated by active control, as shown in [7]. However, vibration occurrences may be effectively studied in fully-constrained manipulators also by simply considering cables as linear or nonlinear springs such as in [1, 5, 10]. For cable-suspended robots, kinetostatic models taking into account hefty elastic cables were proposed with discretized masses in [15] and with approximated cable static configuration description [8, 11].

In this paper, we present an exact model for the direct elasto-static problem of the planar point mass sagged cable-suspended robot with m-elastic cables with distributed mass as insight investigation of the results presented in [16].

2 Elasto-Static Model for Cable-Suspended Robots

In the following we will consider the elasto-static problem of a cable-driven manipulator with m cables and n DOFs, where E denotes a reference point of the moving platform that can be described with respect to a global reference frame OXY attached to the fixed base (Fig. 1).

An approximate solution, which constitutes the first trial for finding the solution of the exact nonlinear problem, is obtained by solving the kinetostatics of cable-suspended robots, based on the assumption of mass less inextensible cables, with the hypothesis that they are always in tension and can thus be treated as line segments representing bilateral constraints. In this case, cable configurations can be assumed as coincident with the chords denoted by vectors $\mathbf{r}_{E0i}$ of length $\overline{L}_{0i}$.

According to the mass less inextensible cable model, cable lengths can be computed as (for $i = 1, \ldots, m$)

$$\overline{L}_{0i}(\boldsymbol{r}_{E0}) = \| \boldsymbol{r}_{E0} - \boldsymbol{a}_i \|; \qquad \boldsymbol{\theta}_i(\boldsymbol{r}_{E0i}) = \cos^{-1}\left(\frac{\boldsymbol{X}_i \cdot \boldsymbol{r}_{E0i}}{\| \boldsymbol{X}_i \| \| \boldsymbol{r}_{E0i} \|}\right) \tag{1}$$

where $\boldsymbol{r}_{E0} = [x_{E0}, y_{E0}]^{\top}$ are the coordinates of center of gravity of the end-effector expressed in the fixed frame, $\overline{L}_{0i}$ and θ_i are respectively the length and angle for the i-th line segment cable, A_i are the attachments points. Accordingly, cable tensions can be evaluated by imposing the equilibrium at the end-effector by using the inverse of Moore–Penrose matrix [2], (when $n < m$).

However, a classical cable modeling approach permits to consider cable elasticity and sagged catenary assuming homogeneous properties and negligible flexural stiffness [9].

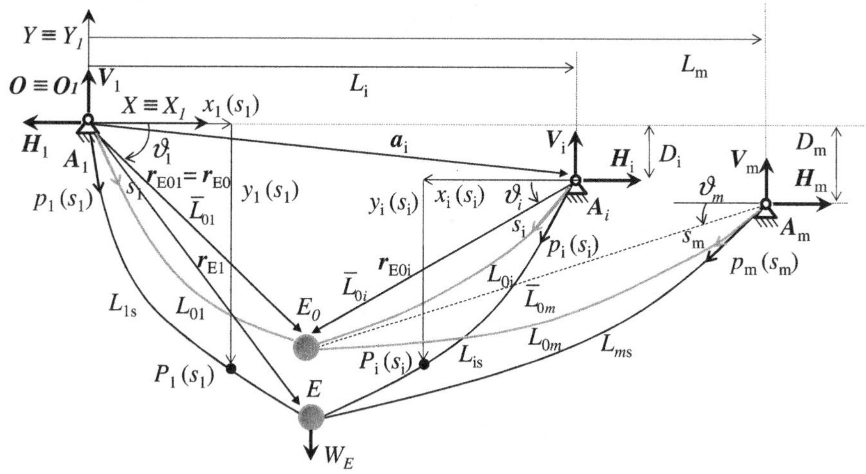

Fig. 1 Static configuration of a planar m-cables point mass manipulator

On this basis, here, a model describing the static configuration of a planar point mass manipulator suspended by m sagged cables, as depicted in Fig. 1, is derived.

The elastic catenary of each cable is described in a local frame $A_iX_iY_i$ through the curve $p_i(s_i)$, being s_i the curvilinear abscissa of the reference unstretched cable configuration going from the support at $s_i = 0$ to the final length $s_i = L_{0i}$.

The elastic catenary curves are described in the vertical plane, satisfying the following constraints

$$\left(\frac{dx_i}{dp_i}\right)^2 + \left(\frac{dy_i}{dp_i}\right)^2 = 1 \tag{2}$$

and the global equilibrium in the elastic configuration of each i-th cable particle point $P_i(s_i)$ at length coordinate $p_i(s_i)$, requires that

$$N_i \frac{dx_i}{dp_i} = H_i; \quad N_i \frac{dy_i}{dp_i} = V_i - w_i s_i \qquad i = 1, \ldots, m \tag{3}$$

being w_i the constant weight per unit natural length with $W_i = w_i L_{0i}$ as the total cable weight, and H_i and V_i are the unknown horizontal and vertical components of the support reaction at A_i, respectively (Fig. 1).

Adopting an exact kinematic formulation for the cable axial strain and linear elasticity of the material, the constitutive relation and finite cable axial strains are

$$N_i = \mathscr{E}_i \mathscr{A}_i \epsilon_i; \quad \epsilon_i = \frac{dp_i}{ds_i} - 1 \qquad i = 1, \ldots, m \tag{4}$$

where $\mathscr{E}$ is the Young's modulus and $\mathscr{A}$ is the cable's area.

Consequently, the two parametric planar curves S_i, exactly describing the i-th cable configurations in the local reference frame, whose map from the s_i-parametric domains to the vertical planes, are defined by the coordinate functions $x_i(s_i)$ and $y_i(s_i)$.

Assuming satisfied the geometric boundary conditions $x_i(0) = 0$ and $y_i(0) = 0$ at each supports $A_i(s_i = 0)$ for $i = 1, \dots m$, the following equations in the $2m$ unknown reactions H_i, V_i, are describing each elastic catenary

$$
\begin{aligned}
x_i(s_i) &= \frac{H_i}{\mathscr{E}_i \mathscr{A}_i} s_i + \frac{H_i}{w_i} \Psi_{1i}(s_i); \\
y_i(s_i) &= \frac{V_i}{\mathscr{E}_i \mathscr{A}_i} s_i - \frac{1}{2} \frac{w_i}{\mathscr{E}_i \mathscr{A}_i} s_i^2 + \frac{H_i}{w_i} \Psi_{2i}(s_i)
\end{aligned} \tag{5}
$$

in which

$$
\begin{aligned}
\Psi_{1i}(s_i) &= \operatorname{arcsinh}[\tan(\varphi_{Si})] - \operatorname{arcsinh}[\tan(\varphi_{si})]; \\
\Psi_{2i}(s_i) &= [1 + \tan^2(\varphi_{Si})]^{1/2} - [1 + \tan^2(\varphi_{si})]^{1/2}
\end{aligned} \tag{6}
$$

where

$$
\tan(\varphi_{Si}(s_i)) = \frac{V_i}{H_i}; \quad \tan(\varphi_{si}(s_i)) = \frac{V_i - w_i s_i}{H_i} \tag{7}
$$

On the basis of the given cable model description, the solution strategy, here adopted, for the static problem of the system follows the direct force method in which the $2m$ reactions H_i and V_i at points A_i are assumed as principal unknowns.

The unknowns V_i, H_i can be found solving simultaneously static and compatibility equations. Consequently, satisfying the equilibrium for the i-th cable as

$$
H_{iE} = H_i; \quad V_{iE} = V_i - w_i L_{0i} \tag{8}
$$

Static equations at the end-effector as described in Fig. 2, according to Eq. (8), are given as

$$
\sum_{i=1}^{m} H_i = 0; \quad \sum_{i=1}^{m} (V_i - w_i L_{0i}) - W_E = 0 \tag{9}
$$

while the relevant compatibility equations for the entire system relates the position of the end-effector in the elastic configuration to the position of each cable support (for $i = 2, \dots, m$, see Fig. 1), and it can be written in the global reference frame as

$$
\boldsymbol{r}_{Ei}(L_{01}) - \boldsymbol{r}_{Ei}(L_{0i}) = \boldsymbol{a}_i \tag{10}
$$

where $\boldsymbol{a}_i$ is the position vector of the support A_i in the global reference frame, and $\boldsymbol{r}_{Ei}$ is the vector describing the end-effector position, as

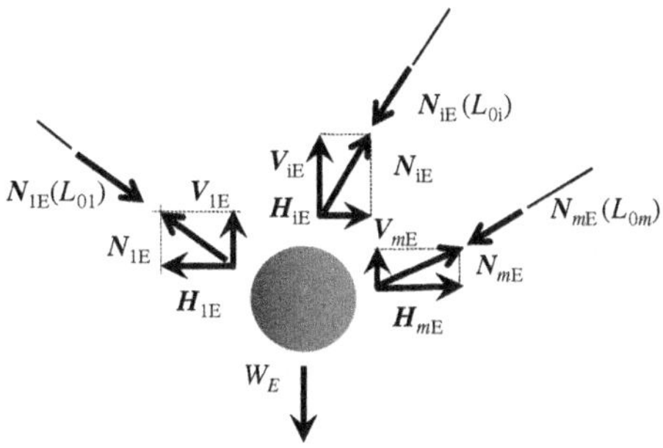

Fig. 2 Static equilibrium at the end-effector of the m-cable robot

$$\boldsymbol{a}_i = [L_i,\ D_i]^\top; \quad \boldsymbol{r}_{Ei}(L_{0i}) = [x_{Ei}(L_{0i}),\ y_{Ei}(L_{0i})]^\top \tag{11}$$

The projection of Eq. (10) for components and the use of the Eq. (5) to express the end effector position as function only of the H_i and V_i unknowns, furnishes the following $2m$-2 scalar equations

$$\begin{aligned}
&\frac{H_1 L_{01}}{\mathscr{E}_1 \mathscr{A}_1} + \frac{H_1}{w_1}\Psi_1(L_{01}) + \frac{H_i L_{0i}}{\mathscr{E}_i \mathscr{A}_i} + \frac{H_i}{w_i}\Psi_1(L_{0i}) = L_i;\\
&\frac{V_1 L_{01}}{\mathscr{E}_1 \mathscr{A}_1} - \frac{1}{2}\frac{w_1}{\mathscr{E}_1 \mathscr{A}_1}L_{01}^2 + \frac{H_1}{w_1}\Psi_2(L_{01}) - \frac{V_i L_{0i}}{\mathscr{E}_i \mathscr{A}_i} + \frac{1}{2}\frac{w_i}{\mathscr{E}_i \mathscr{A}_i}L_{0i}^2 - \frac{H_i}{w_i}\Psi_2(L_{0i}) = D_i
\end{aligned} \tag{12}$$

which complete the set of $2m$ nonlinear algebraic equations in $2m$ unknowns.

Equations (9) and (12) can be solved for any manipulator planar elasto-geometry. The direct problem can be completed determining the elastic catenary by (5)–(7) for each cable providing the end-effector position E and the stretched cable lengths L_{is}.

3 A 3-Cable Planar Point Mass Cable Suspended Robot

According to the proposed model a case of study with 3 cables is considered in Fig. 3 with the following data: $L_2 = 1$ m, $L_3 = 2$ m, $D_i = 0$, and cables made of a textile wire rope with an average nominal diameter of 4 mm and linear density of 1.12 g/m. The Young's modulus is equal to 50 MPa and it was determined experimentally, see [16]. The end-effector mass was set equal to 53.26 g.

The sensitivity of the solution to the cable lengths can be studied by using a monodimensional parameter Λ. Parameter Λ is the ratio between the unstretched cable length L_{0i} and the nominal chord $\overline{L}_{0i}$, which is known as the cable aspect ratio [9, 12].

The Λ parameter was introduced for a cable hanging between two fixed supports to allow a distinction between pre-tensioned ($\Lambda < 1$) and non-pre-tensioned ($\Lambda > 1$) cables as referred in [9]. In the following for cable-suspended robots, the Λ parameter

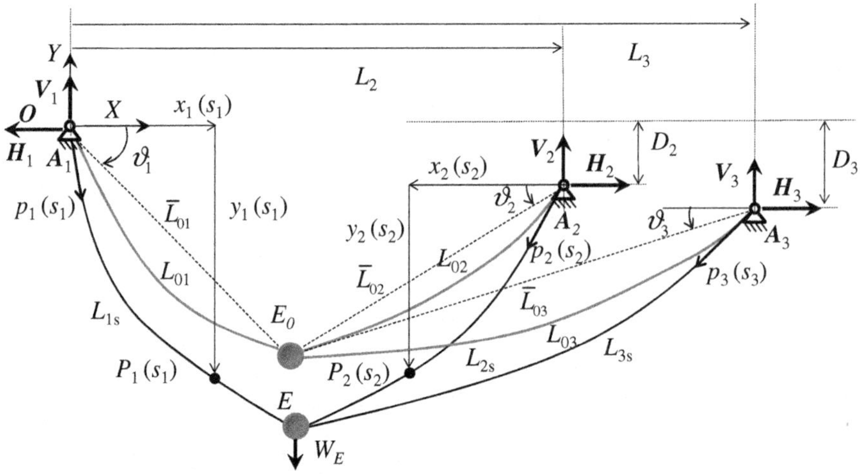

Fig. 3 Static configuration of a planar 3-cables point mass manipulator

has been used to assess the sensitivity of the end-effector positioning to variation of cable lengths with respect to the nominal chord.

It is important to point out that the working cable length may differ from the estimated value during maneuvers due to several factors. Consequently, the studied example is used to evaluate the end-effector position in three different cases. In the first one the input data of the problem (the assumed value of L_{0i} in the Eqs. (9) and (12)) are such that the unstretched cable lengths L_{0i} are equal to the nominal chords $\bar{L}_{0i}$ ($\Lambda = 1$), in the second case cable lengths are greater than the nominal chords ($\Lambda > 1$) and in the third case the lengths are less than the chords ($\Lambda < 1$).

Figure 4 shows selected static configurations in the workspace obtained solving the system of nonlinear equations (9) and (12) in the reaction unknowns and then using their values in the configuration equations (5). It is worth noting that only half of the workspace is spanned due to symmetry of the proposed example. Figure 4a is related to the case of $\Lambda = 1$. For these configurations, going towards to the left boundary of the workspace, center and right cables (2 and 3) show a non-negligible sag.

This feature is more evident for manipulator with cables of lengths greater than the nominal chord evaluated with respect to a pre-selected end effector position (overestimated case with $\Lambda > 1$ Fig. 4b and is less relevant in the case of cable with lengths smaller than the nominal chord (underestimated case with $\Lambda < 1$ in Fig. 4c.

Table 1 reports the cable tensions related to the given configuration in which the unstretched lengths of the cables L_{0i} have been selected on the basis of a nominal configurations of the end-effector within the workspace, as reported in Eq. (1).

In the three different selected cases the cable aspect ratios have used to consider different initial unstretched cable lengths as input data to the problem. The difference in the cable lengths with respect to the chords have been considered equal for each cable.

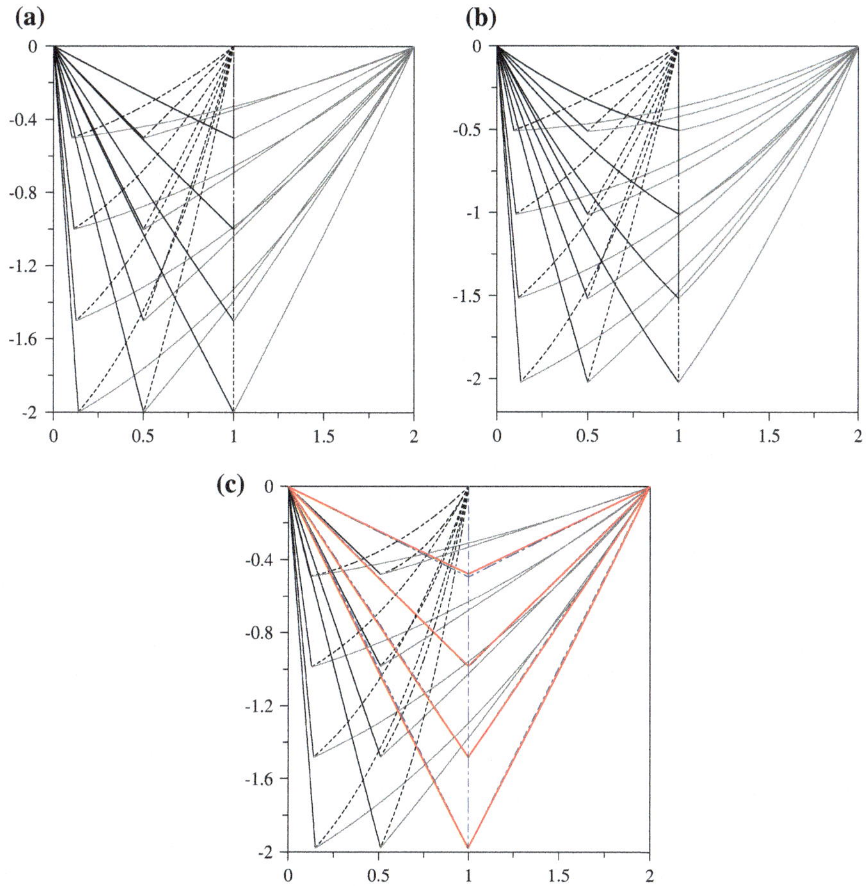

Fig. 4 Simulation results for the planar point mass manipulator with 3 sagged cables when: **a** $\Lambda = 1$; **b** $\Lambda = 1.01$; **c** $\Lambda = 0.99$

Analyzing the solutions of the elasto-static problem a series of consideration can be drawn.

First, the model confirms that close to the left boundary of the workspace the first cable sustains almost all the payload, while the other two cables have vertical components of the tension comparable to the cable weight, as reported in Table 1 (first four rows).

Second, an underestimation of the cable lengths, in the symmetric configurations of the case of study, brings the solution of the system composed by 6-system solving equations to furnish tension vectors for which the central cable results to be compressed. This occurrence is related to the vertical components of the tension vectors, which assume negative values, as it is evidenced in bold letters in Table 1. It is worth noting that $V_{2E} = N_{2E}$ for symmetry.

Table 1 Cable tensions (absolute values) for the 3-cables planar point mass manipulator

		$\Lambda = 1$			$\Lambda = 1.01$			$\Lambda = 0.99$		
x_{E0}	y_{E0}	N_{1E}	N_{2E}	N_{3E}	N_{1E}	N_{2E}	N_{3E}	N_{1E}	N_{2E}	N_{3E}
0.10	0.50	0.525	0.041	0.097	0.521	0.058	0.070	0.527	0.030	0.124
0.10	1.00	0.526	0.041	0.074	0.528	0.043	0.067	0.525	0.038	0.083
0.10	1.50	0.534	0.049	0.075	0.536	0.050	0.071	0.531	0.047	0.080
0.10	2.00	0.542	0.058	0.081	0.545	0.059	0.078	0.540	0.057	0.084
0.50	0.50	0.511	0.146	0.281	0.403	0.352	0.052	0.589	0.018	0.447
0.50	1.00	0.438	0.087	0.207	0.347	0.267	0.065	0.466	0.031	0.259
0.50	1.50	0.429	0.088	0.180	0.353	0.240	0.080	0.449	0.044	0.214
0.50	2.00	0.433	0.094	0.173	0.369	0.223	0.096	0.450	0.057	0.199
1.00	0.50	0.262	0.319	0.262	0.037	0.523	0.037	*0.631*	*0.0*	*0.631*
								17.41	**14.82**	17.41
1.00	1.00	0.281	0.186	0.281	0.052	0.514	0.052	*0.398*	*0.0*	*0.398*
								7.67	**10.19**	7.67
1.00	1.50	0.262	0.319	0.262	0.037	0.523	0.037	*0.34*	*0.0*	*0.34*
								3.86	**5.78**	3.86
1.00	2.00	0.293	0.126	0.293	0.088	0.496	0.088	*0.328*	*0.0*	*0.328*
								2.32	**3.492**	2.32

These results put into evidence that if all three cables possess equal reduced lengths with respect to the cable chord ($\Lambda = 0.99$) the central cable becomes slack for the central symmetric configurations of the workspace (reported as dash-dotted blue lines in Fig. 4c). In this case, the presence of the slack cable needs to be carefully considered in the model. Indeed, the correct solution has to be evaluated considering only two cables (red configurations in Fig. 4c), and new tension vectors should be evaluated as solutions of the 4-system solving equations. The new solutions in terms of cable tensions are reported in italics in Table 1.

4 Conclusions

In this paper, we have proposed a solving approach for elasto-statics of planar point mass cable suspended manipulator with m cables for which is considered elasticity and sag. According to the force method, the solving method utilizes both static and compatibility equations to evaluate the support reactions, which are the only minimal unknowns. In particular, the proposed compatibility conditions express the need of the solution to satisfy the given system geometry (system support relative positions), which is an input data of the problem. Hence, the end-effector positions are not given as an input, but they are used to express the compatibility conditions as function of the force unknowns. The proposed approach to the problem furnishes a set of $2m$ nonlinear equations in the $2m$ support reactions permitting to pursue analytical solutions in both fully- and over-constrained cases with a minimum number of unknowns, as it is typical in elasticity problems solved by force-based methods. Simulation results presented for a planar 3-cable point mass manipulator have shown the ability of the model to determine the cable tensions and consequently the exact system configuration. The relation among the positioning in the workspace, the cable lengths and tensions have evidenced the involvement of each cable in sustaining the payload and the cases in which one cable becomes slack (compressed).

References

1. Behzadipour, S., Khajepour, A.: Stiffness of cable-based parallel manipulators with application to stability analysis. J. Mech. Des. **128**(1), 303–310 (2006)
2. Bruckmann, T., Mikelsons, L., Brandt, T., Hiller, M., Schramm, D.: Wire robots part i kinematics, analysis & design. In: Ryu, J.-H. (ed.) Parallel Manipulators, New Developments. I-Tech Ed. and Pub., Vienna (2008)
3. Bosscher, P., Williams II, R.L., Bryson, L.S., Castro-Lacouture, D.: Cable-suspended robotic contour crafting system. Autom. Constr. **17**(1), 45–55 (2007)
4. Carricato, M., Merlet, J.P.: Stability analysis of underconstrained cable-driven parallel robots. IEEE Trans. Robot. **29**(1), 288–296 (2013)
5. Diao, X., Ma, O.: Vibration analysis of cable-driven parallel manipulators. Multibody Syst. Dyn. **21**, 347–360 (2009)
6. Du, J., Agrawal, S.K.: Dynamic modeling of cable-driven parallel manipulators with distributed mass flexible cables. ASME. J. Vib. Acoust. **137**(2) (2015). doi:10.1115/1.4029486
7. Gattulli, V., Alaggio, R., Potenza, F.: Analytical prediction and experimental validation for longitudinal control of cable oscillations. Int. J. Non-Linear Mech. **43**, 36–52 (2008)
8. Gouttefarde, M., Collard, J.-F., Riehl, N., Baradat, C.: Simplified static analysis of large-dimension parallel cable-driven robots. In: IEEE International Conference on Robotics and Automation (ICRA2012), pp. 2299–2305 (2012)
9. Irvine, H.: Cable Structures. MIT Press, Cambridge (1981)
10. Kawamura, S., Kino, H., Won, C.: High-speed manipulation by using a parallel wire-driven robots. Robotica **18**, 13–21 (2000)
11. Kozak, K., Zhou, Q., Wang, J.: Static analysis of cable- driven manipulators with non-negligible cable mass. IEEE Trans. Robot. **22**, 425–433 (2006)
12. Lepidi, M., Gattulli, V.: Static and dynamic response of elastic suspended cables with thermal effects. Int. J. Solids Struct. **49**(9), 1103–1116 (2012)
13. Merlet, J.-P., Dit-Sandretto, J.A.: The forward kinematics of cable-driven parallel robots with sagging cables. In: 2nd International Conference on Cable-driven Parallel Robots (CableCon), Duisburg, pp. 3–16 (2014)
14. Nguyen, D.Q., Gouttefarde, M.: Study of reconfigurable suspended cable-driven parallel robots for airplane maintenance. In: IEEE/RSJ 2014 Int. Conf. on Intelligent Robots and Systems (IROS 2014), Chicago, USA (2014)
15. Ottaviano, E., Castelli, G.: A study on the effects of cable mass and elasticity in cable-based parallel manipulators. In: Proceedings of the18th CISM-IFToMM Symposium on Robot Design, Dynamics, and Control, Springer Ed. Udine, pp. 149–156 (2010)
16. Ottaviano, E., Gattulli, V., Potenza, F., Rea, P.: Modeling a planar point mass sagged cable-suspended manipulator. In: The 14th IFToMM World Congress. Taipei (2015). doi:10.6567/IFToMM.14TH.WC.OS2.050
17. Skycam, [Online]: http://www.skycam.tv/

Determination of a Dynamic Feasible Workspace for Cable-Driven Parallel Robots

Lorenzo Gagliardini, Marc Gouttefarde and Stephane Caro

Abstract The dynamic equilibrium of the moving platform of a cable-driven parallel robot can be investigated by means of the *Dynamic Feasible Workspace* (DFW), which is the set of dynamic feasible moving platform poses. A pose is said to be dynamic feasible if a prescribed set of moving platform accelerations is feasible, with cable tensions lying in between given lower and upper bounds. This paper introduces an extended version of the DFW with respect to the one usually considered in the literature. Indeed, the improved DFW introduced in this paper takes into account: (i) The inertia of the moving platform; (ii) The external wrenches applied on the moving platform and (iii) The centrifugal and the Coriolis forces corresponding to a constant moving platform twist. Finally, the static, wrench-feasible, dynamic and improved dynamic workspaces of a spatial cable-suspended parallel robot are plotted in order to compare their sizes.

1 Introduction

Several industries, e.g. the naval and renewable energy industries, are facing the necessity to manufacture novel products of large dimensions and complex shapes. In order to ease the manufacturing of such products, the IRT Jules Verne promoted the investigation of new technologies. In this context, the CAROCA project aims at investigating the performance of *Cable Driven Parallel Robots* (CDPRs) to manufacture large products in cluttered industrial environments [5, 6]. CDPRs are a particular class of parallel robots whose moving platform is connected to the robot fixed base

L. Gagliardini
Robotics Team, IRT Jules Verne, Chemin du Chaffault, 44340 Bouguenais, France
e-mail: lorenzo.gagliardini@irt-jules-verne.fr

M. Gouttefarde
CNRS–LIRMM, 161 rue Ada, 34392 Montpellier Cedex 05, France
e-mail: marc.gouttefarde@lirmm.fr

S. Caro (✉)
CNRS–IRCCyN, 1, rue de la Noë, 44321 Nantes Cedex 03, France
e-mail: stephane.caro@irccyn.ec-nantes.fr

J. Lenarčič and J.-P. Merlet (eds.), *Advances in Robot Kinematics 2016*,
Springer Proceedings in Advanced Robotics 4, DOI 10.1007/978-3-319-56802-7_38

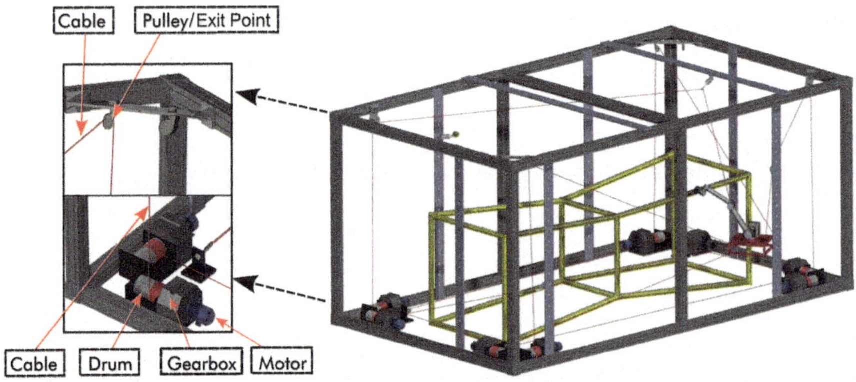

Fig. 1 Example of a CDPR design created in the framework of the IRT JV CAROCA project

frame by a number of cables as illustrated in Fig. 1. CDPRs have several advantages such as a high payload-to-weight ratio, a potentially very large workspace, and possibly reconfiguration capabilities.

The equilibrium of the moving platform of a CDPR is classically investigated by analyzing the CDPR workspace. In serial and rigid-link parallel robots, the workspace is commonly defined as the set of end-effector poses where a number of kinematic constraints are satisfied. In CDPRs, the workspace is usually defined as the set of poses where the CDPR satisfies one or more conditions including the static or the dynamic equilibrium of the moving platform, with the additional constraint of non-negative cable tensions. Several workspaces and equilibrium conditions have been studied in the literature.

The first investigations focused on the static equilibrium and the *Wrench Closure Workspace* (WCW) of the moving platform, e.g. [4, 7, 14–16]. Since cables can only pull on the moving platform, a pose belongs to the WCW if and only if any wrench can be applied by means of non-negative cable tensions. Feasible equilibria of the moving platform can also be analyzed using the *Wrench Feasible Workspace* (WFW) [2, 3, 9]. By definition, the WFW is the set of wrench feasible platform poses where a pose is wrench feasible when the cables can balance a given set of external moving platform wrenches while maintaining the cable tensions in between given lower and upper bounds. The *Static Feasible Workspace* (SFW) is a special case of the WFW, where the sole wrench induced by the moving platform weight has to be balanced [13]. The lower cable tension bound, τ_{min}, is defined in order to prevent the cables from becoming slack. The upper cable tension bound, τ_{max}, is defined in order to prevent the CDPR from being damaged.

The dynamic equilibrium of the moving platform can be investigated by means of the *Dynamic Feasible Workspace* (DFW). By definition, the DFW is the set of dynamic feasible moving platform poses. A pose is said to be dynamic feasible if a prescribed set of moving platform accelerations is feasible, with cable tensions lying in between given lower and upper bounds. The concept of dynamic workspace has already been investigated in [1] for planar CDPRs. Barrette et al. solved the dynamic equations of a planar CDPR analytically, providing the possibility to compute the boundary of the DFW. This strategy cannot be directly applied to spatial CDPRs due to the complexity of their dynamic model. In 2014, Kozlov studied in [12] the possibility to investigate the DFW by using a tool developed by Guay et al. for the analysis of the WFW [11]. However, the dynamic model proposed by Kozlov considers the moving platform as a point mass, neglecting centrifugal and Coriolis forces.

This paper deals with a more general definition of the DFW. With respect to the definitions proposed in [1, 12], the DFW considered in the present paper takes into account: (i) The inertia of the moving platform; (ii) The external wrenches applied on the moving platform; (iii) The centrifugal and the Coriolis forces corresponding to a given moving platform twist. The *Required Wrench Set* (RWS), defined here as the set of wrenches that the cables have to apply on the moving platform in order to satisfy its dynamic equilibrium, is calculated as the sum of these three contributions to the dynamic equilibrium. Then, the corresponding DFW is computed by means of the algorithm presented in [8] to analyze the WFW.

2 Dynamic Model

The CDPR dynamic model considered in this paper consists of the dynamics of the moving platform. A dynamic model taking into account the dynamics of the winches could also be considered but is not used here due to space limitations. Additionally, assuming that the diameters of the cables and the pulleys are small, the dynamics of the pulleys and the cables is neglected.

The dynamic equilibrium of the moving platform is described by the following equation

$$\mathbf{W}\boldsymbol{\tau} - \mathbb{I}_p\ddot{\mathbf{p}} - \mathbf{C}\dot{\mathbf{p}} + \mathbf{w}_e + \mathbf{w}_g = 0 \tag{1}$$

where $\mathbf{W}$ is the wrench matrix that maps the cable tension vector $\boldsymbol{\tau}$ into a platform wrench, and

$$\dot{\mathbf{p}} = \begin{bmatrix} \dot{\mathbf{t}} \\ \boldsymbol{\omega} \end{bmatrix} \quad \ddot{\mathbf{p}} = \begin{bmatrix} \ddot{\mathbf{t}} \\ \boldsymbol{\alpha} \end{bmatrix}, \tag{2}$$

where $\dot{\mathbf{t}} = [\dot{t}_x, \dot{t}_y, \dot{t}_z]^{\mathrm{T}}$ and $\ddot{\mathbf{t}} = [\ddot{t}_x, \ddot{t}_y, \ddot{t}_z]^{\mathrm{T}}$ are the vectors of the moving platform linear velocity and acceleration, respectively, while $\boldsymbol{\omega} = \left[\omega_x, \omega_y, \omega_z\right]^{\mathrm{T}}$ and $\boldsymbol{\alpha} = \left[\alpha_x, \alpha_y, \alpha_z\right]^{\mathrm{T}}$ are the vectors of the moving platform angular velocity and acceleration, respectively.

The external wrench $\mathbf{w}_e$ is a 6-dimensional vector expressed in the fixed reference frame $\mathscr{F}_b$ and takes the form

$$\mathbf{w}_e = \left[\mathbf{f}_e^{\mathrm{T}}, \mathbf{m}_e^{\mathrm{T}}\right]^{\mathrm{T}} = \left[f_x, f_y, f_z, m_x, m_y, m_z\right]^{\mathrm{T}} \tag{3}$$

f_x, f_y and f_z are the x, y and z components of the external force vector $\mathbf{f}_e$. m_x, m_y and m_z are the x, y and z components of the external moment vector $\mathbf{m}_e$, respectively. The components of the external wrench $\mathbf{w}_e$ are assumed to be bounded as follows

$$f_{min} \le f_x, f_y, f_z \le f_{max} \tag{4}$$

$$m_{min} \le m_x, m_y, m_z \le m_{max} \tag{5}$$

According to (4) and (5), the set $[\mathbf{w}_e]_r$, called the Required External Wrench Set (REWS), that the cables have to balance is a hyper-rectangle.

The *Center of Mass* (CoM) of the moving platform, G, may not coincide with the origin of the frame $\mathscr{F}_p$ attached to the platform. The mass of the platform being denoted by M, the wrench $\mathbf{w}_g$ due to the gravity acceleration $\mathbf{g}$ is defined as follows

$$\mathbf{w}_g = \begin{bmatrix} M\mathbf{I}_3 \\ \mathbf{M}\hat{\mathbf{S}}_p \end{bmatrix} \mathbf{g} \tag{6}$$

where $\mathbf{I}_3$ is the 3×3 identity matrix, $\mathbf{MS}_p = \mathbf{R}\left[Mx_p, My_p, Mz_p\right]^{\mathrm{T}}$ is the first momentum of the moving platform defined with respect to frame $\mathscr{F}_b$. The vector $\mathbf{S}_p = \left[x_p, y_p, z_p\right]^{\mathrm{T}}$ defines the position of G in frame $\mathscr{F}_p$. $\mathbf{M}\hat{\mathbf{S}}_p$ is the skew-symmetric matrix associated to $\mathbf{MS}_p$.

The matrix $\mathbb{I}_p$ represents the spatial inertia of the platform

$$\mathbb{I}_p = \begin{bmatrix} M\mathbf{I}_3 & -\mathbf{M}\hat{\mathbf{S}}_p \\ \mathbf{M}\hat{\mathbf{S}}_p & \mathbf{I}_p \end{bmatrix} \tag{7}$$

where $\mathbf{I}_p$ is the inertia tensor matrix of the moving platform, which can be computed by the Huygens-Steiner theorem from the moving platform inertia tensor, $\mathbf{I}_g$, defined with respect to the platform CoM

$$\mathbf{I}_p = \mathbf{R}\mathbf{I}_g\mathbf{R}^{\mathrm{T}} - \frac{\mathbf{M}\hat{\mathbf{S}}_p\mathbf{M}\hat{\mathbf{S}}_p}{M} \tag{8}$$

$\mathbf{R}$ is the rotation matrix defining the moving platform orientation and $\mathbf{C}$ is the matrix of the centrifugal and Coriolis wrenches, defined as

$$\mathbf{C}\dot{\mathbf{p}} = \begin{bmatrix} \hat{\omega}\hat{\omega}\mathbf{MS}_p \\ \hat{\omega}\mathbf{I}_p\omega \end{bmatrix} \tag{9}$$

where $\hat{\boldsymbol{\omega}}$ is the skew-symmetric matrix associated to $\boldsymbol{\omega}$.

3 Dynamic Feasible Workspace

3.1 Standard Dynamic Feasible Workspace

Studies on the DFW have been realised by Barrette et al. in [1]. The boundaries of the DFW have been computed for a generic planar CDPR developing the equations of its dynamic model. Since this method cannot be easily extended to spatial CDPRs, Kozlov proposed to use the method described in [11] in order to compute the DFW of a fully constrained CDPR [12]. The proposed method takes into account the cable tension limits τ_{min} and τ_{max} in checking the feasibility of the dynamic equilibrium of the moving platform for the following bounded sets of accelerations

$$\ddot{\mathbf{t}}_{min} \leq \ddot{\mathbf{t}} \leq \ddot{\mathbf{t}}_{max} \tag{10}$$

$$\boldsymbol{\alpha}_{min} \leq \boldsymbol{\alpha} \leq \boldsymbol{\alpha}_{max} \tag{11}$$

where $\ddot{\mathbf{t}}_{min}, \ddot{\mathbf{t}}_{max}, \boldsymbol{\alpha}_{min}, \boldsymbol{\alpha}_{max}$ are the bounds on the moving platform linear and rotational accelerations. These required platform accelerations define the so-called *Required Acceleration Set* (RAS), $[\ddot{\mathbf{p}}]_r$. The RAS can be projected into the wrench space by means of matrix $\mathbb{I}_p$, defined in (7). The set of wrenches $[\mathbf{w}_d]_r$ generated by this linear mapping is defined as the *Required Dynamic Wrench Set* (RDWS). No external wrench is applied to the moving platform. Accordingly, the DFW is defined as follows

Definition 1 A moving platform pose is said to be dynamic feasible when the moving platform of the CDPR can reach any acceleration included in $[\ddot{\mathbf{p}}]_r$ according to cable tension limits expressed by $[\boldsymbol{\tau}]_a$. The *Dynamic Feasible Workspace* is then the set of dynamic feasible poses, $[\mathbf{p}]_{DFW}$.

$$[\mathbf{p}]_{DFW} = \left\{(\mathbf{t}, \mathbf{R}) \in R^3 \times SO(3) \;:\; \forall \ddot{\mathbf{p}} \in [\ddot{\mathbf{p}}]_r,\ \exists \boldsymbol{\tau} \in [\boldsymbol{\tau}]_a \text{ s.t. } \mathbf{W}\boldsymbol{\tau} - \mathbf{A}\ddot{\mathbf{p}} = 0\right\} \tag{12}$$

In the definition above, the set of *Admissible Cable Tensions* (ACT) is defined as

$$[\boldsymbol{\tau}]_a = \{\boldsymbol{\tau} \mid \tau_{min} \leq \tau_i \leq \tau_{max},\ i = 1, \ldots, m\} \tag{13}$$

3.2 Improved Dynamic Feasible Workspace

The DFW described in the previous section has several limitations. The main drawback is associated to the fact that the proposed DFW takes into account neither the external wrenches applied to the moving platform nor its weight. Furthermore, the model used to verify the dynamic equilibrium of the moving platform neglects the Coriolis and the centrifugal wrenches associated to the CDPR dynamic model.

At a given moving platform pose, the cable tensions should compensate both the contribution associated to the REWS, $[\mathbf{w}_e]_r$, and the RDWS, $[\mathbf{w}_d]_r$. The components of the REWS are bounded according to (4) and (5) while the components of the RDWS are bounded according to (10) and (11).

The dynamic equilibrium of the moving platform is described by (1), where $\mathbf{C}$ is related to the Coriolis and centrifugal forces of the moving platform and $\mathbf{w}_g$ to its weight. These terms depend only on the pose and the twist of the moving platform. For given moving-platform pose and twist, these terms are constant.

Therefore, the DFW definition can be modified as follows.

Definition 2 A moving platform pose is said to be *dynamic feasible* when, for a given twist $\dot{\mathbf{p}}$, the CDPR can balance any external wrench $\mathbf{w}_e$ included in $[\mathbf{w}_e]_r$, while the moving platform can assume any acceleration $\ddot{\mathbf{p}}$ included in $[\ddot{\mathbf{p}}]_r$. The Dynamic Feasible Workspace is the set of dynamic feasible poses, $[\mathbf{p}]_{DFW}$.

$$[\mathbf{p}]_{DFW} \ : \ \forall \mathbf{w}_e \in [\mathbf{w}_e]_r, \forall \ddot{\mathbf{p}} \in [\ddot{\mathbf{p}}]_r, \ \exists \boldsymbol{\tau} \in [\boldsymbol{\tau}]_a \ \text{ s.t. } \ \mathbf{W}\boldsymbol{\tau} - \mathbb{I}_p\ddot{\mathbf{p}} - \mathbf{C}\dot{\mathbf{p}} + \mathbf{w}_e + \mathbf{w}_g = 0 \tag{14}$$

In this definition, we may note that the feasibility conditions are expressed according to three wrench space sets. The first set, $[\mathbf{w}_d]_r$, can be computed by projecting the vertices of $[\ddot{\mathbf{p}}]_r$ into the wrench space. For a 3-dimensional case study (6 DoF case), $[\ddot{\mathbf{p}}]_r$ consists of 64 vertices. The second component, $[\mathbf{w}_e]_r$, consists of 64 vertices as well. Considering a constant moving platform twist, the last component of the dynamic equilibrium, $\mathbf{w}_c = \{\mathbf{C}\dot{\mathbf{p}} + \mathbf{w}_g\}$, is a constant wrench. The composition of these sets generates a polytope, $[\mathbf{w}]_r$, defined as the *Required Wrench Set* (RWS). $[\mathbf{w}]_r$ can be computed as the convex hull of the Minkowski sum over $[\mathbf{w}_e]_r$, $[\mathbf{w}_d]_r$ and $\mathbf{w}_c$, as illustrated in Fig. 2:

$$[\mathbf{w}]_r = [\mathbf{w}_e]_r \oplus [\mathbf{w}_d]_r \oplus \mathbf{w}_c \tag{15}$$

Thus, Definition 2 can be rewritten as a function of $[\mathbf{w}]_r$.

Definition 3 A moving platform pose is said to be *dynamic feasible* when the CDPR can balance any wrench $\mathbf{w}$ included in $[\mathbf{w}]_r$. The Dynamic Feasible Workspace is the set of dynamic feasible poses, $[\mathbf{p}]_{DFW}$.

$$[\mathbf{p}]_{DFW} \ : \ \forall \mathbf{w} \in [\mathbf{w}]_r, \ \exists \boldsymbol{\tau} \in [\boldsymbol{\tau}]_a \ \text{ s.t. } \ \mathbf{W}\boldsymbol{\tau} - \mathbb{I}_p\ddot{\mathbf{p}} + \mathbf{w}_e + \mathbf{w}_c = 0 \tag{16}$$

The mathematical representation in (16) is similar to the one describing the WFW. As a matter of fact, from a geometrical point of view, a moving platform pose will be dynamic feasible if $[\mathbf{w}]_r$ is fully included in $[\mathbf{w}]_a$

$$[\mathbf{w}]_r \subseteq [\mathbf{w}]_a \tag{17}$$

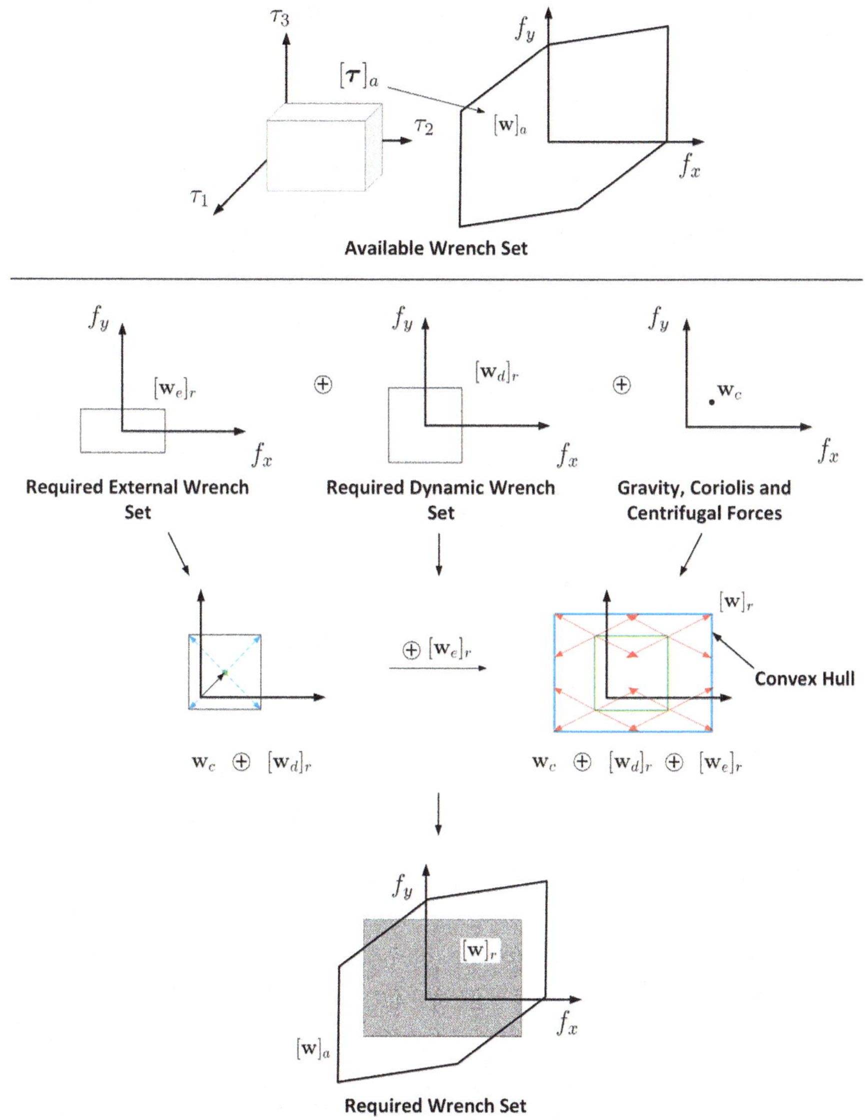

Fig. 2 Computation of the RWS $[\mathbf{w}]_r$. Example of a planar CDPR with 3 actuators and 2 translational DoF

Consequently, the dynamic feasibility of a pose can be verified by means of the hyperplane shifting method [3, 8, 11]. The distances between the facets of the available wrench set, $[\mathbf{w}]_a$, and the vertices of the RWS, $[\mathbf{w}]_r$, is verified according to the following inequality

$$\mathbf{C}\mathbf{w}_r \leq \mathbf{d}_d, \quad \forall \mathbf{w} \in [\mathbf{w}]_r \tag{18}$$

4 Case Study

This section aims at comparing the SFW, WFW, DFW and Improved DFW (IDFW) of the spatial suspended CDPR illustrated in Fig. 3. It has the layout the CoGiRo robot [10] and the size of the IRT JV CAROCA prototype. The robot consists of a moving platform connected to the fixed base frame by $m = 8$ cables. It is 7 m long, 4 m width and 3.5 m high. The maximum cable tension is equal to 6990 N. The moving platform consists of a parallelepiped. Its width, w_p, its length, l_p, and its height, h_p, are equal to 20 cm, 20 cm and 25 cm, respectively. The mass of the moving platform, M, is equal to 100 kg. In the proposed case study, the CoM of the platform, G, does not coincide with the origin O_p of frame $\mathscr{F}_p$, being $\mathbf{S} = [1\text{ cm}, 1\text{ cm}, 1\text{ cm}]^T$.

The volume inside the base frame has been discretized homogeneously into $n_p = 882$ points. Each point has been analysed in order to verify if the corresponding poses of the CDPR belong to the improved DFW. The analysis has been performed assuming that the moving platform is aligned with respect to the axes of frame $\mathscr{F}_b$. The linear velocity of the moving platform is equal to $\dot{\mathbf{t}} = [1\text{ m/s}, 1\text{ m/s}, 1\text{ m/s}]^T$ and its angular velocity is equal to $\boldsymbol{\omega} = [0.05\text{ rad/s}, 0.05\text{ rad/s}, 0.05\text{ rad/s}]^T$. The external wrenches acting on the moving platform are bounded as follows:

$$-100\,\text{N} \leq f_x, f_y, f_z \leq 100\,\text{N} \tag{19}$$

$$-1\,\text{Nm} \leq m_x, m_y, m_z \leq 1\,\text{Nm} \tag{20}$$

Similarly, the range of accelerations of the moving platform is limited according to the following inequalities:

$$-2\,\text{m/s}^2 \leq \ddot{t}_x, \ddot{t}_y, \ddot{t}_z \leq 2\,\text{m/s}^2 \tag{21}$$

$$-0.1\,\text{rad/s}^2 \leq \alpha_x, \alpha_y, \alpha_z \leq 0.1\,\text{rad/s}^2 \tag{22}$$

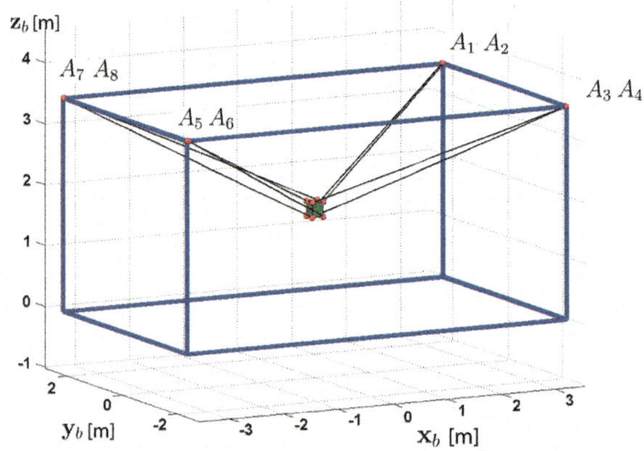

Fig. 3 Layout of the CoGiRo cable-suspended parallel robot [10] with the size of the IRT JV CAROCA prototype

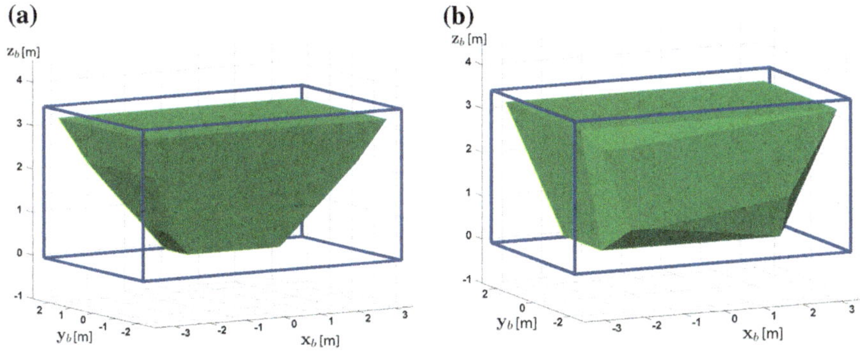

Fig. 4 **a** Improved DFW and **b** DFW of the CDPR under study covering 47.96 and 63.27% of its volume, respectively

Table 1 Comparison of SFW, WFW, DFW and $IDFW$ of the CDPR under study

Workspace type	SFW	WFW	DFW	$IDFW$
Covered volume of the CDPR	99.32%	79.25%	63.27%	47.95%

For the foregoing conditions, the improved DFW of the CDPR covers the 47.96% of its volume. Figure 4a illustrates the improved DFW of the CDPR under study.

The results have been compared with respect to the dynamic feasibility conditions described by Definition 1. By considering only the weight and the inertia of the moving platform, the DFW covers the 63.27% of the volume occupied by the DFW, as shown in Fig. 4b. Neglecting the effects of the external wrenches and the Coriolis forces, the volume of the DFW is 32% larger than the volume of the improved DFW.

Similarly, by neglecting the inertia of the CDPR and taking into account only the external wrenches $\mathbf{w}_e$, the WFW occupies the 79.25% of the CDPR volume. By taking into account only the weight of the moving platform, the SFW covers 99.32% of the CDPR volume. These results are summarized in Table 1.

5 Conclusion

This paper introduced an improved dynamic feasible workspace for cable-driven parallel robots. This novel workspace takes into account: (i) The inertia of the moving platform; (ii) The external wrenches applied on the moving platform and (iii) The centrifugal and the Coriolis forces induced by a constant moving platform twist. As an illustrative example, the static, wrench-feasible, dynamic and improved dynamic workspaces of a spatial suspended cable-driven parallel robot, with the dimensions of a prototype developed in the framework of the IRT JV CAROCA project, are traced. It turns out that the IDFW of the CDPR under study is respectively 1.32 times, 1.65 times and 2.07 times smaller than its DFW, WFW and SFW.

Acknowledgements This research work is part of the CAROCA project managed by IRT Jules Verne (French Institute in Research and Technology in Advanced Manufacturing Technologies for Composite, Metallic and Hybrid Structures). The authors wish to associate the industrial and academic partners of this project, namely, STX, DCNS, AIRBUS and CNRS.

References

1. Barrette, G., Gosselin, C.: Determination of the dynamic workspace of cable-driven planar parallel mechanisms. ASME J. Mech. Design **127**(2), 242–248 (2005)
2. Bosscher, P., Riechel, A., Ebert-Uphoff, I.: Wrench-feasible workspace generation for cable-driven robots. IEEE Trans. Robot. **22**(5), 890–902 (2006)
3. Bouchard, S., Gosselin, C.M., Moore, B.: On the ability of a cable-driven robot to generate a prescribed set of wrenches. In: Proceedings of the ASME International Design Engineering Technical Conferences and Computers and Information in Engineering Conference (IDETC/CIE 2008), pp. 47–58. Brooklyn, NY (2008)
4. Fattah, A., Agrawal, S.K.: Workspace and design analysis of cable-suspended planar parallel robots. In: Proceedings of the ASME International Design Engineering Technical Conferences and Computers and Information in Engineering Conference (IDETC/CIE 2002), pp. 1095–1103. Montreal, QC (2002)
5. Gagliardini, L., Caro, S., Gouttefarde, M., Wenger, P., Girin, A.: A reconfigurable cable-driven parallel robot for sandblasting and painting of large structures. In: Cable-Driven Parallel Robots. Mechanisms and Machine Science, vol. 32, pp. 275–291. Springer, Berlin (2015)
6. Gagliardini, L., Caro, S., Gouttefarde, M., Girin, A.: Discrete reconfiguration planning for cable-driven parallel robots. Mech. Mach. Theory **100**, 313–337 (2016). doi:10.1016/j.mechmachtheory.2016.02.014, http://www.sciencedirect.com/science/article/pii/S0094114X16000513
7. Gouttefarde, M., Gosselin, C.: Analysis of the wrench-closure workspace of planar parallel cable-driven mechanisms. IEEE Trans. Robot. **22**(3), 434–445 (2006)
8. Gouttefarde, M., Krut, S.: Characterization of parallel manipulator available wrench set facets. Advances in Robot Kinematics, pp. 475–484. Springer, Berlin (2010)
9. Gouttefarde, M., Daney, D., Merlet, J.P.: Interval-analysis-based determination of the wrench-feasible workspaceof parallel cable-driven robots. IEEE Trans. Robot. **27**(1), 1–13 (2011)
10. Gouttefarde, M., Collard, J.F., Riehl, N., Baradat, C.: Geometry selection of a redundantly actuated cable-suspended parallel robot. IEEE Trans. Robot. **31**(2), 501–510 (2015). doi:10.1109/TRO.2015.2400253
11. Guay, F., Cardou, P., Cruz, A., Caro, S.: Measuring how well a structure supports varying external wrenches. In: New Advances in Mechanisms, Transmissions and Applications. Mechanisms and Machine Science, vol. 17, pp. 385–392. Springer, Berlin (2014)
12. Kozlov, V.: A graphical user interface for the design of cable-driven parallel robots. Master's thesis, Ecole Centrale de Nantes (2014)
13. Pusey, J., Fattah, A., Agrawal, S., Messina, E.: Design and workspace analysis of a 6–6 cable-suspended parallel robot. Mech. Mach. Theory **39**(7), 761–778 (2004)
14. Roberts, R., Graham, T., Lippitt, T.: On the inverse kinematics, statics, and fault tolerance of cable-suspended robots. J. Robot. Syst. **15**(10), 581–597 (1998)
15. Stump, E., Kumar, V.: Workspaces of cable-actuated parallel manipulators. ASME J. Mech. Design **128**(1), 159–167 (2006)
16. Verhoeven, R., Hiller, M.: Estimating the controllable workspace of tendon-based Stewart platforms. Advances in Robot Kinematics, pp. 277–284. Springer, Berlin (2000)

Adaptive Human Robot Cooperation Scheme for Bimanual Robots

Bojan Nemec, Nejc Likar, Andrej Gams and Aleš Ude

Abstract The paper deals with human robot cooperation, where a bimanual robot and a human are handling highly deformable objects, such as a table cloth. The initial policy is demonstrated by kinesthetic guidance. For safety reasons, the robot operates in high compliance mode, which degrades the performance of trajectory tracking algorithm necessary to perform the demonstrated task. This problem was solved applying iterative adaptation scheme, which successfully diminishes tracking errors in just few adaptation cycles. The proposed approach was verified with a tablecloth placing task involving a human and a bimanual robot composed of two Kuka LWR-4 robot arms.

1 Introduction

Our research applies to dual arm robot manipulators. Nowadays, most of the bimanual control architectures are based on the concept of symmetric control [17], which enables portioning of the task to so-called absolute coordinates and relative coordinates [5]. This formalism allows to easily extend the control and adaptation algorithms developed for single arm systems, to bimanual robotic systems.

Adaptation is one of the key features of new generation of service and humanoid robots, aimed to cooperate with humans. An often applied paradigm for motion adaptation is reinforcement learning (RL) applying probabilistic algorithms [9], which can deal with high dimensionality spaces induced by parameterised policies [16].

B. Nemec (✉) · N. Likar · A. Gams · A. Ude
Humanoid and Cognitive Robotics Lab, Department of Automatics,
Biocybernetics and Robotics, Jožef Stean Institute, Ljubljana, Slovenia
e-mail: bojan.nemec@ijs.si

N. Likar
e-mail: nejc.likar@ijs.si

A. Gams
e-mail: andrej.gams@ijs.si

A. Ude
e-mail: ales.ude@ijs.si

J. Lenarčič and J.-P. Merlet (eds.), *Advances in Robot Kinematics 2016*,
Springer Proceedings in Advanced Robotics 4, DOI 10.1007/978-3-319-56802-7_39

Despite of these advances, humans can still learn much faster and more efficiently. One of promising paradigms to effectively speed up robot learning is also Iterative Learning Control (ILC). The main objective of ILC is to improve the behavior of the control system that operates repeatedly by iterative refinement of the feed-forward control input [3]. Due to its simplicity, effectiveness and robustness when dealing with repetitive operations, ILC is often applied in robotics [12].

The proposed approach aims to contribute to the adaptation aspect during human-robot cooperation (HRC). It is designed to handle the cooperative manipulation of highly deformable object, where the main issue beside the safety for human operators is also to assure that delicate objects made of fabric, such as table cloth, are not damaged during the operation. Therefore, robot arms apply an appropriate control scheme which assures high compliance. On the other hand, the tracking performance of highly compliant robots is often degraded due to the non-modeled robot dynamics (such as friction), as well as environment changes. This problem was solved applying iterative adaptation scheme, which successfully diminishes tracking errors in just few adaptation cycles. Our approach relies on previously proposed adaptation scheme for bimanual peg in hole task [10].

The paper is organized as follow. In Sect. 2 we outline kinematics and dynamics of a bimanual system and propose a solution, which completely decouples absolute and relative tasks at the velocity level. In Sect. 3 we briefly sketch the overall learning and adaptation scheme for a bimanual robot with ability to iteratively adapt to the environment changes and non-modeled robot dynamics. In Sect. 4 the experimental results of the proposed algorithm are presented. Discussion regarding bimanual adaptation properties of the HRC scheme is given in conclusion.

2 Bimanual Task Control

We apply a control scheme for a bimanual system, which is based on previously proposed coordinated task-space control framework [4]. It fully characterizes a cooperative operational space and allows the user to specify the task in terms of geometrically meaningful motion variables defined at the position/orientation level [1, 4]. Our modification of the originally proposed framework decouples both subspaces; motion in relative coordinates does not affect absolute coordinates and vice-versa. First, we define the common base coordinate systems τ_b for both subspaces, as illustrated in Fig. 1. From now on we use the notation where superscript j, $j \in \{1, 2, b\}$ denotes that the given quantity is specified relative to the coordinate system τ_j, while the subscript i, $i \in \{1, 2, a, r\}$ denotes the arm of a bimanual system and relative and absolute coordinates. Absolute and relative task coordinates are specified as

$$\mathbf{p}_a = \frac{1}{2}(\mathbf{p}_1^b + \mathbf{p}_2^b), \tag{1}$$

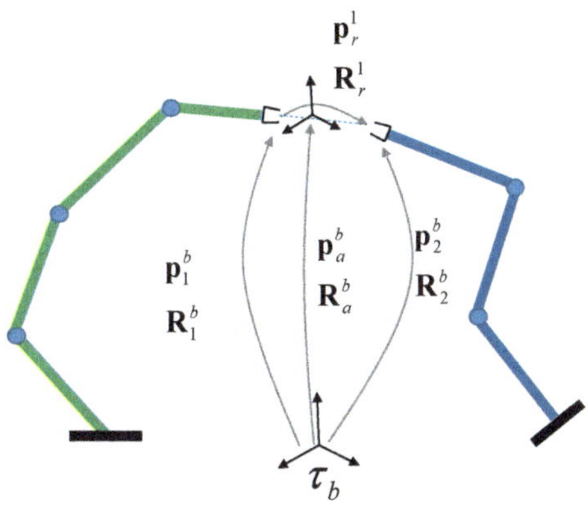

Fig. 1 Dual arm manipulator and the corresponding notation used in the paper

$$\mathbf{R}_a = \mathbf{R}_1^b \mathbf{R}_{\mathbf{k}_{21}^b}^b(\vartheta_{21}/2) \tag{2}$$

$$\mathbf{p}_r = \mathbf{R}_a{}^T(\mathbf{p}_2^b - \mathbf{p}_1^b), \tag{3}$$

$$\mathbf{R}_r = \mathbf{R}_1^{b\,T}\mathbf{R}_2^b, \tag{4}$$

where $\mathbf{p} \in \mathbb{R}^3$ applies to positions vector and $\mathbf{R} \in \mathbb{R}^{3\times 3}$ to rotational matrices. $\mathbf{k}_{21}$ and ϑ_{21} are the axis and angle that realize the rotation $\mathbf{R}_1^b$ to $\mathbf{R}_2^b$. Note that in relative coordinates definition we multiplied $(\mathbf{p}_2^b - \mathbf{p}_1^b)$ multiplied with $\mathbf{R}_a{}^T$. As this rotation compensates for the absolute coordinates rotation, it decouples relative and absolute coordinates. In quaternion notation, (2) and (4) are in the form

$$\mathbf{q}_a = \mathbf{q}_1^b * \mathbf{q}_{\mathbf{k}_{21}^b}^b, \tag{5}$$

$$\mathbf{q}_r = \mathbf{q}_2^1 = \bar{\mathbf{q}}_a^b * \mathbf{q}_2^b, \tag{6}$$

where the quaternion $\mathbf{q}_1^b \in \mathbb{R}^4$ and $\mathbf{q}_2^b \in \mathbb{R}^4$ expresses the rotation of the TCP of the first and the second robot in the common base coordinate frame τ_b, respectively. $\bar{\mathbf{q}}$ denotes conjugate quaternion and operator $*$ denotes quaternion product. $\mathbf{q}_{\mathbf{k}_{21}^b}^b$ denotes the unit quaternion corresponding to $\mathbf{R}_{\mathbf{k}_{21}^b}^b(\vartheta_{21}/2)$, which can be calculated from

$$\mathbf{q}_{\mathbf{k}_{21}^b}^b = \left(\cos\left(\frac{\vartheta_{21}}{4}\right), \mathbf{k}_{21}^b \sin\left(\frac{\vartheta_{21}}{4}\right)\right). \tag{7}$$

Our human robot cooperation scheme uses interactive forces for the demonstration and adaptation of the task. Therefore, it is necessary to calculate the corresponding forces and torques in both absolute and relative coordinates from wrist mounted sensors. The corresponding transformations are given by

$$\mathbf{f}_a = \mathbf{R}_1^b\,\mathbf{f}\,{}_1^1 + \mathbf{R}_2^b\,\mathbf{f}\,{}_2^2, \quad \mathbf{m}_a = \mathbf{R}_1^b\,\mathbf{m}\,{}_1^1 + \mathbf{R}_2^b\,\mathbf{m}\,{}_2^2, \tag{8}$$

$$\mathbf{f}_r = \frac{1}{2}(\mathbf{f}\,{}_1^1 - \mathbf{R}_r\,\mathbf{f}\,{}_2^2), \quad \mathbf{m}_r = \frac{1}{2}(\mathbf{m}\,{}_1^1 - \mathbf{R}_r\,\mathbf{m}\,{}_2^2), \tag{9}$$

where $\mathbf{f}\,{}_i^i \in \mathbb{R}^3$ and $\mathbf{m}\,{}_i^i \in \mathbb{R}^3$ denote the forces and torques measured at the i-th manipulator tool center point (TCP).

In order to control the robot, we have to map the desired relative and absolute task coordinates to the corresponding joint coordinates of both robots, denoted with $\boldsymbol{\theta} = [\boldsymbol{\theta}_1\ \boldsymbol{\theta}_2]^T \in \mathbb{R}^{(N_1+N_2)}$, where N_1 and N_2 is the number of joints of the first and the second robot, respectively. This transformation is obtained through relative and absolute geometrical Jacobian, which maps the corresponding translational and angular velocities to the joint velocities,

$$\begin{bmatrix} \dot{\mathbf{p}}_r \\ \omega_r \end{bmatrix} = \mathbf{J}_r\dot{\boldsymbol{\theta}}, \quad \begin{bmatrix} \dot{\mathbf{p}}_a \\ \omega_a \end{bmatrix} = \mathbf{J}_a\dot{\boldsymbol{\theta}}. \tag{10}$$

Absolute Jacobian is obtained from time derivative of (1)–(2),

$$\mathbf{J}_a = \left[\,\tfrac{1}{2}\mathbf{J}_1\ \tfrac{1}{2}\mathbf{J}_2\,\right]. \tag{11}$$

The derivation of the relative coordinates (3) becomes more complex as they are premultiplied with the inversed absolute rotation, yielding

$$\begin{aligned} \dot{\mathbf{p}}_r &= \mathbf{J}_{r,p}\theta = \dot{\mathbf{R}}_a{}^T(\mathbf{p}2 - \mathbf{p}1) + \mathbf{R}_a{}^T(\dot{\mathbf{p}}_2 - \dot{\mathbf{p}}_1) \\ &= -\mathbf{R}_a{}^T\mathbf{S}^T(\mathbf{p}_2 - \mathbf{p}_1)\omega_a + \mathbf{R}_a{}^T(\dot{\mathbf{p}}_2 - \dot{\mathbf{p}}_1) \\ &= \mathbf{R}_a{}^T(-\mathbf{J}_{1,p}\boldsymbol{\theta}_1 + \mathbf{J}_{2,p}\boldsymbol{\theta}_2 - \mathbf{S}^T(\mathbf{p}_2 - \mathbf{p}_1)\mathbf{J}_{a,\omega}\boldsymbol{\theta}), \end{aligned}$$

where we have taken into account the relation $\dot{\mathbf{R}}^T\mathbf{p} = -\mathbf{R}^T\mathbf{S}^T(\mathbf{p})\omega$ [14]. Subscripts $(.)_p$ and $(.)_\omega$ denote positional and rotational part of the Jacobian. $\mathbf{S}$ is well known skew-symmetric matrix. For the derivation of the rotational part of the relative Jacobian we take into account that the angular velocities are additive as long as they are expressed in the same coordinate frame. Hence, from the definition of relative coordinates (4) it follows $\omega_r = \mathbf{R}_a{}^T(\omega_2 - \omega_1)$ and $\mathbf{J}_{r,\omega} = \mathbf{R}_a{}^T(\mathbf{J}_{1,\omega} - \mathbf{J}_{2,\omega})$. Combining positional and rotational part of the relative Jacobian we obtain

$$\mathbf{J}_r = \begin{bmatrix} \mathbf{R}_a{}^T & 0 \\ 0 & \mathbf{R}_a{}^T \end{bmatrix} \begin{bmatrix} -(\mathbf{J}_{1,p} + \Lambda\frac{\mathbf{J}_{1,\omega}}{2}) & \mathbf{J}_{2,p} + \Lambda\frac{\mathbf{J}_{2,\omega}}{2}) \\ -\mathbf{J}_{1,\omega} & \mathbf{J}_{2,\omega} \end{bmatrix}, \tag{12}$$

where $\Lambda = \mathbf{S}^T(\mathbf{p}_2 - \mathbf{p}_1)$.

To control both absolute and relative coordinates, we define extended task coordinates $\mathbf{x}_e = [\mathbf{p}_a\ \mathbf{q}_a\ \mathbf{p}_r\ \mathbf{q}_r]^T$ and extended Jacobian $\mathbf{J}_e = [\mathbf{J}_a\ \mathbf{J}_r]^T$. The corresponding joint velocities are obtained from

$$\dot{\theta} = \mathbf{J}_e^+(\mathbf{v}_{e,d} + \mathbf{K}_k\mathbf{e}_e) + (\mathbf{I} - \mathbf{J}_e^+\mathbf{J}_e)\dot{\boldsymbol{\theta}}_0, \tag{13}$$

where $\mathbf{J}_e^+$ is the Moore–Penrose pseudo-inverse of the extended Jacobian $\mathbf{J}_e$, $\mathbf{v}_{e,d} \in \mathbb{R}^{12}$ are the desired extended translational and rotational velocities, $\mathbf{I}$ is identity matrix, $\mathbf{K}_k \in \mathbb{R}^{12\times12}$ is a diagonal matrix with the kinematic gains and $\mathbf{e}_e \in \mathbb{R}^{12}$ is the error between the desired and actual extended task coordinates, calculated as

$$\mathbf{e_e} = \begin{bmatrix} \mathbf{p}_{a,d} - \mathbf{p}_a \\ \log(\mathbf{q}_{a,d} * \bar{\mathbf{q}}_a) \\ \mathbf{p}_{r,d} - \mathbf{p}_r \\ \log(\mathbf{q}_{r,d} * \bar{\mathbf{q}}_r) \end{bmatrix}. \tag{14}$$

The rotational part of the error is expressed with logarithmic map log, which maps the quaternion describing the rotation between the desired and current pose to the rotation error vector. This mapping is defined as

$$\log(\mathbf{q}) = \log(v, \mathbf{u}) = \begin{cases} \arccos(v)\dfrac{\mathbf{u}}{\|\mathbf{u}\|}, & \mathbf{u} \neq 0 \\ [0, 0, 0]^{\mathrm{T}}, & \text{otherwise} \end{cases}. \tag{15}$$

Vector $\dot{\boldsymbol{\theta}}_0 \in \mathbb{R}^{(N_1+N_2)}$ is an arbitrary vector of joint velocities that is projected in the null-space of the primary task, selected in such a way that it optimizes an additional secondary task, i.e., obstacle avoidance, joint limit avoidance, singularity avoidance, etc. Note that the dimension of the extended task defined with $\mathbf{x}_e$ can be ≤ 12, which allows to exploit the additional degrees of redundancy for secondary task(s).

The desired joint positions $\boldsymbol{\theta}_d = [\boldsymbol{\theta}_{1,d}\ \boldsymbol{\theta}_{2,d}]^T$ are obtained with the numerical integration of (13) and passed to the joint controller of the robot. The joint controller has the form

$$\rho_{\mathrm{i}} = \mathbf{H}_i(\boldsymbol{\theta}_i)\ddot{\boldsymbol{\theta}}_{i,d} + \mathbf{C}_i(\dot{\boldsymbol{\theta}}_i, \boldsymbol{\theta}_i)\dot{\boldsymbol{\theta}}_{i,d} + \mathbf{G}_i(\boldsymbol{\theta}_i) - \mathbf{K}_p\mathbf{e}_{i,\theta} - \mathbf{K}_d\dot{\mathbf{e}}_{i,\theta}, \tag{16}$$

where $\rho_{\mathrm{i}} \in \mathbb{R}^{N_i}$ are joint torques supplied to the robot torque controller of the i-th robot arm, $i \in \{1, 2\}$, $\mathbf{H}_i \in \mathbb{R}^{N_i\times N_i}$ is the inertia expressed in joint space, $\mathbf{C}_i \in \mathbb{R}^{N_i\times N_i}$ is the matrix of Coriolis and radial forces, $\mathbf{G}_i \in \mathbb{R}^{Ni}$ is the gravitational vector, $\mathbf{K}_p$ and $\mathbf{K}_d \in \mathbb{R}^{N_i\times N_i}$ are diagonal positional and damping matrices and $\mathbf{e}_{i,\theta} \in \mathbb{R}^{N_i}$ denotes the tracking error in the joint coordinates. Gains of $\mathbf{K}_p$ define the desired compliance in joint space and gains of $\mathbf{K}_d$ are chosen in such a way that the close loop system is critically damped. Note that the control law (16) is designed using passivity framework [2].

3 Bimanual Task Learning and Adaptation

Learning by demonstration of a bimanual task can be simplified with the decomposition into relative and absolute coordinates. Both subtasks can be learned independently. The relative task can be demonstrated by moving only one arm, while the other remains fixed. Once the relative task is defined, the absolute part of the task is demonstrated by again moving only one arm, while the other follows accordingly in order to fulfill the previously demonstrated relative task.

In our research we used kinesthetic guidance for initial demonstration of a bimanual task. For this, the robot has to operate in the gravity compensation mode and the task is demonstrated by manual guidance of a single arm, as explained previously. Absolute and relative coordinates are recorded as a time series

$$\mathcal{G}_r = \{\mathbf{p}_r(k), \mathbf{q}_r(k)\}, \; \mathcal{G}_a = \{\mathbf{p}_a(k), \mathbf{q}_a(k)\} \mid_{k=0}^{T}, \tag{17}$$

where T denotes the number of recorded samples. They are calculated form the Cartesian coordinates of both robot arms using (1), (3), (5) and (6).

The learned absolute and relative coordinates are then passed to the bimanual kinematic control (13) as the desired values. Initially demonstrated absolute trajectory might not be perfect or might need additional modification when the human cooperates with the robot. Here we consider the case when the task is repeated more than once. In each repetition cycle the human operator can modify the trajectory from the previous cycle using interactive forces. The modification of absolute task is accomplished with adaptation of the absolute forces and torques (8) using admittance force/torque control law

$$\mathbf{p}_{a,l}(k) = \mathbf{p}_{a,l-1}(k) + \mathbf{K}_f \mathbf{f}_{a,l}(k), \tag{18}$$

$$\mathbf{q}_{a,l}(k) = \mathbf{q}_{a,l-1}(k) * \mathbf{K}_m \mathbf{m}_{a,l}(k). \tag{19}$$

Index l denotes the task repetition cycle and $\mathbf{K}_f, \mathbf{K}_m \in \mathbb{R}^{3\times3}$ are diagonal force and torque adaptation matrices. Note that when multiplying 3-D vectors with quaternions like in (19), vectors are interpreted as quaternions with scalar part equal to zero. Using the above adaptation law, the human operator can modify the previously learned trajectory whenever it exerts forces and torques to the common manipulated object. In order to suppress sensor noise, a threshold is usually applied to the measured force and torque signals.

During the task execution, the tracking of relative coordinates is degraded due to the low gains $\mathbf{K}_p$ in the joint controller (16), which are needed to achieve compliant behavior of the robot. To overcome this problem, we propose iterative controller in the form

$$\mathbf{p}_{r,l}(k) = \mathbf{Q}(\mathbf{p}_{r,l-1}(k) + \mathbf{L}\mathbf{e}_{p,l-1}(k+1)) \tag{20}$$

$$\mathbf{q}_{r,l}(k) = exp(L\mathbf{e}_{q,l-1}(k+1)) * \chi(\mathbf{q}_{r,l-1}(k), Q) \tag{21}$$

where $\mathbf{e}_p(k) = \mathbf{p}_r(k) - \mathbf{p}_{r,m}(k)$, $\mathbf{e}_q(k) = \bar{\mathbf{q}}_r(k) * \mathbf{q}_{r,m}(k)$, $\mathbf{Q} \in \mathbb{R}^{3\times3}$ and $\mathbf{L} \in \mathbb{R}^{3\times3}$ are diagonal matrices with ILC gains. Operator $\chi(\mathbf{q}, Q)$ denotes scaling of unit quaternion $\mathbf{q}$ by scalar Q, implemented by transforming quaternion to the angle-vector notation, scaling the angle and calculating quaternion back from the angle-vector representation. Scalars Q and L in (21) are ILC gains with the same role as diagonal matrices **Q** and **L** in (20). Subscript $(.)_m$ denotes measured signals. Operator exp(.) is defined as

$$\exp(\mathbf{r}) = \begin{cases} \cos(\|\mathbf{r}\|) + \sin(\|\mathbf{r}\|)\dfrac{\mathbf{r}}{\|\mathbf{r}\|}, \ \mathbf{r} \neq 0 \\ 1 + [0, 0, 0]^{\mathrm{T}}, \ \text{otherwise.} \end{cases} \tag{22}$$

4 Experimental Evaluation

The performance of the proposed human robot cooperation scheme was evaluated on the table cloth placing task. Bimanual robot composed of two Kuka LWR-4 robot arms holds one side and the human operator holds the other side of the table cloth. The setup is shown in Fig. 2. The width of the tablecloth was initially demonstrated with kinesthetic guidance of one robot arm; it defines the relative part of the task. Similarly, the absolute part of the task was demonstrated with kinesthetic guidance of one robot arm, while the other maintained the previously demonstrated relative part of the task, i.e., the desired distance between the arms.

Initially demonstrated absolute task was adapted by proprioceptive sensing. In our scheme, absolute forces determine absolute positional velocities and absolute torques determine absolute rotational velocities. During the task execution, the robot adapts its absolute coordinates using (18) and (19) in order to minimize the interaction forces. In our implementation, we added an offset force of 10 N in the -X direction and a small offset force in -Z direction in order to allow displacement of the absolute coordinates in -X and -Z direction, respectively (see Fig. 2). This was necessary since with highly deformable object such as the table cloth we can exert forces only by pulling and not by pushing.

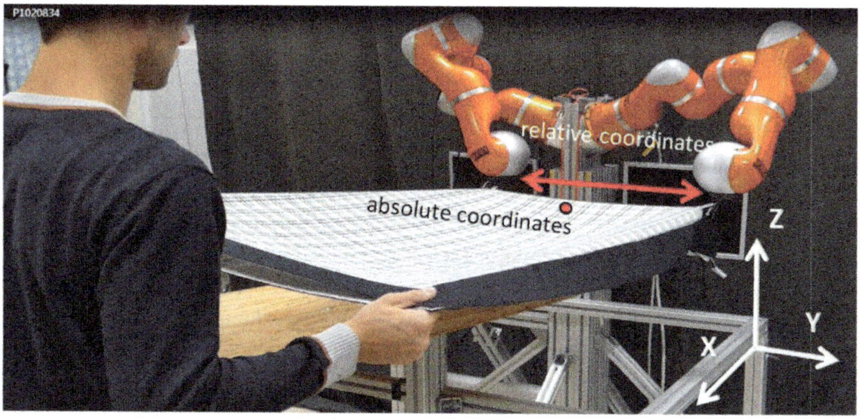

Fig. 2 Experimental platform for bimanual table cloth placing

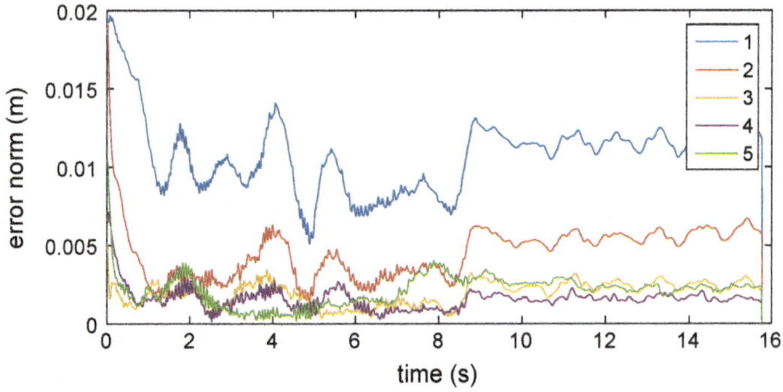

Fig. 3 Evolution of the position tracking error norm of relative coordinates through five repetitions (denoted with 1–5) of the same task

During the execution of the demonstrated absolute task, the robot was not able to precisely maintain the desired relative coordinates due to the arm compliance and the interaction forces. In order to overcome this problem, we applied ILC control scheme (20) and (21) which iteratively modified the controlled relative coordinates, resulting in perfect tracking of the desired relative coordinates. In this experiment, we used the following settings: $\mathbf{Q} = \mathbf{I}$, $\mathcal{Q} = 1, \mathbf{L} = 0.8\,\mathbf{I}$, $\mathcal{L} = 0$. Figure 3 shows how $L2$ norm of the tracking error of the relative coordinates diminished in five iterations. Note that in this experiment we did not explicitly control the internal (relative) forces. Rather, we controlled the relative positions, as the internal forces were very low during this experiment. Moreover, the measurement of the relative forces was not precise enough, since Kuka LWR-4 robot estimates TCP forces from the joint torques measurements. However, the extension of the control law (20) and (21) in order to control also the internal forces is straightforward.

5 Conclusions

In the paper we proposed a new adaptive human-robot cooperation scheme for bimanual robot systems. The main advantage of the proposed algorithms is that the adaptation acts separately in relative and absolute coordinates. This scheme does not require information about interaction forces for motion coordination. However, interaction forces can still be used to compensate for robot and environment model errors. In our previously presented bimanual HRC schemes [6, 11] robot arms were considered as independent agents, where adaptation and coordination of both arms was accomplished using force interaction, similar as in master-slave bimanual approaches [15, 18]. As such, they are not appropriate for handling of highly deformable objects.

In this work, we introduced a modified definition of the relative task (3), which results in an additional compensation term in the relative Jacobian (12) with respect to the one introduced in [5]. We believe that this definition properly decouples absolute and relative coordinates. Consequently, the proposed control law results in improved tracking of relative coordinates and assures proper mapping to the Jacobian null-space, when additional secondary tasks which exploit kinematic redundancy of the overall system are applied.

The adaptation of relative coordinates relies on ILC framework, which effectively improves initial demonstrated policy through task repetitions. An important issue is also the robustness of the control/adaptation scheme. In human-robot cooperation, the controller has to compensate also for the stochastic disturbances, induced by a human. Previous results [13], as well as our experiments, demonstrate the robustness of the ILC based controller against such disturbances. Another beneficial property of ILC is incremental policy adaptation, determined with the gain $\mathbf{L}$ in (20). Although similar policy could be obtained also with classical high gain feedback control [7], ILC is still favorable since the adaptation is independent form the feedback control. Thus, low feedback gains can be applied, as high gain controllers are not appropriate for robots interacting with humans due to the safety reasons [8].

Acknowledgements This work was supported by the Slovenian Research Agency grant J2-7360, "Learning and autonomous adaptation of dual arm assembly and service tasks".

References

1. Adorno, B., Fraisse, P., Druon, S.: Dual position control strategies using the cooperative dual task-space framework. In: IEEE/RSJ International Conference on Intelligent Robots and Systems (IROS), pp. 3955–3960. Taipei, Taiwan (2010)
2. Albu-Schäffer, A., Ott, C., Hirzinger, G.: A unified passivity-based control framework for position, torque and impedance control of flexible joint robots. Int. J. Robot. Res. **26**(1), 23–39 (2007)
3. Bristow, D., Tharayil, M., Alleyne, A.: A survey of iterative learning control. IEEE Control Syst. Mag. **26**(3), 96–114 (2006)
4. Caccavale, F., Chiacchio, P., Chiaverini, S.: Task-space regulation of cooperative manipulators. Automatica **36**, 879–887 (2000)
5. Chiacchio, P., Chiaverini, S.: Kinematic Control of Dual-Arm Systems. Springer, Berlin (1998)
6. Gams, A., Nemec, B., Petric, T., Ude, A.: Coupling of trajectories for human-robot cooperative tasks. In: Lenarcic, J., Khatib, O. (eds.) Advances in Robot Kinematics. Springer, Berlin (2014)
7. Goldsmith, P.B.: Brief on the equivalence of causal LTI iterative learning control and feedback control. Automatica **38**(4), 703–708 (2002)
8. Gribovskaya, E., Kheddar, A., Billard, A.: Motion learning and adaptive impedance for robot control during physical interaction with humans. In: 2011 IEEE International Conference on Robotics and Automation (ICRA), pp. 4326–4332 (2011)
9. Kober, J., Bagnell, D., Peters, J.: Reinforcement learning in robotics: a survey. Int. J. Robot. Res. **11**, 1238–1274 (2013)
10. Likar, N., Nemec, B., Žlajpah, L., Ando, S., Ude, A.: Adaptation of bimanual assembly tasks using iterative learning framework. In: 15th IEEE-RAS International Conference on Humanoid Robots (Humanoids), 2015, pp. 771–776 (2015)
11. Nemec, B., Gams, A., Denisa, M., Ude, A.: Human-robot cooperation through force adaptation using dynamic motion primitives and iterative learning. In: IEEE International Conference on Robotics and Biomimetics (ROBIO), pp. 1439–1444 (2014)

12. Norrlöf, M.: An adaptive iterative learning control algorithm with experiments on an industrial robot. IEEE Trans. Robot. Autom. **18**(2), 245–251 (2002)
13. Norrlöf, M., Gunnarsson, S.: Disturbance aspects of iterative learning control. Eng. Appl. Artif. Intell. **14**(1), 87–94 (2001)
14. Siciliano, B., Sciavicco, L., Villani, L., Oriolo, G.: Robotics: Modelling, Planning and Control. Springer, Berlin (2009)
15. Tarn, T., Bejczy, A., Yun, X.: Coordinated control of two robot arms. In: IEEE International Conference on Robotics and Automation. Proceedings., vol. 3, pp. 1193–1202 (1986)
16. Theodorou, E.A., Buchli, J., Schaal, S.: A generalized path integral control approach to reinforcement learning. J. Mach. Learn. Res. **11**, 3137–3181 (2010)
17. Uchiyama, M., Dauchez, P.: A symmetric hybrid position/force control scheme for the coordination of two robots. In: IEEE International Conference on Robotics and Automation (ICRA), vol. 1, pp. 350–356. Philadelphia, PA (1988)
18. Zheng, Y., Luh, J.: Joint torques for control of two coordinated moving robots. IEEE Int. Conf. Robot. Autom. **3**, 1375–1380 (1986)

Influence of the Wind Load in the Trolley-Payload System with a Flexible Hoist Rope

Jianjie Zhang and Gabriel Abba

Abstract The anti-sway controllers are widely discussed due to the increasing requirements of crane automation in seaports. In this paper, the dynamic model of the trolley-payload pendulum system is put forward considering the flexibility and damping of the hoist rope as well as the wind load as the external excitation. As indicated from the simulation, the wind load increases both the static and fluctuating part of the response of sway angle; the flexibility of the hoist rope cannot be ignored especially near the destination of the final position of the payload. As inferred from the results, the sway angle is the main source of the position error of the payload in both horizontal and vertical direction.

1 Introduction

As the most important equipment in seaports, the Ship-To-Shore quayside container crane (STS) affects the total efficiency for harbor operation. Traditionally, the most skilled staffs are specially chosen to operate the STS in order to obtain the best performance of the equipment.

As shown in Fig. 1, the photo shows STSs co-operating together to load and unload a Post-Panamax container vessel. The cranes move to a particular position to be ready for the loading or unloading operation with the gantry motion only. After the crane is positioned, the gantry stands still and only the hoist and trolley devices are moving.

With the increasing requirements of the seaport automation, the research and development of the crane automatic operating is widely discussed especially in engineering application. As an important aspect, the anti-sway control strategy is the most popular branch in this field. By analyzing latest papers and articles in three years, Singhose W. and his group uses model reference control method to control several

J. Zhang
LCFC, ENSAM, Arts et Métiers ParisTech, Paris, France
e-mail: samonzhang1@gmail.com

G. Abba (✉)
LCFC, National Engineering School of Metz, Metz, France
e-mail: abba@enim.fr

J. Lenarčič and J.-P. Merlet (eds.), *Advances in Robot Kinematics 2016*,
Springer Proceedings in Advanced Robotics 4, DOI 10.1007/978-3-319-56802-7_40

Fig. 1 The photo of Ship-To-Shore quayside container cranes (STS)

types of cranes in anti-sway applications [1–3]. He, B. et al. developed an anti-sway controller for overhead cranes based on the analysis of the Lyapunov method [5]. Liu, X. et al. designed the novel controller in anti-sway application [6]. Viet, L.D. discussed the sway reduction by radial spring and damper [19]. Smoczek, J. et al. introduced fuzzy controller into the anti-sway application [16, 17]. Schaper, U. et al. introduced an load position estimation method in gantry crane application [13]. Suzuki, K. et al. designed an anti-sway controller with the equivalent load as feedback [18]. Sato, K. et al. discussed the influence of the wind load in anti-sway control [12]. Park, J., Kwon et al. developed a method to generate anti-sway trajectory for incompletely restrained system [10].

Although within different application background, the trolley-payload system can be regarded as special under-actuated cable-driven parallel robot and researches on this topic are referenced. Gouttefarde M. et al. modeled and analyzed several types of redundant and reconfigurable cable-driven robots [4, 9], and discussed the influence of cable mass [11]. Pott A. et al. investigated a cable-driven robot and modeled the elastic cable and hysteresis effect in the system [8]. Merlet J.P. analyzed kinematics of cable-driven robots taking sagging cable into account [7].

As mentioned in the above references, regardless what particular control methods the researchers are applied, two assumptions in dynamic models can be inferred: (a) the sway angle is induced mainly from the trolley motion (paying little attention to the environmental load); (b) in anti-sway application, the hoist rope is rigid.

Nevertheless, other papers in civil engineering fields report the elastic elongation and damping is essential in the stability of structures [14]. In mechanical engineering applications, the elastic elongation always needs to be paid attention to as mentioned by many maintenance engineers during their daily work [20]. However, there are rare papers or technical reports discussing the influence of the flexibility of the hoist rope as well as the influence of the wind load in anti-sway control methods.

In this paper, the wind load is applied as an external excitation into the dynamic model of a trolley-payload pendulum system with a flexible hoist rope. Dynamic response of the system is discussed in both the time and frequency domain.

2 Modeling of the Outdoor Anti-sway System

2.1 Problem Specifications and Assumptions

For a crane with bridge structure at large seaports, the operation of loading and unloading a container vessel is always a motion in the vertical plane. Only the hoist motion and the trolley moving are considered in this issue. The rope weights about zero point nine percent of the payload and about four percent of the empty spreader, so in this system, the self-weight of rope is ignored.

The assumption of the model for anti-sway system is that:

1. The concerned motion of the system is in the vertical plane;
2. The hoisting rope is flexible and mass-less;
3. The wind load is source of the external excitation.

As shown in Fig. 2, the system is composed by Trolley and Payload these two motion parts. The payload is connected with a steel hoisting rope to the trolley traveling along the rail. The external excitation is mostly from the wind load on the payload.

As shown in Fig. 2, a global Cartesian coordinate system is defined as $O - xy$, whose origin is at the seaside limit (as shown the left direction in the figure) of the trolley. So, the position of the trolley and the payload are defined in global Cartesian coordinate system as (x_{tr}, y_{tr}) and (x_p, y_p); the length of the hoisting rope is l and the sway angle α which is the angle between hoisting rope and vertical.

The length of connect rope l is the sum of rigid rope length l_r and flexible elongation l_f; l_r may change according to the input of desired position. l_f is the rope elastic

Fig. 2 Trolley-payload coordinate system

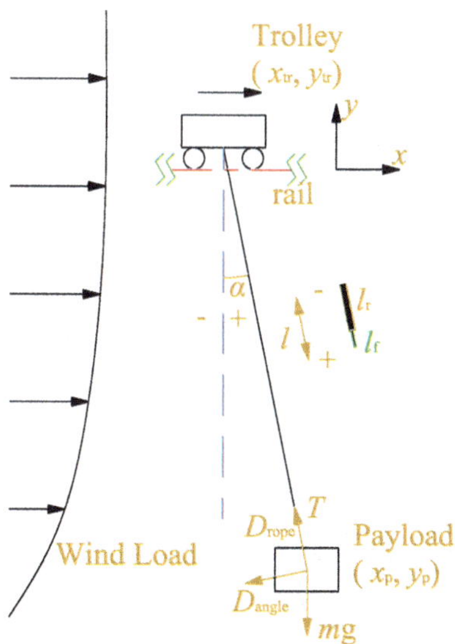

elongation which is a function of tension force T. $D_{\rm rope}$ is the damping force caused by ropes elongation; and $D_{\rm angle}$ is the damping force caused by sway motion.

In this model, we have the assumption that: (1) the rope is flexible so that the length of rope is a function of both the input signal $l_{\rm r}$ and tension force T; (2) the rope damping force is taken into account.

2.2 Wind Load Simulation

In wind engineering, the wind load on cranes and the payloads is concluded as the wind near ground which is affected by the friction on ground. So the wind velocity can be regarded as a sum of average and fluctuating wind as listed in Eq. (1).

$$v_{\rm wind} = \widetilde{v}_{\rm wind} + \bar{v}_{\rm wind} \tag{1}$$

In which, the average wind velocity is a function of height defined in Eq. (2)

$$\bar{v}_{\rm wind} = f(h) = \bar{v}_{\rm wind}({\rm h_{st}}) \cdot ({\rm h/h_{st}})^{\alpha_{\rm wind}} \tag{2}$$

where, $\bar{v}_{\rm wind}({\rm h_{st}})$ is the average wind velocity at a standard height, ${\rm h_{st}} = 10\,{\rm m}$ is used in this case as widely applied in wind engineering researches; h is height of payload; $\alpha_{\rm wind}$ is ground roughness coefficient whose value is saved in design standards.

The fluctuating wind velocity varies during time to time whose average is zero. The time history of the wind load is calculated according to the harmonic superposition method shown in Eq. (3).

$$\widetilde{v}_{\rm wind} = \sum_{n_i=n_{low}}^{n_{up}} \sqrt{S(h, n_i)} \cdot sin(n_i \cdot t + \phi_i) \tag{3}$$

where, n_{low} and n_{up} are the upper and lower limit of the concerned frequency; ϕ_i is the random phase angle; $S(h, n)$ is PSD spectrum of the fluctuating wind velocity; n is the frequency of the wind velocity.

Once the payload height is determined, the wind velocity spectrum is only a function of frequency. The Kaimal spectrum is used in this research [15]. As shown in Fig. 3, is the plot of Kaimal spectrum and the time history of wind velocity at the position of 10 m high.

With the known wind velocity, (both the average part and fluctuating part) the wind pressure $p(h) = C_{\rm P} \cdot 0.625 \cdot v_{\rm wind}^2(h)$. Where, $C_{\rm P}$ is wind coefficient of the object which is related to the aerodynamic shape of the object which is defined in a crane design code.

The wind load is integration of wind pressure on object's windward surface.

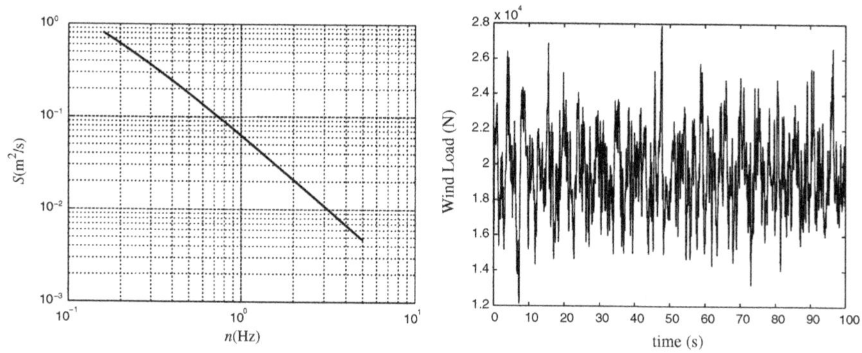

Fig. 3 Kaimal spectrum and time history of wind velocity at height of 10 m

2.3 Dynamic Equation of Trolley-Payload System with Flexible Hoist Rope

With the issues discussed above, the dynamic equations are put forward. Following the analysis in Fig. 2, the equations are based on the force balance function, the constitutive relation function and the geometric function as listed in Eqs. (4) and (5):

$$\begin{cases} -T_{\text{hoist}} sin\alpha - D_{\text{rope}} sin\alpha - m\ddot{x}_p - D_{\text{angle}} cos\alpha + P_{\text{wind}} = 0 \\ T_{\text{hoist}} cos\alpha + D_{\text{rope}} cos\alpha - m\ddot{y}_p - mg - D_{\text{angle}} sin\alpha = 0 \end{cases} \tag{4}$$

$$\begin{cases} x_{\text{p}} = lsin\alpha + x_{\text{tr}} \\ y_{\text{p}} = -lcos\alpha \end{cases} \tag{5}$$

where, D_{angle} is the damping force acting on the payload; D_{rope} is the damping force in the hoist rope. The damping force of the sway angle is $D_{\text{angle}} = C_{\text{angle}}\dot{\alpha}/l$, the damping force in the hoisting rope is $D_{\text{rope}} = C_{\text{rope}}\dot{l}_{\text{f}}$, and the tension force in the hoisting rope is $T_{\text{hoist}} = k_{\text{rope}}(l_{\text{r}})l_{\text{f}}$. In which, $k_{\text{rope}}(l_{\text{r}})$ is a nonlinear function of the rope length and $k_{\text{rope}}(l_{\text{r}}) = A_{\text{rope}}E_{\text{rope}}/l_{\text{r}}$; A_{rope} is the area of the rope section whose unit is m^2; E_{rope} is the equivalent Young modulus of the rope whose unit is Pa.

$\dot{\alpha}$ is the angular velocity of the sway angle; T is the tension force in the hoisting rope; C_{angle} is the damping coefficient of the sway angle; $l = l_{\text{r}} + l_{\text{f}}$, $\dot{l} = \dot{l}_{\text{r}} + \dot{l}_{\text{f}}$ and $\ddot{l} = \ddot{l}_{\text{r}} + \ddot{l}_{\text{f}}$ are the length, the velocity and acceleration of the hoisting rope; P_{wind} is the wind load actuating the payload; m is the mass of the payload. As a result, the position of the payload (x_p, y_p) is calculated according to Eq. (5).

Table 1 Parameters of a STS quayside container crane for simulation

Content	Value	Content	Value
The length of hoist rope:	5–70 m	Modulus of the rope:	1.1 10^{11} Pa
The payload mass (including spreader):	11500–61500 kg	Damping coef. in the rope:	138.23 kg/s
Rated operational wind velocity:	17.1 m/s	Damping coef. of the angle:	10^6 kg·m/s

The dynamic equation of the pendulum system is shown in Eq. (6):

$$\begin{cases} \ddot{\alpha} = -\frac{C_{\text{angle}}}{m}\dot{\alpha}\frac{1}{l^2} + \frac{P_{\text{wind}}}{m}cos\alpha\frac{1}{l} - gsin\alpha\frac{1}{l} - 2\dot{\alpha}\frac{1}{l} - \ddot{x}_{\text{tr}}cos\alpha\frac{1}{l} \\ \ddot{l} = -\frac{A_{\text{rope}}E_{\text{rope}}}{m}\frac{1}{l_{\text{r}}}l_{\text{f}} + \frac{P_{\text{wind}}}{m}sin\alpha + gcos\alpha + l\dot{\alpha}^2 - \ddot{x}_{\text{tr}}sin\alpha - \frac{C_{\text{rope}}}{m}\dot{l} + \frac{C_{\text{rope}}}{m}\dot{l}_{\text{r}} \end{cases} \tag{6}$$

Finally, it can be concluded that the input of the system is the acceleration of the trolley $\ddot{x}_{\text{tr}}$ and the velocity of the rigid term of the hoist rope $\dot{l}_{\text{r}}$, while the out put of the system is the cargo position x_{p} and y_{p}. As defined in Eqs. (5) and (6), the terms in the dynamic equation influence each other and the relationship between the inputs and outputs is nonlinear.

3 Numerical Simulation

3.1 Load Case Specification

As an example, the parameters of a STS quayside container crane are applied to set the model for simulation. The mainly concerned parameters are shown in Table 1.

To discuss the influence of the wind load and the flexibility in the hoist rope to the response of the system, two load cases are used for simulation:

(1) In load Case 1, with wind load only, the trolley and hoist devices are standing still and the only excitation of the system is the wind load which lists below in detail: (a) The length of the hoist rope $l = 40$ m, the payload weight $m = 11500$ kg (which is the weight of a spreader and an empty container), the height of the trolley is 50 m which indicates the height of the payload is 10 m; (b) The simulated wind velocity is 17.1 m/s which is the rated operational wind velocity; (c) The windward area $A_{\text{wind}} = 50\,\text{m}^2$, which is close to a standard 60 ft container area and its spreader.

(2) In load Case 2, a typical operation is simulated. The trolley, hoist device moving caused inertial load and the wind load are all taken into account as excitations of the system. The detail of this load case lists below: (a) The payload weight $m = 36500$ kg, which is the mid-value of the payload which occurs mostly in daily operation; (b) The

maximum hoisting velocity is 5 m/s, the maximum hoisting acceleration is $2\,\text{m/s}^2$; (c) The maximum trolley velocity is 1.2 m/s, the maximum trolley acceleration is $0.2\,\text{m/s}^2$, the jerk of trolley is $1\,\text{m/s}^3$;

3.2 Analysis of the Influence of the Stiffness of the Hoist Rope

As mentioned, in most researches for the anti-sway applications, the hoist rope is regarded as a rigid though in some of the study, the length of the rope changes during operation. However, the stiffness of the hoist rope play a role in actual system.

As listed in most rope producer's user manual, the stiffness of the rope is related to the length of the rope. The stiffness varies from 1.6×10^6 to $2.8 \times 10^6\,\text{N/m}$ according to different length of the rope. It is the reason why in Eq. (6), the flexible length of the rope is also related to the rigid length of the rope.

3.3 Analysis of the Influence of the Wind Load

As described above, the wind load is a lateral load without particular restriction and it extends the sway angle of the system. To discuss the influence of the wind load only, the Load Case 1 is introduced for simulation. The time history of the sway angle response and its Fast Fourier Transform is shown in Fig. 4. As shown in Fig. 4, the time history of the sway angle indicates that the mean value of it is approximately 0.17 rad, when the vibration is about the 7 mrad around the mean value. So, one of the essential influence of wind load is from the average part of it.

After transfer the vibration value of the sway angle into the frequency domain with FFT, it is clear that there are two essential frequencies 0.04 and 0.08 Hz. They

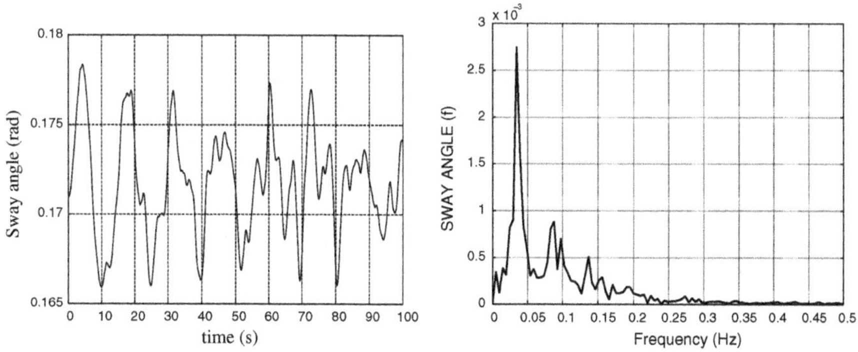

Fig. 4 Time history of sway angle and its FFT actuated by wind load only

are close to the natural frequency of the simulated payload pendulum (about 0.07 Hz) and within the wind loads spectrum range.

3.4 Analysis of a Typical Operation

In above sections, the response of the system with separate excitations is discussed. In this section, the system response during a typical operation which is defined in Load Case 2 is put forward.

(1) The results about the sway angle and the elastic elongation

As a simulation result, the time history of sway angle and its FFT during a typical operation is shown in Fig. 5. As shown in Fig. 5, the sway angle with frequency 0.04 Hz increases in this load case compared with the load case with wind load only. In order to compare the characteristics of the results, the maximum value and the average of the sway angle in both the typical operation with and without wind load as well as with wind load only are listed in Table 2.

As a summary, the wind load amplifies both the maximum and static sway angle of the payload, however, at the same time, it has little influence on the elastic elongation of the hoist rope. And the average of the sway angle which mainly caused by wind load is much larger than the fluctuating part caused by trolley movement. Although,

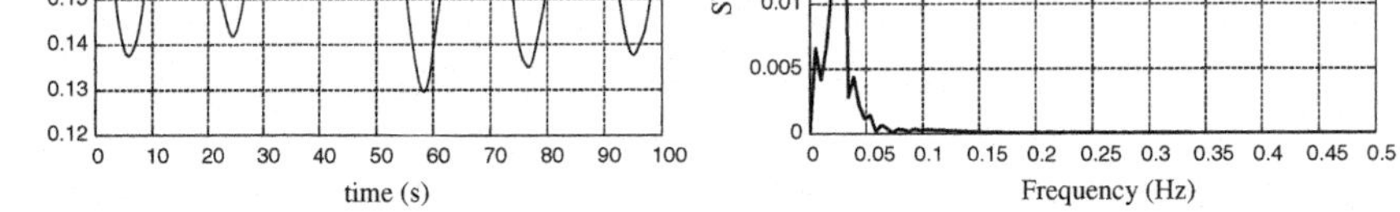

Fig. 5 Time history of sway angle and its FFT during a typical operation with wind load

Table 2 Characteristics of sway angle in different load cases

Content	α_{max} (rad)	$\bar{\alpha}$ (rad)	l_{fmax} (m)	$\bar{l_f}$ (m)
Typical operation with wind load	0.216	0.171	0.155	0.070
Typical operation without wind load	0.045	−8.3e-4	0.151	0.069
With wind load only	0.181	0.172	0.083	0.041

there is fluctuating part in the wind load, the contribution of the fluctuating wind load is really small.

(2) The results about the position of the payload

The most concerned characteristics of the system for actual trolley and hoist system, is the accuracy of the positioning of the cargo. As a result, the time history of the cargo position with and without wind load and the compare with the desired position is shown in Fig. 6. The coordinate definition is shown in Fig. 2. The initial position of the payload with wind load is the balance position under the average wind load which simulates the effect of a continuous wind load, and the initial of the desired position and without the wind load is from zero. To compare the error between the actual position with the desired one in both typical operation with and without wind load as well as with wind load only, the results are listed in Table 3.

It can be inferred from Table 3 that the trolley motion may cause large vibration in x direction as the maximum position error is about 9 m, although, it has little influence of the static position error. At the same time, wind load induced position error is mainly in the static error in x direction. The typical operation with wind load has both the static error mainly caused by the wind load and the fluctuating error mainly caused by the trolley motion. The hoist motion increases the rope elongation in y direction but it is really small compared with the position error in x direction.

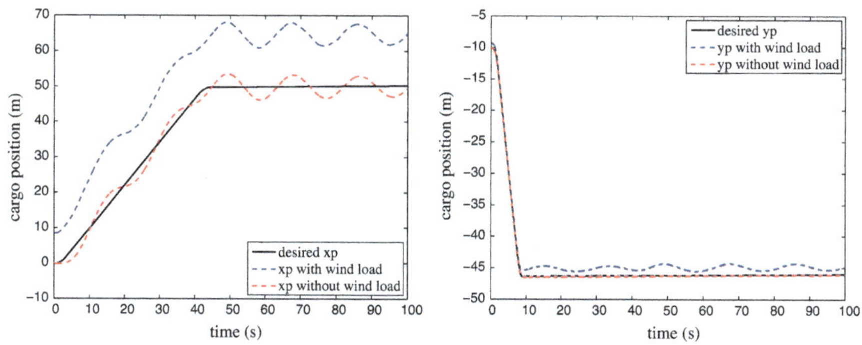

Fig. 6 Time history of payload position compared with the desired position

Table 3 Error between the actual and desired payload position in different load cases

Content	$err^x_{\max}$ (m)	$err^y_{\max}$ (m)	err^x_{static} (m)	err^y_{static} (m)
Typical operation with wind load	18.437	1.962	14.388	1.190
Typical operation without wind load	10.162	0.729	0.040	0.110
With wind load only	8.837	0.784	8.550	0.695

3.5 Results Summary and Discussion

As a summary, the analysis of different Load Cases indicated that:

(1) In the entire working space, the stiffness of the rope changes over two hundred percent due to the variation of rope's current length;

(2) Both the trolley motion and the external excitation, wind load for example, cause payload sway; and the interaction of these two loads amplify the sway angle and change the frequency of the forced vibration;

(3) According to the dynamic equation, the wind load can be regarded as a state variable and observed in the state feedback system.

Based on these results, it can be suggested that:

(1) The trolley-payload system can be regarded as a large scale under-actuated cable-driven robot so robotic kinematics and dynamics analysis can be applied in the future research;

(2) The system is obvious a non-linear system, so the non-linear control strategy should be studied for this specific application.

4 Conclusion

As all the dimensions discussed above, the conclusions list below: (1) The stiffness of the hoist rope changes a lot along with its length and shows nonlinear characteristics; (2) Based on wind spectrum, the dynamic wind load is simulated, and its influences to the positioning accuracy is discussed; (3) The main source of positioning error is from the combination of the wind load induced and trolley motion induced sway angle. And the wind load induced sway angle is mainly a static value while the motion induced one is mainly an oscillation around the balance position of the payload. So, it maybe helpful to decrease the amplitude of sway angle by considering the effects listed above while designing the controller for cargo positioning.

References

1. Adams, C., Potter, J., Singhose, W.: Input-shaping and model-following control of a helicopter carrying a suspended load. J. Guid. Control Dyn. **38**(1), 94–105 (2015)
2. Fujioka, D., Shah, M., Singhose, W.: Robustness analysis of input-shaped model reference control on a double-pendulum crane. In: Proceedings of ACC, pp. 2561–2566 (2015)
3. Fujioka, D., Singhose, W.: Performance comparison of input-shaped model reference control on an uncertain flexible system. In: IFAC Proceedings Volumes, pp. 129–134 (2015)
4. Gouttefarde, M., Lamaury, J., Reichert, C., Bruckmann, T.: A versatile tension distribution algorithm for n-dof parallel robots driven by n + 2 cables. IEEE Trans. Robot. **31**, 1444–1457 (2015)
5. He, B., Fang, Y., Sun, N.: Anti-sway control of an overhead crane with persistent disturbances by lyapunov method. In: Chinese Control Conference, CCC, pp. 889–894 (2015)

6. Liu, X., Zhu, H., Liu, H., Zhang, R.: Design a novel controller with remote upgrading based on gprs of tower crane. In: Proceedings of the World Congress on Intelligent Control and Automation (WCICA), pp. 4945–4950 (2015)
7. Merlet, J.P.: On the inverse kinematics of cable-driven parallel robots with up to 6 sagging cables. In: IEEE International Conference on Intelligent Robots and Systems, pp. 4356–4361 (2015)
8. Miermeister, P., Kraus, W., Lan, T., Pott, A.: An elastic cable model for cable-driven parallel robots including hysteresis effects. Mech. Mach. Sci. **32**, 17–28 (2015)
9. Nguyen, D., Gouttefarde, M.: Study of reconfigurable suspended cable-driven parallel robots for airplane maintenance. In: IEEE International Conference on Intelligent Robots and Systems, pp. 1682–1689 (2014)
10. Park, J., Kwon, O., Park, J.: Anti-sway trajectory generation of incompletely restrained wire-suspended system. J. Mech. Sci. Technol. **27**(10), 3171–3176 (2013)
11. Riehl, N., Gouttefarde, M., Krut, S., Baradat, C., Pierrot, F.: Effects of non-negligible cable mass on the static behavior of large workspace cable-driven parallel mechanisms. In: IEEE International Conference on Robotics and Automation, pp. 2193–2198 (2009)
12. Sato, K., Ohishi, K., Miyazaki, T.: Anti-sway crane control considering wind disturbance and container mass. IEEE Trans. Ind. Appl. **133**(3), 262–271 (2013)
13. Schaper, U., Sawodny, O., Zeitz, M., Schneider, K.: Load position estimation for crane anti-sway control systems. J. Dyn. Syst. Meas. Control Trans. ASME **136**(3), 70–81 (2014)
14. Shen, W.H., Zhao, Z.H., Ren, G.X., Wang, N.Y.: Multi-body dynamic simulation of impact on cross deck pendant. Zhendong yu Chongji/J. Vib. Shock **34**(5), 73–77, 94 (2015)
15. Shinozuka, M., Deodatis, G.: Simulation of stochastic process by spectral representation. Appl. Mech. Rev. **44**, 191–203 (1993)
16. Smoczek, J.: Fuzzy crane control with sensorless payload deflection feedback for vibration reduction. Mech. Syst. Signal Process. **46**(1), 70–81 (2014)
17. Smoczek, J., Szpytko, J.: Fuzzy logic-based adaptive control system prototyping for laboratory scaled overhead crane. In: 2013 18th International Conference on MMAR 2013, pp. 92–97 (2013)
18. Suzuki, K., Murakami, T.: Anti-sway control of crane system by equivalent force feedback of load. In: 2013 IEEE International Conference on Mechatronics, ICM 2013, pp. 688–693 (2013)
19. Viet, L.: Crane sway reduction using coriolis force produced by radial spring and damper. J. Mech. Sci. Technol. **29**(3), 973–979 (2015)
20. Yanping, Z., Zilong, Z., Guoyun, L.: Analysis of steel rope shock absorber amplitude frequency characteristics under the basic excitation. Appl. Mech. Mater. **312**, 287–291 (2013)

Towards a Unified Notion of Kinematic Singularities for Robot Arms and Non-holonomic Platforms

Andreas Müller and Peter Donelan

Abstract Kinematic singularities are classically defined in terms of the rank of Jacobians of associated maps, such as forward and inverse kinematic mappings. A more inclusive definition should take into account the Lie algebra structure of related tangent spaces. Such a definition is proposed in this paper, initially for serial manipulators and non-holonomic platforms. The definition can be interpreted as a change in the number of successive infinitesimal motions required for the system to reach an arbitrary configuration in the vicinity of the given configuration. More precisely, it is based on the filtration of a controllability distribution.

1 Introduction

The singularities of holonomic mechanisms are fairly well-understood, and there is a well-established concept for holonomic mechanisms [4, 10]. There is yet no established notion for non-holonomic systems. It is thus instructive to point out some common features of the output singularities of holonomic serial manipulators (SM) and the input singularities of non-holonomic platforms.

It is known that, in a forward kinematic singularity, the complexity of infinitesimal motions that a SM has to perform in order to reach nearby configurations increases. This is reflected by a drop of rank of the forward kinematics Jacobian. Furthermore, the nesting level of Lie brackets of the instantaneous joint screws necessary to generate the Lie algebra corresponding to the motion subgroup of the kinematic chain increases [6, 12]. This is a non-generic phenomenon, and singularities form closed dense subspaces in the configuration space. The SM can be regarded as a driftless kinematic control system, and the Lie bracketing determines the accessibility

A. Müller (✉)
Johannes Kepler University, Linz, Austria
e-mail: a.mueller@jku.at

P. Donelan
Victoria University of Wellington, Wellington, New Zealand
e-mail: peter.donelan@vuw.ac.nz

J. Lenarčič and J.-P. Merlet (eds.), *Advances in Robot Kinematics 2016*,
Springer Proceedings in Advanced Robotics 4, DOI 10.1007/978-3-319-56802-7_41

algebra of this control system. An unconstrained holonomic SM is always kinematically controllable.

Also a non-holonomic platform can modeled as a driftless kinematic control problem. It is configuration controllable if it is completely non-holonomic, i.e. there is no integral manifold defined by the constraints. Controllability is ensured by the Lie algebra rank condition, i.e. when the nested Lie brackets of the control vector fields span the accessibility algebra. Even if this local property holds, the complexity of the infinitesimal motions necessary to reach nearby configurations may change in configurations that are referred to as singular.

The apparent similarity of the holonomic and non-holonomic systems is discussed in this paper, and a unified definition accounting for both types of systems is proposed. The definition rests on the concept of a filtration of a distribution associated with the kinematic control system, which characterizes the complexity of motion in a singularity. It provides a general framework of SMs and completely non-holonomic systems. This preliminary result can potentially be extended to parallel manipulators.

2 Singularities of Lower-Pair Serial Manipulators

For a serial manipulator (SM) whose joints have n degrees of freedom, denote its joint variables by $\mathbf{q} = (q_1, \ldots, q_n) \in \mathbb{V}^n$, the configuration space (c-space) of the SM. The task of the SM is to position an end-effector (EE) and its task space is a subset of $SE(3)$. The forward kinematic mapping $f : \mathbb{V}^n \rightarrow SE(3)$ relates the configuration $\mathbf{q} \in \mathbb{V}^n$ of the SM to the EE configuration $\mathbf{C} = f(\mathbf{q}) \in SE(3)$. The mapping can be expressed as a PoE:

$$f(\mathbf{q}) = \exp(\mathbf{Y}_1 q_1) \cdot \cdots \cdot \exp(\mathbf{Y}_n q_n)\,\mathbf{A} \tag{1}$$

where $\mathbf{Y}_i \in se(3)$ (the Lie algebra of the Euclidean group) is a twist generating the motion of of joint i, with respect to to the chosen global frame and $\mathbf{A} \in SE(3)$ is the EE pose in the reference configuration $\mathbf{q} = \mathbf{0}$.

The EE twist $\mathbf{V} = (\boldsymbol{\omega}, \mathbf{v}) \in se(3)$ arising from a trajectory $\mathbf{q}(t)$ through configuration $\mathbf{C}$, in spatial representation, is determined by $\widehat{\mathbf{V}} = \dot{\mathbf{C}}\mathbf{C}^{-1} \in se(3)$, where $\widehat{\mathbf{V}}$ is the matrix representation of $\mathbf{V} \in se(3)$. It is determined in terms of joint velocities by the spatial Jacobian $\mathbf{J}(\mathbf{q}) : \mathbb{R}^n \rightarrow se(3)$ as

$$\mathbf{V}(\mathbf{q}, \dot{\mathbf{q}}) = \mathbf{J}_1(\mathbf{q})\,\dot{q}_1 + \cdots + \mathbf{J}_n(\mathbf{q})\,\dot{q}_n = \mathbf{J}(\mathbf{q})\,\dot{\mathbf{q}}. \tag{2}$$

The columns $\mathbf{J}_j, j = 1, \ldots, n$ of $\mathbf{J}$ are the instantaneous joint screws in configuration $\mathbf{C}$, given by

$$\mathbf{J}_j(\mathbf{q}) = \mathbf{Ad}_{g_j}\mathbf{Y}_j. \tag{3}$$

with $g_j(\mathbf{q}) = \exp(\mathbf{Y}_1 q_1) \cdot \cdots \cdot \exp(\mathbf{Y}_{j-1} q_{j-1})$.

Definition 1 (*Singularity—rank criterion*) A *(forward) kinematic singularity* of an SM is a critical point $\mathbf{q} \in \mathbb{V}^n$ of the kinematic mapping f, i.e. a configuration where $\text{rank}\,(\mathbf{J}(\mathbf{q})) < r_{\max} = \min\{n, \dim(\text{task space})\}$.

Since the rank is lower semi-continuous and the assignment is analytic, this is equivalent to saying there is no neighbourhood of $\mathbf{q}$ on which rank $\mathbf{J}$ is constant, whereas at a regular point, the rank will be locally constant and equal to $r_{\max}$.

The vector space of achievable EE twists at a given configuration $\mathbf{q}$ is given by:

$$D_{\mathbf{q}} := \text{im}\,\mathbf{J}\,(\mathbf{q}) = \text{span}_{\mathbb{R}}(\mathbf{J}_1\,(\mathbf{q})\,,\ldots,\mathbf{J}_n\,(\mathbf{q})). \tag{4}$$

The $\mathbf{J}_j$ are analytic right-invariant vector fields on $SE(3)$ so the assignment $\mathbf{q} \mapsto (\mathbf{J}_1\,(\mathbf{q})\,,\ldots,\mathbf{J}_n\,(\mathbf{q}))$ is smooth and so, at a regular point, $\mathbf{q} \mapsto D_{\mathbf{q}} \subseteq se(3)$ is a smooth map to the Grassmannian of subspaces of $se(3)$ of dimension $r_{\max}$.

An important characteristic of an SM is the vector space of all EE twists the SM can generate at a given configuration $\mathbf{q}$. This is a subspace of the involutive closure of $D_{\mathbf{q}}$. The latter is the Lie algebra, denoted $\overline{D}_{\mathbf{q}} \subseteq se(3)$, generated by all Lie brackets of the joint screws. It can be determined by taking all nested Lie brackets of $\mathbf{J}_j(\mathbf{q})$, so constructed by means of the filtration of $D_{\mathbf{q}}$, which is the sequence of vector spaces $D_{\mathbf{q}}^{i+1} := D_{\mathbf{q}}^{i} + \left[D_{\mathbf{q}}, D_{\mathbf{q}}^{i}\right]$, with $D_{\mathbf{q}}^{1} := D_{\mathbf{q}}$. This terminates with $\overline{D}_{\mathbf{q}} = D_{\mathbf{q}}^{\kappa}$ for some κ. It can be shown for the task space $SE(3)$ that $\kappa \leq 4$ [6]. So, for example, $\overline{D}_{\mathbf{0}} = \text{span}_{\mathbb{R}}(\mathbf{Y}_i, [\mathbf{Y}_i, \mathbf{Y}_j], [\mathbf{Y}_i, [\mathbf{Y}_j, \mathbf{Y}_k]], [\mathbf{Y}_i, [\mathbf{Y}_j, [\mathbf{Y}_k, \mathbf{Y}_l]]]])$. The dimensions $\dim D_{\mathbf{q}}^{i}$ are collected in the *growth vector* $\rho \in \mathbb{N}^{\kappa}$.

Moreover, the involutive closure is the same at any $\mathbf{q} \in \mathbb{V}^n$ and we denote this common closure by $\overline{D}$. This follows from the expression (3) for the instantaneous joint screws, invoking the BCH formula. Hence $r_{\max} \leq \min\left(\dim \overline{D}, n\right)$. Moreover, the subgroup G generated by the subalgebra $\overline{D}$ is the smallest $SE(3)$ subgroup comprising all possible EE configurations. Hence $f\,(\mathbf{q}) \in G$ and $\text{im}\,\mathbf{J}\,(\mathbf{q}) \subseteq \overline{D}$ for any $\mathbf{q} \in \mathbb{V}^n$, and f can be regarded as a mapping $f : \mathbb{V}^n \to G$. But note that this restricted forward kinematic mapping may still not be surjective if the EE motions do not form a subalgebra, since then $D_{\mathbf{q}} \subsetneq \overline{D}$.

The filtration at $\mathbf{q}$ locally characterizes the process of manipulating the EE when starting at $\mathbf{q}$. The length κ of the filtration is the maximal number of successive infinitesimal joint motions necessary to produce any given (feasible) EE twist. It seems intuitively clear that this number should change at a singularity. The following can be proved using the BCH formula.

Lemma 1 *The filtrations of $D_{\mathbf{q}}$ are identical at all regular configurations $\mathbf{q} \in \mathbb{V}^n$ of the SM, i.e. when $\mathbf{J}\,(\mathbf{q})$ has full rank. The configuration $\mathbf{q}$ is a kinematic singularity of f if and only if the filtration of $D_{\mathbf{p}}$ is not constant for $\mathbf{p}$ in a neighbourhood of $\mathbf{q}$.*

This leads an alternative definition of kinematic singularities for SM. Denote with κ_0 the length of the filtration at a regular point.

Definition 2 (*Singularity—filtration criterion*) A configuration $\mathbf{q} \in \mathbb{V}^n$ of an SM is a *forward kinematic singularity* if and only if the length $\kappa(\mathbf{q}) > \kappa_0$.

Fig. 1 Kinematic model of a 7 DOF KUKA LWR

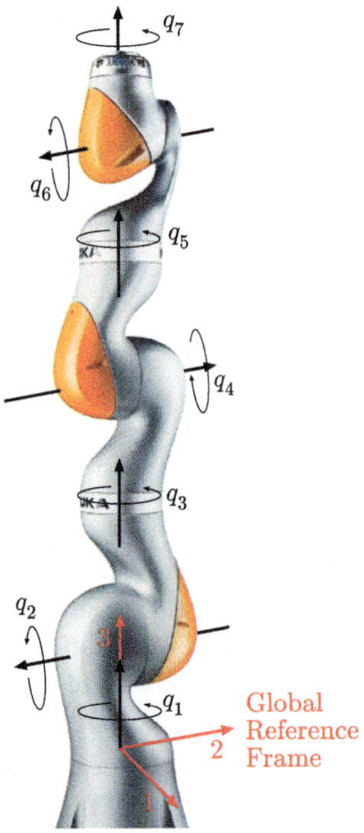

The condition $\kappa > \kappa_0$ is equivalent to say that the filtration of D is not constant in a neighbourhood of $\mathbf{q}$. Since the filtration reveals the effect of higher-order infinitesimal motions it allows for identification of the joint motions that lead the SM out of a singularity [7, 12].

Example 1 Consider the redundant 7 DOF (anthropomorphic) SM in the reference configuration $\mathbf{q} = \mathbf{0}$ in Fig. 1. The SM has 7 revolute joints, and its c-space is $\mathbb{V}^n = T^7$. The joint screw coordinate vectors in the reference configuration w.r.t. to the global frame are

$$\mathbf{Y}_{1,3,5,7} = \begin{pmatrix} 0 \\ 0 \\ 1 \\ 0 \\ 0 \\ 0 \end{pmatrix}, \mathbf{Y}_2 = \begin{pmatrix} 0 \\ -1 \\ 0 \\ -L_3 - L_5 - L_E \\ 0 \\ 0 \end{pmatrix}, \mathbf{Y}_4 = \begin{pmatrix} 0 \\ 1 \\ 0 \\ L_5 + L_E \\ 0 \\ 0 \end{pmatrix}, \mathbf{Y}_6 = \begin{pmatrix} 0 \\ -1 \\ 0 \\ -L_E \\ 0 \\ 0 \end{pmatrix}.$$

The Jacobian $\mathbf{J}(\mathbf{0}) = (\mathbf{Y}_1, \mathbf{Y}_2, \mathbf{Y}_3, \mathbf{Y}_4, \mathbf{Y}_5, \mathbf{Y}_6, \mathbf{Y}_7)$ has rank $\mathbf{J}(\mathbf{0}) = \dim D_{\mathbf{0}} = 3$. In regular configurations $\mathbf{q}$ it is rank $\mathbf{J}(\mathbf{q}) = 6$. The configuration $\mathbf{q} = \mathbf{0}$ is thus a corank 3 singularity, according to Definition 1. The Lie brackets

$$\begin{aligned}
[\mathbf{Y}_1, \mathbf{Y}_2] &= (1, 0, 0, 0, -L_3 - L_5 - L_E, 0)^T \\
[\mathbf{Y}_1, \mathbf{Y}_4] &= (-1, 0, 0, 0, L_5 + L_E, 0)^T \\
[\mathbf{Y}_2, \mathbf{Y}_4] &= (0, 0, 0, 0, 0, -L_3)^T
\end{aligned}$$

for instance, yield linearly independent vectors not in $D_{\mathbf{0}}$. Hence the vector space $D_{\mathbf{0}}^2 = se(3)$ has dim $D_{\mathbf{0}} = 6$. The filtration length at $\mathbf{q} = \mathbf{0}$ is thus $\kappa = 2$, and the growth vector is $\rho(\mathbf{0}) = (3, 6)$. Consequently, the SM may escape from this singularity by first-order motions of joints 1, 2, and 4. The filtration length in regular configurations $\mathbf{q}$ is $\kappa_0 = 1$ since $D_{\mathbf{q}} = SE(3)$. With $\kappa > \kappa_0$ the configuration is a singularity according to Definition 2.

Using the definition $\widehat{\mathbf{V}} = \dot{\mathbf{C}}\mathbf{C}^{-1}$ of spatial velocity, the relation (2) can be written as a right-invariant driftless control system on $SE(3)$

$$\dot{\mathbf{C}} = \left(\widehat{\mathbf{J}}_1 u_1 + \cdots + \widehat{\mathbf{J}}_n u_n\right)\mathbf{C}. \tag{5}$$

The vector space D serves as the right-trivialized controllability distribution of this control system. Necessary and sufficient for (5) to be locally controllable at $\mathbf{q}$ is that $\overline{D}_{\mathbf{q}} = \overline{D}$ [1, 5]. Hence the system (5), and thus the SM, is always locally controllable, even in forward kinematic singularities.

Remark 1 Thus far only holonomic SM were considered. In general, holonomic manipulators are mechanisms comprising closed kinematic loops. This is beyond the scope of this paper, but a note is in order. A mechanism is a physical realization of set of kinematic relations. The mathematical model for the kinematics of a general holonomic mechanism consists of its c-space $V := h^{-1}(\mathbf{0}) \subset \mathbb{V}^n$, where the system of k holonomic constraints arising from closed loops is written as $h(\mathbf{q}) = \mathbf{0} \in \mathbb{R}^k$; the input space $\mathscr{I} \subset \mathbb{R}^m$; and the output space $\mathscr{W} \subset SE(3)$ [8]. These objects are related via the input mapping f_{I} and the output mapping f_{O}

$$\mathscr{W} \xleftarrow{f_{\mathrm{O}}} V \xrightarrow{f_{\mathrm{I}}} \mathscr{I}. \tag{6}$$

The c-space V is an analytic variety when the constraint mapping h is formulated in terms of POEs. Configurations $\mathbf{q}$ where V is not locally a smooth manifold are *c-space singularities*. Configurations $\mathbf{q} \in V$ where the constraint Jacobian $\mathbf{J}_h$ is not constant in a neighbourhood of $\mathbf{q}$ in V are *constraint singularities*. It is important to note that c-space singularities are automatically constraint singularities, but the opposite is not necessarily the case [9]. At c-space singularities the mobility of the mechanism changes. How this affects the possible input and output motions is determined by f_{I} and f_{O}, respectively. *Input (output) singularities* are such that the input (output) Jacobian is not constant in a neighbourhood of $\mathbf{q}$ in V. These three types of singularities can occur simultaneously. All possible combinations and their instantaneous phenomenology were reported in [14]. The consequences for the local finite mobility is yet to be explored.

The associated kinematic control system for a closed loop mechanism can be written in implicit form as:

$$\mathbf{J}_h(\mathbf{q})\,\dot{\mathbf{q}} = \mathbf{0}, \qquad \mathbf{J}_\mathrm{I}(\mathbf{q})\,\dot{\mathbf{q}} = \mathbf{u}, \qquad \mathbf{J}_\mathrm{O}(\mathbf{q})\,\dot{\mathbf{q}} = \mathbf{V} \tag{7}$$

where $\mathbf{u}$ denotes an input vector. Suppose that, locally, rank $\mathbf{J}_h = n - \delta_{\mathrm{loc}}$ (i.e. no c-space singularities, no underconstrained mechanisms). Away from input singularities, the Implicit Function Theorem enables one to locally invert f_I. Then the first two constraint equations in (7) can be rewritten in the form $\mathbf{F}(\mathbf{q})\,\mathbf{u} = \dot{\mathbf{q}}$. The problem of input singularities can be circumvented by working directly with the codistribution defined by the constraints. This will not be pursued further here. Rather the purpose has been to signal the connection to non-holonomic systems, which will be investigated next.

3 Singularities of Non-holonomic Mobile Platforms

Wheeled mobile platforms are frequently used in mobile robotics. The rolling constraint gives rise to non-holonomic constraints so that they can be treated as non-holonomic kinematic control systems. Denote with $\mathbf{x} \in \mathbb{V}^p$ the p coordinates represent the configuration of the system. They are subjected to a system of k non-holonomic Pfaffian constraints

$$\mathbf{A}(\mathbf{x})\,\dot{\mathbf{x}} = \mathbf{0}. \tag{8}$$

Throughout this section the system (8) is assumed to be completely non-holonomic. Then a configuration $\mathbf{q}$ is a *constraint singularity* if $\mathbf{A}(\mathbf{q})$ is not full rank.

Remark 2 If the constraints were not completely non-holonomic, it would be necessary to determine the integral manifold M passing through $\mathbf{x}$. The configuration $\mathbf{x}$ is a *constraint singularity* if and only if rank $\mathbf{A}$ is not constant in a neighbourhood of $\mathbf{x}$ in M, adopting the concept of holonomic mechanisms [10]. For completely non-holonomic constraints this simply means that $\mathbf{A}$ is not full rank.

In the following $\mathbf{A}$ is assumed to have full rank k. Then there are $m = n - k$ vector fields $\mathbf{g}_1, \ldots, \mathbf{g}_m$ that span ker $\mathbf{A}$. They constitute the columns of the orthogonal complement $\mathbf{G}$ of $\mathbf{A}$. This gives rise to the driftless control system

$$\dot{\mathbf{x}} = \mathbf{g}_1(\mathbf{x})\,u_1 + \cdots + \mathbf{g}_m(\mathbf{x})\,u_m = \mathbf{G}(\mathbf{x})\,\mathbf{u}. \tag{9}$$

Usually the inputs $\mathbf{u}$ form a subset of $\mathbf{x}$ consisting of steering and rolling velocities. The associated controllability distribution is $\Delta := \mathrm{span}_{\mathbb{R}}(\mathbf{g}_1, \ldots, \mathbf{g}_m)$, which is also referred to as the constraint distribution [11]. The distribution Δ is regular if it has a locally constant dimension, i.e. if the constraints (8) have locally constant rank.

The similarity to the forward kinematics (2) of holonomic manipulators is obvious, but now (9) describes how input rates affect the system velocity. Now the matrix $\mathbf{G}$

plays the role of an input Jacobian. In case of holonomic mechanisms a drop of rank would be necessary and sufficient for an input singularity, and the definitions 1 and 2 are equivalent. With the assumption that $\mathbf{G}$ is full rank, the control system (9) would not have a singularity according to the classical Definition 1 in terms of the rank of $\mathbf{G}$. However, if one accepts that a singularity is a configuration in which the kinematic accessibility changes then there are further situations that qualify as singular in case of non-holonomic systems.

Definition 3 The configuration $\mathbf{x}$ is a *kinematic singularity* if the filtration of Δ at $\mathbf{x}$ is not constant in a neighbourhood of $\mathbf{x}$. If additionally Δ is regular, i.e. $\mathbf{G}(\mathbf{x})$ has full rank, the configuration $\mathbf{x}$ is a *non-holonomic kinematic singularity*.

The difference to Definition 2 is that $\mathbf{G}$ may have full rank at a non-holonomic singularity. Only for non-holonomic systems can $\mathbf{G}(\mathbf{x})$ be regular but the filtration Δ not be regular at $\mathbf{x}$.

Corollary 1 *The set of non-holonomic singularities, denoted Σ_{nh}, is closed in $\mathbb{V}^p$.*

Example 2 Consider the car with two trailers in Fig. 2. The $p = 7$ system coordinates are $(x, y, \theta_1, \theta_2, \varphi, \alpha) \in \mathbb{V}^7 = \mathbb{R}^2 \times T^5$. The kinematic control system is

$$\begin{pmatrix} \dot{x} \\ \dot{y} \\ \dot{\theta} \\ \dot{\theta}_1 \\ \dot{\theta}_2 \\ \dot{\varphi} \\ \dot{\alpha} \end{pmatrix} = \begin{pmatrix} R\cos\varphi\cos\theta \\ R\cos\varphi\sin\theta \\ \frac{R}{L}\sin\varphi \\ \frac{R}{L}\sin\varphi - \frac{R}{L_1}\cos\varphi\sin\theta_1 \\ \frac{R}{L_1}\cos\varphi\sin\theta_1 - \frac{R}{L_2}\cos\varphi\cos\theta_1\sin\theta_2 \\ 0 \\ 1 \end{pmatrix} u_1 + \begin{pmatrix} 0 \\ 0 \\ 0 \\ 0 \\ 0 \\ 1 \\ 0 \end{pmatrix} u_2. \tag{10}$$

The accessibility distribution $\Delta = \text{span}(\mathbf{g}_1, \mathbf{g}_2)$ is regular, i.e. has constant dimension for all $\mathbf{x} \in \mathbb{V}^7$. Its filtration terminates with the accessibility algebra $\overline{\Delta} = \mathbb{R}^7$, and

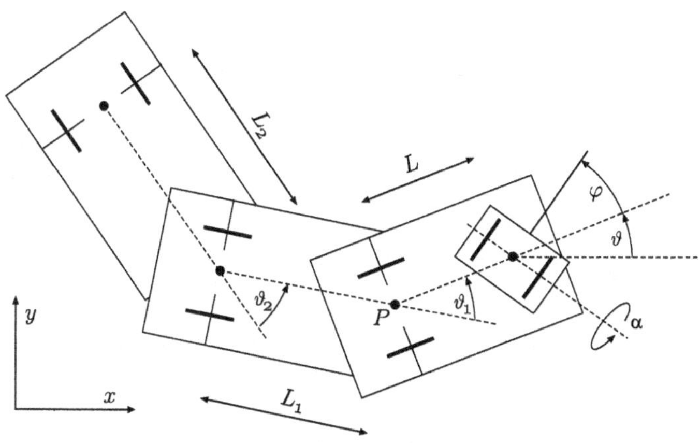

Fig. 2 Kinematic model of a car with two trailers

the system is thus accessible and controllable. As long as $\varphi \neq \pm\frac{n}{2}\pi$ the filtration terminates with growth vector $\rho = (2, 3, 5, 6, 7)$. But if the steering angle attains $\varphi = \pm\frac{n}{2}\pi$, the length of the filtration increases by two and the growth vector is $\rho = (2, 3, 5, 5, 6, 6, 7)$. These are non-holonomic singularities: $\Sigma_{\text{nh}} = \{\mathbf{x}|\varphi = \pm\frac{n}{2}\pi\}$. In these singularities the control of the car with two trailers becomes more complex than it is in regular configurations. This is intuitively clear since a steering motion must be performed first so to reorientate the front axis. It can be shown that for each additional trailer, the length of the filtration at a singularity increases by one.

4 Discussion and Conclusion

Singularities are kinematic configurations where the kinematic properties change. For SMs this is simply reflected by a rank-deficient forward kinematics Jacobian. This means that the complexity of the infinitesimal motion to reach a point in the vicinity of a singularity increases. Non-holonomic systems possess further critical configurations that qualify as singularities although the input Jacobian is full rank. They are also characterized by an increase of the complexity of the motion. It has been proposed here that the complexity in holonomic and non-holonomic cases is characterized and exemplified by the degree of nesting of Lie brackets necessary to generate the screw algebra defined by the joint screws of an SM, respectively the controllability algebra of a mobile platform. The iteration depth of nested Lie brackets (filtration of distribution) is used as defining property of singularities.

Acknowledgements The first author acknowledges partial support of this work by the Austrian COMET-K2 program of the Linz Center of Mechatronics (LCM).

References

1. Cardetti, F., Gordina, M.: A note on local controllability on Lie groups. Syst. Control Lett. **57**, 978–979 (2008)
2. Chitour, Y., Sussmann, H.J.: Motion planning using the continuation method. In: Ballieul, J., Sastry, S.S., Sussmann, H.J. (eds.) Essays on Mathematical Robotics, pp. 91–125. Springer, Berlin (1998)
3. Chitour, Y., Jean, F., Trélat, E.: Singular trajectories of driftless and control-affine systems. SIAM J. Control Optim. **47**(2), 1078–1095 (2008)
4. Donelan, P.: Singularity-theoretic methods in robot kinematics. Robotica **25**, 641–659 (2007)
5. Jurdevic, V., Sussmann, H.J.: Control systems on Lie groups. J. Differ. Equ. **12**, 313–329 (1972)
6. Müller, A.: On the manifold property of the set of singularities of kinematic mappings: genericity conditions. ASME J. Mech. Robot. **4**(1), 1–9 (2012)
7. Müller, A.: A proposal for a unified concept of kinematic singularities for holonomic and non-holonomic mechanisms. In: 36th Mechanisms and Robotics Conference (MECH), ASME 2012 International Design Engineering Technical Conferences, Chicago, IL, pp. 1543–1552 (2012)

8. Müller, A.: On the terminology and geometric aspects of redundantly actuated parallel manipulators. Robotica **31**(1), 137–147 (2013)
9. Müller, A.: Local kinematic analysis of closed-loop linkages—mobility, singularities, and shakiness. ASME J. Mech. Robot. (in press) (2016)
10. Müller, A., Zlatanov, D.: Kinematic singularities of mechanisms revisited. In: IMA Conference on the Mathematics of Robotics, Oxford, UK (2015)
11. Nijmeijer, H., van der Schaft, A.J.: Nonlinear Dynamical Control Systems. Springer, Berlin (1990)
12. Rico, J.M., Gallardo, J., Duffy, J.: A determination of singular configurations of serial non-redundant manipulators, and their escapement from singularities using Lie products. In: Merlet, J.P., Ravani, B. (eds.) Computational Kinematics, pp. 143–152. Kluwer, Dordrecht (1995)
13. Tchoń, K.: On kinematic singularities of nonholonomic robotic systems. In: Proceedings of the RoManSy 13, Zakopane, Poland, pp. 75–84 (2000)
14. Zlatanov, D., Bonev, I.A., Gosselin, C.M.: Constraint singularities as C-space singularities. In: Proceedings of the 8th International Symposium on Advances in Robot Kinematics (ARK 2002), Caldes de Malavella, Spain, pp. 183–192 (2002)
15. Zadarnowska, K., Tchoń, K.: A control theory framework for performance evaluation of mobile manipulators. Robotica **25**, 703–715 (2007)

Dynamic Singularities of Non-holonomic Robotic Systems: An Analytic Approach

Krzysztof Tchoń

Abstract This work develops a concept of singular configurations of non-holonomic systems, rooted in the endogenous configuration space approach. The approach takes as a starting point a control system representation of the equations of motion of a non-holonomic robotic system in the form of a control affine system with output. The input-output map is introduced whose derivative is defined as the system's Jacobian. Dynamic singular configurations are defined as control functions at which the input-output map is not surjective, i.e. the system's Jacobian gets rank deficient. It is shown that the dynamic singularities coincide with the singular optimal controls of the control affine system. As a by-product of the Jacobian setting dynamic dexterity measures of the non-holonomic systems are designed. The concept of dynamic singularities is illustrated with an example of the front wheel driven car.

1 Introduction

The problem of singularities has been creating a challenge for robotics research for decades. For holonomic, serial or parallel manipulators, the basic characteristics of singularities are well established. In the area of non-holonomic (mobile) robots there is even no commonly accepted definition of singularities. Specifically, the kinematics studies of the non-holonomic robots represented by driftless control system concentrate on the control distribution whose properties give rise to two kinds of singularities referred to as the posture and the configuration singularities [8]. Recently, the posture singularities have been examined systematically in [6, 7], within an approach that may be called geometric. As a counterpoint, in [10] we have presented fundamentals of configuration singularities, called an analytic approach, and suggested that both these approaches might be unified based on the recent results of sub-Riemannian geometry concerned with a determination of the distribution by its singular curves [4, 5]. In accordance with the analytic approach the configuration singularities live in an infinite-dimensional space of control functions. A singular configuration is then

K. Tchoń (✉)
Wroclaw University of Technology, Wroclaw, Poland
e-mail: krzysztof.tchon@pwr.edu.pl

J. Lenarčič and J.-P. Merlet (eds.), *Advances in Robot Kinematics 2016*,
Springer Proceedings in Advanced Robotics 4, DOI 10.1007/978-3-319-56802-7_42

identified with a control function at which the input-output map loses rank. This property can be verified by reference to the Gram matrix of the linear approximation to the control system representation of the kinematics. Moreover, it is proved that the configuration singularities coincide with the singular optimal control functions, so they can be computed in a systematic way [3].

This paper attempts at extending the analytic approach of [10] to non-holonomic systems with dynamics, represented by control affine systems. Analogously to [10] we propose a definition of the dynamically singular configuration, establish checkable necessary and sufficient conditions for singularity, and show that dynamically singular configurations are identical with the singular optimal control functions. As a by-product we get a concept of dynamic dexterity measures for the non-holonomic robotic system, originally announced in [11].

The composition of this work is the following. Section 2 introduces basic concepts including the Jacobian. Section 3 defines dynamic singularities. A link between dynamic singular configurations and singular optimal control is established in Sect. 4. Dynamic dexterity is shortly discussed in Sect. 5. An example of the front wheel driven car is presented in Sect. 6. Section 7 contains conclusions.

2 Basics

We shall consider non-holonomic robotic systems with dynamics whose equations of motion can be represented by a control affine system with output, of the form

$$\begin{cases} \dot{q} = f(q) + G(q)u = f(q) + \sum_{i=1}^{m} g_i(q)u_i, \\ y = k(q). \end{cases} \tag{1}$$

Hereabove $q \in R^n$, $u \in R^m$ and $y \in R^r$ denote, respectively, the state, the control and the output variable. The drift vector field $f(q)$ and the control vector fields $g_1(q), \ldots, g_m(q)$ are assumed to be smooth (C^∞). The motion of the system will be examined over a time interval $[0, T]$. The configuration space of the system (1) is defined as the Hilbert space $\mathcal{U} = L^2_m[0, T]$ of Lebesgue square integrable functions on $[0, T]$ assuming values in R^m, with the inner product $\langle u_1(\cdot), u_2(\cdot)\rangle = \int_0^T u_1^T(t)u_2(t)dt$. The space $\mathcal{U}$ will be called the endogenous configuration space of the system [9]. It is assumed that for every admissible control function $u(\cdot) \in \mathcal{U}$ and the initial state q_0 the system's trajectory $q(t) = \varphi_{q_0,t}(u(\cdot))$ exists for every $t \in [0, T]$. The corresponding output trajectory $y(t) = k(\varphi_{q_0,t}(u(\cdot)))$. Further on the control function will be called shortly the control. With a suitable interpretation of system's variables, the representation (1) encompasses non-holonomic mobile robots or mobile manipulators composed of a non-holonomic mobile platform carrying on-board either a holonomic or a non-holonomic manipulator.

For a fixed initial state we define the input-output map of the system (1) as the value at the time T of the system's output, subject to a control $u(\cdot)$,

$$K_{q_0,T} : \mathscr{U} \longrightarrow R^r, \quad K_{q_0,T}(u(\cdot)) = y(T) = k(\varphi_{q_0,T}(u(\cdot))). \tag{2}$$

This map is differentiable with respect to control. Its derivative will be called the system's Jacobian,

$$J_{q_0,T}(u(\cdot)) : \mathscr{U} \longrightarrow R^r, \quad J_{q_0,T}(u(\cdot))v(\cdot) = D\,K_{q_0,T}(u(\cdot))v(\cdot). \tag{3}$$

The Jacobian can be computed by means of the linear approximation to the control system (1), namely given a control $u(t)$ we find the corresponding trajectory $q(t)$, and compute the matrices

$$A(t) = \frac{\partial (f(q(t)+G(q(t))u(t))}{\partial q}, \; B(t) = G(q(t)), \; C(t) = \frac{\partial k(q(t))}{\partial q}. \tag{4}$$

The linear approximation of (1) along $(u(t), q(t))$ is defined as a linear, time-dependent control system

$$\begin{cases} \dot{\xi} = A(t)\xi + B(t)v, \\ \eta = C(t)\xi. \end{cases} \tag{5}$$

Now, the Jacobian is the input-output map at T of (5) initialized at $\xi_0 = 0$,

$$J_{q_0,T}(u(\cdot))v(\cdot) = \eta(T) = C(T)\xi(T) = C(T)\int_0^T \Phi(T,s)B(s)v(s)ds. \tag{6}$$

The matrix $\Phi(t,s)$ appearing above satisfies the evolution equation $\frac{\partial}{\partial t}\Phi(t,s) = A(t)\Phi(t,s)$ initialized at $\Phi(s,s) = I_n$.

3 Dynamic Singularities

Dynamic singularities of the non-holonomic system will be defined with reference to the Jacobian (3).

Definition 1 An endogenous configuration $u(\cdot) \in \mathscr{U}$ is called regular, if the Jacobian $J_{q_0,T}(u(\cdot))$ is a surjection of $\mathscr{U}$ onto R^r, otherwise this configuration is singular.

Below we shall derive a necessary and sufficient condition for regularity. To this objective, let us introduce the Gram matrix of the linear system (5),

$$\mathscr{D}_{q_0,T}(u(\cdot)) = C(T)\int_0^T \Phi(T,s)B(s)B^T(s)\Phi^T(T,s)ds\, C^T(T). \tag{7}$$

It is easy to see that the Gram matrix $\mathscr{D}_{q_0,T}(u(\cdot)) = C(T)\mathscr{M}_{q_0,T}(u(\cdot))C^T(T)$, where the inner factor satisfies the Lyapunov matrix differential equation

$$\dot{\mathscr{M}}_{q_0,t}(u(\cdot)) = B(t)B^T(t) + A(t)\mathscr{M}_{q_0,t}(u(\cdot)) + \mathscr{M}_{q_0,t}(u(\cdot))A^T(t) \quad (8)$$

with initial condition $\mathscr{M}_{q_0,0}(u(\cdot)) = 0$. We have the following

Theorem 1 *An endogenous configuration $u(\cdot)$ is regular if and only if the Gram matrix (7) has full rank r. Moreover, the regularity implies a local input-output controllability of the control system (1).*

Proof Sufficiency: Let $rank\ \mathscr{D}_{q_0,T}(u(\cdot)) = r$, and consider the Jacobian equation

$$J_{q_0,T}(u(\cdot))v(\cdot) = C(T)\int_0^T \Phi(T,s)B(s)v(s)ds = \eta \quad (9)$$

for an $\eta \in R^r$. It is easily checked that this equation has a solution

$$v(t) = B^T(t)\Phi^T(T,t)C^T(T)\mathscr{D}^{-1}_{q_0,T}(u(\cdot))\eta,$$

therefore the Jacobian is surjective.

Necessity: Now assume that $rank\ \mathscr{D}_{q_0,T}(u(\cdot)) < r$, so there exists a non-zero vector $\eta \in R^r$, such that $\mathscr{D}_{q_0,T}(u(\cdot))\eta = 0$. Suppose for a while that there exists a $v(\cdot) \in \mathscr{U}$ such that (9) holds. By definition of $\mathscr{D}_{q_0,T}(u(\cdot))$, we have

$$\eta^T\mathscr{D}_{q_0,T}(u(\cdot))\eta = \int_0^T ||B^T(t)\Phi^T(T,t)C^T(T)\eta||^2 dt = 0,$$

where $||\cdot||$ denotes the Euclidean norm in R^m. This results in

$$\eta^T C(T)\Phi(T,t)B(t) = 0,$$

for every $t \in [0,T]$. Invoking once again the Jacobian equation (9) we conclude that

$$\eta^T J_{q_0,T}(u(\cdot))v(\cdot) = \eta^T C(T)\int_0^T \Phi(T,s)B(s)v(s)ds = ||\eta||^2 = 0,$$

so η must be zero, what leads to a contradiction.

The local input-output controllability means that if a control $u(\cdot) \in \mathscr{U}$ transfers in time T the system's output to a point y_0 then there exist controls that allow to reach in T any point y from an open neighbourhood of y_0. This property in a direct consequence of the local surjectivity of the map $K_{q_0,T}$, see [1], Theorem 2.5.9. □

4 Singular Optimal Control

Consider an optimal control problem in the system (1), consisting in the determination of a control $u(t)$ that transfers the system's state at T to the terminal manifold $M_T = \{q \in R^n | y_d - k(q) = 0\}$, and simultaneously minimizes the Lagrangian objective function

$$\int_0^T L(q(t), u(t))dt.$$

Having introduced the adjoint variables $p_0 \in R$, $p \in R^n$, we define the Hamiltonian

$$H(q, u, p_0, p) = p^T(f(q) + G(q)u) - p_0 L(q, u). \tag{10}$$

The canonical Hamiltonian equations take the form

$$\dot{q} = \frac{\partial H}{\partial p}, \quad \dot{p} = -\frac{\partial H}{\partial q}, \tag{11}$$

where the optimality condition is $\frac{\partial H}{\partial u} = 0$. An extremal $(u(t), q(t), p(t))$ of the problem is called singular if it solves (11) for $p_0 = 0$, $p(t) \neq 0$, satisfies the optimality condition and the transversality condition $p^T(T) = \rho^T C(T)$, for a non-zero vector $\rho \in R^r$. The control included in a singular extremal is referred to as the singular optimal control. The following theorem characterizes dynamic singularities of the non-holonomic robotic system.

Theorem 2 *Dynamic singularities of a non-holonomic robotic system are identical with the singular optimal controls.*

Proof Necessity: Suppose that $(u(t), q(t), p(t))$ is a singular extremal. This means that $p_0 = 0$ and $p^T(t)G(q(t)) = 0$. The adjoint vector $p(t)$ is non-zero and satisfies the Hamiltonian equation

$$\dot{p}^T(t) = -p^T(t)\frac{\partial(f(q(t)) + G(q(t))u(t))}{\partial q} = -p^T(t)A(t), \tag{12}$$

see (4), and the transversality condition. Now, the singularity condition and (12) result in

$$p^T(t)G(q(t)) = p^T(t)B(t) = 0. \tag{13}$$

To proceed we refer to the Gram matrix (7) and look at the Lyapunov equation (8). Having multiplied this equation from the left by $p^T(t)$, after appropriate substitutions from (12) and (13), we obtain

$$p^T(t)\dot{\mathscr{M}}_{q_0,t}(u(\cdot)) = -\dot{p}^T(t)\mathscr{M}_{q_0,t}(u(\cdot)) + p^T(t)\mathscr{M}_{q_0,t}(u(\cdot))A^T(t).$$

This leads to the differential equation

$$\frac{d}{dt}\left(p^T(t)\mathcal{M}_{q_0,t}(u(\cdot))\right) = \left(p^T(t)\mathcal{M}_{q_0,t}(u(\cdot))\right) A^T(t). \tag{14}$$

For initially $\left(p^T(0)\mathcal{M}_{q_0,0}(u(\cdot))\right) = 0$, we deduce that the zero solution satisfies (14), so $\left(p^T(t)\mathcal{M}_{q_0,t}(u(\cdot)))\right) = \left(p^T(T)\mathcal{M}_{q_0,T}(u(\cdot)))\right) = 0$. Using the transversality condition we get $\left(p^T(T)\mathcal{M}_{q_0,T}(u(\cdot))p(T)\right) = \rho^T\mathcal{D}_{q_0,T}(u(\cdot))\rho = 0$, what means a rank deficiency of the Gram matrix. Thus, the endogenous configuration $u(\cdot)$ is singular. We have demonstrated that every singular optimal control is a singular endogenous configuration.

Sufficiency: Choose a singular endogenous configuration $u(\cdot)$, take the Hamiltonian (10) for $p_0 = 0$, and and let the triple $(u(t), q(t), p(t))$ denote the corresponding solution of the canonical equations (11). By singularity of the configuration we get the existence of a non-zero vector $\rho \in R^r$ such that $\rho^T\mathcal{D}_{q_0,T}(u(\cdot))\rho = 0$, i.e.

$$\rho^T C(T)\mathcal{M}_{q_0,T}(u(\cdot))C^T(T)\rho = 0.$$

Taking into account the identity (7) we deduce that

$$\int_0^T ||B^T(s)\Phi^T(T,s)C^T(T)\rho||^2 ds = 0,$$

therefore, for any $t \in [0, T]$,

$$\rho^T C(T)\Phi(T,t)B(t) = 0. \tag{15}$$

Now, by solving the canonical equation $p^T(t) = -\frac{\partial H}{\partial q} = -p^T(t)A(t)$ we obtain

$$p^T(t) = p^T(T)\Phi(T,t), \tag{16}$$

where $\Phi(t,s)$ denotes the fundamental matrix of (5), and $p^T(T) = \rho^T C(T)$. Finally, from (15) and (16), we conclude that $p^T(t)B(t) = p^T(t)G(q(t)) = 0$, i.e. $u(t)$ is a singular optimal control. □

5 Dynamic Dexterity

In accordance with [10, 11], the Gram matrix (7) plays the role of a dynamic dexterity matrix of the non-holonomic system, and can be employed as a means for defining dynamic performance measures. For example, the dynamic dexterity ellipsoid at the endogenous configuration $u(\cdot)$ is obtained as the image by the Jacobian of the unit sphere in $\mathcal{U}$,

$$S_{q_0,T}(u(\cdot)) = \{v(\cdot) \in \mathscr{U} \,|\, ||v(\cdot)|| = 1\},$$

defined with respect to the norm induced by the inner product in $\mathscr{U}$. The dynamic dexterity ellipsoid

$$E_{q_0,T}(u(\cdot)) = J_{q_0,T}(u(\cdot))S_{q_0,T}(u(\cdot)) = \{\eta \in R^r \mid \eta^T \mathscr{D}^{-1}_{q_0,T}(u(\cdot))\eta = 1\} \quad (17)$$

is the image in R^r of the unit sphere by the Jacobian. It is inscribed into the sphere in R^r of radius $\overline{\lambda}^{1/2}_{\mathscr{D}_{q_0,T}(u(\cdot))}$ and circumscribed on the sphere of radius $\underline{\lambda}^{1/2}_{\mathscr{D}_{q_0,T}(u(\cdot))}$, determined, respectively, by the largest and the smallest eigenvalue of the dynamic dexterity matrix. The square root of the determinant

$$d_{q_0,T}(u(\cdot)) = \sqrt{\det \mathscr{D}_{q_0,T}(u(\cdot))},$$

describes the dynamic dexterity of the non-holonomic system at the endogenous configuration $u(\cdot)$. Observe that the dexterity at a singular configuration is zero.

6 Example

Consider a front wheel driven car shown schematically in Fig. 1. The well known control system representation of its kinematics will be extended to a control affine system including the car's kinematics and dynamics. The vector of generalized coordinates of the car $q = (q_1, \ldots, q_6) \in R^6$ describes its position, orientation, the heading angle, the forward and the steering wheel velocities. The controlled inputs will be the forward and the steering forces/torques. The output variable coincides with

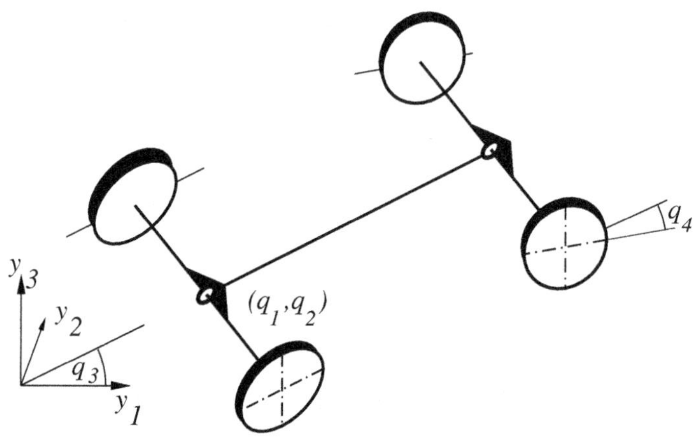

Fig. 1 Front wheel driven car

the state variable, $y = q$. The car moves without the side slip of its front and rear wheels. For simplicity all geometric and dynamic parameters of the car are assumed equal to 1. After a feedback transformation the control affine system (1) comprises the drift vector field

$$f^T(q) = (q_5 \cos q_3 \cos q_4, q_5 \sin q_3 \cos q_4, q_5 \sin q_4, q_6, 0, 0)^T,$$

and a pair of control vector fields $g(q) = e_5$, $g_2(q) = e_6$, e_i denoting the ith unit vector in R^6. It is well known that at the kinematic configuration singularities the car does not move except for arbitrary changes of its heading angle [2].

Following a procedure provided in [3], we shall compute the dynamic configuration singularities of the car. Let $p \in R^6$ denote the adjoint variable. Then, the singular extremal $(u(t), q(t), p(t))$ satisfies for $t \in [0, T]$ the following identities (for the sake of simplicity of notation the argument t will be omitted)

$$p^T f(q) = const, \quad p^T g_1(q) = 0, \quad p^T g_2(q) = 0 \ \text{ and } \ \dot{p}^T = -p^T \frac{\partial f(q)}{\partial q},$$

for $p \neq 0$. Specifically, we get $p_1 = const$, $p_2 = const$, $\dot{p}_3 = q_5 \cos q_4 (p_1 \sin q_3 - p_2 \cos q_3)$, $p_4 = p_5 = p_6 = 0$, $\dot{p}_4 = q_5((p_1 \cos q_3 + p_2 \sin q_3) \sin q_4 - p_3 \cos q_4)$ and $\dot{p}_5 = -(p_1 \cos q_3 + p_2 \sin q_3) \cos q_4 - p_3 \sin q_4$. A combination of the identities $\dot{p}_4 = 0$, $\dot{p}_5 = 0$, $\ddot{p}_4 = 0$, $\ddot{p}_5 = 0$ and $p_5^{(3)} = 0$ results in

$$((p_1 \cos q_3 + p_2 \sin q_3) \sin q_4 - p_3 \cos q_4)^2 \left(u_1^2 + u_2^2\right) = 0,$$
$$q_5(p_1 \sin q_3 - p_2 \cos q_3) = 0 \ \text{ and } \ (p_1 \cos q_3 + p_2 \sin q_3) \sin q_4 - p_3 \cos q_4 = const.$$

For the reason that $u_1^2 + u_2^2 \neq 0$ (otherwise the singular controls become zero), there must be

$$(p_1 \cos q_3 + p_2 \sin q_3) \sin q_4 - p_3 \cos q_4 = 0. \tag{18}$$

Joining this with $\dot{p}_5 = 0$ we deduce that

$$p_1 \cos q_3 + p_2 \sin q_3 = 0 \ \text{ and } \ p_3 = 0.$$

Now, consider the identity $q_5(p_1 \sin q_3 - p_2 \cos q_3) = 0$. Since

$$\frac{d}{dt}(p_1 \sin q_3 - p_2 \cos q_3) = p_1 \cos q_3 + p_2 \sin q_3 = 0,$$

we have $p_1 \sin q_3 - p_2 \cos q_3 = const$, so $q_5 = 0$. But this means that $u_1 = \dot{q}_5 = 0$. Finally, the dynamically singular configurations of the car have the form

$$(u_1(t) = 0, u_2(t) \ - \ \text{arbitrary}).$$

7 Conclusion

We have presented an analytic study of dynamic configuration singularities of non-holonomic robotic systems, parallelizing a similar study of kinematic singularities accomplished in [10]. A comparison reveals that our analytic characterization of dynamic singularities extends in a natural way the previous concept of kinematic configuration singularities. The sub-Riemannian geometry provides a means for computing both the kinematic as well as the dynamic configuration singularities. An extension of the geometric approach to the dynamic singularities of non-holonomic systems seems to be an interesting problem for further research.

Acknowledgements This research was supported by the Wrocław University of Technology under a statutory research project.

References

1. Abraham, R., Marsden, J.E., Ratiu, T.: Manifolds, Tensor Analysis, and Applications. Springer, New York (1988)
2. Chitour, Y., Sussmann, H.J.: Motion planning using the continuation method. Essays on Mathematical Robotics, pp. 91–125. Springer, New York (1988)
3. Chitour, Y., Jean, F., Trélat, E.: Singular trajectories of control-affine systems. SIAM J. Control Optim. **47**, 1078–1095 (2008)
4. Kryński, W.: Singular curves determine generic distributions of corank at least 3. J. Dyn. Control Syst. **11**, 375–388 (2005)
5. Montgomery, R.: A survey on singular curves in sub-Riemannian geometry. J. Dyn. Control Syst. **1**, 49–90 (1995)
6. Müller, A.: A proposal for a unified concept of kinematic singularities for holonomic and non-holonomic mechanisms. In: Proceedings of the 36th Mech. Robotics Conference ASME 2012 International Design Engineering Technical Conferences, pp. 1543–1552. Chicago, IL (2012)
7. Müller, A., Donelan, P.: Towards a unified notion of kinematic singularities for robot arms and non-holonomic platforms (INRIA). Advances in Robot Kinematics, pp. 397–405. Springer, Berlin (2016)
8. Tchoń, K.: On kinematic singularities of nonholonomic robotic systems. In: Proceedings of RoManSy, vol. 13, pp. 75–84. Zakopane, Poland (2000)
9. Tchoń, K., Jakubiak, J.: Endogenous configuration space approach to mobile manipulators: a derivation and performance assessment of Jacobian inverse kinematics algorithms. Int. J. Control **76**, 1387–1419 (2003)
10. Tchoń, K., Sienkiewicz, H.: Kinematic singularities of mobile manipulators: an analytic approach. Advances in Robot Kinematics, pp. 407–415. INRIA (2016)
11. Zadarnowska, K., Tchoń, K.: A control theory framework for performance evaluation of mobile manipulators. Robotica **25**, 703–715 (2007)

A Taylor-Based Continuation Method for the Determination and Classification of Robot Singularities

Gauthier Hentz, Isabelle Charpentier, Lennart Rubbert and Pierre Renaud

Abstract Robot design in contexts such as computer-assisted medical interventions remains challenging. Compact, dexterous mechanisms with particular mobilities are needed, the synthesis of which requires a systematic evaluation of workspace and singular positions. The evaluation of singular positions and their classification are still difficult to perform in a systematic manner. In this paper, an automated method is presented and evaluated on a complex planar mechanism. A higher-order continuation method is used to provide continuous and accurate representation of singular locii. Classification is then performed by testing all the existing singularity types through a direct evaluation. Only the mechanism loop-closure equations are required thanks to automatic differentiation and the Diamanlab software developed for use of continuation. The evaluation of the method shows promising results.

1 Introduction

Robot design in contexts such as computer-assisted medical interventions remains challenging. Compact, dexterous mechanisms with particular mobilities are needed, which synthesis is today still an issue. In order to help the designer, efficient and automated methods are mandatory for mechanism assessment.

Workspace is a first property to be evaluated. We have described in [6] the use of automatic differentiation (AD) and the so-called Diamant higher-order continuation approach to compute the boundaries of reachable workspace in a simple manner. Built as an extension of [5], our approach is of interest for its high accuracy, low computation time and the description of workspace boundaries as Taylor series that yields a continuous representation.

The determination of mechanism singularities is another major requirement in mechanism evaluation. The problem is complex because of the variety of singular situations for a mechanism. Six singularity types exist and up to 21 different singularity classes can be detected during a mechanism analysis [10]. Several methods on

G. Hentz (✉) · I. Charpentier · L. Rubbert · P. Renaud
ICube, Illkirch-Graffenstaden, France
e-mail: gauthier.hentz@icube.unistra.fr

J. Lenarčič and J.-P. Merlet (eds.), *Advances in Robot Kinematics 2016*,
Springer Proceedings in Advanced Robotics 4, DOI 10.1007/978-3-319-56802-7_43

singularity determination have been reported [3, 7, 8]. The determination is however usually not possible for all the singularity classes, and a general manipulator. In [10] algebraic and geometric techniques are given for finding singularity sets. Their application remains however tedious and time consuming. More recently, [1] proposed a general method based on interval analysis to perform an exhaustive numerical determination and classification of the singularities. The latter process relies on a sequential test process that can be affected by numerical accuracy issues, according to the authors.

In this paper, we investigate the extension of our continuation method [6] to propose a complete singularity analysis for general manipulators. Taking advantage of the continuity and accuracy properties of continuation, the method is based on a two-step process: firstly the determination of singularity locations, and secondly their classification. This latter phase is based on a direct evaluation of the 21 possible situations formulated in [10].

The paper is organized as follows. The singularity definition and classification is introduced in Sect. 2. The proposed determination and classification method is then detailed in Sect. 3. The method is tested in Sect. 4 by considering a planar mechanism with complex kinematic behavior. Conclusions on the potential of the method and future developments are detailed in Sect. 5.

2 Formulation of the Singularity Analysis Problem

We consider here a mechanism requiring N independent variables to define its configuration in a unique manner. Following [5], we choose to describe the mechanism *configuration* with a column-array $\mathbf{q} = [\mathbf{u}^\mathsf{T}, \mathbf{v}^\mathsf{T}, \mathbf{w}^\mathsf{T}]^\mathsf{T} \in I\!R^{n_q}$, with $\mathbf{u}$, $\mathbf{v}$ and $\mathbf{w}$ being respectively the output, input and passive coordinates. The mobility of the mechanism is denoted by n. We consider a non-redundant manipulator so that $\dim(\mathbf{u}) = \dim(\mathbf{v}) = n$ and $n_q \geq N + n$. The configuration space is the set of admissible values of $\mathbf{q}$ for the manipulator:

$$\mathscr{C} = \{\mathbf{q} | \mathbf{R}^{\mathscr{C}}(\mathbf{q}) = \mathbf{0}\} , \tag{1}$$

where $\mathbf{R}^{\mathscr{C}}(\mathbf{q}) = \mathbf{0}$ is a system of n_{eq} non-linear equations, typically the loop-closure equations for a parallel mechanism.

Feasible instantaneous motions are characterized by the *velocity equations* [10]:

$$\mathbf{R}^{\mathscr{C}}_{\mathbf{q}}(\mathbf{q})\, \dot{\mathbf{q}} = 0 , \tag{2}$$

with $\dot{\mathbf{q}} = [\dot{\mathbf{u}}^\mathsf{T}, \dot{\mathbf{v}}^\mathsf{T}, \dot{\mathbf{w}}^\mathsf{T}]^\mathsf{T}$ a column-array of velocities and $\mathbf{R}^{\mathscr{C}}_{\mathbf{q}}$ a $n_{eq} \times n_q$ Jacobian matrix. The following two kinematic problems are usually formulated:

- Inverse instantaneous kinematics problem (IIKP): Find $\dot{\mathbf{q}}$ for a prescribed $\dot{\mathbf{u}}$.
- Forward instantaneous kinematics problem (FIKP): Find $\dot{\mathbf{q}}$ for a prescribed $\dot{\mathbf{v}}$.

Singularities can be defined as indeterminations in these two problems. To express that, let sets of variables $\mathbf{z} = [\mathbf{v}^\mathsf{T}, \mathbf{w}^\mathsf{T}]^\mathsf{T}$ and $\mathbf{y} = [\mathbf{u}^\mathsf{T}, \mathbf{w}^\mathsf{T}]^\mathsf{T}$ be two sets of variables issued from $\mathbf{q}$. We note $\mathbf{R}^{\mathscr{C}}_{\mathbf{z}}$, $\mathbf{R}^{\mathscr{C}}_{\mathbf{y}}$ and $\mathbf{R}^{\mathscr{C}}_{\mathbf{w}}$ the submatrices obtained from $\mathbf{R}^{\mathscr{C}}_{\mathbf{q}}$ by removal of the columns related to $\mathbf{u}$, $\mathbf{v}$ and $(\mathbf{u},\mathbf{v})$, respectively [1].

Indetermination situations of the IIKP occur when $\mathbf{R}^{\mathscr{C}}_{\mathbf{z}}$ is rank deficient. The configurations $\mathbf{q} \in \mathscr{S}_{\mathbf{z}} \subset \mathscr{C}$ where these situations occur are called *inverse singularities* and are solutions [5] of:

$$\begin{bmatrix} \mathbf{R}^{\mathscr{C}}(\mathbf{u},\mathbf{z}) \\ \left(\mathbf{R}^{\mathscr{C}}_{\mathbf{z}}(\mathbf{u},\mathbf{z})\right)^\mathsf{T}\,\xi \\ \xi^\mathsf{T}\xi - 1 \end{bmatrix} = 0\,, \tag{3}$$

where $\xi \in I\!R^{n_{eq}}$. The two lower terms express the rank deficiency of the Jacobian $\mathbf{R}^{\mathscr{C}}_{\mathbf{z}}$.

Indetermination situations of the FIKP constitute a set $\mathscr{S}_{\mathbf{y}}$ of solutions to

$$\begin{bmatrix} \mathbf{R}^{\mathscr{C}}(\mathbf{v},\mathbf{y}) \\ \left(\mathbf{R}^{\mathscr{C}}_{\mathbf{y}}(\mathbf{v},\mathbf{y})\right)^\mathsf{T}\,\xi \\ \xi^\mathsf{T}\xi - 1 \end{bmatrix} = 0\,. \tag{4}$$

Solutions to systems (3) and (4) define all singularities of a mechanism [1].

In order to express the nature of singularities for a mechanism, a classification has to be performed. A classification of the singularities in 6 singularity types was introduced in [10] with redundant input (RI), redundant output (RO), redundant passive motion (RPM), impossible input (II), impossible output (IO), and increased instantaneous motion (IIM) singularities. A set of 6 criteria allows such classification [10]:

- $\mathbf{q} \in \{RI\} \Leftrightarrow rank(\mathbf{R}^{\mathscr{C}}_{\mathbf{z}}) < rank(\mathbf{R}^{\mathscr{C}}_{\mathbf{w}}) + n$
- $\mathbf{q} \in \{RO\} \Leftrightarrow rank(\mathbf{R}^{\mathscr{C}}_{\mathbf{y}}) < rank(\mathbf{R}^{\mathscr{C}}_{\mathbf{w}}) + n$
- $\mathbf{q} \in \{RPM\} \Leftrightarrow rank(\mathbf{R}^{\mathscr{C}}_{\mathbf{w}}) < N - n$
- $\mathbf{q} \in \{II\} \Leftrightarrow rank(\mathbf{R}^{\mathscr{C}}_{\mathbf{y}}) < rank(\mathbf{R}^{\mathscr{C}}_{\mathbf{q}})$
- $\mathbf{q} \in \{IO\} \Leftrightarrow rank(\mathbf{R}^{\mathscr{C}}_{\mathbf{z}}) < rank(\mathbf{R}^{\mathscr{C}}_{\mathbf{q}})$
- $\mathbf{q} \in \{IIM\} \Leftrightarrow rank(\mathbf{R}^{\mathscr{C}}_{\mathbf{q}}) < N$

A singular position can be of several types at the same time. A total of 21 classes exists [10], each one corresponding to a particular degenerate kinematic behavior of the mechanism. Indetermination of either the IIKP or the FIKP are the most common and studied cases and correspond respectively to the well known *output* (also designated as serial or (RI,IO)) singularities or *input* (also designated as parallel or (RO,II)) singularities. Configurations satisfying both systems (3) and (4) can belong to any of the other singularity classes. The corresponding singularity sets are typically of lower dimension than the whole singularity set, implying different kinematic behavior [1].

3 A Fully Automated Computation Method

The method we propose is based on computation of singularities followed by their classification.

Determination of the Singularity Set

The solutions to the systems (3) and (4) are composed of several continuous sets, that can be designated as solution branches. We propose to follow solution branches from an initial point by a numerical continuation technique. Intersections between solution branches can occur. This corresponds to so-called bifurcation points. At such points, the continuity of the solution set is broken and rank deficiency of the Jacobian of the system occurs [9]. The detection of such points allows us to locate branch intersections. A branch switching may then be carried out, allowing a complete determination of all the connected branches from one initial point. If the singularity set is one-dimensional, continuation can be directly applied. If not, linear relationships between the coordinates can be added to discretize the determination of solution sets, as described in [5] and used in [6].

With a classical first-order continuation technique, branches are computed step-by-step as a collection of solution points. With *higher-order continuation*, the branches of solutions are computed as Taylor series which constitute continuous and accurate representations. The system to be solved needs however a formulation that usually requires complex and time-consuming additional work. On the contrary, Diamant[1] [2], the numerical method we use, is an implementation of higher-order continuation with integration of automatic differentiation (AD). Using this framework, non-linear equations describing a mechanism behavior are directly implemented in their standard format and are being solved in a fully automated manner. Bifurcation detection and branch switching are carried out automatically as well. In addition, the Tapenade software [4] is used to build the Jacobian $\mathbf{R}_{\mathrm{q}}^{\mathscr{C}}$ from system (1) at no cost, benefiting again from AD. Finally our method builds on AD and Diamant to propose a fully automated framework for singularity analysis, allowing to generate and solve systems (3) or (4) in a fully automated manner. The reader is referred to [6] and the references therein for further explanations on AD and Diamant.

In addition to this exploration technique of the singularity sets, we propose to monitor the rank deficiency of $\mathbf{R}_{\mathrm{y}}^{\mathscr{C}}$ (respectively $\mathbf{R}_{\mathrm{z}}^{\mathscr{C}}$) during the determination of solutions of (3) (respectively (4)). In this way, if a rank deficiency of the monitored Jacobian occurs, the corresponding configuration is located at the intersection of $\mathscr{S}_{\mathrm{z}}$ and $\mathscr{S}_{\mathrm{y}}$. We can therefore identify the input and output singularities and also the other types of singularities. This means we can identify precisely and immediately configurations that can belong to any of the 21 singularity classes. When the described

[1] Stands for *Différentiation Automatique de la Méthode Asymptotique Numérique Typée*.

rank deficiency is detected, we can in addition use the corresponding points to switch from branches of $\mathscr{S}_{\mathbf{z}}$ to branches of $\mathscr{S}_{\mathbf{y}}$ and conversely. There is then no need for another initial point to compute all the connected branches of $\mathscr{S}_{\mathbf{z}} \cup \mathscr{S}_{\mathbf{y}}$.

Thanks to AD, Diamant, monitoring of rank deficiency and automated branch switching, we obtain the whole connected singularity set $\mathscr{S}_{\mathbf{z}} \cup \mathscr{S}_{\mathbf{y}}$ as a grid of solution branches connected by singular points. The determination step can be finally described as:

1. Determination:
   a. Automatic differentiation of system (1) with respect to $(\mathbf{u}, \mathbf{v}, \mathbf{w})$ using Tapenade software [4] to build the Jacobian $\mathbf{R}_{\mathbf{q}}^{\mathscr{C}}$ and the submatrices $\mathbf{R}_{\mathbf{z}}^{\mathscr{C}}$, $\mathbf{R}_{\mathbf{y}}^{\mathscr{C}}$ and $\mathbf{R}_{\mathbf{w}}^{\mathscr{C}}$,
   b. Computation of the singularities by solving systems (3) and (4) with the non-linear solver Diamant [2]. Taylor series allow for the local construction of solution branches. Monitoring of the rank deficiency of $\mathbf{R}_{\mathbf{y}}$ during continuation on system (3), and conversely, allows to detect points of $\mathscr{S}_{\mathbf{z}} \cap \mathscr{S}_{\mathbf{y}}$.

Singularity Classification

The solution branches of either (3) or (4) can be directly classified respectively in (RI,IO) or (RO,II) singularity classes. Other detected singular points can belong to any of the other existing singularity classes, which requires additional classification steps. Thanks to the high accuracy in the estimation of singular points, a direct classification of the solutions of $\mathscr{S}_{\mathbf{z}} \cap \mathscr{S}_{\mathbf{y}}$ can be performed. Each point is tested from the propositions given in Sect. 2. The classification step can be finally described as:

2. Classification of the points detected in $\mathscr{S}_{\mathbf{z}} \cap \mathscr{S}_{\mathbf{y}}$ using the test functions. For each point:
   a. Computation of the ranks of matrices $\mathbf{R}_{\mathbf{q}}^{\mathscr{C}}$, $\mathbf{R}_{\mathbf{w}}^{\mathscr{C}}$, $\mathbf{R}_{\mathbf{y}}^{\mathscr{C}}$ and $\mathbf{R}_{\mathbf{z}}^{\mathscr{C}}$,
   b. Determination of the singularity types from the criteria given in Sect. 2,
   c. Determination of the singularity class (see Table 1 in [10]) by testing in a successive manner membership to (a) combinations of RI, RO, RPM types (rows of the Table), (b) IIM type, (c) IO type and (d) II type.

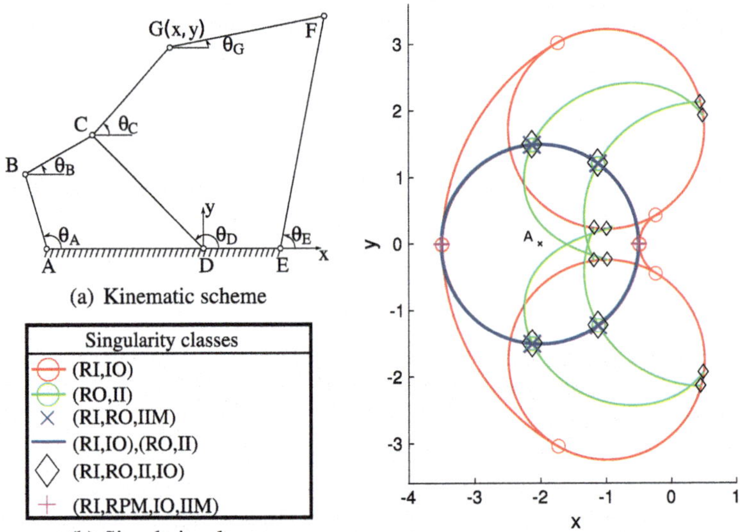

(a) Kinematic scheme

(b) Singularity classes

(c) Set $\mathscr{S}_\mathbf{z} \cup \mathscr{S}_\mathbf{y}$ obtained with our method.

Fig. 1 Singularity set **(c)** of the double-loop mechanism depicted in **(a)** after projection onto the space of the x and y variables. Each singularity in the configuration space $\mathscr{C}$ is classified according to the singularity classes indicated in **(b)**

4 Application to a 2-DOF Double-Loop Manipulator

Manipulator Description

In order to assess the efficiency of the method, the 2-DOF planar manipulator depicted in Fig. 1a is considered. This mechanism is a double-loop parallel mechanism, and it possesses two mobilities ($n = 2$). In terms of singularity analysis, it is very challenging, especially with the geometric parameters issued from [1], for which all existing singularity types occur. Its planar nature allows at the same time a comprehensive interpretation of the kinematic behavior implied by each class of singularity, see [1], which makes it a very interesting case study for the method benchmark.

The constraint equations describing the mechanism kinematics, adapted from [1], are obtained by writing the closure equations for both loops and two equations that constrain the position of point G:

$$\mathbf{R}^{\mathscr{C}}(\mathbf{q}) = \begin{bmatrix} CD\cos(\theta_D) + CG\cos(\theta_C) - x \\ CD\sin(\theta_D) + CG\sin(\theta_C) - y \\ AB\cos(\theta_A) + BC\cos(\theta_B) - CD\cos(\theta_D) - AD \\ AB\sin(\theta_A) + BC\sin(\theta_B) - CD\sin(\theta_D) \\ CD\cos(\theta_D) + CG\cos(\theta_C) + FG\cos(\theta_G) - EF\cos(\theta_E) - DE \\ CD\sin(\theta_D) + CG\sin(\theta_C) + FG\sin(\theta_G) - EF\sin(\theta_E) \end{bmatrix} = \mathbf{0}\,, \quad (5)$$

with θ_A and θ_E the actuated joint coordinates, x and y the coordinates of point G. The selected passive coordinates are θ_B, θ_C, θ_D and θ_G. Geometric parameters are defined as $AB = BC = DE = 1$, $AD = CD = FG = 2$, $CG = 1.5$, $EF = 3$.

Results and Interpretation

The implementation is performed using Taylor series at a truncated order of 20 and a precision of $1e-6$ on the residual [2]. Eigenvalues of square matrices and singular values of rectangular matrices are considered null if inferior to a threshold of $1e-5$. The solution branches corresponding to the output and input singularity locii are defined in the configuration space $\mathscr{C}$. For sake of representation they are projected onto the space of the x and y variables only and plotted in Fig. 1c, using respectively green and red lines.

Bifurcation points detected automatically during the continuation process are indicated with a circle-shape marker ($\circ$). They allow us to detect branch intersections in $\mathscr{S}_\mathbf{z}$ or $\mathscr{S}_\mathbf{y}$. Intersection points between the output and input singularity sets, detected by monitoring the simultaneous rank deficiency of both $\mathbf{R_z}$ and $\mathbf{R_y}$, are indicated by the other markers. As expected, detection of the intersection points and of the bifurcation points allow for the computation of all connected solution branches of $\mathscr{S}_\mathbf{z} \cup \mathscr{S}_\mathbf{y}$. All the singularities other than input or output singularities are then classified by using the second step of the method.

The accuracy in the determination of the branches and the location method allows us to assign a detected point to the class it belongs to in a straightforward manner. All connected singularities can thus be entirely determined and classified. Markers using multiplication symbol ($\times$) designate (RI,RO,IIM) configurations. The points depicted as diamonds ($\diamond$) correspond to simultaneous output and input singularities (RI,RO,II,IO). Two points belonging to (RI,RPM,IO,IIM) class are indicated with plus-shape markers ($+$). Some points on the projected diagram are referenced with several classes. This means they correspond to several configurations in $\mathscr{C}$, because a configuration must belong to a unique class [10]. The blue circle of center A is in particular the superimposition of (RI,IO) and (RO,II) branches. The known singularities are obtained, at the exception of singularities that are superimposed to the blue circle in Fig. 1c (see also [1], Fig. 4). Further analysis shows that the gain of mobility in that situation would require an exploration with continuation along an additional direction. This would require as described earlier the introduction of a discretization and an additional relationship in the system, a development to be investigated.

5 Discussion and Conclusion

In this paper, we have introduced the use of higher-order continuation for accurate and automatic determination of singularity sets. The accuracy of Taylor series computation together with automatic differentiation allows us to easily implement the singularity determination problem from loop-closure equations and perform an unambiguous classification of the obtained singularities. This is of particular interest for the designer since only very little information on the mechanism is needed to conduct the analysis.

The evaluation is promising, with the estimation of all input and output singularities, and more complex situations as described in Fig. 1. Only one set of IIM singular positions is today not determined. Their estimation requires an improvement of the exploration strategy by continuation that will be investigated.

Future work will be focused on the evaluation of the method by considering other 3D mechanisms. It will be also interesting to combine our method with the interval-based method from [1] to get accurate and efficient determination of singularities as well as an exhaustive determination, even in difficult cases of disconnected sets, without any initial knowledge on the singular configurations. Sensitivity analysis also provided by the Diamant framework will then also be exploited to provide the designer with a tool for optimization of the mechanism geometry.

Acknowledgements This work was supported by the French government research program Investissements d'avenir through the Robotex Equipment of Excellence (ANR-10-EQPX-44).

References

1. Bohigas, O., Zlatanov, D., Manubens, M., Ros, L.: On the numerical classification of the singularities of robot manipulators. In: ASME 2012 International Design Engineering Technical Conferences, pp. 1287–1296 (2012)
2. Charpentier, I.: On higher-order differentiation in nonlinear mechanics. Optim. Methods Softw. **27**, 221–232 (2012)
3. Gosselin, C., Angeles, J.: Singularity analysis of closed-loop kinematic chains. IEEE Trans. Robot. Autom. **6**(3), 281–290 (1990)
4. Hascoet, L., Pascual, V.: The Tapenade automatic differentiation tool: principles, model, and specification. ACM Trans. Math. Softw. **39**, 20:1–20:43 (2013)
5. Haug, E.J., Luh, C.M., Adkins, F.A., Wang, J.Y.: Numerical algorithms for mapping boundaries of manipulator workspaces. ASME J. Mech. Design **118**, 228–234 (1996)
6. Hentz, G., Charpentier, I., Renaud, P.: Higher-order continuation for the determination of robot workspace boundaries. Comptes Rendus Mécanique **344**(2), 95–101 (2016)
7. Merlet, J.P.: A formal-numerical approach for robust in-workspace singularity detection. IEEE Trans. Robot. **23**(3), 393–402 (2007)
8. Park, F., Kim, J.W.: Singularity analysis of closed kinematic chains. J. Mech. Design **121**(1), 32–38 (1999)
9. Seydel, R.: Practical Bifurcation and Stability Analysis, 3rd edn. Springer, Berlin (2009)
10. Zlatanov, D., Fenton, R.G., Benhabib, B.: A unifying framework for classification and interpretation of mechanism singularities. ASME J. Mech. Design **117**(4), 566–572 (1995)

Identifying Singularity-Free Spheres in the Position Workspace of Semi-regular Stewart Platform Manipulators

Anirban Nag, Vikranth Reddy, Saurav Agarwal and Sandipan Bandyopadhyay

Abstract This paper presents a method to compute the largest sphere inside the position-workspace of a semi-regular Stewart platform manipulator, that is free of gain-type singularities. The sphere is specific to a given orientation of the moving platform, and is centred at a designated point of interest. The computation is performed in two parts; in the first part, a Computer Algebra System (CAS) is used to derive a set of exact symbolic expressions, which are then used further in a purely numerical manner for faster computation. The method thus affords high computation speed, while retaining the exactness and generic nature of the results. The numerical results are validated against those obtained from an established numerical algebraic geometry tool, namely, `Bertini`, and are illustrated via an example.

1 Introduction

This paper presents a method for finding a sphere inside the position workspace of a semi-regular Stewart platform manipulator (SRSPM), which is free of *gain-type*[1] singularities. The singularity-free sphere (SFS) is derived for a given orientation of the moving platform, and is centred at a designated point of interest. The choice of this point is typically motivated by the intended applications of the manipulator.

[1]Gain-type singularities (also known as *type-II singularities*) occur when the forward kinematic solutions of a manipulator merge. See [1] and the references therein for more details.

A. Nag · S. Agarwal · S. Bandyopadhyay (✉)
Department of Engineering Design, Indian Institute of Technology Madras,
Chennai 600036, India
e-mail: nag.anirban16@gmail.com

S. Agarwal
e-mail: agr.saurav1@gmail.com

S. Bandyopadhyay
e-mail: sandipan@iitm.ac.in

V. Reddy
Systemantics India Pvt. Ltd., Bengaluru, India
e-mail: vikranthiit@gmail.com

J. Lenarčič and J.-P. Merlet (eds.), *Advances in Robot Kinematics 2016*,
Springer Proceedings in Advanced Robotics 4, DOI 10.1007/978-3-319-56802-7_44

The identification of such an SFS facilitates several aspects of path-planning and design of such manipulators. As the sphere describes a convex region in $\mathbb{R}^3$, it is obvious that any path consisting of line segments is free of (gain-type) singularities, so long as the end-points of the segments are *inside* the SFS. For any given manipulator, such a calculation needs to be done *only once*, for any given orientation. If such an SFS is to be identified for a range of orientations of the moving platform, then one can scan the said range (up to some desired resolution), and identify the smallest SFS, which would be free of singularities for the entire range of orientations. Identifying such an SFS forms an important part of computing the "Safe Working Zone (SWZ)" of such a manipulator, where the manipulator can operate without encountering singularities and other issues, as explained in [11]. It is, therefore, possible to think of a design algorithm, to invert the problem, and identify the geometric parameters which would allow the manipulator to be free of singularities over a desired range of orientations, and a spherical region in $\mathbb{R}^3$.

Motivated by such utilities, several attempts have been made in the recent times to obtain such an SFS, or variants of the same. Determination of the maximal SFS in the orientation workspace, parametrised by Euler angles for the Minimal Simplified Symmetric Manipulator (MSSM) has been presented by Jiang et al. [7]. Li et al. [8] have tried to solve this problem in the six-dimensional space of rigid body motions, $\mathbb{SE}(3)$, by finding a sphere that is tangential to the gain-singularity manifold in this space. The formulation, however, seems to lack mathematical rigour for several reasons, as explained below. It is well-known that $\mathbb{SE}(3)$ does not admit a bi-invariant Riemannian metric (see, e.g., [9], Corollary A.5.1, pp. 427), and hence the notion of "distance" or length in $\mathbb{SE}(3)$ is *non-unique*. Thus, the application of the Euclidean metric to define a sphere in $\mathbb{SE}(3)$ is mathematically inaccurate, given that the Euclidean metric is a bi-invariant one. Furthermore, because of the non-existence of a unique "natural/characteristic length" in $\mathbb{SE}(3)$, the results obtained by the application of this method are always subject to the choice of the assumed *characteristic length*, and have therefore limited value in any generic problem. Also, it is not clear as to how the eliminations were implemented to solve the system of equations, and the corresponding computational efforts involved are not mentioned. Finally, in the process of solution, the number of solutions is stated to be 81, which is much higher than the total-degree Bézout's number of 27.

In this paper, the formulation adheres to the standard definition of a sphere in $\mathbb{R}^3$, and accordingly, the SFS is computed only in the position space. Thus, there is an SFS for each point in $\mathbb{SO}(3)$ which is accessible to the moving platform. The formulation is therefore free of any mathematical inaccuracies, and it renders the problem to be solvable analytically. The analytical description of the singularity manifold of an SRSPM is available in [2], which is used in this work. The formulation leads to three cubic equations in the coordinates of the point of tangency between the SFS and the singularity surface. By elimination of two of the coordinates, a univariate polynomial of degree 48 is obtained in the remaining one. It may be noted that the degree of the final polynomial is still higher than the theoretical limit of 27, but is closer to the same. The coefficients of this polynomial are computed *exactly*, via a series of intermediate expressions which are evaluated numerically in the end. Thus, the entire formulation is implemented symbolically and the final univariate polynomial expression obtained in a manner, which can be ported to any numerical programming environment like C or C++, thereby making the steps performed inside CAS a one-time procedure. The roots obtained are validated numerically, as well as against the numerical algebraic geometry (NAG) tool Bertini [3], and the solution are illustrated geometrically.

It may be noted that a complementary formulation of the problem is feasible, i.e., a singularity-free sphere could be identified in $\mathbb{SO}(3)$, for a given position of the end-effector. It is mathematically consistent, when the Euclidean distance is used in conjunction with the quaternion-based representation of $\mathbb{SO}(3)$ [6, 10]. However, the computations required are very demanding in this case, as the problem is defined in terms of four polynomials, one of total degree 2, and the rest of total degree 6 each, resulting in a Bézout number of 432, which puts this problem out of the scope of the present work.

The rest of the paper is organised as follows: in Sect. 2 the mathematical formulation of the problem is described, followed by the solution of the resulting equations. The results are described in Sect. 3. Finally, the paper is concluded in Sect. 4.

2 Mathematical Formulation

This section describes the geometry of the manipulator and the derivation of the equations describing the SFS in the position workspace of the SRSPM.

2.1 Geometry of the Manipulator

The SRSPM has *semi-regular* hexagonal top and bottom platforms, with alternate sides in each platform having equal lengths. The angular spacings between the adjacent pairs of legs are denoted by $2\gamma_t$ and $2\gamma_b$ for the top (see Fig. 1a, b) and the bottom platforms, respectively. Without any loss of generality, the radius of the circum-circle of the bottom platform is scaled to unity, thus rendering all the linear dimensions unit-less in this work. The circum-radius of the top platform is denoted by r_t. The orientation of the top platform is represented by the Rodrigue's parametrisation (see, e.g., [5], pp. 31) of $\mathbb{SO}(3)$, namely, $\{c_1, c_2, c_3\}$.

2.2 Derivation of the SFS Equations

The objective of this work is to find the largest sphere in $\mathbb{R}^3$, centred at a given point of practical interest, say, $\mathbf{p}_0 = \{x_0, y_0, z_0\}^\top$. The formulation is motivated by the observation that such a sphere would be the *smallest* among those tangential to the singularity surface in $\mathbb{R}^3$. Thus, the first and the main task is to find *all* the

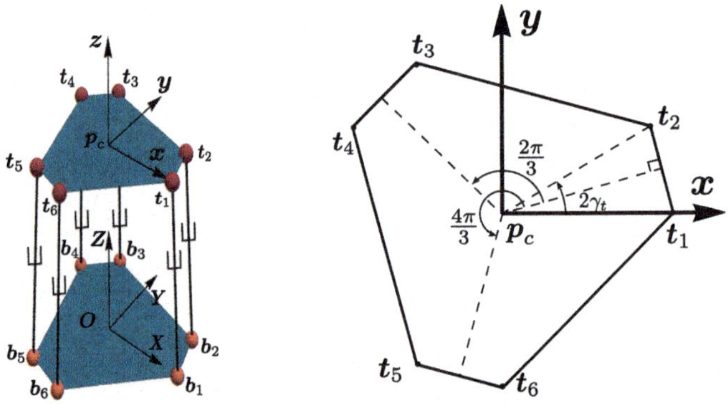

(a) Schematic diagram of SRSPM. (b) Geometry of the top platform.

Fig. 1 Architecture of the SRSPM manipulator

spheres centred at $\mathbf{p}_0$, which are tangential to the singularity surface. In this case, the singularity surface is given by $f(x, y, z) = 0$, where:

$$f(x, y, z) = a_1x^2z + a_2x^2 + a_3xyz + a_4xy + a_5xz^2 + a_6xz + a_7x + a_8y^2z + a_9y^2 + a_{10}yz^2 + a_{11}yz + a_{12}y + a_{13}z^3 + a_{14}z^2 + a_{15}z + a_{16}. \quad (1)$$

The coefficients $a_i \in \mathbb{R}$ depend only on the orientation parameters c_1, c_2, c_3, and the architecture parameters γ_b, γ_t, and r_t [2]. The equation of the sphere is given by:

$$g(x, y, z) = (x - x_0)^2 + (y - y_0)^2 + (z - z_0)^2 - r^2 = 0, \quad (2)$$

where r is the radius of the sphere, and $\mathbf{p} = \{x, y, z\}^\top$ is the point of tangency between the sphere and the singularity surface. Therefore, at $\mathbf{p}$, the normals to these two surfaces should align (see Fig. 2), giving rise to the tangency conditions:

$$\nabla \mathbf{f} \times \nabla \mathbf{g} = \mathbf{0} \Rightarrow h_i(x, y, z) = 0, \; i = 1, 2, 3. \quad (3)$$

As only two of the equations $h_i = 0$ are linearly independent, any two of the three can be taken in combination with the equation defining the singularity surface, namely, Eq. (1), to complete the set of three equations in the three unknowns, x, y, z. Each real root of these equations leads to a sphere that is tangent to the singularity surface. The one with the smallest value of r among these is the SFS.

2.3 Solution Procedure

The degrees of h_1, h_2, h_3 in x, y, z individually are found to be $\{2, 3, 3\}$, $\{3, 2, 3\}$ and $\{2, 2, 2\}$, respectively, while the total degree in x, y, and z equals 3 in each case. In

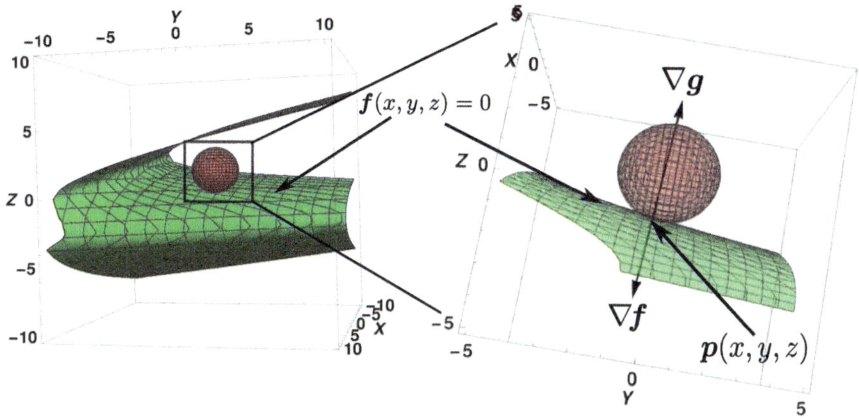

Fig. 2 Tangency of sphere with the singularity surface at the point $\mathbf{p}(x, y, z)$

view of these, $h_1 = 0$, $h_3 = 0$ are chosen for the solution process, alongside $f = 0$. From these equations, x, y are eliminated sequentially,[2] as shown schematically in Eq. (4):

$$\left.\begin{matrix} f(x,y,z)=0 \\ h_1(x,y,z)=0 \\ h_3(x,y,z)=0 \end{matrix}\right) \xrightarrow{\times x} \left.\begin{matrix} g_1(y,z)=0 \\ g_2(y,z)=0 \end{matrix}\right) \xrightarrow{\times y} g_3(z)=0. \tag{4}$$

In the above, "$\xrightarrow{\times x}$" denotes the elimination of the variable x from two or more equations in x, via computation of resultants with respect to x. The functions g_1 and g_2 have degrees $\{4, 8\}$ and $\{6, 7\}$ in y and z, respectively. However, g_2 is of the form yg_2', i.e., g_2' is of degree 5 in y. The variable y is eliminated between $g_1 = 0$ and $g_2' = 0$ (under the assumption $y \neq 0$; the case $y = 0$ is treated separately) using Bézout's method, leading to a Bézout matrix of size 5×5. Direct expansion of the determinant of this matrix leads to a polynomial in the only remaining unknown, z. However, the *size*[3] of the resulting symbolic expression is huge (about 29 GB). The time taken for expanding the determinant symbolically is about 17 min. The time taken for evaluating the determinant and the complexity of the resulting expression, makes this method computationally inefficient and practically eliminates the chance of it being used to find the SFS for a range of orientations, as a part of a larger but more relevant analysis/design problem.

In order to overcome the above-mentioned drawbacks, *a cascaded approach* was adopted to evaluate the 5×5 determinant, wherein it is expanded first in terms of five 4×4 sub-determinants, which, in turn are expanded in terms of 20 (of which

[2]It may be noted that many different elimination sequences are possible. The one presented here resulted in relatively smaller degrees of the intermediate and final polynomials.

[3]The "size" of an expression in this context indicates the amount of memory required to store the expression in the internal format of the computer algebra system (CAS) used, namely, `Mathematica`.

only 10 are distinct) sub-determinants of size 3×3. Thus the coefficients of the final univariate polynomial, $g_3(z)$, are obtained in terms of two stages of intermediate expressions. Firstly, each of the 3×3 determinants are obtained in closed-form, in terms of the coefficients a_j (defined in Eq. (1)):

$$\Delta_{3i} = \sum_{k=1}^{29} b_{ik}(a_j) z^{k-1}, \; i = 1, \ldots, 20; \; j = 1, \ldots, 16. \quad (5)$$

The new sets of coefficients, b_{ik}, are obtained as closed-form expressions in terms of the original coefficients, a_j. In the next step, the five 4×4 determinants are obtained in a similar manner, leading to the new set of coefficients c_{ik}:

$$\Delta_{4i} = \sum_{k=1}^{40} c_{ik}(b_{lm}) z^{k-1}, \; i = 1, \ldots, 5; \; l = 1, \ldots, 20; \; m = 1, \ldots, 29. \quad (6)$$

Finally, the required 5×5 determinant is computed in terms of the 4×4 determinants, and is cast as a polynomial in z:

$$\Delta_5 = \sum_{i=1}^{49} d_i(c_{jk})\, z^{i-1}, \; j = 1, \ldots, 5; \; k = 1, \ldots, 40. \quad (7)$$

Therefore, the final univariate equation in z is obtained as:

$$g_3(z) = \Delta_5 = 0. \quad (8)$$

Equation (8) is solved to find all the 48 solutions of z. The real solutions of z are used to find the values of x and y, and the radius of the desired sphere is obtained. These steps are explained with the help of a numerical example in the next section.

Symbolic expansion of the determinant of each of the 3×3 matrices takes an average of 0.5 s, and their original size is about 30 MB each. However, after symbolic simplification using the built-in `Mathematica` routine `Simplify`, the sizes of these determinants vary from 6.897 to 12.791 MB, with a total size of 93.158 MB (for the ten *unique* determinants). The actual coefficients of the 3×3 determinants are then replaced by the intermediate dummy variables (see Eq. (5)). Proceeding further, the sizes of the five 4×4 determinants (defined as Δ_{4i} in Eq. (6)) are found to be (in MB): 1.083, 0.971, 0.892, 1.067, and 1.267, respectively. The final determinant, Δ_5, is obtained in a similar manner.

These steps of computing the final set of coefficients d_i starting from the inputs a_l allow much faster computation (i.e., 11 s), and also leads to simpler final expressions. The univariate equation, $g_3(z) = 0$, consists of a total of 49 terms, with a cumulative size of 1.842 MB, while the largest term among these is only 160 KB in size. The comparison between the symbolic and the numeric computations of $g_3(z)$, in terms of the computational efforts and sizes of the expressions involved, are presented in Table 1.

Table 1 Comparison between the direct symbolic evaluation of the 5 × 5 determinant and the proposed approach. CPU specifications of the computer used: Intel(R) Core(TM) i7-4790 CPU running at a clock speed of 3.6 GHz, with 16 GB RAM

	Symbolic evaluation	Numerical evaluation
Software used	`Mathematica`, Symbolic mode	`Mathematica`, Numeric mode, with default working precision
Size of expressions	Final univariate polynomial, $g_3(z)$, (obtained by direct expansion of the 5 × 5 determinant): 29 GB	Final univariate polynomial $g_3(z)$ (computed following Eqs. (5)–(7)): 1.842, 100.280 MB inclusive of all intermediate expressions
Time taken	17 min and 33 s	11 s

It is important to note, that the expressions lead to the *exact* values of the final coefficients, subject only to the working precision of the numerical computation environment used. More importantly, it allows for a purely numerical implementation of the solution process (e.g., in `C` or `C++`) without either impacting the exact nature of the computation of the coefficients or restricting the computation to the symbolic computation environment of a CAS. Another point worth noting is that once the coefficients are obtained till the last level, the process need not be repeated, when the point of interest (centre of the sphere) or the architectural parameters of the SRSPM is changed. It also paves the way for computationally efficient scanning of the orientation workspace of the manipulator for finding the smallest SFS.

3 Results

A sample problem was solved in CAS `Mathematica` [12] version 10.4 using the default working precision of the system. The values of the architecture parameters are adopted from [4]: $\gamma_t = 0.0863$ rad, $\gamma_b = 0.0835$ rad, and $r_t = 0.8479$ (after scaling the base circum-radius r_b to 1). The fixed centre of the SFS is taken to be at $\mathbf{p}_0 = \{0, 0, 1.9500\}^\top$. The orientation parameters were taken to be $c_1 = 0.1013$, $c_2 = 0.0368$, and $c_3 = 0.2962$. The monic form of Eq. (8) for these inputs is given below (Fig. 3):

$$z^{48} + 4.4567 \times 10^{15} z^{47} + 3.9157 \times 10^{19} z^{46} + 8.4802 \times 10^{21} z^{45} + 8.8816 \times 10^{23} z^{44} - \cdots$$
$$+ 4.0056 \times 10^{64} z^3 + 3.2054 \times 10^{64} z^2 - 7.7684 \times 10^{64} z + 1.2071 \times 10^{64} = 0. \tag{9}$$

Bézout's limit for the number of solutions in this case was 3 × 3 × 3 = 27. The higher degree of Eq. (9) indicates introduction of spurious solutions in the process of elimination of variables. Therefore, after completing the solutions with the corresponding values of x, y, the original set of equations (i.e., Eq. (4)) are used to filter out any such

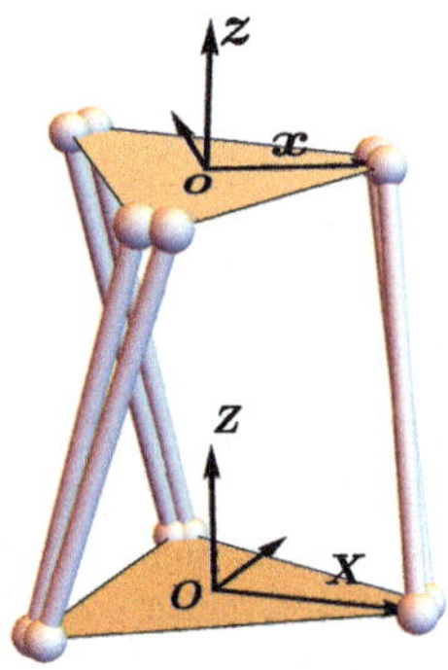

Fig. 3 Manipulator pose for the given input parameters

solutions. Only three sets of real solutions survive this test, producing residues of the order of 10^{-23}: $\{x, y, z\} = \{-0.4384, -0.3125, 0.1696\}$, $\{-0.3996, -6.4295, 2.2232\}$, $\{0.3653, -0.7859, 4.7318\}$. The corresponding values of r are: 1.8599, 6.4477, and 2.9137. Therefore, the SFS has a radius of 1.8599 for the given inputs. The actual tangency is depicted in Fig. 2. For spheres with radii greater than the minimum radius, the sphere may be tangential to the singularity surface at one point, and intersect the surface at another point, thus making them irrelevant for the purpose at hand. Figure 4a depicts the sphere with minimal radius that is tangential to the singularity surface. Figure 4b shows the sphere with radius 2.9137, which, though tangential at one point, actually cuts the singularity surface at several places.

The above solutions were obtained for the case $y \neq 0$. For $y = 0$, the obtained solutions were $\mathbf{p} = \{-66.3514, 0, 5.6010\}^{\top}$, and the corresponding $r = 66.4518$, which is more than the minimum radius already obtained. Hence, the above-reported

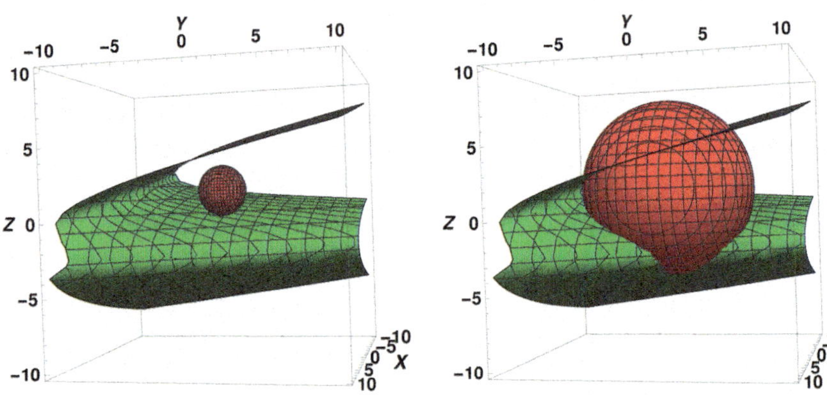

(a) $r = 1.8599$ (b) $r = 2.9137$

Fig. 4 Tangency of the minimal sphere with the singularity surface radius of the SFS holds. The residues obtained on substituting this solution set in the original set of equations was found to be of the order of 10^{-17}.

The results obtained above are prone to numerical errors, due to the high degree of the final equation in z and the huge variations in the order of magnitude of the values of the coefficients in Eq. (9). Thus, it is desirable to solve the system in Eq. (4) using another method, in order to assess the correctness of the solutions obtained. For that purpose, the NAG tool, `Bertini` [3] is used, which is well-known for its capability to compute *all* the solutions of a given polynomial system to a desired level of accuracy. As expected, `Bertini` finds only 27 solutions, of which 22 are finite, and the others escape to infinity. The real solutions match the solutions obtained above up to 10 digits after the decimal point, establishing the correctness of the solutions obtained.

4 Conclusion

A method for computing the largest gain-type singularity-free sphere inside the workspace of the SRSPM has been presented in this paper. The said sphere is a subset of the position workspace of the manipulator, and is derived for a given orientation of the moving platform. The formulation leads to three cubic equations in the coordinates of the point of tangency between the sphere and the singularity surface. A method is presented to derive a univariate equation of degree 48 from these three equations, such that all the coefficients of the intermediate as well as the final polynomials are computed *exactly*, albeit in a numerical manner. This is the main contribution of the paper, which allows, perhaps for the first time, fast computation of these spheres inside a purely numerical computation environment, without losing the accuracy of the solutions obtained. Although `Mathematica` was used to perform the numerical computations, none of the symbolic capabilities of `Mathematica` were made use of in the numerical evaluation of the coefficients.

There are existing numerical techniques, which allow the problem to be solved in a completely numerical framework. For example, Sylvester's dialytic method, leads to a matrix which has polynomial entries in a single variable. This matrix can be used to solve a generalised eigenproblem, where the eigenvalues of the system are the same as the roots obtained by solving the univariate polynomial after expanding the determinant. There exist efficient eigensolvers, which are capable of solving the problem. However, the methods being purely numerical, they have difficulties of their own. It is hard to ensure the numerical accuracy of the solutions, in particular, when fixed precision computational environments are used.

The cascaded approach presented in this paper produces the coefficients of the final univariate in their exact forms, thereby allowing accurate computations of these coefficients in a purely numerical environment. Work is in progress to implement the method presented in `C++`, so as to speed up the computations even more. Furthermore, it is intended to use this method in the computation of the SWZ of SRSPM and more general Stewart platform manipulators, as a part of their design process.

References

1. Agarwal, A., Nasa, C., Bandyopadhyay, S.: Dynamic singularity avoidance for parallel manipulators using a task-priority based control scheme. Mech. Mach. Theory **96, Part 1**, 107–126 (2016). doi:10.1016/j.mechmachtheory.2015.07.013
2. Bandyopadhyay, S., Ghosal, A.: Geometric characterization and parametric representation of the singularity manifold of a 6-6 Stewart platform manipulator. Mech. Mach. Theory **41**(11), 1377–1400 (2006)
3. Bates, D.J., Hauenstein, J.D., Sommese, A.J., Wampler, C.W.: Bertini: software for numerical algebraic geometry. http://bertini.nd.edu. doi:10.7274/R0H41PB5
4. Egner, S.: Semi-numerical solution to 6/6-Stewart-platform kinematics based on symmetry. Appl. Algebra Eng. Commun. Comput. **7**(6), 449–468 (1996). doi:10.1007/BF01293263
5. Ghosal, A.: Robotics: Fundamental Concepts and Analysis, 10th edn. Oxford University Press, New Delhi (2014)
6. Huynh, D.Q.: Metrics for 3D rotations: comparison and analysis. J. Math. Imaging Vis. **35**(2), 155–164 (2009)
7. Jiang, Q., Gosselin, C.M.: Determination of the maximal singularity-free orientation workspace for the Gough-Stewart platform. Mech. Mach. Theory **44**(6), 1281–1293 (2009)
8. Li, H., Gosselin, C.M., Richard, M.J.: Determination of the maximal singularity-free zones in the six-dimensional workspace of the general Gough-Stewart platform. Mech. Mach. Theory **42**(4), 497–511 (2007). doi:10.1016/j.mechmachtheory.2006.04.006
9. Murray, R.M., Li, Z., Sastry, S.S.: A Mathematical Introduction to Robotic Manipulation. CRC Press, Boca Raton (1994)
10. Ravani, B., Roth, B.: Motion synthesis using kinematic mappings. J. Mech. Transm. Autom. Design **105**(3), 460–467 (1983)
11. Srivatsan, R.A., Bandyopadhyay, S.: Determination of the safe working zone of a parallel manipulator. In: Thomas, F., Perez Gracia, A. (eds.) Computational Kinematics: Proceedings of the 6th International Workshop on Computational Kinematics (CK2013), pp. 201–208. Springer, Netherlands (2014). doi:10.1007/978-94-007-7214-4
12. Wolfram, Research: Mathematica, version 10.4 edn. Wolfram Research, Inc., Champaign, Illinois (2016)

Geometric Algebra Based Kinematics Model and Singularity of a Hybrid Surgical Robot

Tanio K. Tanev

Abstract The paper presents a kinematics modelling and singularity analysis of a novel hybrid robot using geometric algebra. The introduced hybrid robot is designed for minimally invasive surgery (MIS). The geometric condition for singularity and the instantaneous uncontrollable motion in a singular configuration are derived in the geometric algebra terms. The singularities within the workspace are graphically presented. Avoiding the singularities of the medical parallel robots is important for the surgical manipulation success and the patient's safety.

1 Introduction

Medical robots enhance the surgeons' capability and encourage introducing novel and more complex surgical techniques. In the robotic assisted Minimally Invasive Surgery (MIS) the small abdominal incision restricts the motion of the end-effector and acts as a pivoting point. Different mechanisms could be applied in order to provide the "Remote Centre-of-Motion" (RCM) [4]. RCM is a point where one or more rotations are centred and located outside the mechanism itself. The analysis of singularities in parallel robots is important not only from the control point of view but also for safety's sake [7]. In singular configurations the parallel manipulator may have undesired behaviour and compromised performance. Since the patient's safety is of paramount importance, the analysis of singular configurations in surgical robots (especially parallel ones) is not only justified but also crucial. The singularities in parallel medical manipulators need to be well analysed and should be avoided during surgical manipulations. Various methods have been used for the analysis of singularities of parallel manipulators, namely, Study's kinematic mapping [8], Grassmann geometry [6], Grassmann-Cayley algebra [3], screw theory [14], geometric algebra [11]. A few papers have been dedicated to the singularity analysis of hybrid medical robots [12, 13].

T.K. Tanev (✉)
Institute of Systems Engineering and Robotics - Bulgarian Academy of Sciences, Sofia, Bulgaria
e-mail: tanev_tk@hotmail.com

J. Lenarčič and J.-P. Merlet (eds.), *Advances in Robot Kinematics 2016*,
Springer Proceedings in Advanced Robotics 4, DOI 10.1007/978-3-319-56802-7_45

In this paper, the geometric algebra (GA) is applied in the modelling of the kinematics and analysing the singularities of a novel type of hybrid robot designed for MIS operations. This robot is the second one of the family of surgical robots, the first one having been previously presented by the author (e.g. [12]).

2 Modelling of the Robot Kinematics

The parallel manipulator has three limbs. Two of them have SPU arrangement and the third one is a double parallelogram (Fig. 1). The double parallelogram has been used in some MIS robots in order to provide remote centre of motion (RCM) [4, 5]. The constraints, enforced by the double parallelogram limb, provide a RCM for the robot. The RCM is at the incision point of the patient's body. Each limb has one driven joint: the prismatic joints for the SPU legs are driven and the first revolute joint of the third leg (double parallelogram), which connects the planar parallelogram mechanism with the base platform, is driven. In addition to these three active joints, an active prismatic joint allowing translation of the end-effector along the line OB_3 is added. The overall degrees of freedom (dof) for the robot are four. The axes of the revolute joints at A_3 and at B_3 are perpendicular to the planes of the base $(A_1A_2A_3)$ and the moving platforms $(B_1B_2B_3)$, respectively. The origins of the reference (base) coordinate system OXYZ and the coordinate systems {1}, {2} and {3} coincide with the incision point of the patient's body. The kinematic modelling could be performed via different geometric algebras. The fundamentals of the geometric algebra can be found in [1, 2]. The kinematics of the considered robot is modelled via geometric algebra $G_{3,0,1}$ and conformal geometric algebra $G_{4,1}$, but here only the model from $G_{3,0,1}$ is presented, since the transformations needed for the singularity analysis are

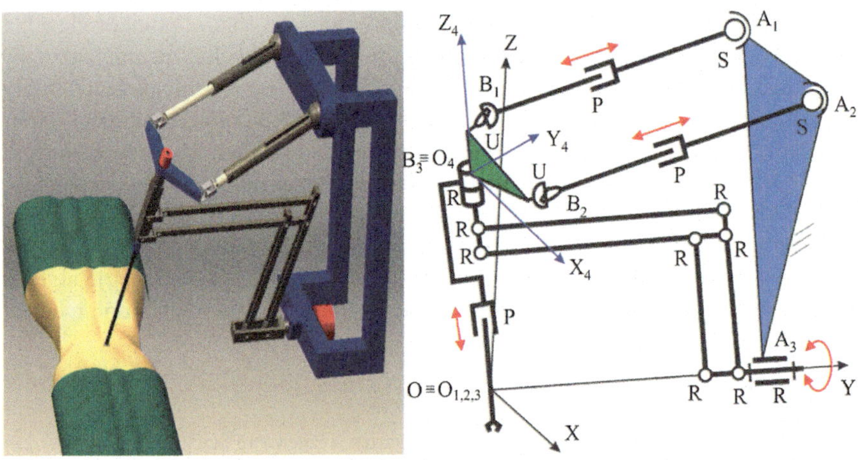

Fig. 1 The cad model and kinematic scheme of the surgical robot

mostly rotations and translations of vectors and screws. The $G_{3,0,1}$ is more convenient for this purpose. The geometric algebra $G_{3,0,1}$ has basis vectors $\mathbf{e}_1$, $\mathbf{e}_2$, $\mathbf{e}_3$, $\mathbf{e}_0$; the first three of them square to $+1$ and the last one squares to 0, i.e., $\mathbf{e}_i \cdot \mathbf{e}_i = 1, i = 1, 2, 3$; $\mathbf{e}_0 \cdot \mathbf{e}_0 = 0$.

The coordinate transformation for the considered robot can be written as

$$\mathbf{Q}_{04} = Rot(Y, \beta)Rot(X, \alpha)Rot(Z, \gamma)Trans(Z, d) \equiv \mathbf{Q}_1\mathbf{Q}_2\mathbf{Q}_3\mathbf{Q}_4, \quad (1)$$

where β, α and γ are angles of rotation about the Y, X and Z axes, respectively; d is a translation along Z axis; $\mathbf{Q}_1 = \cos(\beta/2) - \mathbf{e}_3\mathbf{e}_1 \sin(\beta/2)$, $\mathbf{Q}_2 = \cos(\alpha/2) - \mathbf{e}_2\mathbf{e}_3 \sin(\alpha/2)$, $\mathbf{Q}_3 = \cos(\gamma/2) - \mathbf{e}_1\mathbf{e}_2 \sin(\gamma/2)$, $\mathbf{Q}_4 = 1 - (d/2)\mathbf{e}_3\mathbf{e}_0$.

A vector $\mathbf{x}$ written in G_3 can be represented in $G_{3,0,1}$ as point $x = (\mathbf{1} + \mathbf{e}_0\mathbf{x})I_3$; $I_3 = \mathbf{e}_1\mathbf{e}_2\mathbf{e}_3$ is the unit pseudoscalar of G_3.

A general screw could be written as a multivector in G_3, i.e.,

$$\mathbf{s} = \mathbf{u} + \mathbf{r} \wedge \mathbf{u} + hI_3\mathbf{u} = v_1\mathbf{e}_1 + v_2\mathbf{e}_2 + v_3\mathbf{e}_3 + b_1\mathbf{e}_2 \wedge \mathbf{e}_3 + b_2\mathbf{e}_3 \wedge \mathbf{e}_1 + b_3\mathbf{e}_1 \wedge \mathbf{e}_2, \quad (2)$$

where $\mathbf{u}$ is the direction of the screw axis; $\mathbf{r}$ is the position vector of a point of the screw axis; h is the pitch of the screw; v_i and b_i, $i = 1, 2, 3$ are scalar coefficients.

The same screw could be written as 2-vector in $G_{3,0,1}$, i.e.,

$$S = \mathbf{u}I_3 + (\mathbf{r} \wedge \mathbf{u} + hI_3\mathbf{u})\mathbf{e}_0 I_3 = v_1\mathbf{e}_2\mathbf{e}_3 + v_2\mathbf{e}_3\mathbf{e}_1 + v_3\mathbf{e}_1\mathbf{e}_2 + b_1\mathbf{e}_1\mathbf{e}_0 + b_2\mathbf{e}_2\mathbf{e}_0 + b_3\mathbf{e}_3\mathbf{e}_0. \quad (3)$$

The 2-vector from Eq. (3) is similar to the equation of a line given in [9]. Equation (3) can represent screws with finite and zero (pure rotation) pitches. A screw with infinite pitch (pure translation) can be written as:

$$S_P = \mathbf{n}\mathbf{e}_0 = n_1\mathbf{e}_1\mathbf{e}_0 + n_2\mathbf{e}_2\mathbf{e}_0 + n_3\mathbf{e}_3\mathbf{e}_0, \quad (4)$$

where $\mathbf{n} = n_1\mathbf{e}_1 + n_2\mathbf{e}_2 + n_3\mathbf{e}_3$ is the vector of translation for the prismatic joint.

Then, a point and a screw (M) could be transformed as $M' = \mathbf{Q}M\mathbf{Q}^{\dagger}$, where $\mathbf{Q}$ is a transformation operator and $\mathbf{Q}^{\dagger}$ is reverse of $\mathbf{Q}$. All the screws needed for the singularity analysis are obtained via this model.

3 Singularity Analysis of the Parallel Manipulator and the Double Parallelogram Limb

The geometric algebra approach for singularity analysis of parallel manipulators with limited mobility has been developed and presented by the author in two previous papers [10, 11]. For this reason details of the approach are not given here. In order to perform the singularity analysis, the parallel manipulator could be represented by an equivalent kinematic scheme. In other words, the double parallelogram limb could

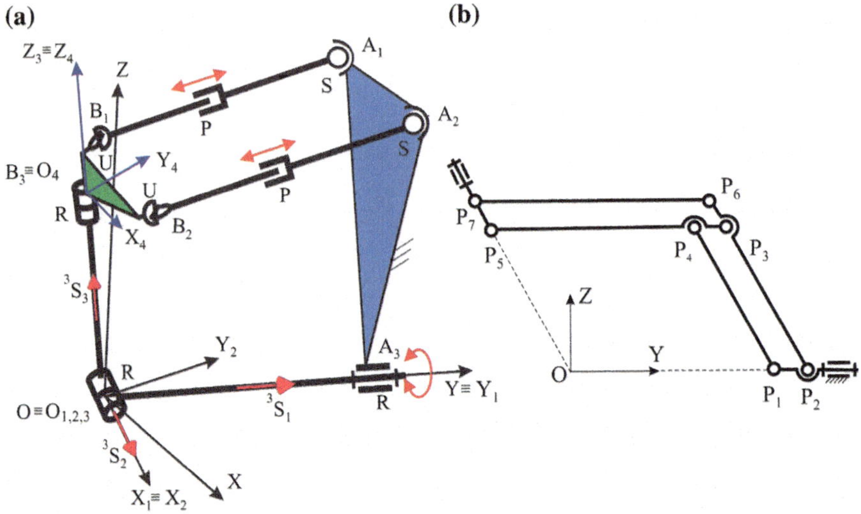

Fig. 2 **a** The equivalent scheme of the parallel manipulator; **b** The double parallelogram limb

be represented by a RRR limb, where the second revolute joint should be located at the incision point (Fig. 2a).

The concept of duality is utilized in the singularity analysis where the mathematical operation involves the outer product of up to six screws. Each screw has six components (Eq. (3)). For this reason, it is convenient to use 6D geometric algebra for the singularity analysis. Therefore, in contrast to Sect. 2, in this section the 6D geometric algebra G_6 with basis vectors $\{\mathbf{e}_1, \mathbf{e}_2, \mathbf{e}_3, \mathbf{e}_4, \mathbf{e}_5, \mathbf{e}_6\}$, $\mathbf{e}_i^2 = +1$ $(i = 1, 2, \ldots, 6)$ is used. Then, the obtained screws in Sect. 2 with finite (or zero) ($\mathbf{S}$) and infinite pitches ($\mathbf{S}_P$), respectively, could be rewritten in G_6 as

$$\mathbf{S} = v_1\mathbf{e}_1 + v_2\mathbf{e}_2 + v_3\mathbf{e}_3 + b_1\mathbf{e}_4 + b_2\mathbf{e}_5 + b_3\mathbf{e}_6; \quad \mathbf{S}_P = n_1\mathbf{e}_4 + n_2\mathbf{e}_5 + n_3\mathbf{e}_6, \tag{5}$$

where the coefficients v_i, b_i and n_i, $i = 1, 2, 3$ are the same as in Eqs. (3) and (4).

If we consider the j-th limb of a parallel manipulator containing active, passive and dummy (in case of a limited mobility limb) joints, then this limb will have full mobility. Then, the subspace of twists (freedom) for five screws with exception of the active (dummy) joint is represented by the following 5-blade for the j-th leg

$${}^j\mathbf{A}_k = {}^j\mathbf{S}_1 \wedge {}^j\mathbf{S}_2 \wedge \cdots \wedge {}^j\mathbf{S}_{k-1} \wedge {}^j\mathbf{S}_{k+1} \wedge \cdots \wedge {}^j\mathbf{S}_6, \tag{6}$$

where the subscript k denotes the active (or dummy) joint of the j-th leg.

The dual of the 5-blade from Eq. (6) is the orthogonal complement to the 5-blade

$${}^j\mathbf{D}_k = {}^j\mathbf{A}_k I_6^{-1}, \tag{7}$$

where $I_6 = \mathbf{e}_1\mathbf{e}_2\mathbf{e}_3\mathbf{e}_4\mathbf{e}_5\mathbf{e}_6$ is the unit pseudoscalar of the G_6 and $I_6^{-1} = \mathbf{e}_6\mathbf{e}_5\mathbf{e}_4\mathbf{e}_3\mathbf{e}_2\mathbf{e}_1$ is its inverse; the subscript k denotes the active or dummy joint of the j-th leg.

The condition for singularity of a general parallel manipulator with fewer than six dof (but with dummy joints) can be expressed in GA terms as [10–12]

$$\mathbf{D}_{a_1} \wedge \cdots \wedge \mathbf{D}_{a_p} \wedge \mathbf{D}_{d_1} \wedge \cdots \wedge \mathbf{D}_{d_r} = 0, \tag{8}$$

where $p + r = 6$; p is the number of the active joints and r is the number of the dummy joints; $\mathbf{D}_{a_i}$ is a dual vector (grade-1 blade) associated to the i-th active joint and $\mathbf{D}_{d_i}$ is a dual vector (grade-1 blade) associated to the i-th dummy joint.

Note that the computation of the dual vectors from Eq. (7) is straightforward and involves only addition and multiplication operations. That is why this geometric algebra approach is computationally more efficient than the screw theory method where the computing of the reciprocal screws could be a complicated operation. This is one of the advantages of the proposed geometric algebra approach.

The RRR leg has one active and three dummy joints. Therefore, four dual vectors associated with these joints can be written as (only ${}^3\mathbf{D}_a$ and ${}^3\mathbf{D}_{d_1}$ are listed here, since the remaining two could be obtained similarly to ${}^3\mathbf{D}_{d_1}$)

$$\begin{aligned} {}^3\mathbf{D}_a &= ({}^3\mathbf{S}_2 \wedge {}^3\mathbf{S}_3 \wedge {}^3\mathbf{S}_{d_1} \wedge {}^3\mathbf{S}_{d_2} \wedge {}^3\mathbf{S}_{d_3}) I_6^{-1}, \\ {}^3\mathbf{D}_{d_1} &= ({}^3\mathbf{S}_1 \wedge {}^3\mathbf{S}_2 \wedge {}^3\mathbf{S}_3 \wedge {}^3\mathbf{S}_{d_2} \wedge {}^3\mathbf{S}_{d_3}) I_6^{-1}, \end{aligned} \tag{9}$$

where ${}^3\mathbf{S}_i$, $(i = 1, 2, 3)$ are joint screws of the R-joints; and ${}^3\mathbf{S}_{d_i}$, $(i = 1, 2, 3)$ are screws of the dummy joints.

The two SPU legs (leg 1 and leg 2) of the parallel manipulator have full mobility and, therefore, have only dual vectors ${}^1\mathbf{D}_a$ and ${}^2\mathbf{D}_a$, respectively, associated with the active joints, i.e.,

$${}^j\mathbf{D}_a = ({}^j\mathbf{S}_1 \wedge {}^j\mathbf{S}_2 \wedge {}^j\mathbf{S}_3 \wedge {}^j\mathbf{S}_5 \wedge {}^j\mathbf{S}_6) I_6^{-1}, \; (j = 1, 2), \tag{10}$$

where the fourth P (prismatic) joint is active and is not included in the formula.

The dual vectors ${}^1\mathbf{D}_a$ and ${}^2\mathbf{D}_a$ are lines (zero pitch screws) and their elliptic polars ${}^1\mathbf{R}_a$ (${}^1\mathbf{R}_a = {}^1\tilde{\mathbf{D}}_a$) and ${}^2\mathbf{R}_a$ (${}^2\mathbf{R}_a = {}^2\tilde{\mathbf{D}}_a$) are lines along the SPU legs, respectively. In this case, there is no need to calculate all joint screws of the SPU legs, because the dual vectors could be obtained by the elliptic polars of the lines along the SPU legs.

The condition for singularity from Eq. (8) can be written as

$${}^1\mathbf{D}_a \wedge {}^2\mathbf{D}_a \wedge {}^3\mathbf{D}_a \wedge {}^3\mathbf{D}_{d_1} \wedge {}^3\mathbf{D}_{d_2} \wedge {}^3\mathbf{D}_{d_3} = 0. \tag{11}$$

Equation (11) involves dummy vectors which could be eliminated as shown in [11, 12]. The duality between inner and outer products is used in the elimination process. Inner and outer products are dual to each other and the following identities could be written

$$(\mathbf{A}\cdot\mathbf{M})I_n = \mathbf{A}\wedge(\mathbf{M}I_n); \quad (\mathbf{A}\wedge\mathbf{M})I_n = \mathbf{A}\cdot(\mathbf{M}I_n), \tag{12}$$

where **A** is a vector; **M** is multivector and I_n is the unit pseudoscalar of n-dimensional space.

The process of elimination of the dummy vectors is explained by Eq. (13), where the identities from Eq. (12) are used. The elimination example is for the outer product of two dual vectors from Eq. (11) (the third leg which contains dummy joints), i.e.

$$\begin{aligned} {}^3\mathbf{D}_a \wedge {}^3\mathbf{D}_{d_1} &= -{}^3\mathbf{D}_{d_1} \wedge {}^3\mathbf{D}_a = -{}^3\mathbf{D}_{d_1} \wedge [({}^3\mathbf{S}_2 \wedge {}^3\mathbf{S}_3 \wedge {}^3\mathbf{S}_{d_1} \wedge {}^3\mathbf{S}_{d_2} \wedge {}^3\mathbf{S}_{d_3})I_6^{-1}] \\ &= -[{}^3\mathbf{D}_{d_1} \cdot ({}^3\mathbf{S}_2 \wedge {}^3\mathbf{S}_3 \wedge {}^3\mathbf{S}_{d_1} \wedge {}^3\mathbf{S}_{d_2} \wedge {}^3\mathbf{S}_{d_3})]I_6^{-1} \\ &= -({}^3\mathbf{D}_{d_1} \cdot \mathbf{S}_{d_1})({}^3\mathbf{S}_2 \wedge {}^3\mathbf{S}_3 \wedge {}^3\mathbf{S}_{d_2} \wedge {}^3\mathbf{S}_{d_3})I_6^{-1} = c({}^3\mathbf{S}_2 \wedge {}^3\mathbf{S}_3 \wedge {}^3\mathbf{S}_{d_2} \wedge {}^3\mathbf{S}_{d_3})I_6^{-1} \end{aligned} \tag{13}$$

where $c = -{}^3\mathbf{D}_{d_1} \cdot \mathbf{S}_{d_1} \neq 0$ is a scalar, while referring to Eq. (9) it can be seen that the following inner products are zero: ${}^3\mathbf{D}_{d_1} \cdot {}^3\mathbf{S}_i = 0$, $(i{=}1, 2, 3)$ and ${}^3\mathbf{D}_{d_1} \cdot {}^3\mathbf{S}_{d_i}{=}0$, $(i = 2, 3)$; these values are used in the expansion of the inner product between a vector and a blade in Eq. (13).

Then, after elimination, the 4-blade formed by the outer product of the four dual vectors associated with the RRR leg (the third leg) can be written as

$${}^3\mathbf{D}_a \wedge {}^3\mathbf{D}_{d_1} \wedge {}^3\mathbf{D}_{d_2} \wedge {}^3\mathbf{D}_{d_3} = \lambda({}^3\mathbf{S}_2 \wedge {}^3\mathbf{S}_3)I_6^{-1}, \tag{14}$$

where λ is a scalar coefficient and is irrelevant to the geometric condition.

The 4-blade from Eq. (14) represents a blade of non-freedom for the RRR leg. This is another advantage of the method, i.e., the twists of non-freedom (the wrenches of constraint, respectively) could be represented by a blade and therefore, there is no need to obtain each single twist of non-freedom. Then, the condition for singularity (Eq. (11)) becomes

$${}^1\mathbf{D}_a \wedge {}^2\mathbf{D}_a \wedge [({}^3\mathbf{S}_2 \wedge {}^3\mathbf{S}_3)I_6^{-1}] = 0. \tag{15}$$

Now, another advantage of the geometric algebra approach could be applied here. Equation (15) could be further manipulated by applying the identities from Eq. (12), i.e.

$${}^1\mathbf{D}_a \wedge [{}^2\mathbf{D}_a \cdot ({}^3\mathbf{S}_2 \wedge {}^3\mathbf{S}_3)]I_6^{-1} = {}^1\mathbf{D}_a \wedge (\mathbf{V}I_6^{-1}) = ({}^1\mathbf{D}_a \cdot \mathbf{V})I_6^{-1} = 0. \tag{16}$$

The results from Eq. (16) could be geometrically interpreted and the drawn conclusions are explained below. The inner product between the vector ${}^2\mathbf{D}_a$ and the bivector ${}^3\mathbf{S}_2 \wedge {}^3\mathbf{S}_3$ (Eq. (16)) is the vector $\mathbf{V} = {}^2\mathbf{D}_a \cdot ({}^3\mathbf{S}_2 \wedge {}^3\mathbf{S}_3)$. This implies that **V** is perpendicular to ${}^2\mathbf{D}_a$ in the 6-D space and **V** belongs to the bivector space ${}^3\mathbf{S}_2 \wedge {}^3\mathbf{S}_3$. Since the screws ${}^3\mathbf{S}_2$ and ${}^3\mathbf{S}_3$ are lines passing through the origin of the coordinate system, therefore the vector **V** represents a line passing through the origin, too (Fig. 3). Thus, the line **V** lies in the plane defined by the lines ${}^3\mathbf{S}_2$ and ${}^3\mathbf{S}_3$. The elliptic polar of ${}^2\mathbf{D}_a$ is the line ${}^2\mathbf{R}_a$ along the second SPU leg (A_2B_2). Since the vectors **V** and ${}^2\mathbf{D}_a$ are mutually perpendicular in the 6-D space, then the elliptic

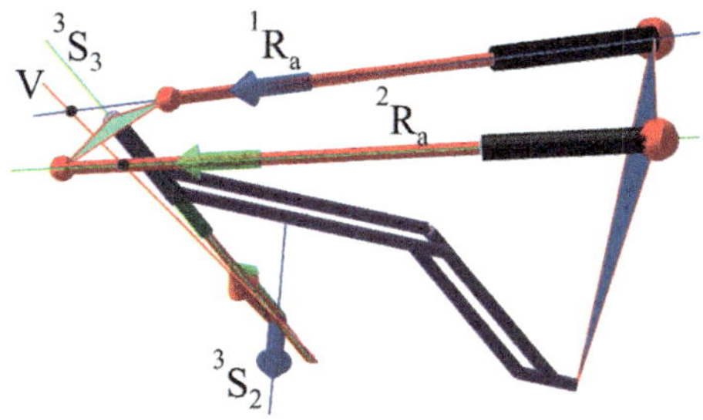

Fig. 3 A singular configuration of the surgical robot

polar $^2\mathbf{R}_a$ (line) of the screw $^2\mathbf{D}_a$ ($^2\mathbf{R}_a =^2 \tilde{\mathbf{D}}_a$) and the line $\mathbf{V}$ should intersect at a common point, i.e., the screws $\mathbf{V}$ and $^2\mathbf{R}_a$ are reciprocal. It can be concluded from the last part of Eq. (16) that the manipulator is in singular configuration when the inner product of vectors $^1\mathbf{D}_a$ and $\mathbf{V}$ is zero ($^1\mathbf{D}_a \cdot \mathbf{V} = 0$). This means that the elliptic polar $^1\mathbf{R}_a$ (line) of the screw $^1\mathbf{D}_a$ ($^1\mathbf{R}_a =^1 \tilde{\mathbf{D}}_a$) and the line $\mathbf{V}$ should intersect, i.e., the screws $\mathbf{V}$ and $^1\mathbf{R}_a$ are reciprocal. Thus, the geometric condition for singularity could be stated as: the considered parallel manipulator is in singular configuration if the vector (line) $\mathbf{V}$ intersects both lines $^1\mathbf{R}_a$ and $^2\mathbf{R}_a$ which are along the SPU legs, respectively. Similar results are obtained for the other robot (with different kinematic scheme) of the family of MIS robots, presented in [12].

For completeness, the singularity of the parallelogram limb is considered. Since the double parallelogram limb (Fig. 2b) consists of two connected parallelograms, the singularity of each of them determines the singularity of this limb. The revolute joints attached to the base and moving platforms do not affect the limb singularity. Thus, considering the 1-dof planar mechanism $P_1P_2P_3P_4P_5P_6P_7$ is enough for the analysis. Let us assume that the active joint is at P_2. The first parallelogram $P_1P_2P_3P_4$ can be considered as a parallel manipulator with two legs P_2P_3 and P_1P_4. The dual vectors associated with the active and dummy joints of the two legs are

$$\mathbf{D}_a = (\mathbf{P}_3 \wedge \mathbf{P}_d \wedge \mathbf{e}_{234})I_6^{-1};\, \mathbf{D}_1 = (\mathbf{P}_2 \wedge \mathbf{P}_3 \wedge \mathbf{e}_{234})I_6^{-1};\, \mathbf{D}_2 = (\mathbf{P}_1 \wedge \mathbf{P}_4 \wedge \mathbf{e}_{234})I_6^{-1}, \quad (17)$$

where $\mathbf{P}_i$, $(i = 1, 2, 3, 4)$ are joint screws; $\mathbf{e}_{234} = \mathbf{e}_2 \wedge \mathbf{e}_3 \wedge \mathbf{e}_4$ is a 3-blade associated with dummy joints, which restrict the mechanism to move only in planes parallel to $Y - Z$ plane. Then, referring to [11], the condition for singularity leads to

$$\mathbf{D}_a \wedge \mathbf{D}_1 \wedge \mathbf{D}_2 = c_3(\mathbf{P}_1 \wedge \mathbf{P}_4 \wedge \mathbf{P}_3 \wedge \mathbf{e}_{234})\mathbf{e}_{234} = 0, \quad (18)$$

where c_3 is a scalar coefficient which is irrelevant to the geometric condition.

Therefore, the condition for singularity is $\mathbf{P}_1 \wedge \mathbf{P}_4 \wedge \mathbf{P}_3 = 0$, and the mechanism is in singular configuration if these parallel lines are linearly dependent, i.e., lie in a single plane. In case of the parallelogram mechanism this occurs when lines P_2P_3 and

P_1P_4 coincide. The similar condition can be obtained for the second parallelogram $P_3P_6P_7P_5$. It could be concluded that the limb is in singular configuration when all lines (P_2P_3, P_1P_4, P_6P_7, P_4P_5) coincide.

3.1 The Instantaneous Uncontrollable Motion in a Singular Configuration

In a singular configuration the dual vectors from Eq. (11) are linearly dependent. The instantaneous uncontrollable twist could be obtained by the dual of a 5-blade. This 5-blade includes all the constituent members of Eq. (15) with the exception of one vector. If we exclude ${}^1\mathbf{D}_a$ from Eq. (15), we will obtain the following expression for the uncontrollable twist $\mathbf{U}$

$$\begin{aligned}\mathbf{U} &= [{}^2\mathbf{D}_a \wedge ({}^3\mathbf{S}_2 \wedge {}^3\mathbf{S}_3)I_6^{-1}]I_6^{-1} = \\ &\{[{}^2\mathbf{D}_a \cdot ({}^3\mathbf{S}_2 \wedge {}^3\mathbf{S}_3)]I_6^{-1}\}I_6^{-1} = {}^2\mathbf{D}_a \cdot ({}^3\mathbf{S}_2 \wedge {}^3\mathbf{S}_3) \equiv \mathbf{V}.\end{aligned} \tag{19}$$

Equation (19) proves that the uncontrollable twist $\mathbf{U}$ coincides with the vector $\mathbf{V}$ from Eq. (16). Since $\mathbf{V}$ is a line, therefore the uncontrollable motion in this singular configuration is a pure rotation about the line $\mathbf{V}$ (Fig. 3), which implies that this instantaneous rotation does not affect (change) the RCM. In this case, the instantaneous uncontrollable motion is obtained geometrically in a coordinate-free manner.

4 Singularities Within the Workspace

The algebraic formulation of the singular condition could be derived by the scalar part from Eq. (15), which is a function of the design and input parameters. This function is parametrized by the three angles α, β and γ. Then, the obtained singular surface is shown in Fig. 4 together with the workspace (in terms of the angles α, β and γ) of the robot. Also, several slices of the workspace and the singularities are presented in the same figure. Examining the workspace and the singular surface it can be observed that the singularities are outside the workspace for the range $\gamma = -15° \div 15°$ (approximately). The singularity surface and workspace are obtained for the following design parameters: $\mathbf{OA}_1 = (-0.2, 0.52, 0.56)$; $\mathbf{OA}_2 = (0.2, 0.52, 0.56)$; $\mathbf{OA}_3 = (0, 0.52, 0)$; $\mathbf{O}_4\mathbf{B}_1 = (-0.1075, 0.05, 0)$; $\mathbf{O}_4\mathbf{B}_2 = (0.1075, 0.05, 0)$; $OP_1 = 0.422$; $P_1P_2 = P_3P_4 = 0.06$; $P_1P_4 = P_2P_3 = 0.33$; $P_3P_6 = P_5P_7 = 0.06$; $P_3P_5 = P_6P_7 = 0.482$; $OO_4 = 0.52$ (all values are in meters). The workspace is obtained by imposing a range [0.30, 0.58] of the motion of the two SPU legs and a restriction of the motion of the planar double parallelogram ($\alpha = [-80° \div 80°]$).

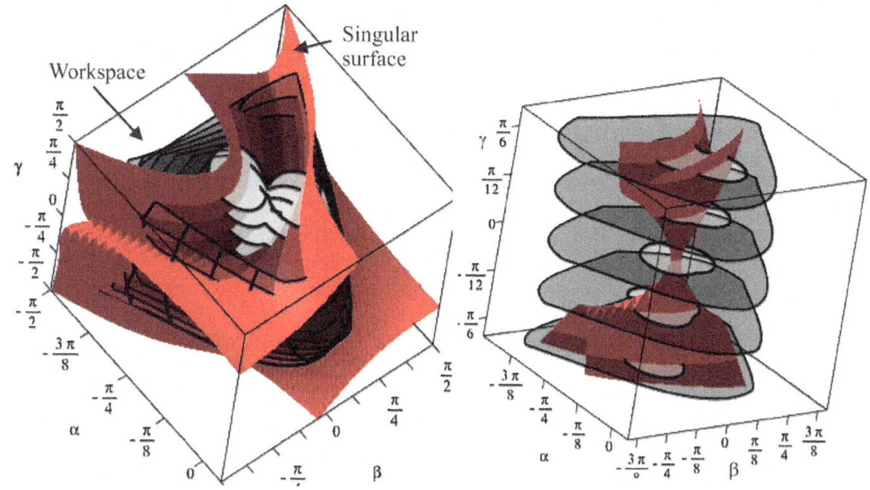

Fig. 4 Workspace, singular surface, and workspace slices and singularities

5 Conclusions

The geometrical condition for singular configuration of the proposed MIS robot is obtained in a basis-free form in terms of the geometric algebra. The approach provides a good geometrical insight into the singularity of this robot. The singularities are presented within the robot workspace. The analysis revealed that a singularity-free workspace exists for a given range of orientations, i.e. in this case the singularities are outside the workspace. The results from the presented analysis could be applied in the robot path planning, where the singular configurations should be avoided. The outcome of this study also suggests that the singularities could be completely avoided if the robot is endowed with an additional redundant rotation around the axis of the surgical instrument.

References

1. Doran, C., Lasenby, A.: Geometric Algebra for Physicists. Cambridge University Press, Cambridge (2007)
2. Hestenes, D.: New Foundations for Classical Mechanics, 2nd edn. Kluwer Academic Publishers, Dordrecht (1999)
3. Kanaan, D., Wenger, P., Chablat, D.: Singularity analysis of limited-DOF parallel manipulators using Grassmann-Cayley algebra. In: Lenarčič, J., Wenger, P. (eds.) Advances in Robot Kinematics, Analysis and Design, pp. 59–68. Springer, Berlin (2008)
4. Kuo, C.H., Dai, J.S.: Robotics for minimally invasive surgery: a historical review from the perspective of kinematics. In: Yan, H.-S., Ceccarelli, M. (eds.) International Symposium on History of Machines. Springer Science+Business Media B.V., Dordrecht (2009)

5. Li, J., Xing, Y., Liang, K., Wang, S.: Kinematic design of a novel spatial remote center-of-motion mechanism for minimally invasive surgical robot. ASME. J. Med. Devices **9**(1) (2015)
6. Merlet, J.-P.: Singular configurations of parallel manipulators and Grassmann geometry. Int. J. Robot. Res. **8**(5), 45–56 (1989)
7. Merlet, J.-P.: Parallel Robots, 2nd edn. Springer, Berlin (2006)
8. Schadlbauer, J., Walter, D.R., Husty, M.L.: A complete kinematic analysis of the 3-RPS parallel manipulator. In: 15th National Conference on Machines and Mechanisms (NaCoMM2011), 30 Nov–2 Dec, Madras, India (2011)
9. Selig, J.: Clifford algebra of points, lines and planes. Robotica **18**, 545–556 (2000)
10. Tanev, T.K.: Singularity analysis of a 4-DOF parallel manipulator using geometric algebra. In: Lenarčič, J., Roth, B. (eds.) Advances in Robot Kinematics, Mechanism and Motion, pp. 275–284. Springer, Berlin (2006)
11. Tanev, T.K.: Geometric algebra approach to singularity of parallel manipulators with limited mobility. In: Lenarčič, J., Wenger, P. (eds.) Advances in Robot Kinematics, Analysis and Design, pp. 39–48. Springer, Berlin (2008)
12. Tanev, T.K.: Singularity analysis of a novel minimally-invasive-surgery hybrid robot using geometric algebra. In: Wenger, P., et al. (eds.) New Trends in Medical and Service Robots, 4th International Workshop on Medical and Service Robots, 8–10 July, Nantes, France. Springer (to appear) (2015)
13. Vaida, C., Pisla, D., Schadlbauer, J., Husty, M., Plitea, N.: Kinematic analysis of an innovative medical parallel robot using study parameters. In: Wenger, P., et al. (eds.) New Trends in Medical and Service Robots, 4th International Workshop on Medical and Service Robots, 8–10 July, Nantes, France. Springer (to appear) (2015)
14. Zlatanov, D., Bonev, I.A., Cosselin, C.M.: Constraint singularities of parallel mechanisms. In: Proceedings of the IEEE International Conference on Robotics and Automation, Washington, DC, vol. 1, pp. 496–502 (2002)

Kinematic Singularities of a 3-DoF Planar Geared Robot Manipulator

S. Vahid Amirinezhad and Peter Donelan

Abstract By incorporating gearing into a planar 3R mechanism, one obtains a family of mechanisms in which the gear ratios play a central kinematic role. Special choices of these parameters result in interesting simplifications of the kinematic mapping. An explicit expression for the mapping can be derived using the 'matroid method' of Talpasanu et al. [6]. We use this relatively simple mechanism to illustrate singularity analysis for geared mechanisms.

1 Introduction

The use of gear pairs in a mechanism may confer a number of advantages. For example, they can enable more efficient placement of the actuators thereby reducing their mass and inertia. Epicyclic gear trains (EGTs), in which the centre of one gear wheel revolves around that of another, are the simplest form and therefore play an important role in geared mechanisms (GMs). By utilising EGTs, we can easily place actuators close to the base of a GM and rotation of inputs can be efficiently transmitted to the end-effector. Careful choice of gear ratios can also enable end-effector motion to be tailored to specific inputs.

The fundamental kinematic equation for an epicyclic gear is due to Willis [10]. Subsequent authors have introduced methods of global analysis for GMs that ensure the equations are correctly formulated for a given mechanism topology and design. Notably, Buchsbaum and Freudenstein [2] introduced combinatoric methods to represent the topology of the mechanism. This approach was later developed by Tsai [8], Hsu and Lam [3]. In order to enhance the computational effectiveness of the method, Talpasanu et al. [5, 6] refined and to some extent recast the approach, introducing the 'incidence and transfer method' that uses the cycle matroid of the mechanism's

S.V. Amirinezhad (✉) · P. Donelan
School of Mathematics and Statistics, Victoria University of Wellington,
PO Box 600, Wellington, New Zealand
e-mail: vahid.amirinezhad@msor.vuw.ac.nz

P. Donelan
e-mail: peter.donelan@vuw.ac.nz

J. Lenarčič and J.-P. Merlet (eds.), *Advances in Robot Kinematics 2016*,
Springer Proceedings in Advanced Robotics 4, DOI 10.1007/978-3-319-56802-7_46

directed graph. A comparison of Talpasanu's method with that of Tsai–Tokad was made in [1]. In this paper, we illustrate Talpasanu's method for a simple geared version of a planar 3R mechanism in order to determine its kinematic mapping and thereby its singularities. The goal is to develop a systematic approach to the kinematic analysis of GMs and the determination of their singularities. While the example presented is straightforward, it is intended to provide a model for extending singularity analysis of GMs to more complex cases, including those which are genuinely spatial in their kinematics and to parallel mechanisms incorporating gearing.

2 The Mechanism

A simple planar GM consists of $n + 1$ links, $L_0, \ldots, L_n$, and m joints that include t revolute (turning) pairs, $T_1, \ldots, T_t$, and g gear pairs, $G_1, \ldots, G_g$, so that $m = t + g$. Note that the number of links, excluding the base L_0, is assumed equal to the number of simple pairs, i.e. $t = n$. In effect, the mechanism without gears contains no closed chains.

By placing three actuator joints at the base and using simple spur-gear pairs to transmit motion to the end-effector, one obtains a geared mechanism based on a simple serial planar 3R mechanism (see Fig. 1). One EGT, consisting of three gear wheels and using link L_1 as carrier, transmits motion to the link L_7, while a second EGT of five gear wheels with links L_1 and L_7 as carriers transmits motion to the end-effector.

A functional schematic for the mechanism is illustrated in Fig. 2. The inputs, which are attached to the base L_0, are via T_i, $i = 1, 2, 3$ while link L_9 is the output planet gear or end-effector. Note that the carrier arms L_1 and L_7 that form the first two links in the underlying planar 3R are also gear wheels. Other links are intermediate (idler) gear wheels.

In its directed graph (digraph) representation, Fig. 3, the links (including gear wheels) are vertices ($L_0, \ldots, L_9$) while joints are edges. Specifically, the revolute pairs ($T_1, \ldots, T_9$) are solid edges and gear pairs ($G_1, \ldots, G_6$) are dashed. Note that the solid edges form a spanning tree for the graph; put another way, each simple cycle contains at least one gear pair as an edge. The direction of an edge connecting vertices (links) L_i and L_j is $L_i \rightarrow L_j$ if the transmission from input to output flows from L_i to L_j.

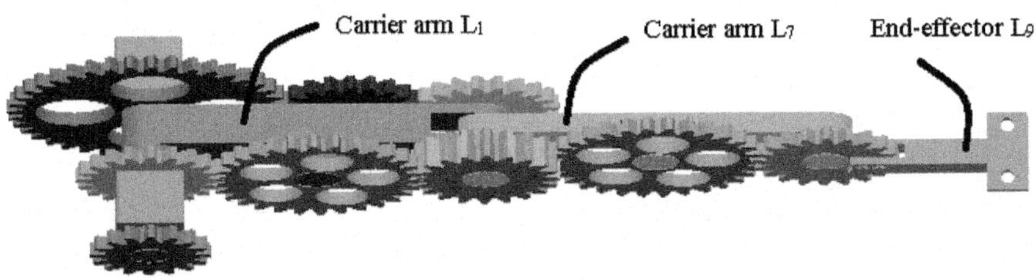

Fig. 1 3-DoF geared planar manipulator

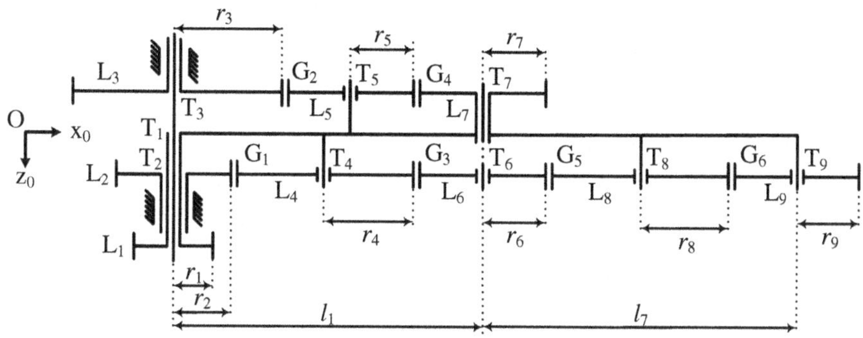

Fig. 2 Functional schematic of manipulator in Fig. 1

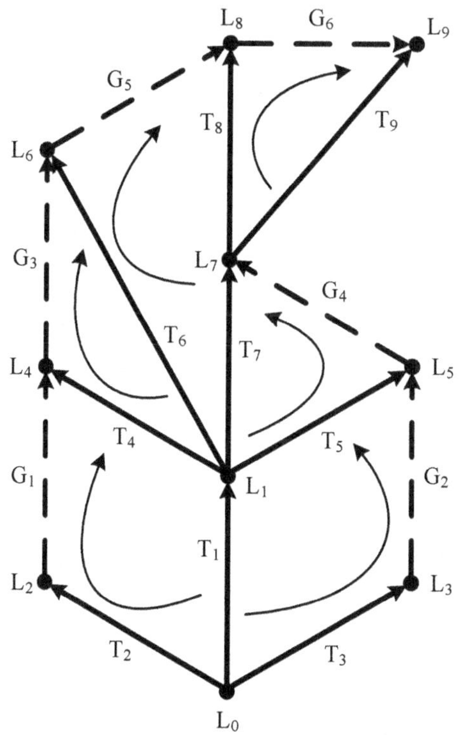

Fig. 3 Associated digraph

Application of the Chebyshev-Grübler-Kutzbach (CGK) formula shows that the GM has three degrees of freedom (3 dof), noting that a gear pair has 2 dof. Alternatively, Talpasanu [5] and Tsai [8] observe that there is a relation between the degrees of freedom f of the GM, the number of links and the number of gear pairs:

$$f = n - g \tag{1}$$

again yielding $f = 3$.

3 Constraint Analysis via the Matroid Method

To perform the kinematic analysis, we apply the matroid method of Talpasanu [6] to obtain the Willis kinematic equations for all gear pairs and solve these equations to express all joint variables in terms of the input (sun) variables. This enables us to express the kinematic mapping as a product of exponentials (PoE) in terms of input variables alone and consequently to undertake the singularity analysis. It is worth noting that the Willis equations usually express the relation between angular velocities in a gear-pair/carrier cycle but since the relations between the joint variables themselves are linear, the same equations hold between the underlying variables as between their velocities.

There are essentially three stages to the matroid or incidence–transfer method: the first stage codifies the topology of the digraph representation of the GM in matrix form. The second stage builds the specific design on to this by introducing dimensions that can then be interpreted as gear ratios. The method insures that we obtain a minimal set of linear (Willis) equations and the third stage is to solve these for the joint variables in terms of the input variables.

Associated to the digraph are two matrices. The *incidence matrix* $\mathbf{\Pi}^0$ has rows labelled by the vertices and columns by edges and its entries π^0_{ij} are -1 or 1 according as edge j leaves or enters vertex i, or else is 0. In this setting, the base L_0 is fixed and its row (containing only -1 and 0) is linearly dependent on the other rows. So, for the purpose of analysis we omit this row and arrive at the *reduced* incidence matrix $\mathbf{\Pi}$ as follows:

$$\mathbf{\Pi} = \begin{array}{c} \\ L_1 \\ L_2 \\ L_3 \\ L_4 \\ L_5 \\ L_6 \\ L_7 \\ L_8 \\ L_9 \end{array} \begin{array}{c} \begin{array}{ccccccccc|cccccc} T_1 & T_2 & T_3 & T_4 & T_5 & T_6 & T_7 & T_8 & T_9 & G_1 & G_2 & G_3 & G_4 & G_5 & G_6 \end{array} \\ \left[\begin{array}{ccccccccc|cccccc} 1 & 0 & 0 & -1 & -1 & -1 & -1 & 0 & 0 & 0 & 0 & 0 & 0 & 0 & 0 \\ 0 & 1 & 0 & 0 & 0 & 0 & 0 & 0 & 0 & -1 & 0 & 0 & 0 & 0 & 0 \\ 0 & 0 & 1 & 0 & 0 & 0 & 0 & 0 & 0 & 0 & -1 & 0 & 0 & 0 & 0 \\ 0 & 0 & 0 & 1 & 0 & 0 & 0 & 0 & 0 & 1 & 0 & -1 & 0 & 0 & 0 \\ 0 & 0 & 0 & 0 & 1 & 0 & 0 & 0 & 0 & 0 & 1 & 0 & -1 & 0 & 0 \\ 0 & 0 & 0 & 0 & 0 & 1 & 0 & 0 & 0 & 0 & 0 & 1 & 0 & -1 & 0 \\ 0 & 0 & 0 & 0 & 0 & 0 & 1 & -1 & -1 & 0 & 0 & 0 & 1 & 0 & 0 \\ 0 & 0 & 0 & 0 & 0 & 0 & 0 & 1 & 0 & 0 & 0 & 0 & 0 & 1 & -1 \\ 0 & 0 & 0 & 0 & 0 & 0 & 0 & 0 & 1 & 0 & 0 & 0 & 0 & 0 & 1 \end{array}\right] \end{array} \quad (2)$$

Further, this is partitioned as indicated into submatrices: $\mathbf{\Pi}_{n\times m} = [\mathbf{P}_{n\times t}|\hat{\mathbf{P}}_{n\times g}]$.

A *cycle basis matrix* $\mathbf{\Gamma}$ for a digraph consists of a maximally independent set of rows $G_i, i = 1, \ldots, g$, each corresponding to a cycle, whose entries γ_{ij}, $j = 1, \ldots, m$ are -1 or 1 according as edge j appears in that cycle directed with, or opposed to, a given vertex order for the cycle, or otherwise 0. The cycle space is in fact the nullspace of the incidence matrix so, according to Euler's formula, its dimension is $m - n$. Given the special structure of the digraph for a GM, we have $m - n = g$ and a basis for the cycle space can be indexed by the gear pairs,

$G_1, \dots, G_g$. For the given GM, with the vertex order as indicated in Fig. 3 by arrows in each basis cycle, we have:

$$\boldsymbol{\Gamma} = \begin{array}{c} \\ G_1 \\ G_2 \\ G_3 \\ G_4 \\ G_5 \\ G_6 \end{array} \begin{array}{c} \begin{array}{ccccccccc|cccccc} T_1 & T_2 & T_3 & T_4 & T_5 & T_6 & T_7 & T_8 & T_9 & G_1 & G_2 & G_3 & G_4 & G_5 & G_6 \end{array} \\ \left[\begin{array}{ccccccccc|cccccc} -1 & 1 & 0 & -1 & 0 & 0 & 0 & 0 & 0 & 1 & 0 & 0 & 0 & 0 & 0 \\ -1 & 0 & 1 & 0 & -1 & 0 & 0 & 0 & 0 & 0 & 1 & 0 & 0 & 0 & 0 \\ 0 & 0 & 0 & 1 & 0 & -1 & 0 & 0 & 0 & 0 & 0 & 1 & 0 & 0 & 0 \\ 0 & 0 & 0 & 0 & 1 & 0 & -1 & 0 & 0 & 0 & 0 & 0 & 1 & 0 & 0 \\ 0 & 0 & 0 & 0 & 0 & 1 & -1 & -1 & 0 & 0 & 0 & 0 & 0 & 1 & 0 \\ 0 & 0 & 0 & 0 & 0 & 0 & 0 & 1 & -1 & 0 & 0 & 0 & 0 & 0 & 1 \end{array}\right] \end{array} \tag{3}$$

Again, this can be partitioned into submatrices: $\boldsymbol{\Gamma}_{g\times m} = [\mathbf{C}_{g\times t}|\mathbf{I}_{g\times g}]$ where the second block is the identity matrix. Note that $\boldsymbol{\Gamma}$ is the cycle basis matrix corresponding to the specific spanning tree for the digraph, which one can obtain by deleting the dashed lines in Fig. 3. In any graph with edge set E, the collection I of subsets of E that do not include a cycle form a *matroid*, mathematical objects that capture the abstract idea of independence. Spanning trees are maximally independent while simple cycles are minimally dependent objects.

The second step is to introduce design parameters into the matrices. The constraint imposed by the cycles on the motion of the GM is captured by the *joint position matrix* $\boldsymbol{\Delta}$ whose entries are $\delta_{ij} = c_{ij}d_{ij}$, where c_{ij} are components of the (reduced) cycle basis matrix $\mathbf{C}$ and $d_{ij} = y_{T_j} - y_{G_i}$ where $y_{T_j}, j = 1, \dots, t$ and $y_{G_i}, i = 1, \dots, g$ are distances of the axes of turning joint T_j and meshing joint G_i from the base in home configuration. These distances are the radii of the various gear wheels $r_k, k = 2, \dots, t$ so that:

$$\boldsymbol{\Delta} = \begin{array}{c} \\ G_1 \\ G_2 \\ G_3 \\ G_4 \\ G_5 \\ G_6 \end{array} \begin{array}{c} \begin{array}{ccccccccc} T_1 & T_2 & T_3 & T_4 & T_5 & T_6 & T_7 & T_8 & T_9 \end{array} \\ \left[\begin{array}{ccccccccc} r_2 & -r_2 & 0 & -r_4 & 0 & 0 & 0 & 0 & 0 \\ r_3 & 0 & -r_3 & 0 & -r_5 & 0 & 0 & 0 & 0 \\ 0 & 0 & 0 & -r_4 & 0 & -r_6 & 0 & 0 & 0 \\ 0 & 0 & 0 & 0 & -r_5 & 0 & -r_7 & 0 & 0 \\ 0 & 0 & 0 & 0 & 0 & -r_6 & r_6 & -r_8 & 0 \\ 0 & 0 & 0 & 0 & 0 & 0 & 0 & -r_8 & -r_9 \end{array}\right] \end{array} \tag{4}$$

For an oriented gear pair $G_i, i = 1, \dots, g$ connecting link (gear wheel) L_p to L_q, denote the corresponding gear ratio $\rho_i = -r_p/r_q$. The rows of the matrix represent equations that hold between the joint variables at each revolute pair (or equivalently their angular velocities) so that each row can be independently scaled by one of the radii to realise the *gear ratio matrix*:

$$\boldsymbol{\Lambda} = \begin{array}{c} \\ G_1 \\ G_2 \\ G_3 \\ G_4 \\ G_5 \\ G_6 \end{array} \begin{array}{c} \begin{array}{ccccccccc} T_1 & T_2 & T_3 & T_4 & T_5 & T_6 & T_7 & T_8 & T_9 \end{array} \\ \left[\begin{array}{ccccccccc} \rho_1 & -\rho_1 & 0 & -1 & 0 & 0 & 0 & 0 & 0 \\ \rho_2 & 0 & -\rho_2 & 0 & -1 & 0 & 0 & 0 & 0 \\ 0 & 0 & 0 & -\rho_3 & 0 & -1 & 0 & 0 & 0 \\ 0 & 0 & 0 & 0 & -\rho_4 & 0 & -1 & 0 & 0 \\ 0 & 0 & 0 & 0 & 0 & -\rho_5 & \rho_5 & -1 & 0 \\ 0 & 0 & 0 & 0 & 0 & 0 & 0 & -\rho_6 & -1 \end{array}\right] \end{array} \tag{5}$$

To arrive finally at a complete set of Willis equations for the GM, it is necessary to incorporate the component $\mathbf{P}$ of the reduced incidence matrix that provides the connection between the angles of rotation θ_i for each link L_i and the joint variables ϕ_j at each revolute pair T_j, $i, j = 1, \ldots, n$ ($= t$ as noted in Sect. 2). Specifically, set:

$$\mathbf{\Sigma}_{g\times t} = \mathbf{\Lambda}_{g\times t}.\mathbf{P}^T_{t\times t}, \tag{6}$$

then the Willis equations have the matrix form:

$$\mathbf{\Sigma}.\boldsymbol{\theta} = \mathbf{0} \tag{7}$$

where $\boldsymbol{\theta}$ is the vector of link rotations.

We can partition $\boldsymbol{\theta}$ between *input* variables and *passive* variables. Following Eq. (1), there are three input variables and six passive variables. Explicitly:

$$\boldsymbol{\theta} = \left[\boldsymbol{\theta}_f \mid \boldsymbol{\theta}_g\right]^T = \left[\theta_1\ \theta_2\ \theta_3 \mid \theta_4\ \theta_5\ \theta_6\ \theta_7\ \theta_8\ \theta_9\right]^T \tag{8}$$

Partitioning $\mathbf{\Sigma}$ in a similar way, and expanding the product gives:

$$\mathbf{\Sigma} = \left[\mathbf{Z}_f \mid \mathbf{Z}_g\right] = \begin{array}{c} \\ G_1 \\ G_2 \\ G_3 \\ G_4 \\ G_5 \\ G_6 \end{array} \begin{array}{c} \begin{array}{ccc|cccccc} L_1 & L_2 & L_3 & L_4 & L_5 & L_6 & L_7 & L_8 & L_9 \end{array} \\ \left[\begin{array}{ccc|cccccc} \rho_1+1 & -\rho_1 & 0 & -1 & 0 & 0 & 0 & 0 & 0 \\ \rho_2+1 & 0 & -\rho_2 & 0 & -1 & 0 & 0 & 0 & 0 \\ \rho_3+1 & 0 & 0 & -\rho_3 & 0 & -1 & 0 & 0 & 0 \\ \rho_4+1 & 0 & 0 & 0 & -\rho_4 & 0 & -1 & 0 & 0 \\ 0 & 0 & 0 & 0 & 0 & -\rho_5 & \rho_5+1 & -1 & 0 \\ 0 & 0 & 0 & 0 & 0 & 0 & \rho_6+1 & -\rho_6 & -1 \end{array}\right] \end{array} \tag{9}$$

Now, we can rewrite Eq. (7) in the form of $[\mathbf{Z}_f \mid \mathbf{Z}_g].[\boldsymbol{\theta}_f \mid \boldsymbol{\theta}_g]^T = \mathbf{0}$. From which it follows that, provided $\mathbf{Z}_g$ is non-singular which is easily verified in this case:

$$[\boldsymbol{\theta}_g] = -[\mathbf{Z}_g]^{-1}.[\mathbf{Z}_f].[\boldsymbol{\theta}_f] \tag{10}$$

Solving Eq. (10), gives passive variables in terms of input variables as follows:

$$\begin{bmatrix} \theta_4 \\ \theta_5 \\ \theta_6 \\ \theta_7 \\ \theta_8 \\ \theta_9 \end{bmatrix} = \begin{bmatrix} \rho_1+1 & -\rho_1 & 0 \\ \rho_2+1 & 0 & -\rho_2 \\ 1-\rho_1\rho_3 & \rho_1\rho_3 & 0 \\ 1-\rho_2\rho_4 & 0 & \rho_2\rho_4 \\ \rho_1\rho_3\rho_5-\rho_2\rho_4(1+\rho_5)+1 & -\rho_1\rho_3\rho_5 & \rho_2\rho_4(1+\rho_5) \\ -\rho_1\rho_3\rho_5\rho_6+\rho_2\rho_4(\rho_5\rho_6-1)+1 & \rho_1\rho_3\rho_5\rho_6 & -\rho_2\rho_4(\rho_5\rho_6-1) \end{bmatrix} . \begin{bmatrix} \theta_1 \\ \theta_2 \\ \theta_3 \end{bmatrix} \tag{11}$$

4 Kinematic and Singularity Analysis

The forward kinematic map of the mechanism can be written in terms of the revolute pair rotations as a PoE in the relevant Euclidean group (see Murray et al. [4]). In this case as the mechanism is planar, the group is $SE(2)$. The form of PoE derives from the corresponding *open-loop chain* (see Tsai [9]) as follows:

$$\boldsymbol{T}(\boldsymbol{\phi}) = e^{\boldsymbol{X}_1\phi_1} e^{\boldsymbol{X}_7\phi_7} e^{\boldsymbol{X}_9\phi_9} \boldsymbol{T}(\boldsymbol{0}) \tag{12}$$

where $\boldsymbol{T}(\boldsymbol{0})$ is the transformation between base and end-effector frames at the rest position $\boldsymbol{\phi} = \boldsymbol{0}$ and $\boldsymbol{X}_i$ denote the infinitesimal rotations of revolute joints T_i, $i = 1, 7, 9$ about their centres of rotation. Explicitly, we can use homogeneous representations as follows:

$$\boldsymbol{X}_i = \begin{bmatrix} 0 & -1 & 0 \\ 1 & 0 & -\xi_i \\ 0 & 0 & 0 \end{bmatrix} \tag{13}$$

where (with respect to appropriate choices of body coordinates) $\xi_1 = 0, \xi_7 = l_1, \xi_9 = l_1 + l_7$ with $l_1 = r_2 + 2r_4 + r_6 = r_3 + 2r_5 + r_7$ and $l_7 = r_6 + 2r_8 + r_9$ the lengths of the carrier arms L_1 and L_7 (see Fig. 2); and:

$$\boldsymbol{T}(\boldsymbol{0}) = \begin{bmatrix} 1 & 0 & l_1 + l_7 \\ 0 & 1 & 0 \\ 0 & 0 & 1 \end{bmatrix} \tag{14}$$

Then the homogeneous form of the forward kinematic map is

$$\boldsymbol{T}(\boldsymbol{\phi}) = \begin{bmatrix} \cos(\phi_1 + \phi_7 + \phi_9) & -\sin(\phi_1 + \phi_7 + \phi_9) & l_1 \cos\phi_1 + l_7 \cos(\phi_1 + \phi_7) \\ \sin(\phi_1 + \phi_7 + \phi_9) & \cos(\phi_1 + \phi_7 + \phi_9) & l_1 \sin\phi_1 + l_7 \sin(\phi_1 + \phi_7) \\ 0 & 0 & 1 \end{bmatrix} \tag{15}$$

This can be more simply expressed in terms of link rotation variables using $\theta_1 = \phi_1$, $\theta_7 = \phi_1 + \phi_7$, and $\theta_9 = \phi_1 + \phi_7 + \phi_9$. Moreover, for purposes of singularity analysis it is preferable to work with a local representation of the kinematic mapping $\mathbf{T}$. Simply using the angle θ_9 to parametrise the rotation matrix that constitutes the top left 2×2 block of the homogeneous transformation, the local representation is:

$$(\theta_1, \theta_7, \theta_9) \mapsto \begin{bmatrix} \theta_9 \\ l_1 \cos\theta_1 + l_7 \cos\theta_7 \\ l_1 \sin\theta_1 + l_7 \sin\theta_7 \end{bmatrix} \tag{16}$$

In this form, we have simply made use of the passive variables that describe the kinematics of the underlying 3R mechanism. These can now be expressed using Eq. (11) in terms of the input variables. Hence, the kinematic mapping for the GM can be expressed as a function $\mathscr{F} : \mathbb{R}^3 \to \mathbb{R}^3$, where:

$$\mathscr{F}(\theta_1, \theta_2, \theta_3) = \begin{bmatrix} \beta_1\theta_1 + \beta_2\theta_2 + \beta_3\theta_3 \\ l_1 \cos\theta_1 + l_7 \cos(\alpha_1\theta_1 + \alpha_3\theta_3) \\ l_1 \sin\theta_1 + l_7 \sin(\alpha_1\theta_1 + \alpha_3\theta_3) \end{bmatrix} \tag{17}$$

and $\alpha_1 = 1 - \rho_2\rho_4$, $\alpha_3 = \rho_2\rho_4$, $\beta_1 = -\rho_1\rho_3\rho_5\rho_6 + \rho_2\rho_4(\rho_5\rho_6 - 1) + 1$, $\beta_2 = \rho_1\rho_3\rho_5\rho_6$, and $\beta_3 = -\rho_2\rho_4(\rho_5\rho_6 - 1)$. It is worth noting that by judicious choice of gear ratios the rotation of the end-effector can be made independent of one or more input variables. For example, setting $\rho_1\rho_3 = \rho_5\rho_6 = 1$ (equivalently $r_2 = r_6 = r_9$) ensures the rotation is independent of θ_1, θ_3 and is directly equal to θ_2. This is nicely illustrated by Thang [7].

Finally, to find singularities we need to investigate the *Jacobian* of the kinematic mapping $\mathscr{F}$. From Eq. (17) we obtain:

$$\mathscr{J} = \begin{bmatrix} \beta_1 & \beta_2 & \beta_3 \\ -l_1 \sin\theta_1 - \alpha_1 l_7 \sin(\alpha_1\theta_1 + \alpha_3\theta_3) & 0 & -\alpha_3 l_7 \sin(\alpha_1\theta_1 + \alpha_3\theta_3) \\ l_1 \cos\theta_1 - \alpha_1 l_7 \cos(\alpha_1\theta_1 + \alpha_3\theta_3) & 0 & \alpha_3 l_7 \cos(\alpha_1\theta_1 + \alpha_3\theta_3) \end{bmatrix} \tag{18}$$

For a singularity, we require:

$$\det(\mathscr{J}) = \alpha_3\beta_2 l_1 l_7 \sin(\theta_1 - \alpha_1\theta_1 - \alpha_3\theta_3) = 0. \tag{19}$$

The design parameters $\alpha_3, \beta_2, l_1, l_7$ are assumed non-zero so the GM is singular if and only if $\sin(\theta_1 - \alpha_1\theta_1 - \alpha_3\theta_3) = 0$ and hence:

$$\theta_1 - \theta_3 = \frac{n\pi}{\rho_2\rho_4}, \quad \text{for any integer } n.$$

Thus, the singular configurations of mechanism in Fig. 1 are strictly contingent on the difference between input variables θ_1 and θ_3. It can be concluded that increasing the product of gear ratios $\rho_2\rho_4$ connecting gear wheels L_3 and L_7 can cause more singular points in the joint space, while keeping it close to zero will reduce singularities. It must be noticed that having $\rho_2\rho_4 \ll 1$ may have dynamic consequences. The images of the singularity set in the workspace of course correspond to the expected singular configurations in which carrier arms L_1 and L_7 are collinear.

5 Conclusion

We have illustrated the constraint analysis of a geared mechanism involving two epicyclic gear trains, using the matroid method of Talpasanu. This, in turn, leads to an explicit determination of its passive variables in terms of input variables and thereby to a representation of the kinematic mapping in those variables. Determination of the singularity set, not only in terms of specific geometric configurations in the workspace—collinearity of the carrier arms—but also in the input space, is a straightforward consequence. This demonstrates the dependency on gear ratios as design parameters for the GM.

References

1. Amirinezhad, S.V., Uyguroğlu, M. K.: Kinematic analysis of geared robotic mechanism using matroid and T-T graph methods. Mech. Mach. Theory **88**, 16–30 (2015)
2. Buchsbaum, F., Freudenstein, F.: Synthesis of kinematic structure of geared kinematic chains and other mechanisms. J. Mech. **5**, 357–392 (1970)
3. Hsu, C.H., Lam, K.T.: A new graph representation for the automatic kinematic analysis of planetary spur-gear trains. J. Mech. Design **114**, 196–200 (1992)
4. Murray, R.M., Sastry, S.S., Zexiang, L.: A Mathematical Introduction to Robotic Manipulation. CRC Press Inc, Boca Raton (1994)
5. Talpasanu, I.: Kinematics and Dynamics of Mechanical Systems Based on Graph-Matroid Theory. Ph.D. thesis, University of Texas at Arlington (2004)
6. Talpasanu, I., Yih, T.C., Simionescu, P.: Application of matroid method in kinematic analysis of parallel axes epicyclic gear trains. J. Mech. Design **128**, 1307–1314 (2006)
7. Thang, N.D.: Planar manipulator 3. https://www.youtube.com/watch?v=trtH7Id7HSA (2015). Accessed 15 Jan 2016
8. Tsai, L.W.: The kinematics of spatial robotic bevel-gear trains. IEEE J. Rob. Autom. **4**(2), 150–156 (1988)
9. Tsai, L.W.: Robot Analysis: the Mechanics of Serial and Parallel Manipulators. Wiley, New York (1999)
10. Willis, R.: Principles of Mechanism. Green and Co., London (1870)

Index

J. Lenarčič and J.-P. Merlet (eds.), *Advances in Robot Kinematics 2016*,
Springer Proceedings in Advanced Robotics 4, DOI 10.1007/978-3-319-56802-7

T

U

W